MAGNETOSPHERIC PHYSICS

ASTROPHYSICS AND
SPACE SCIENCE LIBRARY

A SERIES OF BOOKS ON THE RECENT DEVELOPMENTS

OF SPACE SCIENCE AND OF GENERAL GEOPHYSICS AND ASTROPHYSICS

PUBLISHED IN CONNECTION WITH THE JOURNAL

SPACE SCIENCE REVIEWS

VOLUME 44

MAGNETOSPHERIC PHYSICS

PROCEEDINGS OF THE ADVANCED SUMMER INSTITUTE
HELD AT SHEFFIELD, U.K., AUGUST 1973

Edited by

B. M. McCORMAC

*Lockheed Palo Alto Research Laboratory,
Palo Alto, Calif., U.S.A.*

D. REIDEL PUBLISHING COMPANY

DORDRECHT-HOLLAND/BOSTON-U.S.A.

Library of Congress Catalog Card Number 74-76472

ISBN-13:978-94-010-2216-3   e-ISBN-13:978-94-010-2214-9
DOI: 10.1007/978-94-010-2214-9

Published by D. Reidel Publishing Company,
P.O. Box 17, Dordrecht, Holland

Sold and distributed in the U.S.A., Canada, and Mexico
by D. Reidel Publishing Company, Inc.
306 Dartmouth Street, Boston,
Mass. 02116, U.S.A.

TABLE OF CONTENTS

PART IV: WAVE-PARTICLE INTERACTIONS

PART V: SUBSTORM PHENOMENA

PART VI: CONCLUSIONS

PREFACE

This book contains the lectures presented at the Summer Advanced Study Institute, 'Earth's Particles and Fields' which was held at the University of Sheffield, England, during the period August 13–24, 1973. One hundred thirty nine persons from sixteen different countries attended the Institute.

The authors and publisher have made a special effort for rapid publication of an up-to-date status of the particles, fields, and processes in the Earth's magnetosphere, which is an ever changing area. Special thanks are due to the lecturers for their diligent preparation and excellent presentations. The individual lectures and the published papers were deliberately limited; the authors' cooperation in conforming to these specifications is greatly appreciated. The contents of the book are organized by subject area rather than in the order in which papers were presented during the Institute. Many thanks are due to Drs Rolf Boström, J. Ronald Burrows, Robert W. Fredricks, Thomas R. Kaiser, Bernt N. Maehlum, Christopher T. Russell, and Martin Walt who served as session chairmen during the Institute and contributed greatly to its success by skillfully directing the discussion period in a stimulating manner after each lecture.

Many persons contributed to the success of the Institute. Drs J. Ronald Burrows, James Dungey, Carl-Gunne Fälthammar, Roger E. Gendrin, Thomas R. Kaiser, Reimar Lüst, Bernt N. Maehlum, Atsuhiro Nishida, J. Ortner, Erwin R. Schmerling, and Martin Walt were especially helpful in preparing the technical program. Dr Thomas R. Kaiser, University of Sheffield, played a most important role in helping to arrange the facilities in Sheffield and in providing visual aids, recording equipment, office support, and projectionists. The assistant editor, Mrs Diana R. McCormac, checked the manuscripts and proofs, and worked hard to achieve a uniform style in this book.

Direct financial support was provided to the Institute by The Advanced Research Projects Agency, Air Force Office of Scientific Research, Defense Nuclear Agency, Lockheed Palo Alto Research Laboratory, and the Office of Naval Research.

Palo Alto,   BILLY M. MCCORMAC
November 1973

PART I

INSTITUTE SUMMARY

INSTITUTE SUMMARY

MARTIN WALT

Lockheed Palo Alto Research Laboratory, 3251 Hanover Street, Palo Alto, Calif. 94304, U.S.A.

The purpose of the final session was to provide a concise review and interpretation of the material presented during the Institute.

However, any review of the present Institute should also reflect on the overall trends in particles and field research and clarify how the various aspects of this subject are evolving. A convenient way of identifying trends is to recall the previous Institutes of this series and observe the growth or decline in popularity of various topics presented at these Institutes.

In the first conference at Bergen in 1965, considerable effort was devoted to the theory of particle motion. Interest quickly divided into considerations of either adiabatic or non-adiabatic motion. The adiabatic theory was quickly completed and is being utilized continuously in almost all trapped particle studies. The development of theories for non-adiabatic motion has been a much slower process, and various aspects have been discussed progressively at Freising, Santa Barbara, Cortina and here at Sheffield. There have also been notable changes in interest in trapped particle populations, and in the time period since the first Institute at Bergen attention has shifted from the near-Earth region to the ring current, magnetosheath, and other features of the more distant magnetosphere. At this Institute there was not a single paper devoted to inner belt protons. There was also no mention of albedo neutrons with respect to the Earth, although the contribution of this source to the trapped protons at Jupiter was described. Artificial radiation belts, which were of major interest at Bergen, were also largely neglected at Sheffield. It is worth remembering, however, that the radiation belts formed by nuclear weapons tests contain a unique body of information which, from time to time, may be valuable in testing a new concept. In the future, the production of artificial belts may be important again when accelerators powerful enough to inject appreciable flux are available.

The growth of research activity in the convection electric fields has been quite marked during the last decade. At the Freising Institute the first measurements were reported. These early measurements were made with Ba clouds, but recently data obtained from rocket and satellite instruments have increased the coverage enormously. Although the gross flow patterns of convected plasma are becoming clear, much of the detail remains obscure, particularly at large distances from the Earth. Associated with the dynamics of the convection electric field are the substorms, a transient phenomena resulting in a rearrangement of magnetospheric fields and plasmas. These events are being studied extensively in programs involving simultaneous measurements with ground based, airborne, and satellite carried instruments. Another field undergoing rapid growth in both theory and observations comes under the general heading

B. M. McCormac (ed.), Magnetospheric Physics, 3–19. All Rights Reserved.
Copyright © 1974 by D. Reidel Publishing Company, Dordrecht-Holland.

of wave-particle interactions. Included are all the situations in which energy is exchanged between charged particle populations and electromagnetic fields.

The summaries of the Institute have been organized to proceed from the world-wide aspects of magnetospheric physics to the microscopic. The global considerations, which embody the interaction of the solar wind with the geomagnetic field and the subsequent convection of magnetosphere plasma, are summarized by Vasyliunas. A consequence of this interaction is the magnetospheric substorm with its various manifestations in auroras, magnetic field fluctuations, and particle perturbations. Several lectures were devoted entirely to substorms, and these are reviewed by Rostoker.

Through the two week Institute, there were numerous reports on particle measurements, and the sorting of various particle populations by type, energy, and location, continues to be a major theme of space research. The recent experimental work on magnetospheric particle distributions is summarized by Burrows.

Lastly, Cornwall describes the status of work on wave-particle interactions. This important subject has played a major role during this conference as well as throughout the magnetosphere. The wave-particle processes are important in forming the bow shock, in driving pitch-angle diffusion, and have generally been invoked to explain or make plausible a number of otherwise mysterious phenomena.

1. Magnetospheric Structure, Field Line Merging and Plasma Convection *

This conference is the first of this series in which the fact that the magnetosphere is open has been taken for granted. In previous conferences there were prolonged debates on whether all the magnetic field lines from the Earth are contained within the magnetosphere or whether they extend into interplanetary space. Two years ago most of the controversy had been resolved in favor of the open magnetosphere, and at this meeting it was recognized that one must try to understand the magnetosphere within the framework of this concept.

Schematically, an open magnetosphere means that magnetic field lines from the Earth connect with the interplanetary magnetic field lines. Among the evidence for an open magnetosphere are solar particles entering the polar caps, magnetospheric effects dependent on the direction of the interplanetary magnetic field, mapping of electric fields from the solar wind to the polar caps, and others. Individually, some of these features could perhaps be explained without resorting to an open magnetosphere; however, taken as a whole, the evidence is overwhelming that the magnetosphere is open. Unfortunately, most of our concepts have been developed with a closed magnetosphere in mind, and in this meeting for the first time there was considerable discussion and attempts to formulate quantitative theories of an open magnetosphere.

A necessary consequence of the open magnetosphere is that there exist so-called X-type neutral points both in the dayside and in the tail of the magnetosphere. (Topologically, the two neutral points lie on a single X-line that encircles the entire

* Presented by V. M. Vasyliunas.

magnetosphere.) The flow of plasma into the vicinity of an X-type neutral point (for example, the flow of heated solar wind plasma toward the dayside neutral point) constitutes the process variously called magnetic field line reconnection, magnetic merging, or magnetic field annihilation. Once one admits that the magnetosphere is open, this process must occur; thus all the evidence for an open magnetosphere is also evidence for the existence of the reconnection process.

Sonnerup reviewed the reconnection process itself, which has already been studied for a long time in connection with solar flares. There is partial agreement that, in the vicinity of the neutral point, there is a series of standing MHD waves which produces the changes in the plasma required by the process. The agreement is not complete since there are other ways of approaching the problem, for example, the 'vacuum merging' advocated by Alfvén, Dessler, and others. This model, which assumes a true neutral sheet rather than a configuration with a non-zero normal field component, seems to work only for a very low density plasma, so that it would be of importance, if at all, only in the tail of the magnetosphere. For the dayside, the type of model developed by Petschek and others is generally accepted. However, there are wide divergences of opinion on some important aspects of the model, in particular with regard to the merging rate (the velocity with which the plasma flows toward the neutral point). There are basically two schools of thought on this rate. One view, strongly advocated by Axford and defended at this meeting by Vasyliunas, is that the rate is determined by the overall plasma flow and field configuration and that, whatever the processes that take place in the vicinity of the neutral point, they will adjust themselves so that the flow can proceed as required. The opposite view is taken by Haerendel who believes that the processes in the near vicinity of the neutral point control the merging rate. To what extent the reconnection is controlled by local or by global processes is important because we are becoming interested in the whole system. It is a significant step forward that people are starting to worry about what happens to the entire magnetosphere, how the configuration might appear, and what happens when the interplanetary field is not conveniently oriented southward. Sonnerup reviewed a considerable amount of material on these factors, and Vasyliunas gave some theoretical estimates for the values of polar cap potentials associated with the convection system. The level of sophistication of these calculations is somewhat analogous to the first attempts of Chapman and Ferraro to compute a closed magnetosphere by considering image dipoles in a plane magnetosphere.

At the neutral point itself finite resistivity is necessarily important; one can no longer assume that $\bar{E} = -\bar{V} \times \bar{B}$ since both $\bar{V}$ and $\bar{B}$ go to zero. Ordinary collisional conductivity is completely insignificant and one has to invoke various types of wave-particle interaction processes. Which process might be important depends on how thick the interaction layer can be. Sonnerup maintains that the layer cannot be smaller than an ion gyro-radius. Other theorists require it to be as thin as an electron gyro radius, so that this point remains an important area of disagreement. There is also the important question of what is the magnetopause for an open magnetosphere. On theoretical grounds the boundary in the open model is a very complicated object, a variety of different discontinuities, waves, and changes in the plasma. Which of these changes is to be identified with 'the' magnetopause remains an open question with which both experimenters and theorists will have to be concerned in the next few years.

The tail of the open magnetosphere also presents many problems. From topological considerations there must necessarily be a neutral point there. However, we do not know where it is located. The solar particle data indicate that it is rather far downstream, maybe several hundred earth radii. Nishida, on the other hand, reported that during substorms the neutral point was located at some 10 to 20 R_E. Whether the observations of Nishida are related to the open magnetosphere or whether they represent a local structure is presently unsettled; the weight of opinion seems to be leaning in favor of a local structure.

Another aspect of the open magnetosphere is the flow pattern of the plasma. In the boundary region one might expect the plasma to flow back from the sub-solar point directly over the poles. Since we do not yet have a quantitative model for the magnetic field within an open magnetosphere, we do not know how to map this flow from the boundary to the polar cap. For lack of any argument to the contrary people have generally assumed, without much thought, that a uniform boundary flow implies also a uniform flow over the polar caps and hence a uniform electric field directed from dawn to dusk. When the electric field measurements of Gurnett, Heppner, and others indicated instead a concentration of flow toward the edges of the polar cap (or toward one or the other depending on the external field orientation). Frank developed a particular model of the magnetosphere in which the flow over the boundary itself was assumed to follow this non-uniform pattern. In this model the plasma flows essentially around the sides of the magnetosphere rather than over the polar cap, so that it is quite easy, topologically at least, to have the polar cusp plasma flowing over the flanks and into the plasma sheet. Until recently much of the particle evidence presented by Frank was consistent with this picture. At this meeting the first definite evidence to the contrary was presented by Paschmann and his group: their measurements of the plasma mantle are most readily interpreted in terms of plasma flowing directly over the polar cap. These two viewpoints lead to rather different concepts of the origin of the plasma sheet. (One point which bears on this is the extent of the polar cusp in local time. In the picture of Frank there is no reason why there should be sharp local time boundaries in the polar cusp; it should just extend from noon and gradually merge into the plasma sheet. Heikkila reports a fairly sharp termination of the polar cusps, so there is a disagreement which requires further clarification.)

Within the magnetosphere itself, the electric field imposed on the polar cap will extend throughout the ionosphere and form the magnetospheric convection pattern. The theory for this is well understood and was reviewed by Boström. If one knows certain quantities, one can calculate the electric field inside the magnetosphere. The key quantities, knowledge of which limits the accuracy of such calculations at present, are electric fields at high latitudes, magnetic fields, and ionospheric conductivities.

All of these quantities are rather poorly known so that the statement that one can do the computation is not as powerful as it might seem.

One major unresolved question is to what extent are there electric fields parallel to magnetic field lines. No one seriously doubts any more that such fields do exist somewhere under some conditions. However, are these large scale fields, extending over hundreds of kilometers and due essentially to anomalous resistivity, or are they concentrated in so-called double layers, the theory for which has been developed primarily by the Swedish school? There is at present little information with which to decide between these alternatives.

In regard to convection in the inner magnetosphere, a controversy has developed essentially at this meeting. It is well known that the plasmasphere has a bulge on the dusk side and that its surface seems to shrink and expand with geomagnetic activity. A reasonable interpretation is that during periods of magnetic activity the outer layers of the plasmasphere are removed; later the region is gradually refilled from the ionosphere, then maybe removed again, leading to a complicated step structure in the density profiles. The measurements supporting this interpretation have been discussed previously by Chappell and many others. The simplest model for removing the plasma is by means of a convection electric field, making use of the fact that the size of the region where plasma flows in closed paths around the Earth varies inversely with the magnitude of the electric field. With this concept Chappell and co-workers have found good qualitative agreement with changes in plasma density as a function of time. McIlwain has now proposed a different way of accounting for the process. Instead of removing the cool plasma by a convection field he proposes that, at the time of substorms, the particles are accelerated to high energies and form the hot plasma sheet. The step structure at the plasmapause results because there is a boundary – beyond which all the cool plasma is accelerated and becomes hot plasma, then the region refills with plasma from the ionosphere until removed by another substorm and so on – and the position of this boundary varies with geomagnetic activity. Again, there appears to be no clear resolution of the controversy at the present time. One will have to look carefully at how these pictures fit other observations.

In summary, there has been considerable progress in the theory of the magnetosphere. The recognition that the magnetosphere is open and the consequent beginning of a search for a quantitative description have been major steps forward. This has made the magnetosphere more complicated, but also, perhaps, more interesting. Within the magnetosphere, the processes of convection and particle transfer are becoming understood in greater detail. There are several unresolved controversies, the most important being about the flow over the polar cap, since what happens in the tail depends crucially on the polar cap flow.

2. Substorms*

Although the ionospheric signatures of substorms were studied for many years the

* Presented by G. Rostoker.

physics of the magnetospheric substorms did not begin to unravel until satellite data became available. The work of Axford and Hines in 1961, which stressed the role of magnetospheric convection, is a historical turning point for the understanding of the physics. They claimed that the energy input occurred through some form of viscous interaction, which was really quite undefined. They also envisioned a closed magnetosphere which, as was summarized by Vasyliunas above, is a view not held at present. Dungey, on the other hand, who was developing his ideas on reconnection at roughly the same time, pictured an open magnetosphere which obtained this configuration through merging of the interplanetary magnetic field lines with Earth dipole field lines on the dayside. The subsequent formation of the magnetotail occurred through reconnection of the dipole field lines in the tail. An interesting thing about magnetospheric physics is that it developed, particularly in the early sixties, in a series of attempts to prove one of these concepts as the expense of the other.

The concept that sharply enhanced tail reconnection was associated with storms and substorms was initiated first by Axford, Petschek and Siscoe back in 1965, and was further developed by Atkinson in a series of papers in 1966–67. Of course, one must relate the magnetic and auroral measurements to the large scale dynamics of the magnetosphere and to the substorm itself. The current flow associated with the entity known as the geomagnetic bay, was for a long time thought to be purely ionospheric. Equivalent current systems were drawn on polar plots. Later the word 'equivalent' was dropped and the plots were regarded as real current systems. However, as far back as 1908 Birkeland suggested that field aligned currents were flowing in the magnetosphere during what he called polar elementary storms. This very old idea was revived in about 1964 by Boström who showed that indeed there was a three dimensional current system. Boström's work was followed not too long thereafter by Atkinson, Akasofu, Hones, and others who developed precise pictures of the current systems flowing and particles moving. The only trouble was with the morphology of the substorms itself since it refused to be as orderly and simple as the theories which were proposed for it. Even at the present time, the state of the morphology of the magnetospheric substorm is in a state of some confusion and disagreement.

This summary reviews some of the observations reported at the Institute and emphasizes points which will help to judge the interpretations and the theory on which they are based.

Rostoker described the magnetic and auroral effects in the auroral zone accompanying the expansion phase of the substorm. He pointed out that during the expansion phase there was a rapid poleward motion of the poleward border of the auroral electrojet, while the equatorward edge of the electrojet was pretty well pinned to one position during the course of the substorm. After the maximum poleward position had been reached by the poleward border, the electrojet could remain relatively stationary, being very wide, very intense, with loops and surges being generated at the poleward border. This behavior really implied a continued energy input and suggested that some sort of maintenance phase might be an integral part of the substorm process. One should perhaps edge away from the term 'recovery' [which legitimately applies in the Akasofu framework] because 'recovery' generally implies a decay in the level of activity, which is not really warranted. Also, Rostoker showed that the northward and westward expansion of the electrojet did not occur continuously. It occurred in jumps with new current elements being formed progressively northward and westward of each preceding one. These elements blend together into a diffuse electrojet region.

Hargreaves was also looking at what was happening to the electrojet when he presented his work on the spread of auroral absorption in the auroral zone. In the polar plots he showed it is apparent that the region of auroral absorption develops very rapidly eastward and less rapidly westward. There is a definite asymmetry in how the auroral absorption pattern expands; eventually the eastward part literally circulates all the way around the pole almost meeting the westward part. One tends to associate the westward motion with a westward surge because a stationary observer at the auroral zone will see increased absorption as the surge passes overhead. The eastward drifting part has been associated with drifting energetic electrons. For example Jelly and Brice back in 1967 suggested that something like 40 or 70 keV energetic electrons were responsible, and the drift time depends not only on the azimuthal drift velocity one would normally predict but also on any electric fields which are present.

Going back to the substorm current itself, McPherron studied the growth of the substorm system using a three dimensional current loop which involved current flowing down field lines into the ionosphere, westward through the ionosphere in the form of a westward electrojet and then back up into the magnetosphere. The basic idea is that this path represents a short circuiting of the cross tail current. McPherron studied this particular loop in both a qualitative and quantitative fashion through the magnetic signature at midlatitudes. He found that there was quite an outstanding magnetic signature of positive H bays at midlatitudes in conjunction with what he called magnetospheric substorms. He also found that when there was a good mid-latitude response the tail lobe field was also very significantly decreased.

The growth phase of the magnetic substorm deserves special mention since there was a great deal of discussion by McPherron and Rostoker and some difference of opinion on the interpretation of magnetic records in terms of a growth phase. The point was brought out from the work of Rostoker and Wiens that a magnetic signature characteristic of growth can occur in the near magnetotail due to an expansion phase in an adjacent sector to the east. In the magnetograms the key thing to note is that to the east one can identify a positive bay which McPherron would claim indicates a substorm expansion phase. To the west however, there is a depression which is the signature of growth. There are however qualifications involved in this interpretation. On rare occasions, a signature is observed which occurs as energy is being added to the magnetosphere, apparently while no substorm activity can be seen. This signature appears to be caused by the double vortex current system SD or S_qP in the polar cap. Rostoker felt however, that the magnetosphere is rarely in a state of pure growth without some localized but significant substorm activity taking place in some azimuthal sector of the night side. Regarding this question, Anger pointed out a rather significant fact in regard to his ISIS satellite photometer data; he rarely observed passes where he could not identify at least the presence of the auroral signatures indicative of substorm activity. This confirmed a growing body of opinion that truly quiet times are very, very rare and that a moderate state of activity is the norm.

McPherron agreed that a growth signature did indeed appear in the low latitude evening sector in conjunction with the midnight sector substorm, but that these H-component changes were relatively sharp. The disagreement centered around the time of a slowly decreasing midlatitude H component and whether there is in fact magnetospheric substorm activity going on during these periods. McPherron contended that small auroral zone bays, which have no pronounced midlatitude response, are a different phenomenon and do not necessarily constitute a magnetospheric substorm of the conventional type. This is a very intriging idea, and Vasyliunas amplified on this feeling by suggesting that perhaps there was not one-to-one correspondence between the phases of the magnetospheric substorm *per se* and the polar substorm signatures in the ionosphere. This interesting attempt to decouple the magnetosphere and the ionosphere warrants much further study.

The question then arises as to what threshold does one chose to distinguish a small bay from a large bay: how can one distinguish between a magnetospheric substorm and what might be termed a small polar substorm? The threshold has not been established as yet but will have to be before further progress can be made.

The response of the magnetospheric tail to substorms was reported in a number of papers. A new concept was introduced suggesting that while there was a neutral point far down the tail through which steady reconnection took place, a magnetospheric substorm involved the creation of an X type neutral point close to the Earth but outside synchronous orbit. McPherron presented data of Russell which showed an X type and an O type neutral point configuration developing. The important feature is that the neutral point was developing in a region of closed field lines. Nishida also presented strong evidence along the same line, confirming similar conclusions reached earlier by Hones, that there was a neutral point at least Earthward of the Vela satellite (i.e., inside $\sim 18 \, R_E$). The key to Nishida's argument is the appearance of negative B_z in the magnetic tail. For a satellite about 25 R_E down the tail the presence of negative B_z is certainly indicative of a neutral point Earthward of that satellite. There is still the question as to whether the old neutral point has moved inward inside the observing satellite or whether a new one had formed Earthward of the satellite. Multisatellite studies are needed to resolve this question.

The localized character of substorms received some comment, and a concensus has begun to develop that even magnetospheric substorms were in some sense localized in azimuthal extent. Both Rostoker and McPherron claimed that this appeared to be the case, and they were supported by Anger on the basis of his auroral data and by Rossberg on the basis of his multisatellite studies. Basically if you were in different locations you would see different things during the course of a magnetospheric substorm. In addition, Boström in his treatment of three dimensional current loops claimed that the large inductance of a loop prevented it from growing too rapidly, suggesting that at least in the beginning the current loop must occupy a confined longitudinal sector.

Nishida, however, felt that the neutral line structure which he was seeing in the tail really stretched across a large portion of the tail although he suggested that the energy input close to the Earth might be dissipated in narrow sectors. This suggestion does not explain some observations which Rostoker has made of a zero tail effect during large substorms. However, Nishida asserted that even when his two satellites were reasonably far apart they saw concurrent responses associated with substorms. His comments thus far apply only to very large events with pronounced midlatitude responses, and these would certainly involve a large current loop and one might expect two satellites to respond together. It is certain that large substorms affect a large portion of the tail. This result appears in the study of Nishida and Hones as well as in the study of tail lobe responses by McPherron and Caan. For these events, the three dimensional current wedge has a large longitudinal extent and the result is going to be a large midlatitude response. Again, one has the question of defining a threshold whereby one can separate these large or magnetospheric substorms from small substorms and ask whether they are indeed different kinds of events.

In the theoretical side of substorm studies Kennel concentrated on the coupling of the magnetosphere and ionosphere through the mechanism of field aligned current flow. Kennel firmly emphasized that he felt the bright arcs were the seat of field aligned currents which coupled the ionosphere to some magnetospheric feature. He invoked polarization effects from the ionosphere to regulate the field aligned current flow at such places as the equatorward and poleward edges of the auroral oval. He further postulated that anomalous resistivity along the field lines penetrating the equatorward border caused energetic protons to precipitate, thus accounting for the region of the hydrogen arc – the H_β glow.

Kennel also commented upon the sequence of events leading up to the onset of substorm activity, concentrating upon the aspects of convection and of reconnection in the magnetotail. Based on the order of magnitude calculations, he claimed that convection was initiated in the magnetosphere shortly after the appearance of a southward interplanetary magnetic field, although it took something of the order of 10^3 s before the maximum convection rate could be achieved. He also claimed that if the arcs were drifting equatorward at roughly the $\bar{E} \times \bar{B}$ drift velocity this fact implied that no significant reconnection was going on in the magnetotail. This is a very interesting and important conclusion. In addition, he favored the concept of a near Earth neutral point consistent with the observations of Nishida and Hones and also the work of McPherron and Russell.

It is very easy to conclude what must be done to resolve the morphological controversies, most of which result because of inadequate spatial coverage of the phenomena. If one wishes to observe a phenomena whose scale size is 1000 km one does not put observatories 2000 km apart. In many ways the people who are studying the morphology are forced into this position by the inadequate distribution of ground based observatories over the Earth's surface. Similarly in the magnetosphere if one wants to study something which varies spatially and temporally, it is difficult to do so with one satellite unless by some means you can define where that satellite lies with respect to the disturbed region. Until you can do that all you can produce are statistical summaries; individual event analyses are difficult if not impossible. In conclusion more and better data, and a better distribution of the observing points are needed to clarify the overall morphology of substorms.

3. Observations of Particles*

It is worth commenting that the particle observations and the models of particle processes which have been presented in this conference are in widely varying states of maturity or development. If one were to present only those problems that have been solved, the conference would have to be short and rather uninteresting. On the other hand some observations and interpretations should be considered more working hypotheses than theories. In reviewing the observations of different features of the magnetospheric particles that were discussed at this conference I will try to emphasize each subject's stage of evolution.

In the high latitude region Paschmann reported some recent, important preliminary HEOS 2 observations which should be valuable in developing the theory of the open magnetosphere. This new 'mantle' region is adjacent to the magnetopause, extending 2 to 4 R_E inside of it downstream from the cleft. The streaming protons, which are present there more than one-half of the time, are flowing outward from the Earth along the magnetic field direction with about one-half of the energy and velocity of the adjacent magnetosheath protons. In the cleft, two proton populations with different energies and flow characteristics were observed. Earthward flowing magnetosheath-like protons occurred in a narrow region near the polar cusp's low latitude side while colder upward flowing protons were found throughout most of the cleft region.

Other observations of cleft fluxes are somewhat more mature than Paschmann's results. They were discussed by Frank using OGO-5 and INJUN-5 data and by Heikkila using ISIS-1 data. Heikkila noted that fluxes had magnetosheath-like spectra and extended about 3° in latitude at low altitude. Frank, on the other hand, empha-sized the spatial separation of the proton and electron flux maxima and the narrower, intense feature in the electron fluxes occurring coincident with the electric field reversal. Anger reported new photometric observations from ISIS-2 in which narrow arcs of 5577 Å emission due to more energetic electrons were observed embedded in a broader 6300 Å region caused by lower energy electrons. Armstrong's TRIAD ob-servations of field aligned currents are also important in modeling the magnetopause-ionosphere current systems emphasized by Heikkila and may help to establish distinctive characteristics of the cleft within the auroral oval. For example he noted the frequent occurrence of small scale structures in the dayside oval having large current densities ($\sim 5 \times 10^{-5}$ A m^{-2}).

---

* Presented by J. R. Burrows.

There are many unresolved questions about the cleft region. We now know that it is there, but how does its structure vary as a function of the solar wind's magnetic and kinetic parameters and as a function of substorm erosion and recovery? One might compare the cleft structure with the bow shock structure reported by Fredricks, who showed that the structure was strongly influenced by the angle between the interplanetary field and the shock normal. Similar work with cleft fluxes should reveal comparable spatial diversity in the cleft which will contribute to the improvement of large scale convection models of the type discussed by Kennel. Such models should also be influenced by further study of the polar cap's auroral morphology reported by Anger.

In the plasma sheet, there now exist good data sets at geostationary orbits (6.6 R_E), at the Vela satellites ($\sim 18\ R_E$), at the orbit of the Moon (60 R_E), and from the many radial cuts that have been obtained from elliptically orbiting satellites. All of these orbits still suffer from temporal-spatial ambiguities, although multisatellite studies are resolving some of these in the evolution of substorms. Thus the data set is generally more mature for the plasma sheet, and modeling is developing qualitatively and in some cases quantitatively. Synoptic plasma observations in the tail were only briefly reviewed and average plasma sheet spectral parameters were not discussed. Proton bulk flow measurements, such as those reported by Frank, are important to further our understanding of tail convection patterns and particle access into the magnetosphere from the magnetosheath. Substorm development in the deep-tail part of the plasma sheet still needs much study in spite of the progress made by Russell, McPherron, Hones and co-workers in analyzing its radial and latitudinal (i.e., X_{GSM} and Z_{GSM}) dependence. For example, where is the quiet time X neutral line located in the tail? How is it related to bulk plasma flow? What influences the formation of a new neutral line closer to the Earth? What regulates its longitudinal (Y_{GSM}) extent?

In the inner plasma sheet, McIlwain described the quantitative modeling of the quiescent convection electric field derived from fitting plasma dispersion curves measured at geostationary orbit. At present this is the only method in use for deriving large scale electric fields synoptically at the equatorial plane. The potential contours obtained indicate a very weak electric field in the evening sector with all sunward convection occurring on the morning side. The conflict between this model and the evidence for sunward convection found at low altitudes probably reflects the limitations inherent in the modeling procedure nearer the magnetopause. The dispersion curves are consistent with particles initially being injected along a spiral shaped inner edge of the plasma sheet. McIlwain also deduced that these particles resulted from *in situ* heating of the local cold plasma rather than from inward convection of hot plasma. This conclusion deserves further study because of its basic implications for plasma sheet dynamics.

The study of the plasma sheet's extension to low altitudes near the auroral oval provides a different type of information about magnetospheric morphology and processes which was treated by several speakers using particle detectors, photometers and magnetometers. Both Frank and Gurnett concluded from INJUN 5 data that the quiet time evening plasma data should be divided into two major regions – the plasma sheet and the higher latitude inverted V region. The latter contained the more intense electron fluxes which were field aligned. They occurred at the electric field reversal on open field lines outside the trapping region. This reversal was a clear feature in the early evening but it was poorly defined near midnight. Gurnett described a mechanism for the generation of the inverted V structures and proposed that they mapped out to the magnetopause and the magnetosheath. Burrows, using ISIS-2 data, also studied the inverted V structures and the premidnight sector. By comparing them with Anger's photometric data obtained simultaneously, he found that they corresponded to arcs extended along the auroral oval, and he also showed that these structures sometimes occur where the magnetic field is still closed. One must conclude that the problem of mapping quiet time low altitude data out to the equatorial plane remains largely unsettled because of the tail's complex inflated geometry.

The Norwegian group studied arcs near the poleward edge of the night side auroral zone. They used a mother-daughter rocket to separate spatial from temporal effects in the small scale structure of the arcs. Moestue reported that isotropic fluxes occurred in the harder core of these arcs while field aligned fluxes occurred in the soft edges.

Anger, in discussing the night side 'diffuse zone' which corresponds roughly to the plasma sheet as described by Frank and Gurnett, emphasized that its total energy input is greater than that in the arcs. Thus one should not ignore this ionization source when modeling ionospheric conductivity patterns.

Armstrong's study of field aligned currents showed that they occur on all passes through the auroral oval, primarily in the form of east-west oriented sheets with current densities usually between 10^{-6}–10^{-5} A m^{-2}. In particular, in the evening sector the current flow is outward at the poleward boundary of the field aligned current region with inward flowing currents at lower latitude giving closure to a varying degree (up to 100%). The necessary synthesis of these various low altitude observations with equatorial observations should be achieved after further careful comparative studies.

In the outer zone, a broad understanding of the particles' dynamic processes has been achieved but the microscopic processes responsible for the loss of particles via precipitation are poorly understood. In the L range from 5 to 7, scattering associated with the generation of electrostatic modes at 3/2 of the electron gyrofrequency appears to be important. Another unsatisfactorily treated problem is that of coupled radial diffusion and precipitation, particularly for off-equatorial mirroring particles. Fritz derived a pitch angle diffusion coefficent at low altitudes for electrons assuming small angle scattering. Further contributions in this area can be made with good pitch angle distribution measurements.

Near the equatorial plane, West reported some very good quantitative pitch angle measurements of energetic electrons ($E \gtrsim 80$ keV) from which one can deduce the geomagnetic field configuration and regions of noise turbulence. The transition from butterfly to isotropic distributions near midnight indicates strong wave scattering there.

Measurements of ring current protons are now being obtained with the S^3 satellite in the equatorial plane out to 5.1 R$_E$. They provide much more detailed energy spectra and pitch angle distributions than were previously available as well as magnetometer measurements of the field inflation. Fritz discussed the dispersion characteristics of storm-time ring current protons arriving at S^3. With further analysis this data set should supplement McIlwain's plasma sheet studies and provide detailed information about various proton loss processes which recent theories have proposed.

Cornwall's treatment of the quiet time ring current as a three dimensional transport problem considered protons mirroring at all altitudes in an attempt to provide a framework for the interpretation of low altitude proton observations.

Particle phenomena occurring in the plasmasphere or associated with the plasmapause received little discussion at the conference, although the plasmapause is a critical boundary in magnetospheric convection models, in the formation of SAR arcs and in relativistic electron precipitation in the 'slot'. The ground-based VLF techniques discussed by Kaiser and others remain a powerful tool in analyzing the plasmapause and the motion of whistler ducts in the plasmasphere.

The ionosphere was not given detailed treatment at this meeting, but it is of course important because it forms the near-Earth boundary of the magnetosphere. As such, its influence is felt in several ways. First, it provides current paths across the magnetic field via Pedersen and Hall conductivities to close the field aligned currents as discussed by Boström. Second, it is the source of cold plasma for the plasmasphere. Third, it is probably the source for the 1 to 12 keV O^+ ions reported by Shelley. Fourth, it is a sink for magnetospheric energy, and the associated conductivity inhomogeneities may regulate magnetospheric convection patterns. Lastly, it is the source for the polar wind which may provide particles for the inner plasma sheet by the heating of cold plasma as suggested by McIlwain.

While it is apparent that much progress has been made in the study of particles in the magnetosphere, it is expected that even greater progress towards a synthesis will be made in the next few years as further intercomparisons are made with the sophisticated data sets which are now becoming available for analysis.

4. Wave-Particle Interactions *

Wave-particle interactions have gone through a history in the last decade very much like the evolution of a wave itself. From a small signal of a few papers the field has exponentiated, and now there is a multitude of papers. Some of the words which apply to non-linear wave-particle interactions may also apply to the literature in the field. Certainly some fields are becoming saturated, and some authors are becoming trapped in their own interactions. Many side bands have been generated, and there is a certain tendency to decay into noise, but still the signal to noise ratio is very healthy.

The wave-particle interaction session had more papers than any other session. Essentially everything that was discussed involved non-linear processes. A good example is the bow shock, an area of magnetospheric physics which has many of these

* Presented by J. M. Cornwall.

non-linear features. In describing the bow shock Fredricks observed that the OGO 5 data now allow one to resolve fine-scale space-time structures (e.g., noise in electrostatic fields in the range of a few kilohertz). More importantly perhaps for the development of the theory, one can now see fine-scale spatial structure in the shock transition. The time resolution of a second or less allows one to observe spatial variations in the magnetic field as one goes through the shock, and such fluctuations are critical to developing the theory of the shock. The first question one might ask is, what is the nature of such a fluctuation? If one idealizes to a single hump, one has a soliton wave structure, which by virtue of non-linearities in the equations can exist by itself and move for quite some period of time. However, it is non-dissipative and cannot by itself lead to a shock structure. In the real world the magnetic field is very jagged (i.e., there are many solitons, roughly speaking), but the main point for consideration is the gradient structure of a soliton. These gradients will be associated with currents such that $\Delta B/\Delta x \approx (4\pi/c)\,j$, where j is the current density. Obviously the larger the change in B and the smaller the change in distance x, the larger the current. If there is a sufficiently large current, it can drive various instabilities (e.g., ion acoustic, ion cyclotron). The instabilities dissipate energy and can lead to a conventional shock structure on a gross spatial scale.

Fredricks observed that at one time he and others believed that in the bow shock the scale length was the electron inertial length, namely the velocity of light divided by the electron plasma frequency. Now he believes that it is the ion inertial length which determines the scale length, or possibly the ion Larmor radius, but these quantities are basically the same thing if the β of the plasma is unity, β being the ratio of particle pressure to magnetic pressure. This result means that the scale length has been reduced somewhat and therefore a class of instabilities that might otherwise have been available for use, such as ion acoustic instabilities, is not going to go, because there is not enough current to drive it. There is one mode (and perhaps others) involving electron cyclotron coupling with gradient drift motion, which could be unstable at harmonics of the electron cyclotron frequency of a few hundred hertz. When these frequencies are doppler shifted upwards one gets into the kilohertz range and this process can therefore explain the noise.

The next topic of concern is the dissipation of this energy. It is certainly observed in the bow shock that the ions are heated, and it does not seem very plausible that the ions can be heated with an instability that works on electrons. While there was little discussion of this point, the possibility is that these modes undergo decay instabilities involving an ion wave mode which could then heat the ions. This question is not settled yet and is one of the most important problems in collisionless shock theory which remains to be settled.

At present one of the most difficult problems in electron non-linear wave-particle interactions is that of triggered emissions. This problem has been around for many years, and everybody knows that the solution somehow involves non-linearities, trappings of particles in the waves, and inhomogeneities. However, it is not likely that the problem will yield to an analytic approach in view of the difficulties encountered in sophisticated computer simulations, even on a computer the size of a CDC 6600. Nunn had a routine in which for reasons of computational inadequacies of the computer he had to suppress side bands. One characteristic feature was that the wave energy kept growing even well into the non-linear regime, which ordinarily does not happen when trapping occurs, in a homogeneous medium. He had to damp the growth phenomenologically. Rising frequency emissions were obtained with the model, and several falling tones, although the latter are difficult to get. I really cannot see any hope of solving this very difficult problem with a brilliant insight; it is going to require larger computers and more effort.

Wave-particle interactions which could lead to finite wave intensity and particle precipitation were described by several authors. Abdalla described a calculation which found that *if* ~ 2 keV electrons were injected isotropically at $L = 10$ and then convected to the plasmasphere, the anisotropy would not be large enough to produce instability. The result of the calculation of course depends critically upon the initial assumptions about energy spectrum, anisotropy, and cold plasma density. In view of the observed wave turbulence, and in view of the observed particle turbulence one might challenge one of the assumptions. Which of these assumptions needs challenging is not clear at the moment, but one thing that seems obvious is to find out what kind of initial anisotropy would be required to produce instability. One could simply redo the calculations with different amounts of initial anisotropy. In that connection it is worth pointing out that the final anisotropy which is developed by any kind of transport process conserving μ and J is determined not so much by the initial pitch angle anisotropy as by the initial energy distribution; the faster the distribution falls off with energy the more anisotropy it will pick up. The average energy is also important; it must be considered in connection with the cold plasma density. If one raises the cold plasma density one can make things more unstable rather readily, and the cold plasma density may possibly be somewhat higher than was considered in those calculations. Those assumptions should be examined to see which one of them if changed most readily yields instability.

Let us briefly recall three statements which have become virtually dogmatic for electromagnetic cyclotron instabilities; they are based on linear theory but have been applied to essentially non-linear problems as well. This dogma can itself be challenged in detail but it is useful as a rapid and quick way of stating several qualitative facts about certain kinds of instabilities. Three conditions are required for instability.

$$E > \frac{B^2}{8\pi N} \frac{1}{A_c^2 (1 + A_c)} \quad \text{(for protons)} \tag{1}$$

$$> \frac{B^2}{8\pi N} \frac{1}{A_c^2 (1 + A_c)^2} \quad \text{(for electrons)}$$

$$A > A_c = \frac{\omega}{\Omega - \omega}. \tag{2}$$

Stably trapped flux: flux $\sim 5 \times 10^{10} \, L^{-4} \, \text{cm}^{-2} \, \text{s}^{-1}$ \hfill (3) where E is the particle energy, A_c is the critical anisotropy of the trapped particles, and N is the plasma density. The quantity $B^2/8\pi N$ is the magnetic energy per particle, including both cold and energetic particles. The first condition is merely a restatement of the cyclotron resonance conditions; the energy of a single particle must exceed this value to interact with the wave. The second condition of the central dogma is that the actual anisotropy of the distribution must be greater than this critical anisotropy which depends on frequency. The closer the wave frequency is to the gyrofrequency the higher the critical anisotropy must be. The anisotropy is defined most simply by stating that $2A$ is the exponent of $\sin \alpha$ in the pitch angle distribution.

The third condition is a rather fuzzy concept introduced by Cornwall and Kennel and Petschek in 1966 and has to do with how much flux is needed to reach an equilibrium with finite wave amplitude. Electromagnetic cyclotron instabilities are usually convective, so waves escape the region of growth. Therefore one must produce waves at least as fast as they are lost, and since the growth rate is proportional to the number of fast particles, one must have enough fast particles to balance growth and loss.

A number of authors have pointed out that if one increases the cold plasma density, thereby lowering the magnetic energy per particle, condition 1 can be satisfied by lower-energy particles. Thus the plasmasphere may be a favorable region for wave-particle interactions which would not be possible in the low-density region outside. It has been suggested that regions of high plasma density may lead both to enhanced wave energy and to increased precipitation flux. Gendrin's group has studied certain aspects of the problem, and Etcheto reported on the generation of plasmaspheric hiss. A self-consistent, quasilinear treatment of equilibrium wave-particle dynamics shows that the wave intensity is roughly proportional to the cold plasma density, all other things being equal, but that the precipitation flux is virtually independent of density. These and other considerations have led the French group to criticize vigorously some of the conclusions drawn by other authors on the basis of the three conditions above, as well as the conditions themselves. Clearly, the three conditions cannot substitute for detailed non-linear calculations. In the future, many more such detailed calculations will have to be done, especially the initial value problem; until then, the three conditions can be profitable, if used judiciously.

Part of the dogma reviewed above states that protons have strong wave-particle interactions just inside the plasmapause. N increases by a factor of as much as a thousand inside the plasmapause, thereby greatly reducing the parallel energy required, so low energy protons which were stable outside suddenly become unstable. Fritz presented evidence from S^3 which showed such intriguing features as low energy particles penetrating furthest into the plasmapause because they need the highest density before they can become unstable. By combining measurements of these flux profiles and anisotropy with the magnetic field, Fritz et al. were able to construct a plausible plasmapause. This result is indirect evidence for the existence of electromagnetic cyclotron modes, but the big question is where are the ultra-low-frequency waves which would be required. The absence of waves is an embarrassment although it is not clear whether theoreticians or the experimentalists should be embarrassed.

Energetic protons are precipitating in the low-density region outside the plasma-pause which, according to the central dogma, suggests that the precipitation is not caused by electromagnetic cyclotron waves. On the other hand, these modes are not the only possible wave-particle interactions. There may be an electrostatic loss-cone instability which goes on outside the plasmapause and causes the energetic proton precipitations. Søraas presented some very interesting data in this regard in which the protons precipitating outside are isotropic in the loss cone. Inside the plasmapause, the proton distributions show mirroring. These data could be compared with Kennel's model of proton arcs containing $H\beta$ emission. Are H arcs produced by the precipitation of energetic protons observed by Søraas, or are they produced as Kennel would have it by 1 to 10 keV protons precipitated by parallel electric fields? This question could be investigated by studying such things as whether there are always associated southward electric fields in auroral electrojets as Kennel's model would demand or whether $H\beta$ arcs are in fact always in the region of the particle precipitation as seen by ESRO 1A and 1B and various other low altitude polar orbiting satellites.

One factor that is most aggravating to the wave-particle interaction theorist is the lack of information on the anisotropy of these protons. Frank has not published such data, and although S^3 is getting some information, there are sometimes detector saturation problems. Fritz had some pitch angle distributions, and hopefully many more can be generated. The net result of this lack of data is that the role of the cyclotron instability for the protons is now uncertain, especially as far as the dynamics of protons in the main phase are concerned. Now what should the anisotropy be? One can make progress with the calculation program outlined by Cornwall and the work of Abdalla, all of which shows that radial diffusion tries to promote this anisotropy. The non-linearities for the protons are fortunately not as forbidding as they are for the electrons, mainly because protons are so heavy it takes a long time to move them around. As a result it takes them much longer to enter into the non-linear regime. Considering the observed wave amplitudes, trapping effects do not seem to be dominant for Pc 1 ULF waves. According to Roux the effects of inhomogeneities might be more important. The idea is that inhomogeneities lead to finite wave spread and thus some energy in the sidebands even if the original wave is monochromatic. The wave hits the protons very hard, flattens out their pitch angle distribution but of course has to leave it anisotropic at the two ends. The sharp anisotropy at the ends then produces rapid growth in side bands. While much work needs to be done on this subject, the experimental results of the French group showing the generation of side bands in Pc 1 emission are very intriguing. This is a most important field for the future study of non-linear proton interactions.

PART II

STRUCTURE AND CHARACTERISTICS OF
THE MAGNETOSPHERE

THE RECONNECTING MAGNETOSPHERE

BENGT U. Ö. SONNERUP

Dartmouth College, Hanover, N.H., U.S.A.

1. Introduction

For present purposes the terms closed and open magnetosphere are defined as follows:
In the former, the net magnetic flux crossing any substantial portion of the magnetopause is too small to generate any magnetospheric dynamo effect of consequence.
In the latter the net crossing flux is sufficient to account directly for such observed
effects. Open models require field reconnection, closed ones do not. In ideal closed
models no flux whatsoever crosses the magnetopause. In ideal open ones the crossing
magnetic flux points consistently outward in one magnetopause region, inward in
another. In reality, substantial deviations from these ideal conditions are expected.

Mounting evidence indicates that some kind of open model is operative during
dynamically active periods. i.e., during substorms, storms, and in general during periods
of magnetic flux adjustment from the magnetospheric front lobe to the tail. But as
yet it has not been demonstrated that the closed magnetosphere does not occur during
quiet periods.

The paper focusses on theoretical aspects of the open magnetosphere, the closed
model being considered only briefly. Primary emphasis is on field topology, reconnection, and magnetopause structure. Magnetospheric convection, steady and nonsteady, is dealt with to a minor extent. References are given without claim to completeness.

2. The Closed Model

The closed magnetosphere model (Figure 1a) usually incorporates the interplanetary

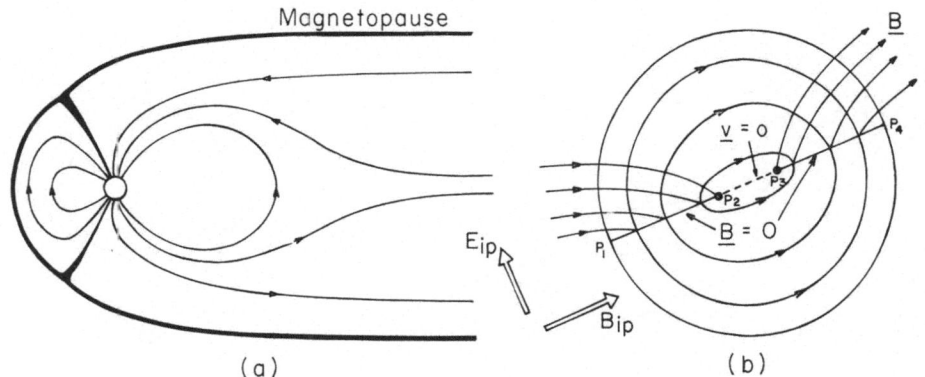

Fig. 1. The closed magnetosphere; (a) sideview; and (b) front view of topology of magnetosheath
magnetic field on the magnetopause surface. P_1–P_2 and P_3–P_4 are magnetic null lines, P_2–P_3 is
a stagnation line. Also shown in (b) are magnetosheath field lines connected to the magnetic null lines.
These lines lie in a plane perpendicular to the paper but have been folded up into the plane of the paper.

B. M. McCormac (ed.), Magnetospheric Physics, 23-33. All Rights Reserved.
Copyright © 1974 by D. Reidel Publishing Company, Dordrecht-Holland.

field as a passive ingredient which renders the solar plasma a continuum but has no other important dynamical effects (Spreiter *et al.*, 1968). No interconnection between terrestrial and interplanetary magnetic field lines occurs and there is no plasma penetration into the magnetosphere.

The absence of plasma penetration implies that the magnetopause is an electrical equipotential surface. The short-circuiting of the impressed interplanetary electric field, E_{ip}, is accomplished along an infinite number of magnetic lines of force on the magnetopause surface. This is illustrated in Figure 1b, which shows the field on the magnetosheath side of that surface. This short-circuiting ability of the magnetopause is the most important topological feature of the closed magnetosphere. It makes the closed model invulnerable to magnetic field reconnection.

The surface field topology in Figure 1b contains a stagnation line rather than a stagnation point and two magnetic null lines. This is compatible with the steady state frozen-field condition written in the form $\mathbf{v} \cdot \nabla (\mathbf{B}/\varrho) = (\mathbf{B}/\varrho) \cdot \nabla \mathbf{v}$, but not with the concept that the magnetosheath field stresses may be neglected. The latter assumption leads to a contradiction, namely an infinite field over the entire magnetopause surface! Thus, the field stresses play an instrumental role in establishing the pattern in Figure 1b.

3. Steady-State Open Magnetosphere

The open magnetosphere model (Dungey, 1961, 1963), shown in Figure 2, operates

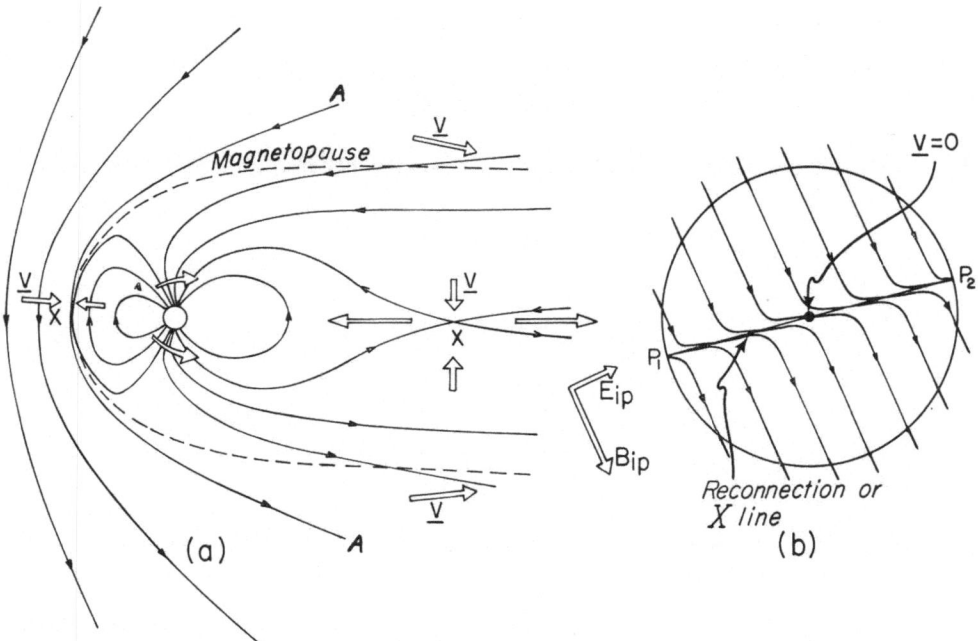

Fig. 2.   The open magnetosphere; (a) sideview; (b) front view of topology of magnetosheath magnetic field on the surface AXA. P_1-P_2 is the reconnection or X line.

most efficiently when the interplanetary magnetic field, B_{ip}, is due south. A reconnection model for northward B_{ip} has also been suggested (Dungey, 1963). In contrast to Figure 1b, Figure 2b shows that on the magnetopause surface there is now but one singular field line responsible for the crow-barring of the interplanetary electric field, E_{ip}. This line closes in the tail, and degenerates to a magnetic null line of the hyperbolic or X type when B_{ip} opposes the equatorial geomagnetic field exactly. Along this line magnetic-field reconnection occurs. Accordingly it is referred to as the reconnection or X line. The line is thought to coincide with a streamline through the subsolar stagnation point.

The short circuiting of E_{ip} along the X line is not expected to be 100% effective. As a result of the hyperbolic field configuration, charged particles accelerating along the X line are deflected away from it rather quickly. Further, should the current density along the line become large, current-driven instabilities will come into play causing a substantial reduction in effective conductivity. The net result of the incomplete short circuit is an electric field component, E_{mp}, parallel to the magnetopause and directed along the singular line. Regardless of the orientation of B_{ip}, E_{mp} always has a component from dawn to dusk. But, as demonstrated later, the magnitude of E_{mp} is a maximum when B_{ip} is due south.

Away from the X line the frozen condition remains approximately valid and implies the presence of a magnetic field component, $B_{\perp}$, perpendicular to the magnetopause according to the relation $\mathbf{E}_{mp} + \mathbf{v}_{mp} \times \mathbf{B}_{\perp} = 0$, where $\mathbf{v}_{mp}$ is the flow speed along the magnetopause. It follows from this formula that $\mathbf{B}_{\perp}$ must point outward from the Earth below the X line, inward above.

In the above discussion the electric field E_{mp} is considered a primary quantity, the presence of which implies a nonzero $B_{\perp}$. Alternatively, $B_{\perp}$ may be considered primary. The solar plasma flow v_{mp} across $B_{\perp}$ is then thought of as an MHD generator producing the charge separation responsible for E_{mp}. The total voltage produced by the generator between the dawn and dusk side of magnetosphere is $\Delta V_{mp} = E_{mp} D$, D being the effective length of the X line.

The terrestrial field lines, which cross the magnetopause, originate in the polar regions. In a steady state the magnetopause potential distribution is mapped along these lines into the polar ionosphere. The resulting magnetospheric electric fields lead to antisolar convection of plasma over the poles with an associated return flow at lower latitudes in the familiar twin-vortex pattern (e.g., Nishida and Obayashi, 1972). In terms of field line transport, this convection implies removal of flux from the magnetospheric front lobe by converting closed field lines to open ones at the X line. The flux on open lines is then carried over the poles and into the tail, there being converted back to closed lines at the tail X line, and finally being returned to the front lobe by convection in the equatorial plane.

Heppner (1972) reports measured potential differences across the polar cap in the range $\Delta V_{pc} = 20$–100 kV. In a steady state $\Delta V_{pc} = = \Delta V_{mp} = E_{mp} D = v_{mp} B_{\perp} D$. With $v_{mp} = 300$ km s^{-1} and $D = 20$ R$_E$ the above voltage range corresponds to $B_{\perp} = 0.5$ to 2.5γ, indicating that the average magnetic field component perpendicular to the magnetopause is so small that it is difficult to measure directly. The interplanetary potential difference ΔV_{ip} across a distance D usually is 300 kV or more so that on the average $\Delta V_{mp} \ll \Delta V_{ip}$ and hence $E_{mp} \ll E_{ip}$. During substorms and storms these inequalities may not be strong ones (Gurnett and Frank, 1973) and measurable values of $B_{\perp}$ may occur.

On account of the small value of $B_{\perp}$ one expects the geometrical features of the open magnetosphere to be nearly the same as those of the closed one. But certain systematic differences are expected and have been observed (Meng, 1970; Fairfield, 1971; Burch, 1972). In a circuit model (Coroniti and Kennel, 1973), the magnetopause generator is connected to a resistor, the polar-cusp ionosphere. The current flows from the magnetopause into the ionosphere along dawn-side field lines, returns to the magnetopause in a similar manner on the dusk side, and finally closes along the front lobe magnetopause, there causing a diminution of the total Chapman-Ferraro current. The solar wind pressure remains unchanged so that this latter effect requires the magnetospheric front lobe to be somewhat smaller than it would be otherwise. The front-lobe shrinkage implies a relocation of magnetic flux, with less flux in the front lobe, more in the tail. Associated effects include lower cusp latitudes and increased tail flare. Coroniti and Kennel predict that the steady-state change in lobe radius is proportional to the generator electromotance ΔV_{mp}, inversely proportional to the net ionospheric resistance and to the intensity of the Chapman-Ferraro currents. Estimated displacements are in the range 1 to 2 R_E.

4. Nonsteady Effects

An important recent advance in open magnetospheric physics is the realization that nonsteady effects play a dominant role.

When magnetopause reconnection is enhanced, in response to a southward turning of B_{ip}, the electromotance ΔV_{mp} of the magnetopause generator increases. This requires a decrease in front lobe size, i.e., a net transport of magnetic flux from that lobe into the tail, an effect associated with the substorm growth phase. Neglecting inductive effects, Coroniti and Kennel (1973) analyzed this nonsteady process. They found a time constant of about 20 min for the establishment of a new equilibrium, regardless of the reconnection rate, and in reasonable accord with the substorm growth phase duration. A single instance of front lobe erosion, at the rate of 1 R_E h^{-1}, has been observed (Aubry et al., 1971). During this event, $B_{\perp}$ remained small most of the time. But occasional large values did occur and suggest patchy reconnection, i.e., a highly time and space variable distribution of $B_{\perp}$.

Erosion of the magnetospheric front lobe, following the onset of magnetopause reconnection occurs because flux transport into the tail initially exceeds the convective return flux, the latter being impeded by the ionosphere. Indications are (Atkinson, 1967) that in a substorm, the return convection initially occurs without tail reconnection, the result being a gradual depletion of closed field lines on the nightside, a thinning of the plasma sheet (Hones et al., 1967), and an earthward displacement of the tail current sheet. When tail reconnection is initiated at the onset of the substorm expansion phase, one expects the night side auroral ionosphere to impede ionospheric motion so that the tail X line initially moves in the antisolar direction with an associated poleward motion of the boundary between open and closed field lines. It also appears that reconnection does not occur simultaneously across the entire tail.

The lumped parameter approach, of which the Coroniti-Kennel model is a simple example, should prove extremely useful in future studies of several of the aforementioned nonsteady effects.

5. Field Reconnection Geometries

The most important element in the open magnetosphere model is the magnetic field reconnection process, which occurs at the magnetopause as well as in the tail, albeit in somewhat different configurations and not necessarily concurrently.

Reconnection occurs between two oppositely magnetized plasmas separated by a thin current sheet. Its initiation is marked by the creation of an X-type magnetic null line and the switching on of a small magnetic field component $B_\perp$ perpendicular to the sheet, to form the geometries apparent near the nose, and in the tail of the open magnetosphere in Figure 2a. An electric field E_{mp} is established parallel to the sheet but perpendicular to the original magnetic field, and a plasma flow ensues, described by $E + v \times B = 0$, except near the X line. The model involves slow inflow with speed v_1 from both sides toward the null point and rapid ejection perpendicular to the inflow and to E_{mp}. The ejection speed is of the order of the Alfvén speed in the inflow regions. In a symmetrical (tail) model the inflow has high magnetization whereas the ejected flow contains only the weak magnetic field $B_\perp$. For this reason the process is sometimes referred to as magnetic field annihilation.

An alternate annihilation model, proposed by Alfvén (1968) and pursued by Dessler (1971) and Cowley (1971), postulates $B_\perp = 0$ and particle ejection along E_{mp}. For reasonable plasma densities, E_{mp} and the annihilation rates are exceedingly small. Hence, the Alfvén model seems more relevant to nonreconnecting than to reconnecting conditions. The model will not be discussed further here.

As modelled by Petschek and coworkers (Levy *et al.*, 1964; Petschek, 1966), magnetospheric reconnection is dominated by MHD waves in most of the flow, with resistive effects limited to a narrow channel around the X line, the diffusion region. A similar incompressible model (Figure 3a) has been proposed by Sonnerup (1970). It forms the only nonsingular member of a family of self-similar flows studied by Yeh and Axford (1970). In the wave regime, this model satisfies the equations of motion exactly, and, being analytically simple, it lends itself readily to the study of special effects. The model contains a set of waves, A–A, that will not occur in the real flow. These waves merely represent a mathematical lumping of MHD interactions in the inflow regions. The remaining waves, B–B, represent the slow shocks in Petschek's model. To date, the latter remains the only reasonably satisfactory compressible model.

Figure 3b shows qualitatively the modification of the symmetrical model when

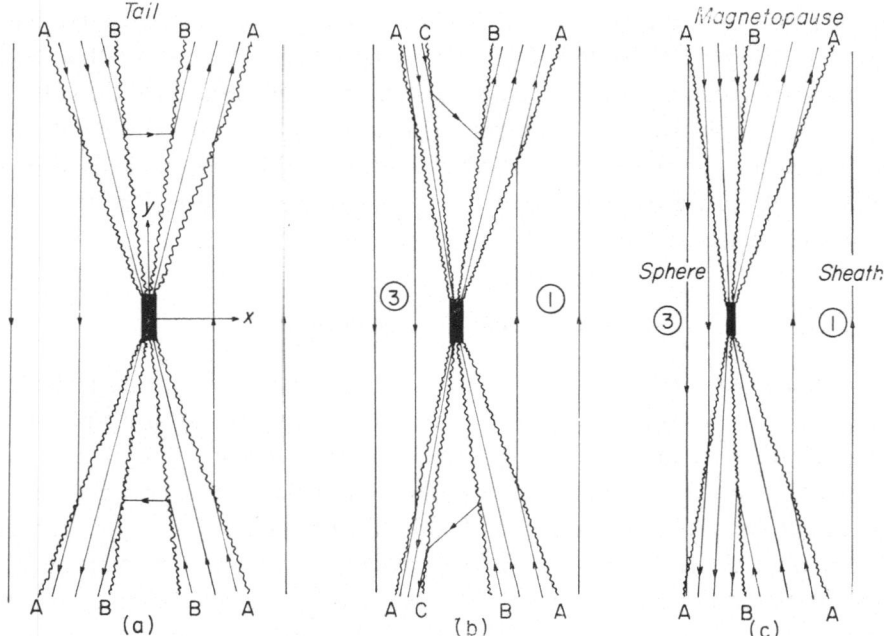

Fig. 3. Qualitative reconnection geometries; (a) symmetrical (tail) reconnection; (b) Alfvén speed in region 3 greater than in region 1; and (c) magnetopause reconnection model which occurs if $\sqrt{\varrho_1/\varrho_3} = 1 + B_1/B_3 + \sqrt{1 + 2(1 + B_1/B_3)B_1/B_3}$. Inflow is assumed to be along x axis, outflow along y axis. Solid lines are magnetic field lines, wavy ones denote standing wave fronts. Diffusion region is at center of each figure.

the left-hand inflow region has a larger Alfvén speed than the right-hand one. If this asymmetry is enhanced, the waves C–C ultimately disappear, and the magnetopause model, shown in Figure 3c, results. In this model the annihilation aspect of the reconnection process is absent. The leading waves (A–A) are unphysical, as before, while the Petschek waves (B–B) resolve into rotational discontinuities and slow expansion fans in compressible flow (Levy *et al.*, 1964).

Another modification of Figure 3a is obtained by inclusion of a velocity component along the magnetic fields in the inflow. The resulting field pattern is asymmetric about the x axis. But indications are that a considerable amount of sidewind is acceptable, so that the X line is unlikely to be blown off to infinity either at the magnetopause or in the tail. This result is of particular importance for the Dungey (1963) model for northward reconnection.

6. Limits on Reconnection

Petschek's (1964) reconnection model predicts a maximum flow speed, v_1, toward the X line, given by $v_1 < k v_{A_1}$, or equivalently $B_\perp < k B_1$, where v_{A_1} is the Alfvén speed based on the field B_1 in the inflow region. The coefficient k depends logarithmically upon conductivity and overall length scale and has values in the range 0.1–0.2.

The model in Figure 3 has a similar restriction but with $k=1+\sqrt{2}$. The discrepancy in k value probably is due to a different definition of v_1 and v_{A_1} in the two models (Sonnerup, 1973).

Petschek's calculation requires the magnetic field in the inflow region to become uniform at large distances from the X line. Under conditions of maximum reconnection, he finds the field at infinity to be considerably higher, and v_1 correspondingly lower, than they are adjacent to the diffusion region. Petschek evaluates v_1 and v_{A_1} at infinity, while the k value of the similarity solution corresponds to the evaluation of these quantities in the inflow but immediately adjacent to the diffusion region. This accounts for the difference in k value and also for the fact that Petschek's k value depends on overall length scale while the k value resulting from the similarity solution depends only on conditions in the diffusion region. Unless observations of v_1 and B_1 adjacent to the diffusion region are available, one is restricted to the use of Petschek's version of the upper limit on reconnection. Recent polar cusp observations (Burch, 1973) appear compatible with the Petschek limit, while direct measurements of $B_\perp$ (Sonnerup and Cahill, 1967) as well as values of $B_\perp$ calculated from ΔV_{pc} indicate a somewhat smaller k value.

At the magnetopause the two reconnecting fields usually are not antiparallel, and it

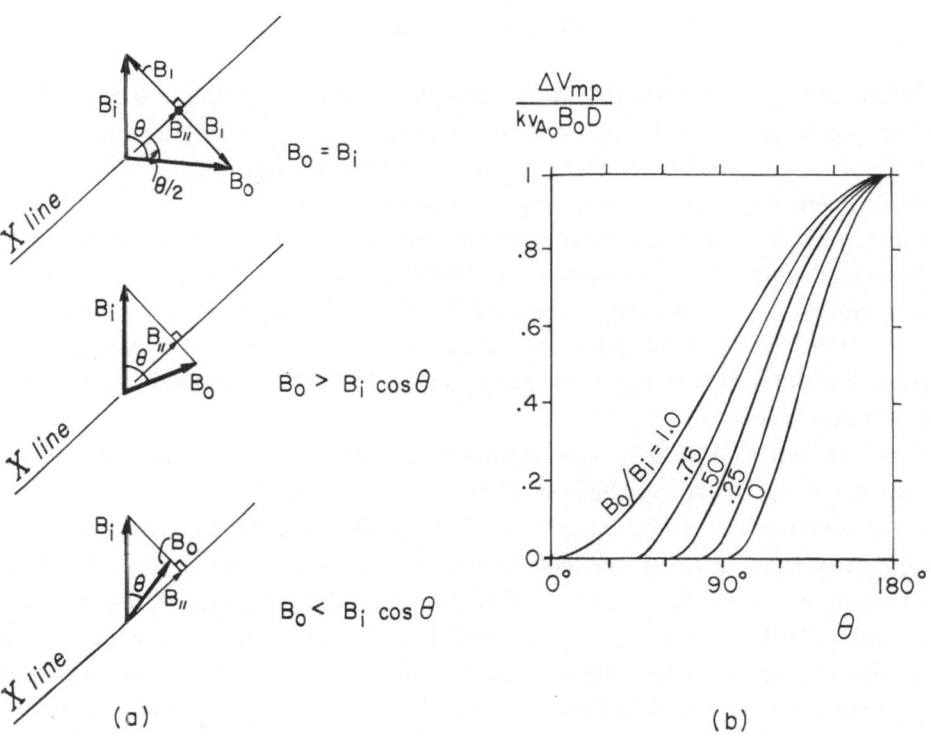

Fig. 4.   Reconnection when fields B_1 and B_0 are not antiparallel; (a) the X line is perpendicular to the vector $(B_1 - B_0)$; and (b) impressed magnetopause voltage versus the angle, θ, between B_1 and B_0. Reconnection does not appear possible for $B_0 < B_1 \cos\theta$.

is often asserted (Petschek, 1966) that the maximum reconnection rate is proportional to $\sin(\theta/2)$, θ being the angle between the two fields. This result is obtained simply by adding a constant magnetic field component, $B_{\parallel}$, along the X line to the basic configuration in Figure 3a. The flow remains unchanged but v_{A_1} in the formula $v_1 < k v_{A_1}$ is based, not on the total field, B_t, in the inflow region, but on the component $B_1 = $ $= B_t \sin(\theta/2)$ perpendicular to the X line. Thus the maximum value of v_1 becomes proportional to $\sin(\theta/2)$. However, the quantity of major geophysical interest is $\Delta V_{mp} = E_{mp} D$. Since $E_{mp} = v_1 B_1$ it follows that the maximum value of ΔV_{mp} is proportional to $\sin^2(\theta/2)$.

For unequal field magnitudes B_i and B_o (just inside and outside the magnetopause, respectively), the angular dependence of ΔV_{mp} becomes more complicated. The X line orientation is constructed as in Figure 4a. For $B_o < B_i$, a configuration with two *opposing* magnetic fields perpendicular to the X line and a constant field $B_{\parallel}$ along that line is possible only when $\cos\theta < B_o/B_i$. When $\cos\theta > B_o/B_i$ reconnection in the simple geometry of Figure 2 is impossible. Either the magnetosphere is closed or, less likely, the reconnection model for northward fields (Dungey, 1963) applies.

For $\cos\theta < B_o/B_i$, and with the subscript o denoting conditions outside the magnetopause, the impressed voltage obeys the inequality (Sonnerup, 1974)

$$\Delta V_{mp} < k v_{A_o} B_0 D \frac{(B_0/B_i - \cos\theta)^2}{[1 + (B_0/B_i)^2 - 2(B_0/B_i)\cos\theta]}.$$

In reality, k may depend not only on various plasma parameters but also on θ, B_i/B_o, and on overall geometry. From a plot of the above relationship (Figure 4b) it is seen that for B_o substantially less than B_i, the impressed voltage remains small unless $\theta > \pi/2$, in agreement with a fairly large body of observations (e.g., Arnoldy, 1971).

Current ideas are that magnetopause reconnection sets in gradually as B_{ip} turns increasingly south $(\theta > \pi/2)$ while tail reconnection, which always has $\theta = \pi$, occurs rather more abruptly and only after a substantial build up of the tail field energy. This difference between symmetric and asymmetric reconnection is not understood at present, but the following argument suggests that the explanation might be found in the diffusion region:

Consider first symmetrical (tail) reconnection under the assumption that gyro-resistivity effects (Dungey and Speiser, 1969; Sonnerup, 1970; Speiser, 1970) dominate. The gyro-conductivity σ is estimated by taking the effective 'collision' time in the usual conductivity formula to be one-half the gyro-period in the ejection field $B_\perp$. Thus $\sigma \approx \frac{1}{2}\pi n e/B_\perp$ where n is the particle density. The halfwidth, x^*, of the diffusion region must adjust itself so that the magnetic Reynolds number $\mu_o \sigma v_1 x^* \approx 1$ which means a smaller x^* value the larger the reconnection speed v_1. But x^* must remain greater than some minimum value, of the order of an ion gyroradius in the field B_1. Smaller values of x^* would be difficult to account for since the tail sheet (and magnetopause) are thought to have a thickness not less than an ion gyroradius. From the above relations, along with the formula $v_1/v_{A_1} \approx B_\perp/B_1$, and after elimination of σ and x^*, the following condition for symmetrical reconnection emerges (Sonnerup, 1973)

$$\beta_p < 4/\pi^2.$$

Here β_p is the ratio of perpendicular ion pressure to magnetic pressure in the inflow regions. The value $4/\pi^2$ is uncertain by a factor of 2 or more but should remain of order unity. A restriction on the β value of the plasma results also when the ion-acoustic two-stream instability dominates the resistivity (Sonnerup, 1973) but the electron to ion temperature ratio is unlikely to be sufficiently high in the tail (or at the magnetopause) for this instability to be of importance. Other current-driven instabilities are likely to yield similar restrictions on the β value. But such instabilities may not have to be invoked, since the gyro-resistivity, associated with the magnetic scattering of particles off the X line, appears sufficient to account for required reconnection rates, as long as β_p remains sufficiently low.

If a $1/\beta$ threshold effect in fact exists, it is natural to assume that stable tail conditions should be such that $1/\beta$ lies slightly below the threshold value. Two known attributes of the substorm growth phase are an increase in tail-field intensity and a cooling of the plasma sheet particles (Hones, 1972), both resulting in an increase of $1/\beta$. Further, if the plasma sheet thickness shrinks to zero (Russell, 1972) a very large increase in $1/\beta$ would occur. When the threshold value is reached tail reconnection, and with it the substorm expansion phase, sets in. Since the gyro-conductivity is inversely proportional to the reconnection rate, the latter will reach its maximum permitted value extremely rapidly.

The apparent absence of a $1/\beta$ threshold at the magnetopause may perhaps be explained in terms of geometrical differences in the diffusion regions of symmetrical and asymmetrical reconnection, including the absence or presence of a field component $B_\parallel$. For instance, the effective conductivity may be substantially lower in the asymmetrical case on account of the higher field intensity in the ejection flow. This and other effects may make the threshold sufficiently low to remain unnoticed in normal circumstances. And we do not know for certain that the threshold is unimportant at the magnetopause.

The preceding discussion indicates that reconnection is poorly understood at present. In particular, further research into diffusion-region processes is needed.

7. Magnetopause

The MHD reconnection models in Figure 3b and 3c predict the appearance of a rotational discontinuity whenever the waves B–B are required to reverse the field component tangential to the wave. Hence, this type of discontinuity should be a prominent feature of the reconnecting magnetopause. A few observations of such structure, including substantial values of $B_\perp$, have been reported (Sonnerup and Cahill, 1967; Sonnerup, 1971; Sonnerup and Ledley, 1974). But the data interpretation is hampered by a lack of understanding of the structure of the rotational discontinuity when its thickness is comparable to the ion gyroradius. Indeed, the factors that determine the structure and thickness of the magnetopause in open as well as closed models remain largely unknown. Observations suggest that the structure often is two or three dimensional, and that reconnection is likely to be a patchy phenomenon (Aubry *et al.*, 1971). Even so, in the open model the component $B_\perp$ on the average should point toward the Earth in the northern hemisphere, away in the southern. A possible open but patchy magnetopause structure is topologically similar to the tail structure proposed by Schindler and Ness (1972).

Finally, it is proposed that a synoptic survey of the magnetopause surface-field might help distinguish between the closed and open topologies in Figures 1b and 2b.

Acknowledgment

The research was supported by National Aeronautics and Space Administration under Grant NGR30-001-040 to Dartmouth College.

References

Alfvén, H.: 1968, *J. Geophys. Res.* **73**, 4379.
Arnoldy, R. L.: 1971, *J. Geophys. Res.* **76**, 5189.
Atkinson, G.: 1967, *J. Geophys. Res.* **72**, 1491.
Aubry, M. P., Kivelson, M. G., and Russell, C. T.: 1971, *J. Geophys. Res.* **76**, 1673.
Burch, J. L.: 1972, *J. Geophys. Res.* **77**, 6696.
Burch, J. L.: 1973, *Radio Sci.* **8**, 955.
Coroniti, F. V. and Kennel, C. F.: 1973, *J. Geophys. Res.* **78**, 2837.
Cowley, S. W. H.: 1971, *Cosmic Electrodyn.* **2**, 90.
Dessler, A. J.: 1971, *J. Geophys. Res.* **76**, 3174.
Dungey, J. W.: 1961, *Phys. Rev. Letters* **6**, 47.
Dungey, J. W.: 1963, in C. DeWitt, J. Hieblot, and A. Lebeau (eds.), *Geophysics, The Earth's Environment*, Gordon and Breach, New York, N.Y., p. 503.
Dungey, J. W. and Speiser, T. W.: 1969, *Planetary Space Sci.* **17**, 1285.
Fairfield, D. H.: 1971, *J. Geophys. Res.* **76**, 6700.
Gurnett, D. A. and Frank, L. A.: 1973, *J. Geophys. Res.* **78**, 145.
Heppner, J. P.: 1972, in E. R. Dyer (ed.), *Critical Problems of Magnetospheric Physics*, National Academy of Sciences, Washington, D.C., p. 107.
Hones, E. W. Jr.: 1972, in B. M. McCormac (ed.), *Earth's Magnetospheric Processes*, D. Reidel Publishing Company, Dordrecht-Holland, p. 365.
Hones, E. W., Jr., Asbridge, J. R., Bame, S. J., and Strong, I. B.: 1967, *J. Geophys. Res.* **72**, 5879.
Levy, R. H., Petschek, H. E., and Siscoe, G. L.: 1964, *AIAA J.* **2**, 2065.
Meng, C. I.: 1970, *J. Geophys. Res.* **75**, 3252.
Nishida, A. and Obayashi, T.: 1972, in E. R. Dyer (ed.), *Critical Problems of Magnetospheric Physics*, National Academy of Sciences, Washington, D. C., p. 179.
Petschek, H. E.: 1964, in W. N. Hess (ed.), *The Physics of Solar Flares*, NASA SP-50, Washington, D.C., p. 425.
Petschek, H. E.: 1966, in R. J. Mackin, Jr., and M. Neugebauer (eds.), *The Solar Wind*, Pergamon Press, New York, N.Y., p. 257.
Russell, C. T.: 1972, in E. R. Dyer (ed.), *Critical Problems of Magnetospheric Physics*, National Academy of Sciences, Washington, D.C., p. 1.
Schindler, K. and Ness, N. F.: 1972, *J. Geophys. Res.* **77**, 91.
Sonnerup, B. U. Ö: 1970, *J. Plasma Phys.* **4**, 161.
Sonnerup, B. U. Ö.: 1971, *J. Geophys. Res.* **76**, 6717.
Sonnerup, B. U. Ö.: 1973, in R. Ramaty and R. G. Stone (eds.), *High Energy Phenomena on the Sun*, NASA X-693-73-193, p. 357.

Sonnerup, B. U. Ö.: 1974, *J. Geophys. Res.*, to appear.
Sonnerup, B. U. Ö. and Cahill, L. J., Jr.: 1967, *J. Geophys. Res.* **72**, 171.
Sonnerup, B. U. Ö. and Ledley, B. G.: 1974, *J. Geophys. Res.*, to appear.
Speiser, T. W.: 1970, *Planetary Space Sci.* **18**, 613.
Spreiter, J. R., Alksne, A. Y., and Summers, A. L.: 1968, in R. L. Carovillano, J. F. McClay, and H. R. Radoski (eds.), *Physics of the Magnetosphere*, D. Reidel Publishing Company, Dordrecht-Holland, p. 301.
Yeh, T. and Axford, W. I.: 1970, *J. Plasma Phys.* **4**, 207.

NEUTRAL LINE IN THE MAGNETOTAIL

ATSUHIRO NISHIDA

Institute of Space and Aeronautical Science, University of Tokyo, Komaba, Tokyo 153, Japan

1. Introduction

In the open model of the magnetosphere, it is thought that magnetic field lines are transported from the dayside of the magnetosphere to the magnetotail as a consequence of the reconnection process operating on the dayside magnetopause (Dungey, 1963). In order that the flux content of the tail does not grow indefinitely it is required that the reconnection process operates also in the magnetotail and allows the field lines to return to the dayside. It has been suggested that this nightside reconnection does not proceed continuously but occurs explosively giving rise to the major magnetospheric phenomenon known as the substorm (Atkinson, 1966; Axford, 1967; Dungey, 1966).

This paper briefly summarizes our recent analysis on the X-type neutral line that is formed in the near tail region during the substorm expansion phase. The work is based on the magnetic field observations by Explorers 33 and 34 in the magnetotail and the interplanetary space, and also on the low energy electron observations by Vela 4A in the magnetotail. Details of the analysis are given in Nishida and Nagayama (1973a) and Nishida and Hones (1974).

2. Neutral-Line Formation During Substorms

The ideal way of detecting the neutral line is to cover the magnetotail by a dense network of satellites equipped with magnetic probes. This being practically impossible, we have chosen to combine the observations made at different locations for different substorms on the assumption that the tail behavior is essentially the same for every substorm. In Figure 1 the tail magnetic field records for two instances of substorms (left and right) are examined. On the top is the projection of the satellite position to the xy plane. (Throughout this paper the solar magnetospheric coordinate system is employed.) Data from two satellites are available for each instance, and F and D are abbreviations for the satellite designations IMP-F (Explorer 34) and AIMP-D (Explorer 33). In the next panel ~ 3 min averages of the B_z (north-south) component of the magnetic field obtained by these satellites are shown. Records in the last panel are tracings of ground magnetograms from equatorial (top) and auroral-zone (bottom) stations that happened to be located near the geomagnetic midnight meridian.

In the ground magnetograms the onset of the expansion phase of the substorm is recognized as onsets of a low latitude positive bay and of an auroral zone negative bay (cf., Iijima and Nagata, 1972). The vertical lines in the Figure are drawn with reference to the onset time noted in the equatorial record which is less subject to local modifications. In the February 21 event (left) both satellites are located at $x \simeq -10\ R_E$

B. M. McCormac (ed.), Magnetospheric Physics, 35–44. All Rights Reserved.
Copyright © 1974 by D. Reidel Publishing Company, Dordrecht-Holland.

ATSUHIRO NISHIDA

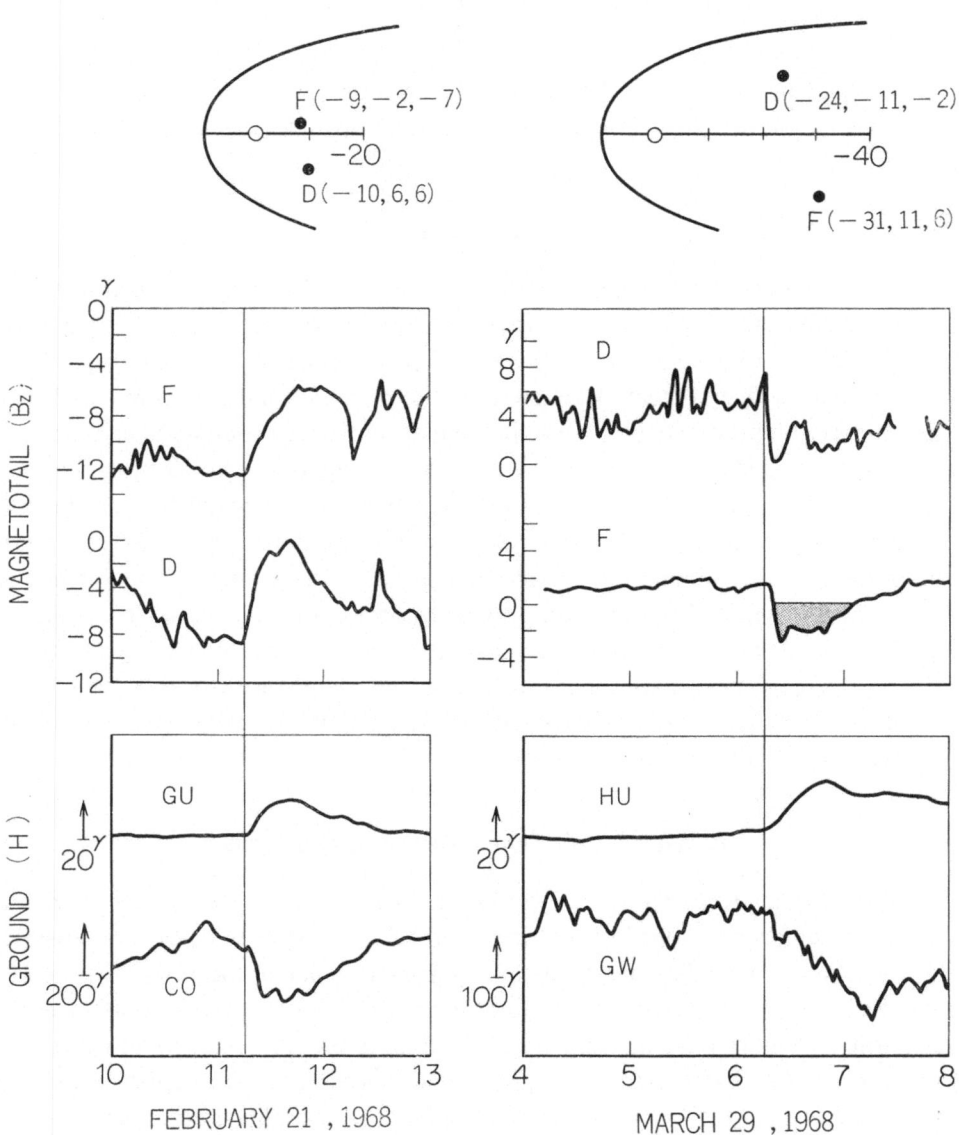

Fig. 1.   Tail magnetic field records (middle panel) obtained during substorms at $x \simeq -10$ R$_E$ (left) and $x \simeq -30$ R$_E$ (right). Observing sites are noted in the top panel in the unit of R$_E$.

and they have recorded the increase in B_z at the expansion phase onset. On the other hand, in the March 29 event (right) where both satellites are around the $x \simeq -25$ to -30 R$_E$ range the decrease in B_z is observed at the expansion phase onset. In particular, at IMP-F located at $y \simeq 11$ R$_E$ B_z turns clearly southward about 10 min after the expansion-phase onset time determined by the ground data. Thus, the magnetotail B_z variation observed in association with the substorm expansion phase changes sign somewhere between $x \simeq -10$ and -25 R$_E$.

This tendency is confirmed for a greater number of cases in Nishida and Nagayama (1973a), and the sequence of magnetic field structural change in the near midnight meridian is shown on left hand side of Figure 2. The top is the familiar quiet time field structure observed about 100 min before the expansion phase onset. The second is the field structure observed toward the end of the growth phase, and it is characterized by the severe stretching of low latitude field lines in the $x \simeq -10$ to -20 R$_E$

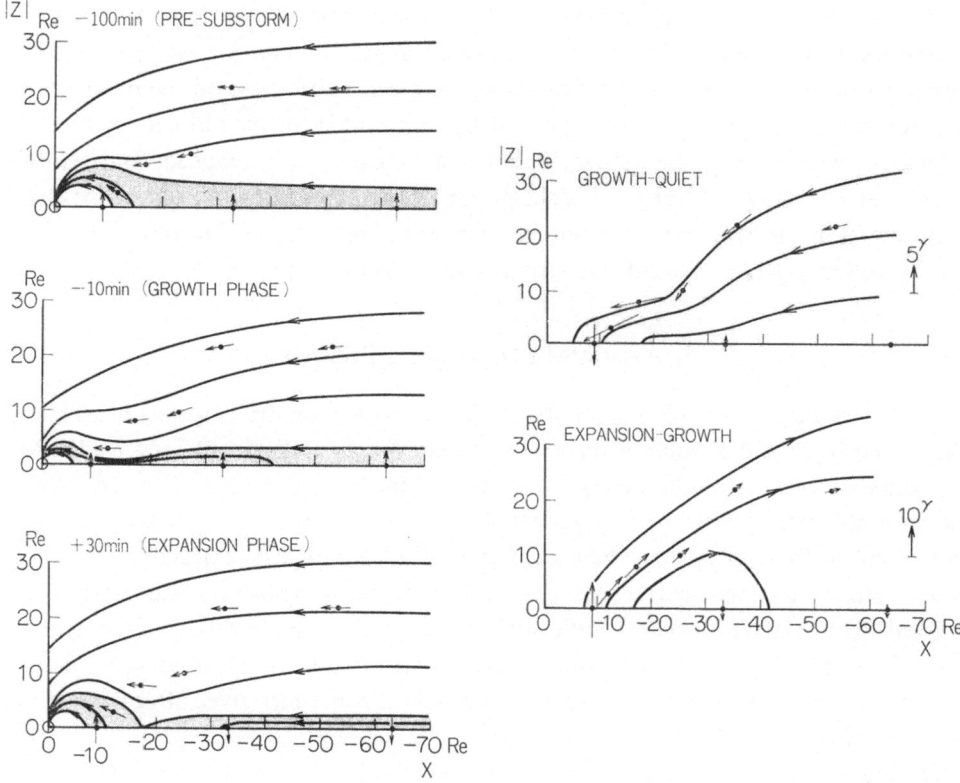

Fig. 2. Left: Structural change of the magnetotail during substorms. Time is counted from the onset time of the low latitude positive bay in the near midnight meridian. The shading expresses the plasma sheet location. Right: Difference in the magnetic field distribution between successive phases of substorms (Nishida and Nagayama, 1973a).

range and also by the increased flaring angle (with respect to the xy plane) and magnitude of the high latitude magnetic field (cf., McPherron, 1972). Then in the expansion phase the field structure turns to the one shown in the bottom which involves a neutral line located in the near tail region.

Unfortunately, the low latitude range at $x \simeq -10$ to -25 R$_E$ where the neutral line is expected to be located is not adequately covered by the available satellite magnetometer records, but the inferences made from the plasma analysis (Hones *et al.*, 1973) locate the neutral line at $x \simeq -15$ R$_E$. The time delay of about 10 min noted in the March 29 event between the expansion phase onset on the ground and the southward field turning at IMP-F would be attributable, at least in part, to the X-type field configuration to spread from $x \simeq -15$ R$_E$ to the IMP-F location of $x \simeq -31$ R$_E$. As exemplified by the March 29 event the B_z decrease at $x \lesssim -25$ R$_E$ is observed in a wide range of y (viz., local time) but a clear turning of B_z to the southward polarity is seen in a limited range of $|y| \lesssim 10$ to 15 R$_E$. The neutral line seems to have a length of 20 to 30 R$_E$ and does not appear to reach the magnetopause.

The right hand side of Figure 2 shows the difference of the magnetic field distributions between the growth phase and the quiet state (above) and between the expansion phase and the growth phase (below). It is seen that during the growth phase the intensification of the currents flowing in the plasma sheet and on the magnetotail surface in the $x \gtrsim -30$ R$_E$ range is principally responsible for the tail field distortion. When the expansion phase sets in these currents are sharply reduced, and the tail current around the earthward end of the plasma sheet drops below the quiet time level, resulting in the neutral line formation in the near tail region. The duration of the field configuration with the neutral line usually ranges from 0.5 to 1 h.

3. Associated Plasma Sheet Behavior

The plasma sheet responds sensitivity to the formation of the neutral line. This is illustrated in Figure 3 where a record of the tail magnetic field obtained by IMP-F is compared with the simultaneous observation of the low energy (63 eV to 18.5 keV) electron pressure by Vela 4A (top panel). The satellite coordinate at 1000 UT is given in the unit of R$_E$ after the satellite code. (In this Figure only, z is replaced by dz which is the distance from the neutral sheet (where B_x changes sign) position estimated by the Russell-Brody (1967) formula.) Both satellites are in the neighborhood of the neutral sheet. The bottom panel presents the ground magnetograms from low latitude (above) and auroral zone (below) stations around midnight. Local times given after the station code are also for 1000 UT.

During the interval of Figure 3 the low latitude tail field enters the southward range three times indicating the repeated formation of the neutral line earthward of the satellite position. Vertical lines are drawn at the timings of these southward turnings. The ground records testify that each of these are associated with the expansion phase of a substorm. According to the plasma data the latter two of these events are associated with clear depressions in the electron pressure lasting roughly throughout the duration of the southward tail field. The same correlation as the above has been noted in several other occasions where simultaneous field and plasma observations at low |dz| are available, and it can be concluded that the rapid thinning of the plasma sheet occurs beyond the neutral line (Nishida and Hones, 1974). When the plasma reappears after the depression it tends to be rich in the energetic electron ($\gtrsim 36$ keV) component, and occasionally a burst of energetic electrons is observed in association with the southward turning of the low latitude tail field.

As for the first event in Figure 3 the electron pressure does not drop but the energetic electron flux has decreased at the corresponding timing. Since the distance

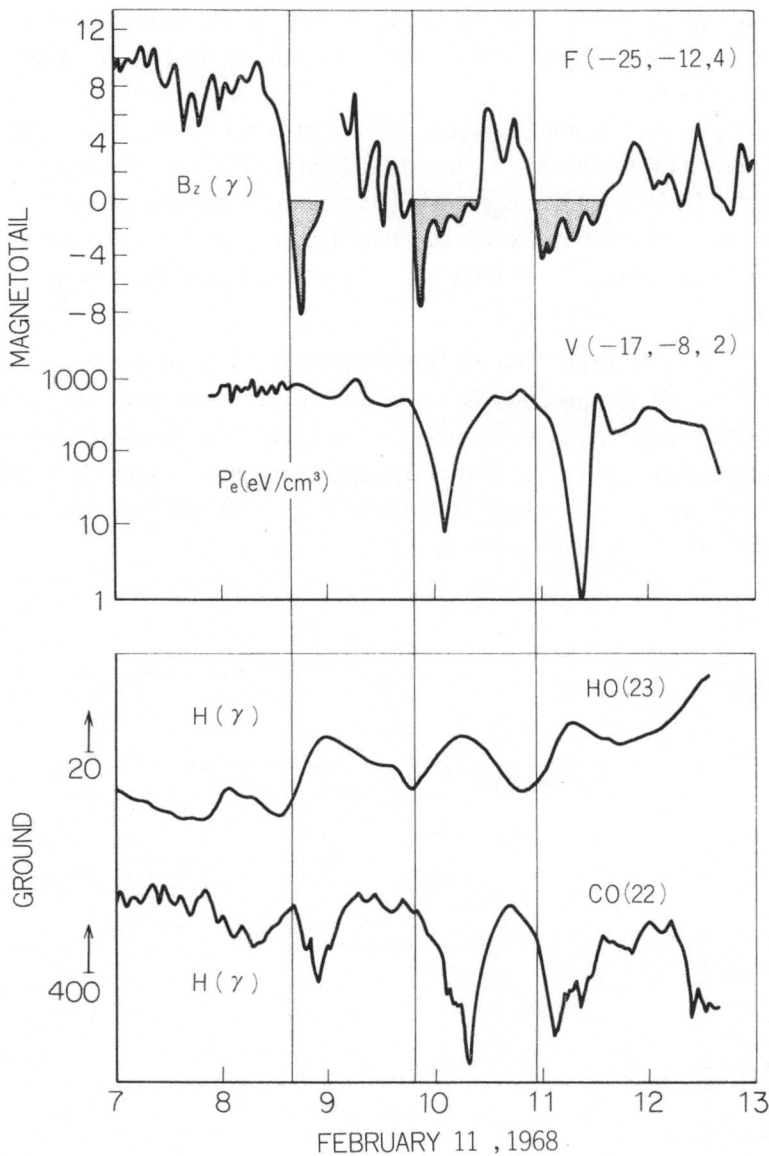

Fig. 3. Comparison of a tail field record (top panel, above) with a simultaneous electron pressure
data (top panel, below) during substorms (Nishida and Hones, 1974).

|dz| of Vela 4A is steadily increasing during this interval from 1.3 R_E (at 0800) to
2.4 R_E (at 1100), this event may be taken to demonstrate that while the plasma sheet
thins it does not entirely disappear. To make the picture of the substorm-associated
plasma sheet behavior complete it should also be noted that in high latitudes of the tail
the plasma sheet thinning is observed to start prior to the expansion phase onset
(Hones et al., 1971). In addition, the inward motion of the earthward boundary of

the plasma sheet is observed at the geosynchronous orbit (6.6 R_E) for ~1 h before
the expansion phase (Mende *et al.*, 1972), viz, before the formation of the neutral line
in the near tail region. The plasma sheet structure in Figure 2 is drawn taking these
points into account.

By way of passing it is noted that two more substorm expansion phases are recognized in Figure 3 to start around 0750 and 1203 but the southward turning of the field
is not detected by IMP-F. This is probably because the satellite is brought to the high
latitude region of the tail due to the irregular flapping or tipping motion of the tail.
Also, the observation is made at large | y | where the effect of the neutral line is not
always clear.

4. Dependence on the Interplanetary Condition

Since the nightside reconnection is considered to be the consequence of the flux
build-up in the magnetotail, the occurrence of the tail neutral line is expected to depend
on the interplanetary condition that is thought to govern the dayside reconnection
rate. The relevant interplanetary parameter is the north-south component B_z of the
interplanetary magnetic field (Dungey, 1963).

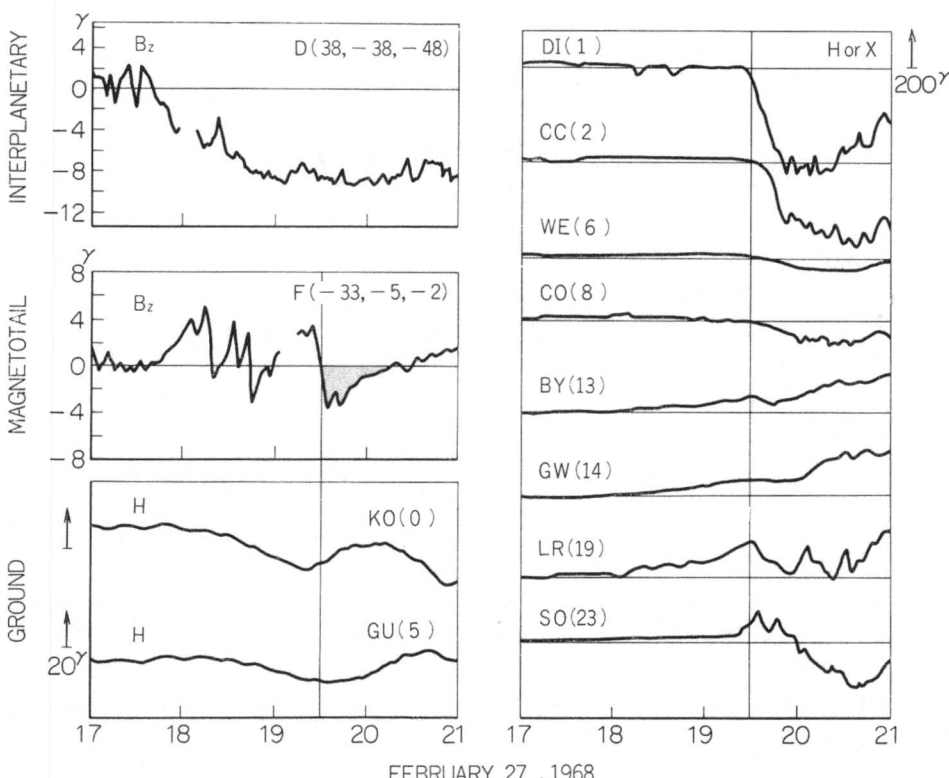

Fig. 4.   Response of the tail field (left, middle) and the ground field at nightside low latitude (left,
bottom) and at nightside auroral zone (right) to a southward movement of the interplanetary magnetic
field (left, top). Satellite positions (solar magnetospheric) and station local times are given for the
expansion phase onset time.

Figure 4 shows the tail and ground field response for a particularly simple case in which the interplanetary field develops a large ($\sim 8\gamma$) southward component after an interval of $|B_z| \lesssim 2\gamma$ lasting for about 1 h (left, top). In this instance the solar wind speed is about 330 km s^{-1} and the transit time correction (estimated by dividing the AIMP-D x coordinate by the solar wind speed) is roughly 11 min. Following the southward movement of the interplanetary magnetic field, B_z in the low latitude tail enters the southward range several times but the most persisting and most intense southward tail field is seen to start about 100 min after the arrival of the southward interplanetary field (left, middle). In the rest of the records (which are ground magnetograms) it can be seen that the beginning of the above interval of the southward tail field (indicated by a vertical line) corresponds to the intensification of the disturbance field that is identifiable as the onset of the substorm expansion phase.

Thus the present sample suggests that the principal reconnection process on the nightside of the magnetosphere starts to operate 1 to 2 h after the onset (or, intensification) of the dayside reconnection. During this 1 to 2 h interval the flux build up is supposed to be proceeding in the lobe region of the tail, and indeed the lobe field has been seen to intensify for 1 to 2 h following well defined southward movements in the interplanetary magnetic field (e.g., McPherron, 1972; Nishida and Nagayama, 1973b). The 'growth phase' duration of this magnitude has also been deduced from particle and field observations at various locations in the magnetosphere during isolated substorm events (e.g., Mende et al., 1972). Although brief intervals of the southward tail field noted in Figure 4 prior to the expansion phase might be taken to be an indication of the occasional formation of neutral lines, such premature neutral lines obviously do not develop to cause a major release of energy from the magnetotail. (These brief southward tail fields may actually be due to flapping or tipping motions of the tail and do not necessarily reflect the neutral-line formation.)

In the sample presented above the interplanetary field continues to have a southward polarity for several hours until ~ 0100 of the next day (February 28), but notably the southward tail field does not last throughout that interval. About 45 min after its southward turning the tail field polarity returns to northward, and correspondingly the ground magnetic activity in the near midnight region begins to subside. This suggests that the nightside reconnection is not a continuous process, even when the interplanetary magnetic condition indicates the persistence of the dayside reconnection. Another case in favor of this inference is Figure 3 which corresponds to a large and relatively steady interplanetary field with a southward polarity ($B_z \simeq -12\gamma$). However, the southward field in the low latitude tail is observed to occur only intermittently every hour with a duration of ~ 0.5 h. Thus it seems that the nightside reconnection ceases as soon as the magnetic flux content of the magnetotail is reduced to some threshold, and is not resumed until the tail flux content becomes significantly larger than that threshold value.

Figure 5 shows the response of ground and tail fields to interplanetary B_z fluctuations (top panel, second record) with a time scale of about 1 h. In the interval concerned the dynamic pressure $P_d = nmv^2$ of the solar wind (top panel, first record) is kept at

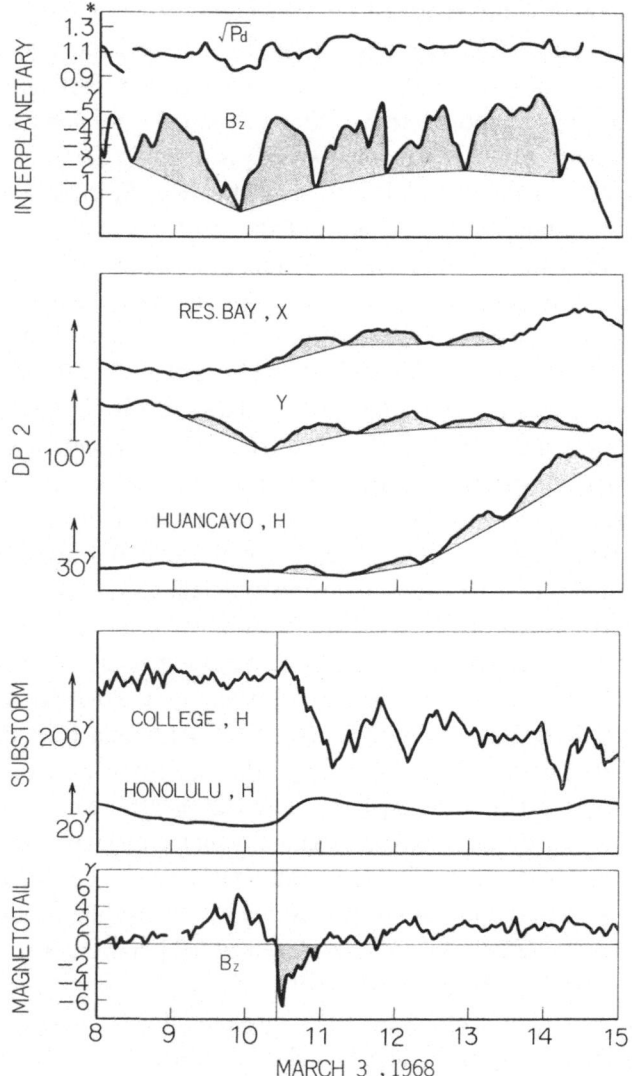

Fig. 5.   Occurrence of DP 2 (second panel) and substorm expansion phase (third and fourth panels) in the interval of fluctuating interplanetary magnetic field (top panel). At 1000, AIMP-D is monitoring the interplanetary space at $(56, -31, -36)$ R_E and IMP-F is in the low-latitude tail at $(-33, -2, -2)$ R_E. *The unit of $\sqrt{P_d}$ is $(10^{-4} \text{ dyn cm}^{-2})^{1/2}$.

a relatively steady level. The solar wind speed is about 400 km s^{-1} and the transit time correction is about 14 min. These B_z fluctuations are reflected in the ground magnetograms from polar cap (Resolute Bay) and dayside equatorial (Huancayo) stations (reproduced in the second panel) as DP 2 fluctuations. The time delay between interplanetary B_z fluctuations (with southward taken as positive) and the ground DP 2 (with H increase at the dayside equator taken as positive) is about 15 min after the transit time correction. On the other hand, the magnetogram records from the nightside auroral zone (College) and low latitude (Honolulu) stations (in the third panel) are dominated by the substorm expansion phase (DP 1) whose onset is associated with the formation of the neutral line in the near tail region as manifested by the southward turning of the low latitude tail field observed by IMP-F (last panel). In this way dayside and nightside reconnection processes are reflected in the ground magnetic field as different types of disturbance fields.

5. Conclusion

We have seen that the formation of the neutral line in the near tail region is due to the severe weakening or disruption of the cross tail current flowing through the earthward end of the plasma sheet. In the stage that immediately precedes the neutral line formation the nightside field lines are stretched to the tail-like configuration even at a distance as close to the Earth as $x \simeq -10$ R_E, and the cross tail current is highly intensified in the region where it is soon to disappear. Hence it would be natural to consider that the instability of this current flow is involved, and the search for plasma waves that have grown out of the instability is of much interest. Scarf and Fredricks (1972) have reported a couple of instances in which the growth of the electric field oscillations in the ~ 1 kHz range is observed by OGO-5 at the distance of 8 to 10 R_E following the onset of the substorm expansion phase. In order to pursue this point we asked Scarf to look at the OGO-5 VLF electric field record at the time of 12 more expansion phase onsets (i.e., cases included in Figures 2 and 6 of Nishida and Nagayama, 1973a). According to Scarf, however, the results of the examination were rather negative. Of the 12 cases examined only 2 were found to accompany the VLF electric field excitation of any kind, and it was pointed out that the prevailing condition $(T_e \ll T_i)$ in the plasma sheet was not favorable for the growth of electrostatic waves. In any event, the observing sites of OGO-5 are probably still ~ 5 R_E earthward of the neutral line position, and the advent of the vehicle which can make *in situ* observations of the neutral line characteristics is highly desirable.

As the formation of the neutral line is associated with the intensification of the electrojet flowing in the auroral zone ionosphere it may be thought that the current that disappeared from the tail is transposed to the ionosphere. Such a model, however, seems to have some difficulties. First, the auroral electrojet persists after the disappearance of the neutral line from the near tail region. All the samples presented here show that the return of the low latitude tail field from the southward to the northward range is roughly coincident with the peak of the auroral zone negative bay, but not the end of it. Second, while the intensification of the westward auroral electrojet does not seem to occur simultaneously all over the nightside ionosphere, the magnetotail collapse tends to occur nearly simultaneously all across the tail. Specifically, the negative bay onsets in the evening sector tend to show a delay that reflects the transit time of the westward traveling surge. However, the B_z increase observed by OGO-5 on the earthward side of the neutral line is within ~ 5 min of the expansion phase onset time registered in the near midnight region even when OGO 5 is in the 19 LT (solar magnetospheric) meridian (Nishida and Nagayama, 1973a). The auroral electrojet, thus, is not likely to be the mere transposition of the cross tail current in the near-tail region. During the reconnection the tail field energy is converted to the kinetic energy of the tail plasma and the heated plasma is convected to the nightside inner magnetosphere. The energy of such plasma would act to create a variety of disturbances observed on the ground as the substorm phenomena.

Finally, there have been suggestions regarding the nightside reconnection process that do not involve the neutral line in the near tail region. Examples are the one that can proceed under $B_z = 0$ (Dungey, 1972) and the other process which is associated with a neutral line located at a greater distance (c.f. Hones *et al.*, 1973). According to our analysis these processes are not likely to be the principal agent of the substorm, but a more detailed analysis is needed to assess the relevance of these suggestions to the overall magnetospheric process.

Acknowledgments

It is a pleasure to acknowledge the collaboration with Mr N. Nagayama, Dr E. W. Hones, Jr., and Dr F. L. Scarf in the course of this work. The satellite magnetometer data are observations by Dr N. F. Ness and are obtained from WDC-A, Rockets, and Satellites.

References

Atkinson, G.: 1966, *J. Geophys. Res.* **71**, 5157.
Axford, W. I.: 1967, *Space Sci. Rev.* **7**, 149.
Dungey, J. W.: 1963, in C. DeWitt, J. Hieblot, and A. Lebeau (eds.), *Geophysics, The Earth's Environment*, Gordon and Breach, New York, p. 503.
Dungey, J. W.: 1966, in B. M. McCormac (ed.), *Radiation Trapped in the Earth's Magnetic Field*, D. Reidel Publishing Company, Dordrecht-Holland, p. 389.
Dungey, J. W.: 1972, in B. M. McCormac (ed.), *Earth's Magnetospheric Processes*, D. Reidel Publishing Company, Dordrecht-Holland, p. 210.
Hones, E. W., Jr., Singer, S., Lanzerotti, L. J., Pierson, J. D., and Rosenberg, T. J.: 1971, *J. Geophys. Res.* **76**, 2977.
Hones, E. W., Jr., Asbridge, J. R., Bame, S. J., and Singer, S.: 1973, *J. Geophys. Res.* **78**, 109.
Iijima, T. and Nagata, T.: 1972, *Planetary Space Sci.* **20**, 1095.
McPherron, R. L.: 1972, *Planetary Space Sci.* **20**, 1521.
Mende, S. B., Sharp, R. D., Shelly, E. G., Haerendel, G., and Hones, E. W.: 1972, *J. Geophys. Res.* **77**, 4682.
Nishida, A. and Nagayama, N.: 1973a, *J. Geophys. Res.* **78**, 3782.
Nishida, A. and Nagayama, N.: 1973b, *Astrophys. Space Sci.* **20**, 459.
Nishida, A. and Hones, E. W., Jr.: 1974, *J. Geophys. Res.* **79**, 535.
Russell, C. T. and Brody, K. I.: 1967, *J. Geophys. Res.* **72**, 6104.
Scarf, F. L. and Fredricks, R. W.: 1972, in B. M. McCormac (ed.), *Earth's Magnetospheric Processes*, D. Reidel Publishing Company, Dordrecht-Holland, p. 329.

IONOSPHERE-MAGNETOSPHERE COUPLING

ROLF BOSTRÖM

Department of Plasma Physics, Royal Institute of Technology, S-100 44 Stockholm, Sweden

Abstract. A review is given of some aspects of the electric coupling between the ionosphere and magnetosphere. Topics discussed are the mapping of time-dependent electric fields, the effects of Birkeland currents on the magnetospheric plasma convection, the relation of Birkeland currents to ionospheric electric field structures such as field reversals, secondary Birkeland currents of ionospheric origin and triggering of auroral particle precipitation, double-layers causing voltage drops along the magnetic field lines, and an equivalent electric circuit for the substorm currents in the coupled ionosphere-magnetosphere system.

1. Introduction

The electric coupling between the ionosphere and magnetosphere is caused by the normally high conductivity along the geomagnetic field lines that connect these regions. The *ionosphere* as the term is used here, is the region of maximum electrical conductivity transverse to the geomagnetic field lines caused by numerous ion-neutral collisions. The *magnetosphere* is characterized by collision-free charged particle motion or sometimes MHD processes. Ionosphere-magnetosphere coupling is the interaction of different physical processes occurring in regions that are partly overlapping although often considered as spatially separated. This interaction is important for many phenomena such as substorms, magnetospheric plasma convection, auroral particle acceleration, and the formation of auroral structures (for an earlier review, see Boström, 1972). The coupling occurs through the flow of Birkeland (field-aligned) currents. Under certain conditions substantial voltage drops may occur along the geomagnetic field lines so that the ionosphere to some extent is decoupled from the magnetosphere.

2. Coupling of Time-Dependent Electric Fields

An electric field generated at one altitude normally spreads to other altitudes due to the high conductivity along the field lines. Small scale fields are damped, especially in the E layer of the ionosphere, but large scale fields are mapped with little damping from the outer magnetosphere to the ionosphere and even into the lower atmosphere. When we perform such a mapping the electric field pattern will be distorted due to the different scale factors for east-west and north-south fields (Mozer, 1970) caused by the magnetic field geometry. Here we will study the damping caused by the transverse conduction and displacement currents.

Considering a two-dimensional problem where the electric field is a wave traveling horizontally in the x direction

$$\mathbf{E} = (U(z), V(z), W(z)) \exp[j(\omega t - kx)] \tag{1}$$

we want to determine the variation of the complex amplitude functions U, V, and W with the altitude coordinate z. The magnitudes of U, V, and W show the field intensity in the directions $\hat{\mathbf{x}}$, $\hat{\mathbf{y}}$, and $\hat{\mathbf{z}}$ and the arguments of the complex numbers show the phase shifts relative to the source field. Using Maxwell's equations curl $\mathbf{E} = -\partial \mathbf{B}/\partial t$ and curl $\mathbf{B} = \mu_0 \mathbf{i} = \mu_0 \sigma \mathbf{E}$, where $\sigma(z)$ is the complex conductivity tensor taking the displacement current into account, and assuming the geomagnetic field lines to be vertical, we find that these functions are governed by the equations

$$\frac{d^2 U}{dz^2} - \frac{k^2}{j\omega\mu_0 + k^2/\sigma_\parallel} \left(\frac{d}{dz}\frac{1}{\sigma_\parallel}\right)\frac{dU}{dz} = \left(j\omega\mu_0 + k^2/\sigma_\parallel\right)\left(\sigma_p U + \sigma_H V\right) \qquad (2)$$

$$\frac{d^2 V}{dz^2} = \left(j\omega\mu_0\sigma_p + k^2\right) V - j\omega\mu_0\sigma_H U \qquad (3)$$

$$W = \frac{jk}{j\omega\mu_0 + k^2/\sigma_\parallel}\frac{1}{\sigma_\parallel}\frac{dU}{dz}. \qquad (4)$$

Except for the contribution from the vacuum displacement current the frequency dependent complex conductivities (Akasofu and De Witt, 1965) are derived from the momentum equations, which, for sinusoidal temporal variations and with usual notation, read

$$j\omega m_e \mathbf{v}_e = -e\left(\mathbf{E} + \mathbf{v}_e \times \mathbf{B}\right) - m_e \nu_{en}\left(\mathbf{v}_e - \mathbf{v}_n\right) - m_e \nu_{ei}\left(\mathbf{v}_e - \mathbf{v}_i\right) \qquad (5)$$

$$j\omega m_i \mathbf{v}_i = e\left(\mathbf{E} + \mathbf{v}_i \times \mathbf{B}\right) - m_i \nu_{in}\left(\mathbf{v}_i - \mathbf{v}_n\right) - m_e \nu_{ei}\left(\mathbf{v}_i - \mathbf{v}_e\right) \qquad (6)$$

$$j\omega \frac{n_n}{n_e} m_n \mathbf{v}_n = -m_e \nu_{en}\left(\mathbf{v}_n - \mathbf{v}_e\right) - m_i \nu_{in}\left(\mathbf{v}_n - \mathbf{v}_i\right). \qquad (7)$$

Working through some tedious algebra to obtain the electric current density $\mathbf{i} = n_e e(\mathbf{v}_i - \mathbf{v}_e)$, introducing the approximations $m_e \ll m_i$ and $m_e \nu_{en} \ll m_i \nu_{in}$ valid at all altitudes, we find

$$\mathbf{i} = \sigma_p \mathbf{E} + \sigma_H \hat{\mathbf{B}} \times \mathbf{E} + \sigma_\parallel \mathbf{E}_\parallel \qquad (8)$$

$$\sigma_p = \frac{a}{1 + a^2}\frac{n_e e}{B} + j\omega\varepsilon_0 \qquad (9)$$

$$\sigma_H = \frac{1}{1 + a^2}\frac{n_e e}{B} \qquad (10)$$

$$\sigma_\parallel = \frac{n_e e^2}{m_e\left(j\omega + \nu_{en} + \nu_{ei}\right)} + j\omega\varepsilon_0 \qquad (11)$$

$$a = \frac{\omega_i}{j\omega + \nu_{in}\dfrac{1}{1 + \dfrac{n_e m_e \nu_{in}}{j\omega n_n m_n}}} + \frac{j\omega + \nu_{en} + \nu_{ei}}{\omega_e} \qquad (12)$$

where the term for the vacuum displacement current, important only in the lowest atmosphere, has been added to σ_P and $\sigma_\parallel$. The complex conductivities are shown in Figure 1 for a typical nighttime ionosphere. Note that $\sigma_P = \sigma_H = 0$ for $\omega = 0$, as all particles are accelerated to the same velocity that is $\mathbf{E} \times \mathbf{B}/B^2$. For a frequency range from roughly $\omega = 10^{-6}$ to $\omega = 1$ the conductivities of the E layer are real frequency independent numbers given by the commonly used formulas. However, at higher altitudes or other frequencies the effect of the inertia of the charged and neutral particles is important and must be included as done in our derivation. The magnetospheric plasma effectively acts as a capacitor for the frequency range studied here.

We have assumed a normalized source field in the outer magnetosphere $\mathbf{E} = \cos(kx - \omega t)\,\hat{\mathbf{x}}$, that is $U = 1 + j0$ and $V = 0 + j0$ at $z = 35000$ km (the equatorial plane for a dipolar field line from the auroral zone). The ground is a good conductor so no horizontal fields should exist there, thus $U = 0 + j0$ and $V = 0 + j0$ at $z = 0$. Equations (2) and (3) have been solved numerically for U and V with these boundary conditions, and W is derived using Equation (4). For $\omega < 1$ rad s^{-1} and a wavelength shorter than 1000 km ($k > 2\pi \times 10^{-6}$) the field in the y-direction (V) caused by the Hall effect is negligibly small compared to U, so for this range of parameters the equations may be simplified.

Figure 2 shows the magnitude of the amplitude of the field component in the x-direction, that is $|U|$, for some wavelengths ($2\pi/k$) and angular frequencies (ω) of the traveling wave. For the limiting case of low frequencies when a scalar potential can be used to represent the electric field, U is proportional to this potential.

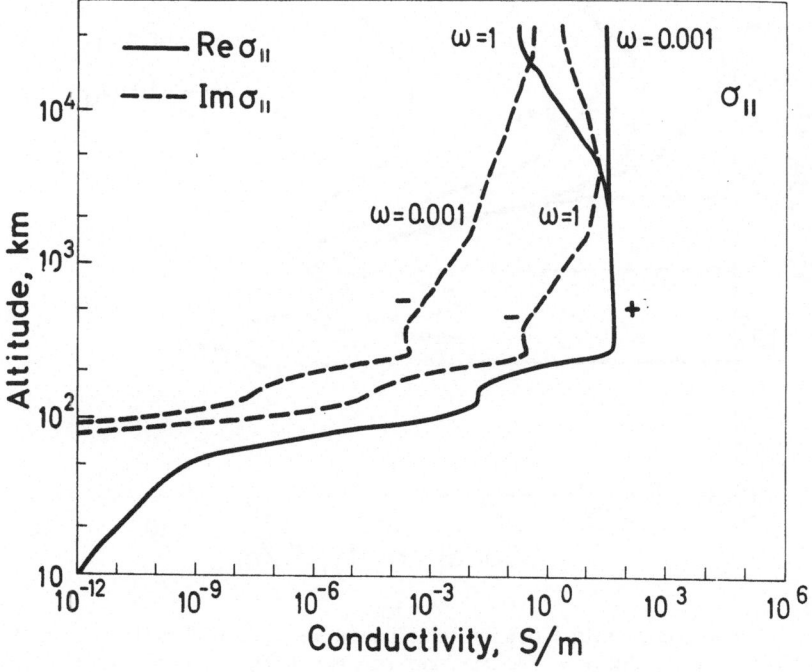

Fig. 1a.

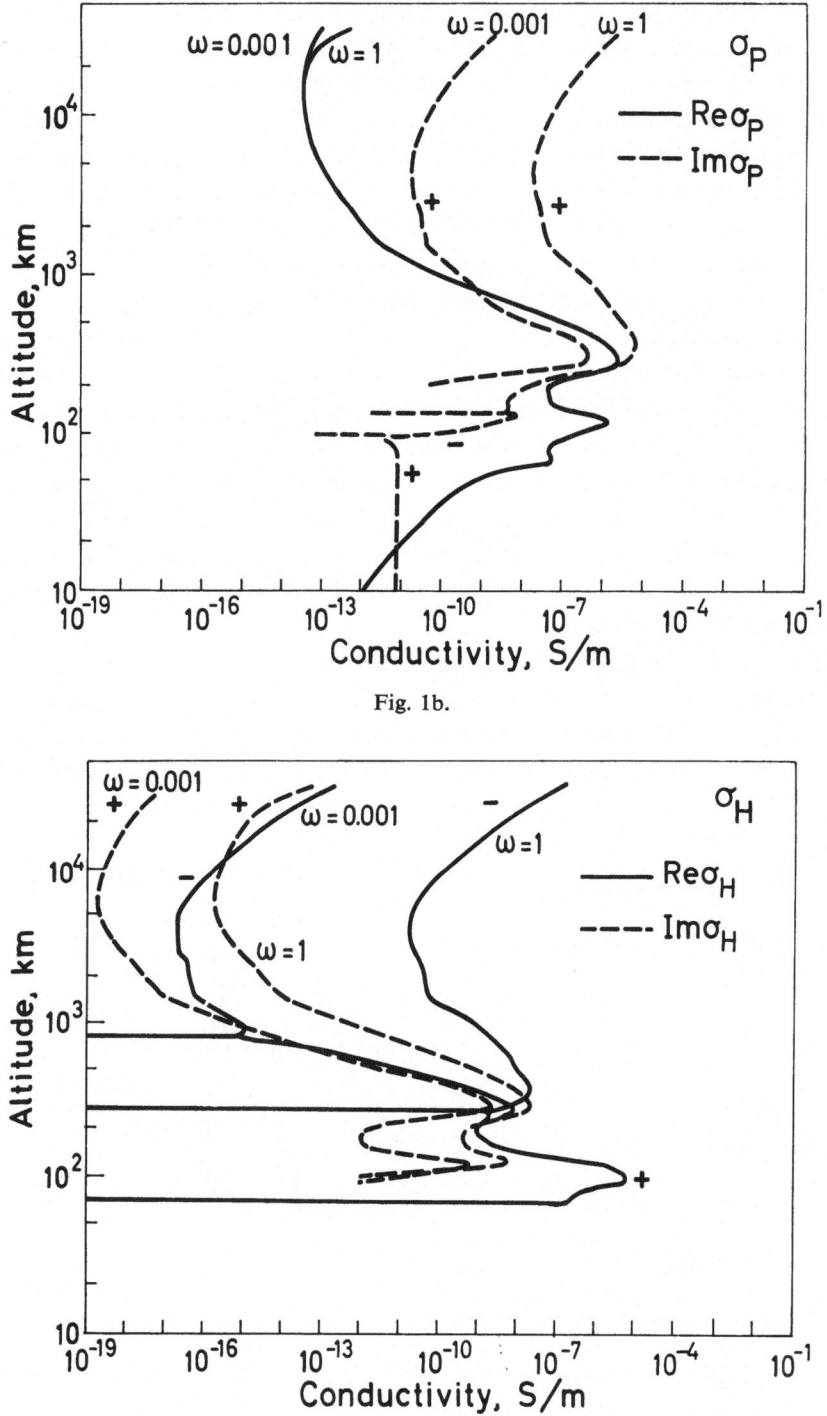

Fig. 1b.

Fig. 1c.

Fig. 1a–c.   Complex conductivities for a nighttime ionosphere. (a) Parallel conductivity. (b) Pedersen conductivity. (c) Hall conductivity. The imaginary part is not shown for some regions where it is much smaller than the real part.

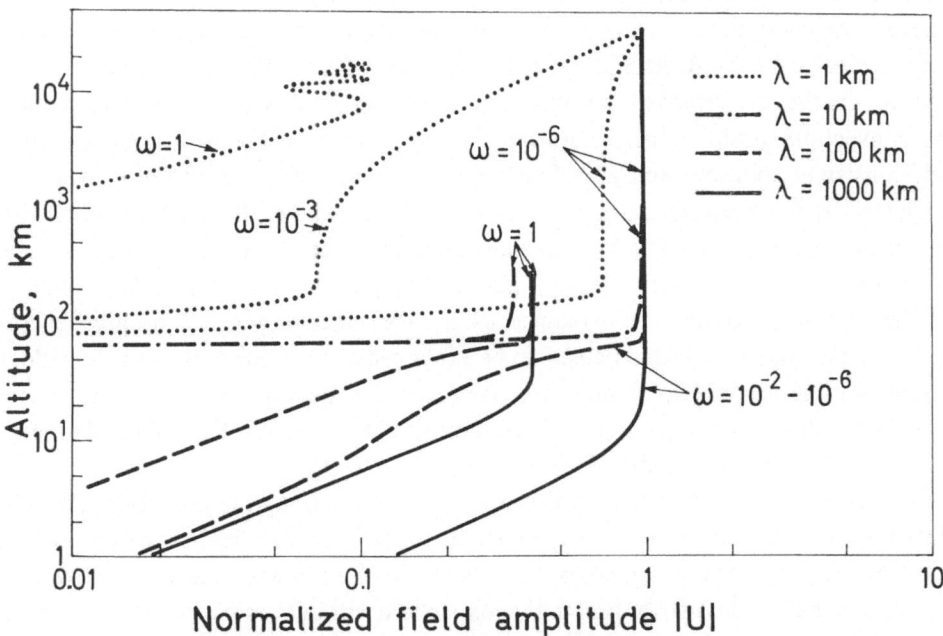

Fig. 2. Damping of the horizontal electric field amplitude with altitude for a magnetospheric source in the form of a traveling wave, for some various wavelengths and angular frequencies. Variations in field amplitude due to convergence of the field lines not included.

For high frequencies and altitudes ($z > 1000$ km for $\omega = 1$ rad s^{-1}), that is when the damping distance is large compared to the hydromagnetic wavelength, U oscillates rapidly. Figure 2 shows only the smoothly damped solutions in the lower region. Fields of a wavelength larger than about 100 km penetrate far down into the atmosphere, but for shorter wavelengths or higher frequencies the damping increases. Thus it should be possible to measure at balloon altitudes (30 to 40 km) electric fields of magnetospheric origin if their spatial scale is larger than about 100 km and they do not vary on a time scale shorter than about 10 s (Boström et al., 1973). Fields of a wavelength of 1 km penetrate into the E layer without much damping, for $\omega = 10^{-6}$ but for $\omega = 10^{-3}$ the amplitude is very much reduced. Thus even for rather slow temporal variations the field mapping is frequency dependent. Note that Figure 2 can be also applied to cases of a source at lower altitudes as long as the Hall field V is not of importance at the source altitude. Of course, our assumption of vertical parallel field lines is not fully applicable all the way out to the equatorial plane. Still our results may be used to indicate whether the damping at high altitudes is small or large.

3. Effects of Birkeland Currents on Plasma Convection

The plasma convection in the magnetosphere has been subject to numerous theoretical studies, but as yet it has not been possible to take the coupling to the ionosphere fully into account. The differences in the motion of ions and electrons give rise to space charges strong enough to change the field distribution completely, unless they are totally discharged by Birkeland currents. However, the coupling cannot be so strong that all charges are removed, because then all electric fields would be discharged and no convection would occur. Thus, space charges must be taken into account in a self-consistent solution, and as a first approximation Birkeland currents have been neglected in some earlier models. Karlson (1971) has shown that if we assume that the plasma sheet contains two components, one high energy (keV) plasma (e.g. solar wind origin) and one low energy (eV) plasma (e.g., ionospheric origin) the inner edge of the plasma sheet may be explained as the forbidden region of the high energy component, while the plasma pause can be explained as the boundary of the forbidden region for the low energy component. As forbidden regions of a different size for particles of different energy play an important role it is obvious that the MHD theory would not suffice for a study of these problems.

If there is a perfect coupling, that is no voltage drop along the field lines, then the distribution of the electric field perpendicular to the magnetic field in the ionosphere must be an image of the magnetospheric field distribution and vice versa. However, if we analyze the field distribution in the ionosphere without taking Birkeland currents into account we would find a distribution that does not match the magnetospheric one. We would then have to assume that parallel electric fields occur in a way exactly to match the potential differences, which seems to be a rather unnatural assumption. As we do not know enough about how parallel fields develop, but still want to take the coupling to the ionosphere into account, we may study the opposite extreme case where there are no parallel electric fields. The real case should be something inbetween these two solutions. For the model with Birkeland currents we can set up a closed set of equations describing the coupled magnetospheric-ionospheric system, as done by Fejer, Swift and Vasyliunas (see Vasyliunas, 1970). To solve the system we may try an iterative procedure, analyzing one region without taking the coupling to the other region into account, then mapping the resulting field distribution onto the other region and finding the resulting Birkeland currents from that region that are fed back to the first region, etc. As shown by Karlson (1972) it is essential to start with as good a model as possible for the iterative procedure to converge. The question is then whether we obtain a more unbalanced state if we start with a solution for the ionosphere or the magnetosphere.

Karlson shows that imposing an arbitrary field distribution from the ionosphere on the magnetosphere, the difference in the forbidden regions for electrons and protons would lead to a Birkeland current of the order of 2×10^6 A. Feeding such a high current back to the ionosphere would change the field configuration there considerably. Thus this approach would hardly converge. The inertia effects of the magnetospheric plasma also give a Birkeland current of the order 2×10^5 A which is of importance mainly in the outer magnetosphere. On the other hand, if we start from a magnetospheric field distribution, derived for a case with no Birkeland currents, and apply this to the ionosphere we would get Birkeland currents from the ionosphere of the order 5×10^5 A with most of the contribution coming from the dayside. Since this current is larger than the inertia current this magnetospheric field model is neither the very best starting point. Karlson showed that a useful first approximation would be a model where the field distribution outside a certain radius was based on the self-consistent solution for the magnetosphere with no Birkeland currents due to inertia effects or differences in forbidden regions, while inside this radius a model field derived for no Birkeland currents from the ionosphere is used. This first order approximation has a forbidden region, common to electrons and protons, which is a circular equipotential contour. The similarity of this solution with the earlier known solution for the other extreme case of no Birkeland current shows that it should be possible to make predictions about the intermediate case found in nature.

4. Birkeland Currents and Ionospheric Electric Field Structures

In the ionosphere the vertical Birkeland currents are coupled to horizontal Pedersen and Hall currents. From the condition that the total current should be divergence-free we find an expression for the Birkeland current flowing into the ionosphere

$$i_{\parallel} = \Sigma_P \operatorname{div} \mathbf{E}_{\perp} + \mathbf{E}_{\perp} \cdot \operatorname{grad} \Sigma_P + \hat{\mathbf{B}} \times \mathbf{E}_{\perp} \cdot \operatorname{grad} \Sigma_H. \tag{13}$$

We have assumed that there are no significant voltage drops along the field lines within the most conductive part of the ionosphere so that we can use the height-integrated conductivities Σ_P and Σ_H. This relation can be interpreted as showing that we can have two kinds of Birkeland currents: a) *primary* Birkeland currents 'of magnetospheric origin' occurring because of spatial variations of the imposed electric field (the first term on the right hand side of Equation (13)) b) *secondary* Birkeland

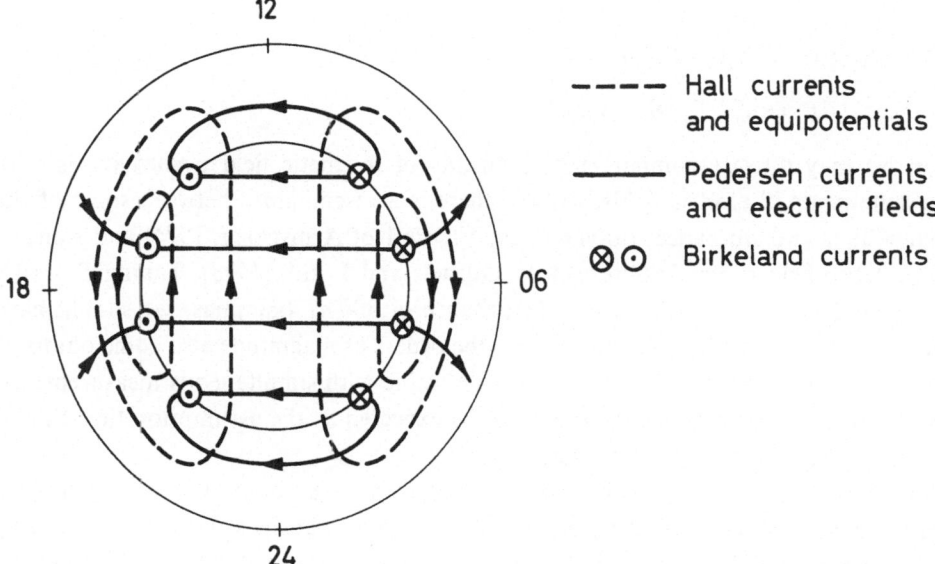

Fig. 3. Idealized pattern for currents and electric fields in the polar ionosphere.

currents 'of ionospheric origin' occurring because of spatial variations in the ionospheric conductivity (the last two terms of Equation (13)).

However, when we discuss time-independent problems the cause and effect relations
are not obvious. As well as saying that a magnetospheric field variation (div $\mathbf{E}_\perp$) gives
rise to a Birkeland current ($i_\parallel$) we can say that it is the Birkeland current which, when
forced to flow to the ionosphere, produces the electric field there, which then can be
regarded as being of ionospheric origin.

Figure 3 shows an idealized pattern for the average Birkeland current and electric
field topology in the polar ionosphere. A sheet of Birkeland current flows into the
auroral regions on the dawn side and out from the dusk side. Assuming a uniform
conductivity they connect to the Pedersen currents only (cf. Equation (13)), while
the Hall currents flow in a divergence-free system. Only the latter contribute to the
magnetic field perturbations observed on the ground, which is of the two cell (DP2)
form. Associated with these currents is an electric field and plasma convection pattern
with field reversals in the regions where the Birkeland currents flow. Over the polar
cap the convection is antisunwards, and at lower latitudes sunwards.

For a model with no variations in the east-west (y) direction we find from Equation (13) an expression for the north-south electric field associated with the Birkeland
sheet currents (assuming uniform conductivities)

$$E_x(x) = \Sigma_p^{-1} \int i_\parallel \, dx + E_0. \tag{14}$$

Experimentally, the current sheets are observed by measurements of the east-west
component of the magnetic field

$$B_y(x) = \mu_0 \int i_\parallel \, dx + B_0. \tag{15}$$

Thus there is a linear relation

$$E_x = \mu_0^{-1} \Sigma_p^{-1} B_y + \text{const}. \tag{16}$$

and we may directly compare measurements of magnetic field perturbations with
measurements of electric fields. In fact, there is a resemblance between some of the
magnetic field disturbances observed (see Figure 1 of Armstrong, 1974) and some of
the electric field reversals observed by Gurnett and Frank (1973). Taking $\Sigma_P = 10$ S
the scale factor is such that for a disturbance of $300 \, nT$ (gammas) we should have
electric field variations of 25 mV m^{-1}, so the orders of magnitude also come out to be
reasonable. Obviously, it would be most desirable with simultaneous measurements.
However, an exact agreement should not be expected as the assumption that Σ_P and
Σ_H are constant may not be satisfied.

Assuming an infinitely sharp field reversal from $-E_0$ at $x<0$ to $+E_0$ at $x>0$
Equation (13) would predict infinitely intense current densities. However, following
Lennartsson (1973b) we can show that if we take the finite conductivity along the
field lines $\sigma_\parallel$ into account the Birkeland currents will be spread out and have a finite

density. The Birkeland current density is independent of the vertical coordinate z above the altitude $z = z_i$ where the horizontal currents flow, thus

$$i_z(x) = - \Sigma_P \frac{\mathrm{d}}{\mathrm{d}x} E_x(x, z_i) = \sigma_{\parallel} E_z(x). \tag{17}$$

We assume that Σ_P and $\sigma_{\parallel}$ are constants, and thus E_z is a function of x only. Furthermore, as curl $\mathbf{E} = 0$,

$$\frac{\mathrm{d}E_z(x)}{\mathrm{d}x} = \frac{\partial E_x(x, z)}{\partial z} = \frac{E_x(x, z_0) - E_x(x, z_i)}{z_0 - z_i} \tag{18}$$

since E_x is a linear function of z as the left hand equality shows. Combining Equations (17) and (18) we obtain a second order ordinary differential equation for $E_x(x, z_i)$ and we find

$$E_x(x, z_i) = \mathrm{sign}\,(x) \cdot E_0 \left[1 - \exp\left(- |x|/L\right) \right] \tag{19}$$

$$i_z(x) = - i_0 \exp\left(- |x|/L\right) \tag{20}$$

$$L = \sqrt{(z_0 - z_i)\,\Sigma_P/\sigma_{\parallel}} \tag{21}$$

$$i_0 = \Sigma_P E_0/L. \tag{22}$$

Equation (20) shows that the Birkeland current is distributed over a region with a width of the order of L and has a maximum current intensity i_0. Assuming the source altitude z_0 to be at least 1000 km above the ionosphere, taking $\Sigma_P = 10$ S, $\sigma_{\parallel} = 30$ S m^{-1}, and $E_0 = 20$ mV m^{-1} we find $L > 600$ m, $i_0 < 3 \times 10^{-4}$ A m^{-2}. Thus we should not expect smaller structures or higher current densities of the Birkeland currents if their origin is a magnetospheric field reversal of this kind. The maximum current densities observed by satellites (Armstrong, 1974) are smaller than our estimate of i_0.

The relations between field reversals and Birkeland currents have been studied by Lennartsson (1973a, b) also for cases when the ionospheric conductivity varies spatially and for cases when the field reversal moves so that displacement currents flow in addition to the resistive currents. The analysis can be based on the results of Section 2.

5. Secondary Birkeland Currents and Triggering of Auroral Precipitation

Secondary Birkeland currents are produced when horizontal ionospheric currents flow through regions of varying conductivity (cf. Equation (13)). If the flow of Birkeland currents to the magnetosphere is restricted then charges will accumulate and secondary polarization electric fields will be produced so that the electric field of Equation (13) will also have a component of ionospheric origin. Models of the auroral electrojets, including polarization electric fields and Birkeland currents, have been discussed by Boström (1964), De Witt (1968), and Coroniti and Kennel (1972). This polarization electric field is opposite to the ionospheric current that connects to the

Birkeland currents, thus the ionosphere acts as a generator while the magnetosphere is the load for this secondary current system. We can also show that the Poynting vector $\mathbf{E} \times \mathbf{B}/\mu_0$, where $\mathbf{E}$ is the secondary polarization electric field and $\mathbf{B}$ the magnetic field disturbance due to the secondary currents, is direct upwards. The current system inferred from the rocket experiment of Park and Cloutier (1971) is of this nature (see Haerendel, 1972; Boström, 1972). Polarization of the auroral regions may also explain the development of a southward electric field during substorms from a westward field before the substorms, as observed by Mozer (1971).

If the secondary Birkeland currents, caused by ionospheric conductivity variations, are carried at least partly by energetic precipitating particles which cause conductivity variations, we may get a feedback leading to an instability. An enhanced ionization gives an enhanced Birkeland current and a still more enhanced ionospheric conductivity. This is the basis of some theories for auroral arc formations, and also for substorm theories (see Atkinson, 1970, 1973; Oguti, 1971).

It has been shown by Stoffregen (1970) that enhanced ionization can trigger auroral precipitation. During a Ba ion cloud experiment the conductivity of the E layer is locally increased, even if the cloud is released in the F layer, as Birkeland currents flow to the E layer to discharge polarization electric fields. There will be an accumulation of charges in the E layer where Birkeland currents flow upwards, because the Birkeland current is carried mainly by electrons moving downwards while the Pedersen current is carried mainly by ions. Stoffregen found an enhanced emission of auroral lines from the region of the atmosphere threaded by field lines from the ion cloud, just as the Ba ion cloud developed. Also on a rocket flight an increased flux of energetic particles was recorded on field lines connected to the ion cloud (Köhn and Page, 1972; Kelley et al., 1973).

6. Double Layers

When the current density of the Birkeland currents exceed a critical limit instabilities may occur which drastically increase the voltage drop along the field lines. The current density needed for this to occur at an altitude around 1000 km is estimated to be a few times 10^{-5} amp m^{-2}, which is a realistic number for observed Birkeland currents. With the ordinary conductivity determined by electron-ion and electron-neutral collisions such a current density would produce a voltage drop along the field lines of at most 10 V, but with the instabilities the voltage drop may well be tens of kilovolts. Then the ionosphere and magnetosphere may to some extent be electrically decoupled.

Two main possibilities have been proposed. One (Swift, 1965; Kindel and Kennel, 1971) is that an instability gives rise to anomalous resistivity and a voltage drop extended over a considerable distance. The other possibility (Akasofu, 1969; Carlqvist and Boström, 1970; Block, 1972) is that the instability gives rise to a double layer (also called space charge region) with a high voltage drop confined to a *short* distance. In the former case most of the energy released is converted into thermal energy, while for the case of double layers the energy is used to accelerate all the charged particles that carry the current into beams. Thus double layers offer the most efficient mechanism for auroral particle acceleration.

The double layer is a region void of plasma where electric quasi-neutrality no longer applies. There is an excess of positive space charge at one end of the layer and negative charge at the other end, giving rise to an electric field much stronger than in the surrounding medium.

As yet we have only indirect evidence for the occurrence of double layers in the magnetosphere (Albert and Lindstrom, 1970; Carlqvist and Boström, 1970). Thus we have to rely largely on laboratory experiments or theoretical predictions. In laboratory plasmas there are two kinds of double layers that we must distinguish between: (a) Stationary layers with a voltage drop of the order of $kT e^{-1}$ occur at boundaries between plasmas of different temperatures and densities, at discharge tube constrictions and walls; (b) Transient or periodically reappearing double layers with much higher voltage drops may occur in a discharge if the current exceeds some critical limit, and it is this kind of double layers that is of interest to magnetospheric physics.

It is interesting to note that the formation of double layers may be described as a development of the two-stream instability (Carlqvist, 1973). Although disturbances of a particular wavelength grow most rapidly, disturbances of other wavelengths will also grow given the time. By applying the theory of the two-stream instability to the Fourier components of a disturbance in the form of a dip in the plasma density Carlqvist indicated that this would develop into a double layer. Thus the critical current density for double layers to appear should be the same as for the two-stream instability.

As the magnetic field lines are nearly equipotential lines in the plasma surrounding the double layer, the vertical voltage drop across the double layer will lead to a horizontal voltage difference between the center and the outer parts of the column above the double layer. Below the double layer only weak fields occur because of the short-circuiting action of the ionosphere. Thus, the double layers are associated with horizontal field reversals, with the fields pointing inwards to the center, if the vertical voltage drop is directed such as to accelerate electrons towards the ionosphere. The strong horizontal fields should thus be found above the layer and energetic electrons below the layer and there are some difficulties in using double layers as a direct explanation of the observations (Gurnett and Frank, 1973) of concurrent strong electric field irregularities and inverted V events.

7. A Substorm Circuit

In this section we will show how we can describe the behavior of the coupled ionosphere-magnetosphere system during a substorm by means of an equivalent electric circuit.

During the expansive phase of the substorm the magnetic field lines of the near tail change from a tail-like form to a dipolar form associated with the release of magnetic energy stored during the growth phase. In terms of currents this change of field topology implies that the neutral sheet current has disappeared. However, the current which is a part of the huge current system that encloses the tail lobes cannot suddenly cease to flow because the current system has a large inductance and small resistance and thus a long time constant. Instead the current is redirected and flows along the field lines to the ionosphere and closes there forming the westward auroral electrojet as proposed by Atkinson (1967, 1973) (Figure 4). This change of currents is represented

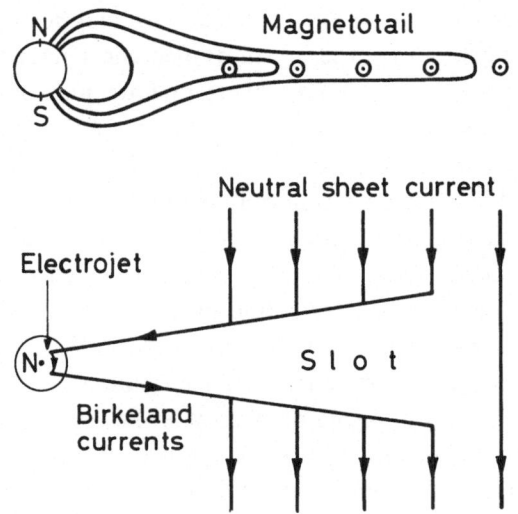

Fig. 4.   Substorm currents caused by magnetotail current interruption.

by the *addition of an eastward* current in the tail, Birkeland currents and auroral electrojets, which should be the current system responsible for the expansive phase of the magnetic substorm (Boström, 1971). The loop to the ionosphere also has a certain time constant due to its inductance, but this can be small if the loop has a small east-west extent initially. Thus it seems likely that the collapse of tail field lines to dipolar form starts in a limited slot, but it may spread longitudinally and radially during the course of the substorm. Evidence of such a localized character of the tail field variations has been given by Rostoker and Camidge (1971). From ground observations we also have indications of a lengthening of the auroral electrojets during substorms (Boström, 1971). The Birkeland currents are carried mainly by electrons while the transverse currents are carried mainly by protons. Thus, at the westward edge of the slot there will be a deficit of charged particles. It seems likely that this will lead to a further decrease of the neutral sheet current there, and a further westward spread of the slot region and the corresponding westward traveling surge at the ionospheric end of the flux tubes. Assuming there are no currents in the slot region a neutral sheet current of 30 mA m^{-1} outside of the slot distributed over a height of 5000 km corresponds to a net loss of 4×10^{10} protons m^{-2} s^{-1}. With a number density of 10^6 protons m^{-3} this gives a westward propagation speed of 40 km s^{-1}, or projected to the auroral zone about 1 km s^{-1}, which seems reasonable.

Figure 5 shows the equivalent circuit (Fälthammar, 1971; Boström, 1972). The 100 H inductance of the current system flowing around the tail is estimated to be that of a single-loop coil with a radius of 10 R_E and a length along the tail of 10 R_E. The total current is about 10^6 A, obtained from half of the neutral sheet current of 30 mA m^{-1} over 10 R_E. Thus the magnetic energy stored in the inductance is $\frac{1}{2}L_{tail}I^2 = 5 \times 10^{13}$ J, which is of the right order for the energy released during a

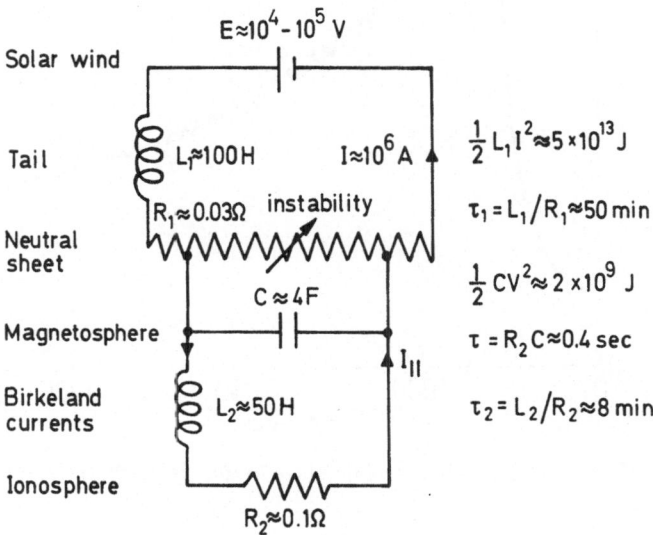

Fig. 5. Equivalent circuit for the substorm current system. At the substorm onset the resistance of the neutral sheet increases and the tail current is redirected to the ionosphere.

substorm. When the resistance of the tail increases, due to some unspecified instability, more current flows through the ionospheric branch. The loop along the field lines has an inductance L ≈ 50 H and the ionospheric resistance R ≈ 0.1 Ω (estimated from electrojet dimensions and ionospheric conductivities or from a voltage drop of 30 kV for an electrojet current of 3×10^5 A). The magnetospheric plasma effectively acts as a capacitor where a displacement current flows. Charging the capacitor is equivalent to setting the plasma into motion, and we estimate the capacitance from the kinetic energy of plasma motions in the magnetosphere. Assuming a velocity of 10 km s^{-1}, a number density of 10^6 m^{-3} and a volume of 100 R_E^3, the kinetic energy is $2 \times \times 10^9$ J. A capacitor charged to 30 kV has the same energy if its capacitance is 4 F. The energy is small compared to the magnetic energy of the system. Thus the capacitor, or the kinetic energy of the plasma, does not play any major role in the substorm process, except for the production of some oscillations.

Figure 6 shows the current in the ionospheric branch of the circuit when the tail resistor is increased considerably so that most of the current is forced to flow through the ionospheric branch. The magnetic disturbance on the ground should be proportional to this current. The build-up of the substorm current depends on how fast

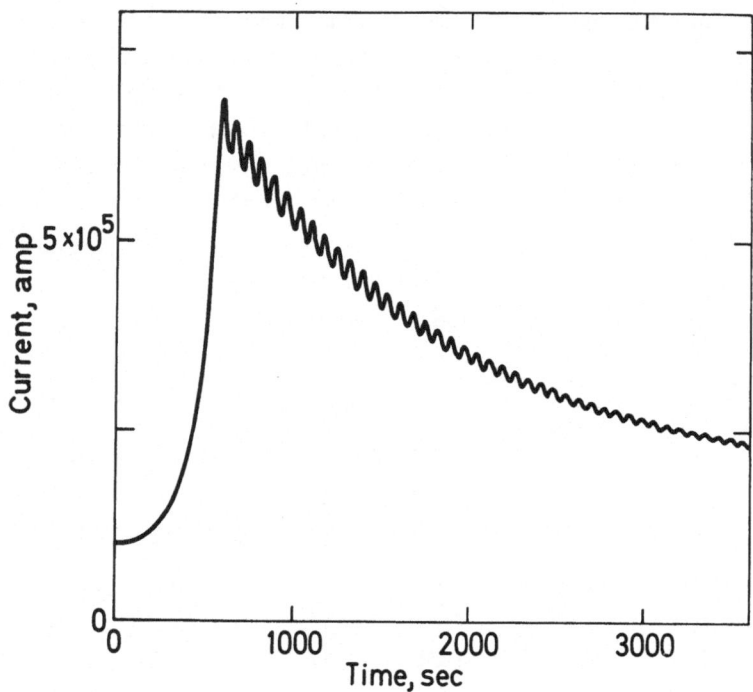

Fig. 6. Current in the loop to the ionosphere for the circuit shown in Figure 5.

the tail resistance increases. The rate of the ensuing current decay depends on the L/R time constant of the circuit, for our estimates of the circuit elements the substorm decay time comes out to be of the right size. Furthermore we find that the oscillations that occur fall in the Pi2 range. Pulsations in this range are typical for substorms. Thus we feel that we could put some confidence in our estimates of the circuit parameters. Of course, a model like this must be only a crude approximation. In reality the circuit elements should be distributed and not lumped, and they should be time dependent. The ionospheric resistance also depends on the amount of energetic particle precipitation. Furthermore, the model does not take into account the thermal energy of the energetic particles in the magnetosphere. However, the behavior of the system still resembles that of the real one.

References

Akasofu, S.-I.: 1969, *Nature* **221**, 1020.
Akasofu, S.-I. and DeWitt, R. N.: 1965, *Planetary Space Sci.* **13**, 737.
Albert, R. D. and Lindstrom, P. J.: 1970, *Science* **170**, 1398.
Armstrong, J. C.: 1974, this volume, p. 155.
Atkinson, G.: 1967, *J. Geophys. Res.* **72**, 5373.
Atkinson, G.: 1970, *J. Geophys. Res.* **75**, 4746.
Atkinson, G.: 1973, preprint, Communications Research Center, Ottawa.
Block, L. P.: 1972, *Cosmic Electrodyn.* **3**, 349.
Boström, R.: 1964, *J. Geophys. Res.* **69**, 4983.

Boström, R.: 1971, in B. M. McCormac (ed.), *The Radiating Atmosphere*, D. Reidel Publishing Company, Dordrecht-Holland, p. 357.

Boström, R.: 1972, in E. R. Dyer (ed.), *Critical Problems of Magnetospheric Physics*, IUCSTP Secretariat, c/o National Academy of Sciences, Washington, D.C., p. 139.

Boström, R., Fahleson, U., Olausson, L., and Hallendal, G.: 1973, Report TRITA-EPP-73-02, Dept. of Plasma Physics, Royal Inst. of Techn., Stockholm.

Carlqvist, P.: 1973, Report TRITA-EPP-73-05, Dept. of Plasma Physics, Royal Inst. of Techn. Stockholm.

Carlqvist, P. and Boström, R.: 1970, *J. Geophys. Res.* **75**, 7140.

Coroniti, F. V. and Kennel, C. F.: 1972, *J. Geophys. Res.* **77**, 2835.

DeWitt, R. N.: 1968, *J. Geophys. Res.* **73**, 6307.

Fälthammar, C.-G.: 1971, private communication.

Gurnett, D. A. and Frank, L. A.: 1973, *J. Geophys. Res.* **78**, 145.

Haerendel, G.: 1972, in E. R. Dyer (ed.), *Solar Terrestrial Physics/1970*, D. Reidel Publishing Company, Dordrecht-Holland, Part IV, p. 87.

Karlson, E. T.: 1971, *Cosmic Electrodyn.* **1**, 474.

Karlson, E. T.: 1972, Report TRITA-EPP-72-22, Dept. of Plasma Physics, Royal Inst. of Techn., Stockholm.

Kelley, M. C., Pedersen, A., Fahleson, U., Jones, D., and Köhn, D.: 1973, to be published.

Kindel, J. M. and Kennel, C. F.: 1971, *J. Geophys. Res.* **76**, 3055.

Köhn, D. and Page, D. E.: 1972, *J. Geophys. Res.* **77**, 4888.

Lennartsson, O. W.: 1973a, Report TRITA-EPP-73-03, Dept. of Plasma Physics, Royal Inst. of Techn., Stockholm.

Lennartsson, O. W.: 1973b, Dept. of Plasma Physics, Royal Inst. of Techn., Stockholm, in preparation.

Mozer, F. S.: 1970, *Planetary Space Sci.* **18**, 259.

Mozer, F. S.: 1971, *J. Geophys. Res.* **76**, 7595.

Oguti, T.: 1971, *Cosmic Electrodyn.* **2**, 164.

Park, R. J. and Cloutier, P. A.: 1971, *J. Geophys. Res.* **76**, 7714.

Rostoker, G. and Camidge, F. P.: 1971, *J. Geophys. Res.* **76**, 6944.

Stoffregen, W.: 1970, *J. Atmos. Terrest. Phys.* **32**, 171.

Swift, D. W.: 1965, *J. Geophys. Res.* **70**, 3061.

Vasyliunas, V. M.: 1970, in B. M. McCormac (ed.), *Particles and Fields in the Magnetosphere*, D. Reidel Publishing Company, Dordrecht-Holland, p. 60.

ON THE TOPOLOGY OF THE GEOMAGNETIC FIELD

MANFRED SCHOLER and GREGOR MORFILL

Max-Planck-Institut für Physik und Astrophysik, Institut für extraterrestrische Physik,
8046 Garching bei München, Germany

1. Introduction

From various magnetic field and plasma experiments on-board highly eccentric satellites, moon orbiters, and interplanetary probes, as well as theoretical considerations of the solar wind interaction with the geomagnetic field, the topology of the magnetosphere is fairly well understood now, at least in as much as the quiet time macroscopic structure is concerned (e.g., reviews by Ness, 1969, 1972; Fairfield, 1970; Cain, 1971; Sugiura, 1971).

There are a number of controversies and disputes remaining however, for instance the stability and microstructure of the plasma sheet region, the transfer of solar wind energy into the magnetosphere (reconnection, viscous interaction), the magnetopause microstructure and dynamic changes in the magnetosphere during geomagnetic storms and substorms.

In this paper we shall try to combine recent measurements of particles and fields as well as theoretical considerations in order to help understand some of these problems.

2. Static Quiet-Time Topology

A current problem in the plasma-physical description of the magnetotail is the applicability of the fluid approximation, in particular the magnetohydrostatic approximation for the plasma sheet (e.g., Cole and Schindler, 1972).

The momentum balance equation becomes in the static limit

$$- \operatorname{grad} p + \frac{1}{c} \mathbf{j} \times \mathbf{B} = 0 \tag{1}$$

where the plasma pressure p is assumed to be isotropic (Bame, 1968). Writing $\mathbf{B} = \operatorname{curl} \mathbf{A}$ where the vector potential $\mathbf{A} = A(x, z) \hat{\mathbf{y}}$ so that we may treat the problem in two dimensions we obtain $\mathbf{B} = (B_x, 0, B_z)$. Here $\hat{\mathbf{y}}$ is a unit vector in the y-direction.

$$\mathbf{j} \times \mathbf{B} = (\operatorname{grad} A) j_y - \hat{\mathbf{y}} (\mathbf{j} \cdot \operatorname{grad} A). \tag{2}$$

From Maxwell's equation

$$\operatorname{curl} \mathbf{B} = 4\pi \mathbf{j} \tag{3}$$

we obtain $\mathbf{j} = (0, j_y, 0)$. Since $(\partial/\partial y) A(x, z) = 0$, Equation (2) becomes

$$\mathbf{j} \times \mathbf{B} = (\operatorname{grad} A) j_y. \tag{4}$$

B. M. McCormac (ed.), Magnetospheric Physics, 61–72. All Rights Reserved.

Thus one of the conditions for Equation (1) to apply is that currents flow only perpendicular to B, or, more weakly, $j_\perp \gg j_\parallel$, where the subscripts $\perp$ and $\parallel$ are with respect to **B**.

Inserting Equation (4) into Equation (1) and taking the curl, we obtain

$$\operatorname{grad} j_\perp \times \operatorname{grad} A = 0 \qquad (5)$$

where $j_y = j_\perp$. Thus a second requirement for Equation (1) to apply is that $\operatorname{grad} j_\perp$ is parallel to $\operatorname{grad} A$ or $j_\perp = j_\perp(A)$. Since A is constant along a magnetic field line, $j_\perp$ must be constant along a field line, too.

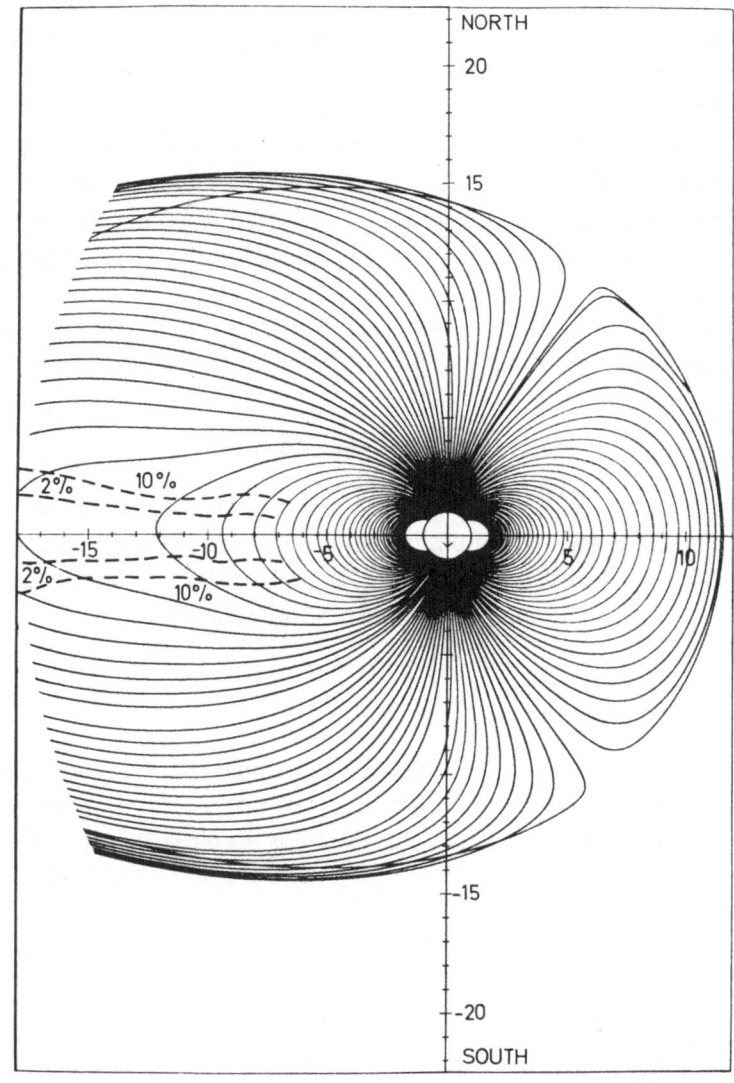

Fig. 1.   Projection of field lines of the Mead-Fairfield model onto the noon-midnight meridian plane.
The region, where $j_\perp \gg j_\parallel$ and $j_\perp \approx$ const within 2% and 10%, respectively, is indicated.

These conditions, which might apply at quiet times, have been checked experimentally by Morfill and Schindler (1973) using magnetic field models constructed from measurements. Mead and Fairfield (1971) have produced a field model based on satellite observations. The analytic expression they fitted to these measurements is a second order polynomial. Figure 1 shows, in the noon-midnight meridian plane, the projection onto that plane of the model field lines, as well as the region on the nightside, where $j_\perp \gg j_\parallel$ and $j_\perp \approx$ constant (to within 2% and 10% respectively) along the model field lines. As can be seen, this region is approximately 4 R_E wide, and thus Morfill and Schindler (1973) conclude that the notion of a thin current sheet is not supported by the results of this model.

Another possible interpretation of the broad reduced field strength region in the plasma sheet (eg. Behannon, 1970) is a thin moving current sheet. The observed field strength is then given by the probability of observing this current sheet at a distance Δz from the mean location. Any agreement with the magnetohydrostatic approximation would then be purely accidental.

It is necessary, however, to enquire into the validity of the Mead-Fairfield model in the region of interest. Since these authors represented the magnetospheric field by a second order polynomial expansion, it is clear that the curl **B** only results in a first

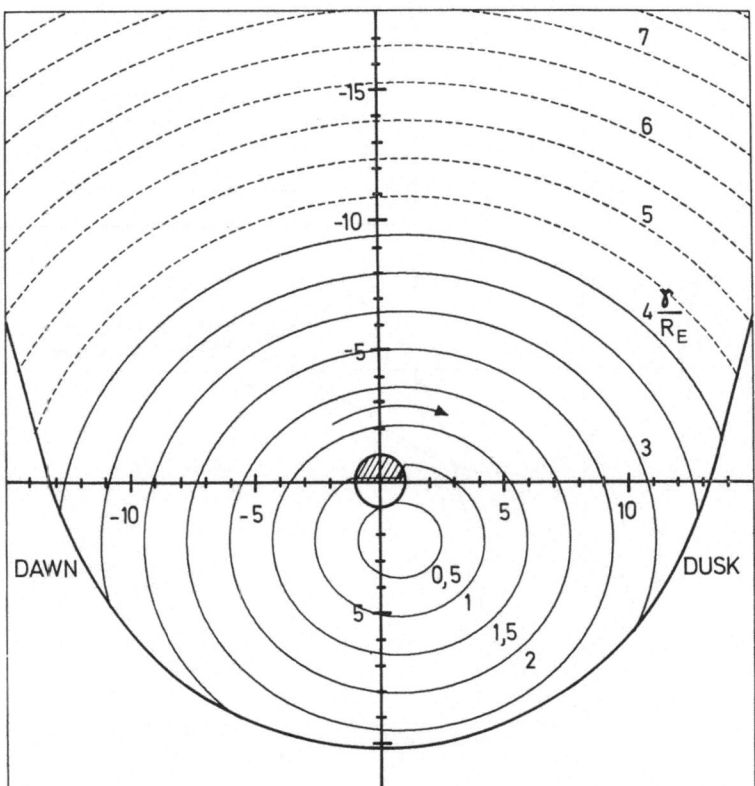

Fig. 2.   Contours of constant current density in the equatorial plane.

order form. In the noon-midnight meridian plane, most of the currents are flowing perpendicular to that plane, so that contours of constant current density are, in fact, straight lines almost parallel to the z axis. Such a result is unrealistic, certainly outside the plasma sheet where no sizeable distributed currents can flow. The problem is then to determine how representative the model is in the region near the equatorial plane.

Figure 2 shows contours of constant current density in the equatorial plane; the arrow points in the direction of the current. Such a picture is consistent, qualitatively, with the existence of a quiet time ring current, and a gradual transition to a plasma sheet with a curved front 'edge'. Olson (1973) has produced a magnetic field model using a sixth order expansion for the magnetic field. His method was to assemble distributed currents due to the magnetic effect of the plasma near the region where particle gradients exist, eg., at the plasma pause and near the front edge of the plasma sheet, and gradient B-drift and curvature drift in the outer magnetosphere.

The Olson model was fitted to the ΔB contours of Sugiura et al. (1971) and can be compared to the Mead-Fairfield model in the region of interest here, i.e., within 2 R_E of the geomagnetic equator in the midnight meridian plane at radial distances greater than $\sim 10\ R_E$. Figure 3 shows the comparison taken from Olson and Pfitzer (1973)) where the 2-term representation of the Olson model is in fact identical (to within $\sim 1\ \gamma$)

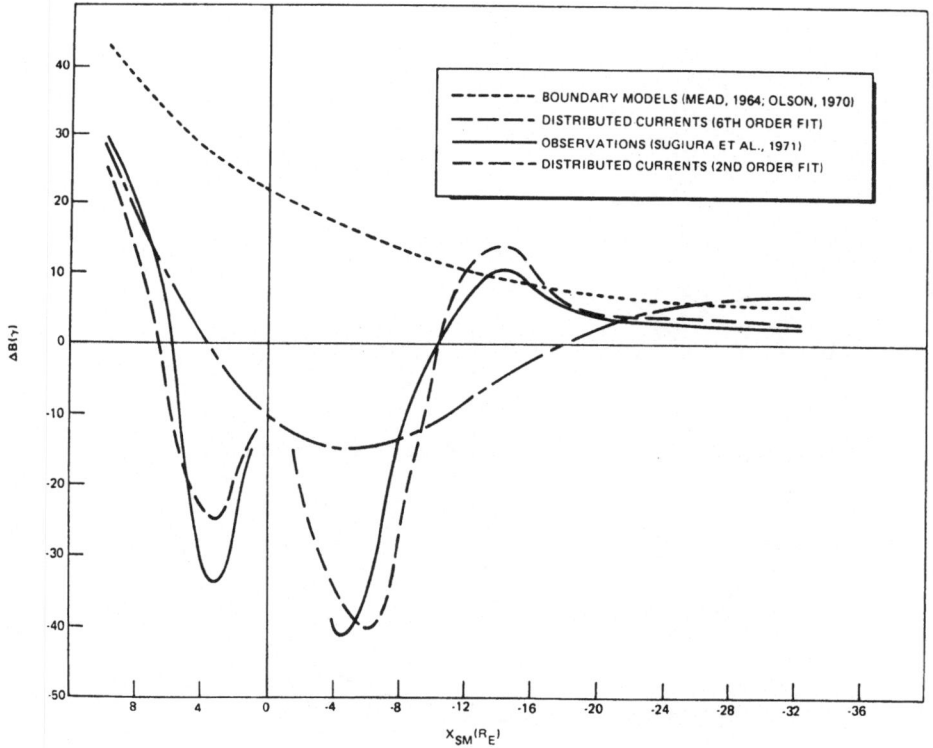

Fig. 3. ΔB along the Sun-Earth line for three models and the observations of Sugiura et al. (1971). (From Olson and Pfitzer, 1973.)

(Pfitzer, 1973) to the Mead-Fairfield result. Figure 3 also shows that the two models are not too dissimilar, as far as ΔB is concerned, in the region of the plasma sheet.

Thus there are some grounds to suppose that at quiet times the plasma sheet region can indeed be thought of as being in magnetohydrostatic equilibrium, although perhaps some further field model work is still required to define the thickness of the current carrying layer.

Figure 4 shows the contribution of the current parallel to the magnetic field of the total current, i.e., the curves are curves of constant $j_{\parallel}/j_{\text{total}}$. Dashed lines are drawn when the current $j_{\parallel}$ is antiparallel to **B** and the lines are continuous when the current $j_{\parallel}$ is parallel to **B**. As can be seen from Figure 4, there is almost no contribution from parallel currents to the magnetic field in the tail region, whereas considerable $j_{\parallel}$ currents are flowing near the dayside neutral cusp. These currents are flowing away from the Earth in both the northern and southern hemisphere. This is consistent with recent plasma measurements on HEOS II in the cusp region reported by Paschmann *et al.* (1973). The observed angular distribution of the arrival of particles in the cusp region results in a current parallel to B away from the Earth.

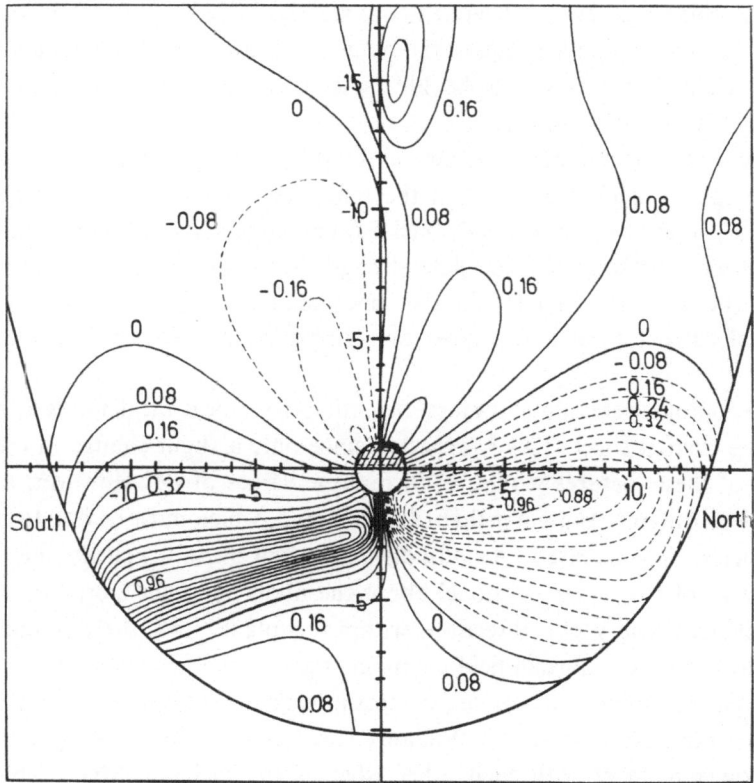

Fig. 4.   Contribution of the current parallel to B to the total current in the Mead-Fairfield model. The curves are curves of constant $j_{\parallel}/j_{\text{total}}$ and are either dashed or continuous when the current is antiparallel or parallel to B, respectively.

3. Comparison of Field Models

It may be of some use at this stage to compare field models which have been used
hitherto; for instance, in the calculations of cosmic ray entry into the magnetosphere
(eg. Gall *et al.*, 1968; Morfill and Quenby, 1971; Morfill and Scholer, 1973a) in order
to see how the use of much simpler models than the Mead-Fairfield (1971) or distributed current (Olson, 1973) models might modify earlier interpretations.

In order to make such a comparison quantitatively, we have chosen the Mead-
Fairfield model as a reference and computed ΔB and $\Delta \alpha$ contours, where $\Delta B =$
$= |\mathbf{B}| - |\mathbf{B_{MF}}|$ and $\Delta \alpha = \cos^{-1}(\mathbf{B} \cdot \mathbf{B_{MF}} / |\mathbf{B}| \|\mathbf{B_{MF}}|)$. Comparisons of trajectory calculations for these models have been shown to yield closely similar results (Morfill and
Scholer, 1973b) so that the above comparison using ΔB and $\Delta \alpha$ should help to distinguish the physical differences between the thin current sheet models and the distributed current models.

Figure 5 shows ΔB and $\Delta \alpha$ contours for (from top to bottom) the Williams-Mead
6-term model (where the field due to currents flowing in the magnetopause has been
expressed by a sixth order harmonic expansion), the Williams-Mead 2-term model
(second order harmonic expansion), and the Voigt (1972) model where the Chapman-
Ferraro problem has been solved self-consistently, including the addition of a thin
current sheet in the nightside equatorial plane under a given magnetopause geometry.
Tail field strength was chosen to be 40 γ, since this value has been most widely used
for the Williams-Mead models.

Contours of constant ΔB are drawn as solid lines when the model under test has
greater magnetic field strength than the reference (Mead-Fairfield) model, and as
dotted lines when the field strength is less. It can be seen from Figure 5 that the thin
current sheet models have larger field strength in the plasma sheet region (z within
± 5 R_E) and near the equator in the inner magnetosphere. Addition of a thick
distributed current sheet and a quiet time ring-current is required to reconcile the
models.

The $\Delta \alpha$ contours generally show large deviations near the neutral points and the location of the current sheet. This is expected since only a slight change in the location
of either of these features would result in large values of $\Delta \alpha$. However, away from
these regions differences greater than 20° are found, which must be due to real model
discrepancies. The increase in $\Delta \alpha$ with distance down the tail is partly due to the lack
of symmetry of the reference model about the noon midnight meridian, and partly
due to different field vector directions within this plane, since the thin current sheet
models do not have a normal field component across the neutral sheet.

Although the differences between the test models do not appear to be major ones,
it is clear from Figure 5 that the Williams-Mead 6-term and the Voigt model are in
fairly good agreement with each other whereas the Williams-Mead 2-term model
deviates somewhat. The thickness of the plasma sheet can be defined as the distance
between the $+10$ γ contours in the tail, and turns out to be ~ 10 R_E for both the
6-term and the Voigt model, and ~ 19 R_E for the 2-term model which is somewhat

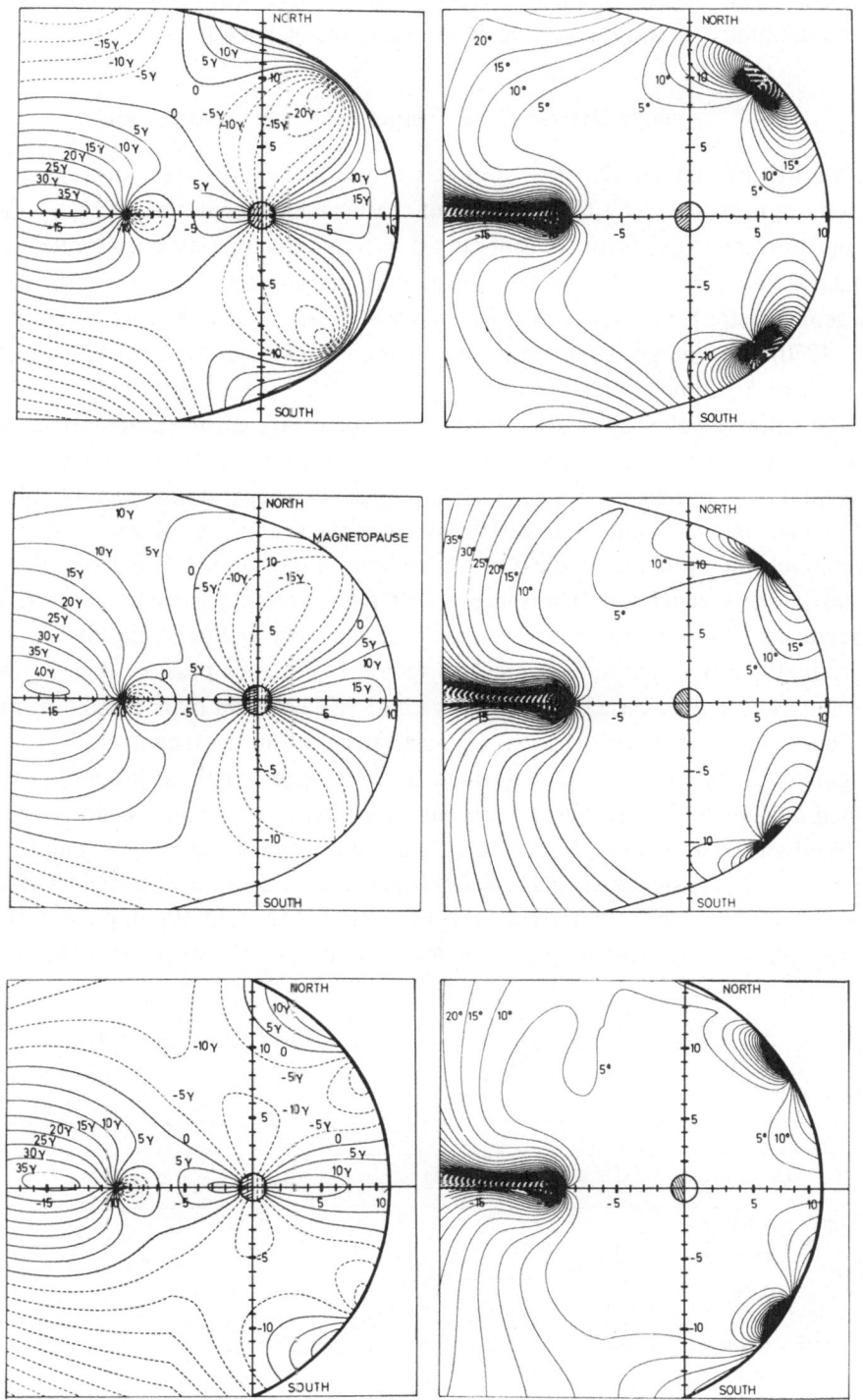

Fig. 5. ΔB and $\Delta \alpha$ contours for (from top to bottom) the Williams-Mead 6-term model, the Williams-Mead 2-term model and the Voigt model. Reference model is the Mead-Fairfield model.

unrealistic. This unrealistic thick 'plasma sheet' has the advantage of leading to a better quantitative description of the field close to the magnetopause at high latitudes.

4. Topology Derived From Energetic Particle Observations

In 1965 Michel and Dessler first proposed to use low energy solar protons as test particles in order to study the topology of the magnetospheric tail. Access of solar protons into the magnetosphere is discussed in terms of either an open model of the magnetosphere (Dungey, 1961) with substantial merging between the interplanetary and geomagnetic field lines or a closed model (Dessler, 1964; Michel and Dessler, 1965, 1970) with the geomagnetic tail extending $\sim 10^4$ R_E in the antisolar direction and with no field line interconnection.

Observation of polar cap structures due to interplanetary particle gradients supports the access mode in an open magnetosphere with reconnected field lines (Scholer, 1972). Tangential discontinuities in interplanetary space can separate regions of different proton flux. As such distinct flux variations are convected by the solar wind along the magnetotail, they provide an ideal means to study the proton access from different regions of the magnetotail. The sharpness of the resulting polar cap structure and the relatively short time delay between the interplanetary and polar cap flux increase is incompatible with any diffusive access mode. It becomes possible then to make quantitative estimates of such quantities as length of a coherent magnetotail, the mean convection electric field and the particle mean free path in the magnetotail.

Figure 6 shows a cross section through the geomagnetic tail. A discontinuity has reached a distance L down the tail and particles entering at a point M down the tail can be observed at a point P located at a distance r from the magnetopause. The magnetic field is assumed to make a small angle θ with the magnetopause due to reconnection with the interplanetary magnetic field. The normal component of the magnetic field at the magnetopause is $B_N = B_0 r \cos\theta/L$. If the interplanetary dis-

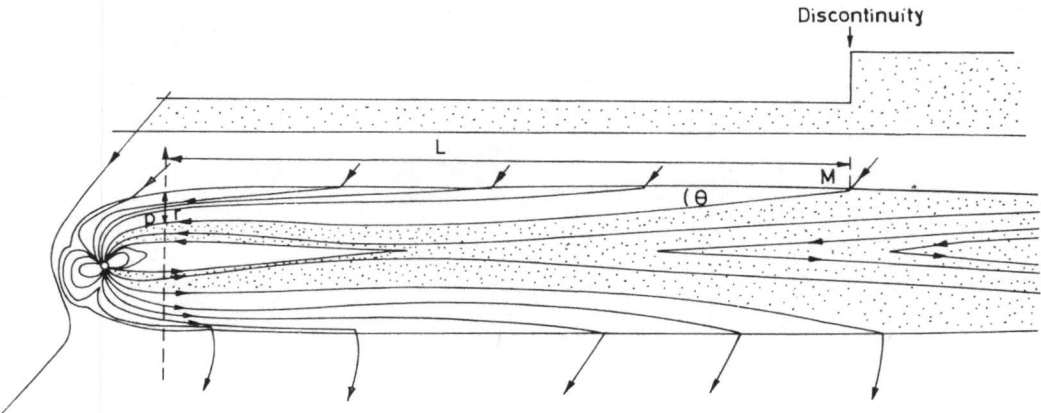

Fig. 6.   Schematic cross section through the geomagnetic tail in order to demonstrate the effect of a discontinuity in the particle intensity caused by interplanetary magnetic field structure.

continuity travels the distance L in time τ, the average normal field $\langle B_N \rangle$ at the magnetopause becomes:

$$\langle B_N \rangle \approx \frac{B_0 r}{V_s \tau}$$

where V_s is the solar wind speed and θ has been assumed to be small. In order to maintain a stationary reconnection between interplanetary and tail field, the magnetic flux has to reconnect across the neutral sheet and field lines must move inwards to the Earth. This inward motion leads to a $\mathbf{V} \times \mathbf{B}$ cross tail electric field

$$E_y = -\frac{1}{c} \langle V_{TN} B_{TN} \rangle$$

where V_{TN} is the velocity of the field lines moving to the Earth and B_{TN} is the perpendicular component of the magnetic field in the neutral sheet. Since in a steady state the rate of flux loss out of the tail must be equal to the flow into the neutral sheet one gets on the average:

$$\langle V_{TN} B_{TN} \rangle (:) \langle B_N \rangle \cdot V_s,$$

so that the cross tail electric field is

$$E_y (:) B_0 r / \tau.$$

The time τ can be determined by multisatellite observations, the distance r can be obtained by projecting the observed polar cap structure of the discontinuity via trajectory calculations onto a cross section of the tail. The quantity $B_0 r$ gives the amount of magnetic flux (or field lines) between the measuring position and the magnetopause and has to be replaced in a more realistic magnetospheric model by the 'magnetic depth' $D = \int \mathbf{B} \times d\mathbf{l}$. The integral is taken from a point inside the tail which connects via a trajectory with the corresponding measurement position over the polar cap to the magnetopause along a path which is perpendicular to B everywhere (Morfill, 1972).

Detailed multisatellite observations of such discontinuities were reported by Scholer (1972) and Rampling (1972). Scholer (1972) obtained from the measured time delays for two discontinuities tail lengths of $< 550 \, R_E$ and $790 \, R_E$, respectively, where the tail length is understood as the distance from the Earth over which the field line, originally at the center of the tail lobe, reaches the magnetopause. The corresponding cross tail electric field is $2.2 \times 10^{-4} \, \text{V m}^{-1}$ and $2.6 \times 10^{-4} \, \text{V m}^{-1}$ (Morfill and Scholer, 1972). From Rampling's (1972) measurements a cross tail electric field of $\sim 1.4 \times \times 10^{-4} \, \text{V m}^{-1}$ can be obtained. Despite the fact that these latter observations were made during times with a southern interplanetary magnetic field component the convection electric field was found to be smaller than that obtained by Morfill and Scholer (1972) during a time with a northern interplanetary magnetic field. These measurements illustrate the variability of the magnetospheric topology and emphasize the need for further detailed measurements of this kind.

5. The Magnetospheric Topology at Disturbed Times

The magnetopause undergoes great changes in its topology during geomagnetic storms and substorms. These changes affect large regions of the magnetosphere and are therefore difficult to measure and interpret with *in situ* observations of the plasma and magnetic field. Changes in the trapped and incident energetic particle distributions reflect such macroscopic space time variations. In the following we concentrate on the variations during geomagnetic storms.

The equatorial H-depression during geomagnetic storms has been shown to be due to the development and decay of a symmetric ring current located between about 2 and 7 R_E. The presence of the ring current not only alters the topology of the geomagnetic field but the increased plasma density gives rise to wave growth, which in turn can lead to wave-particle interaction (e.g., Williams and Heuring, 1973). The

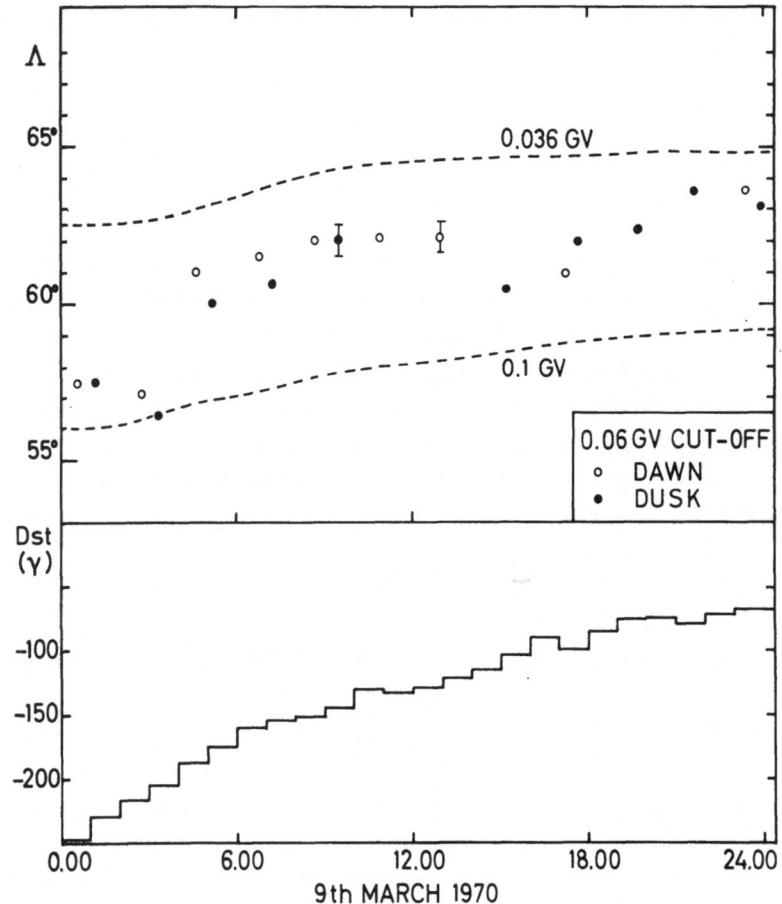

Fig. 7.   Solar proton penetration cut off latitudes during a geomagnetic storm (March 9, 1970). Cut-off latitudes for ~ 0.06 GV protons are shown, as well as the theoretically calculated adiabaticity contours for 0.030 GV and 0.1 GV protons for comparison. The lower panel shows the hourly averaged D_{ST} values during the decay of the ring current.

purely geomagnetic changes in the magnetospheric field, caused by the presence of symmetric ring currents, were analyzed by Morfill (1973). Ring current models of Sckopke (1972) were superimposed onto the Williams-Mead (1965) model and the particle rigidity was calculated at which the Alfvén criterion $X_r = \varrho\,(\mathbf{B} \times \nabla B)/B^2 = 0.1$ holds (ϱ is the particle gyroradius). This was done for different D_{ST} values and it was shown that the addition of a ring current does not alter the position of the radial iso-rigidity contours very much. However, due to the ring current inflation of the magnetosphere, field lines originating at the same position on the Earth reach further out so that the latitude iso-rigidity contours are affected considerably more. In order to apply such calculations the correct magnetopause location has to be chosen, which is affected by the increased interplanetary plasma pressure and the ring current field inside the magnetosphere.

In Figure 7 actual measured positions of particle cut-offs are compared with the contours of adiabatic breakdown. Data were obtained onboard the GRS-AZUR satellite during the geomagnetic storm of March 1970 (Scholer *et al.*, 1972). Measured values of interplanetary plasma pressures (Palmiotto, 1971) were inserted into the field model parameters and thus the contours of breakdown of adiabatic motion shown for comparison were computed. Apart from two time periods, 0.00 to 4.00 UT and 15.00 to 17.30 UT, when strong substorm activity distorted the measurements, good correlation can be seen between the location of experimental cut-off and the non-adiabatic motion contour, as well as their temporal variations.

References

Behannon, K. W.: 1970, *J. Geophys. Res.* **75**, 743.

Cain, J. C.: 1971, *Rev. Geophys. Space Phys.* **9**, 259.

Cole, G. H. A. and Schindler, K.: 1972, *Cosmic Electrodyn.* **3**, 275.

Dessler, A. J.: 1964, *J. Geophys. Res.* **69**, 3913.

Dungey, J. W.: 1961, *Phys. Rev. Letters* **6**, 47.

Fairfield, D. H.: 1970, in *Solar Terrestrial Physics Part III*, D. Reidel Publishing Company, Dordrecht-Holland, p. 1.

Gall, R., Jimenez, J., and Comacho, L.: 1968, *J. Geophys. Res.* **73**, 1593.

Mead, G. D. and Fairfield, D. H.: 1971, Paper, XV General Assembly of IUGG, Moscow.

Michel, F. C. and Dessler, A. J.: 1965, *J. Geophys. Res.* **70**, 4305.

Michel, F. C. and Dessler, A. J.: 1970, *J. Geophys. Res.* **75**, 6061.

Morfill, G.: 1972, *J. Geophys. Res.* **77**, 4010.

Morfill, G.: 1973, Paper, Seventh ESLAB Symposium, Saulgau, Germany.

Morfill, G. and Schindler, K.: 1973, Paper, Chapman Memorial Symposium, Boulder, Colorado.

Morfill, G. and Scholer, M.: 1972, *J. Geophys. Res.* **77**, 4021.

Morfill, G. and Scholer, M.: 1973a, *Space Sci. Rev.*, **15**, 267.

Morfill, G. and Scholer, M.: 1973b, *J. Geophys. Res.* **78**, 5449.

Morfill, G. and Quenby, J. J.: 1971, *Planetary Space Sci.* **19**, 1541

Ness, N. F.: 1969, *Rev. Geophys.* **7**, 97.

Ness, N. F.: 1972, in B. M. McCormac (ed.), *Earth's Magnetospheric Processes*, D. Reidel Publishing Company, Dordrecht-Holland, p. 189.

Olson, W. P.: 1973, McDonnell Douglas, preprint.

Olson, W. P. and Pfitzer, K. A.: 1973, Paper, Seventh ESLAB Symposium, Saulgau, Germany.

Paschmann, G., Grünwaldt, H., Montgomery, M. D., Rosenbauer, H., and Sckopke, N.: 1973, Paper, Seventh ESLAB Symposium, Saulgau, Germany.

Palmiotto, F.: 1971, Techn. Report LPS-71-15, University of Rome.
Pfitzer, K. A.: 1973, private communication.
Rampling, R.: 1972, Ph.D. thesis, Imperial College, London, England.
Scholer, M., Häusler, B., and Hovestadt, D.: 1972, *Planetary Space Sci.* **20**, 271.
Scholer, M.: 1972, *J. Geophys. Res.* **77**, 2762.
Sckopke, N.: 1972, *Cosmic Electrodyn.* **3**, 330.
Sugiura, M.: 1971, in A. J. Zmuda (ed.), *The World Magnetic Survey 1957–1969*, IUGG Publication's Office, Paris.
Sugiura, M., Ledley, B. G., Skillman, T. L., and Heppner, J. P.: 1971, *J. Geophys. Res.* **76**, 7552.
Voigt, G. H.: 1972, *Z. Geophys.* **38**, 319.
Williams, D. J. and Mead, G. D.: 1965, *J. Geophys. Res.* **70**, 3017.
Williams, D. J. and Heuring, F. T.: 1973, *J. Geophys. Res.* **78**, 37.

THE SPEED, DENSITY, AND FLUX VARIATIONS IN LARGE-SCALE SOLAR WIND DISTURBANCES

A. J. HUNDHAUSEN

*High Altitude Observatory, National Center for Atmospheric Research,
Boulder, Col., U.S.A.*

1. Introduction

The large-scale structure of the solar wind is a topic of considerable interest in the field of solar-terrestrial physics. In fact, the existence of such a structure was inferred from studies of geomagnetic activity long before direct interplanetary observations became possible. Two distinct classes of interplanetary 'disturbances' were suggested by these studies – (a) transient, flare-associated shock waves that produced the classical SSC geomagnetic storms, and (b) long-lived, localized plasma streams that produced recurrent geomagnetic activity (see Chapman and Bartels, 1950). The existence of such disturbances has been confirmed by modern *in situ* observations and the study of the detailed properties of these structures, their solar origins, and their magnetospheric effects has been an important and active area of solar wind research. This paper will concentrate on some recent results in this area that suggest a unified view of large-scale solar wind structures and has some interesting implications regarding the energetics of the related magnetospheric processes.

2. The Large-Scale Dynamical Structure of the Solar Wind

Both the transient shock waves and long-lived plasma streams mentioned above were commonly held to stem from the emission of fast-moving plasma from some part of the solar corona. The major difference between these two classes of solar wind disturbances (in addition, of course, to their specific solar origins) lay in their spatial configurations; the shock wave was traditionally thought of as advancing along a broad, nearly spherically-symmetric front, while the plasma streams were expected to extend outward from the sun in a spiral pattern, similar to that now known to describe the average interplanetary magnetic field (e.g., Parker, 1963; Hundhausen, 1972). In the presence of a fluid-like interplanetary medium, the fast-moving plasma in either class of disturbance would be expected to overtake the slower-moving ambient solar wind in front of it, and to recede from the ambient solar wind that might exist behind it. Implicit then, in any such speed inhomogeneity, is an interplanetary evolution of the solar wind density structure (Parker, 1963; Sarahbai, 1963), producing a compression at the leading edge of the solar wind disturbance (where the solar wind speed would be seen to increase with time by a stationary observer) and a rarefaction in the trailing portion of the disturbance (where the solar wind speed would be seen to decrease with time by the same observer).

B. M. McCormac (ed.), Magnetospheric Physics, 73–83. All Rights Reserved.
Copyright © 1974 by D. Reidel Publishing Company, Dordrecht-Holland.

Within the past few years, considerable progress has been made in developing quantitative models for the traditional classes of solar wind speed inhomogeneities. Figure 1 illustrates the results of such a model for a transient shock wave (Hundhausen and Gentry, 1969). A thin, spherical shell of dense, fast-moving plasma was introduced at 1/10 AU into an ambient solar wind flow; the Figure shows the predicted variations with heliocentric distance r of the solar wind speed u and number density n at a later instant of time when the front edge (a shock front) of the resulting disturbance is at 1 AU. The buildup of large densities near the front of the outwardly propagating wave, and the development of a rarefaction (a feature *not* included in the ori-

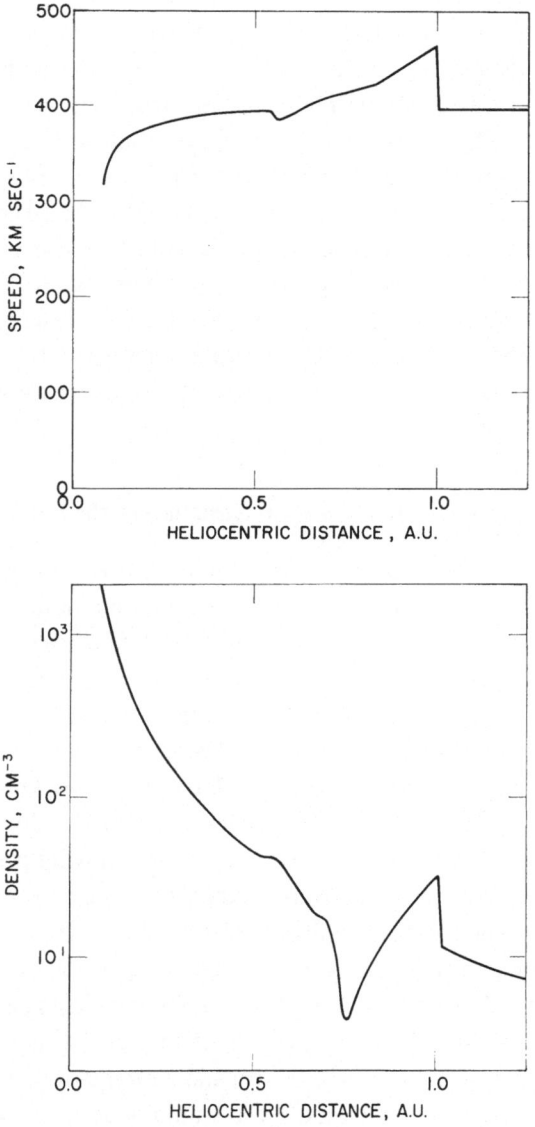

Fig. 1.   The speed and density as functions of heliocentric distance predicted by a model of a transient, spherically-symmetric shock wave (Hundhausen and Gentry, 1969).

ginal perturbation of ambient conditions) in the trailing portion of the wave are clearly demonstrated. The details of this pattern depend upon the details of the assumed perturbation at 1/10 AU (e.g., the maxima in both speed and density occur immediately behind the shock front only for thin shell, or short duration, perturbations); the basic nature of the pattern is general.

Figure 2 shows the results of a model for a localized, long-lived stream of high-speed plasma in the solar wind (Goldstein, 1971). The stream is introduced at 1/20 AU

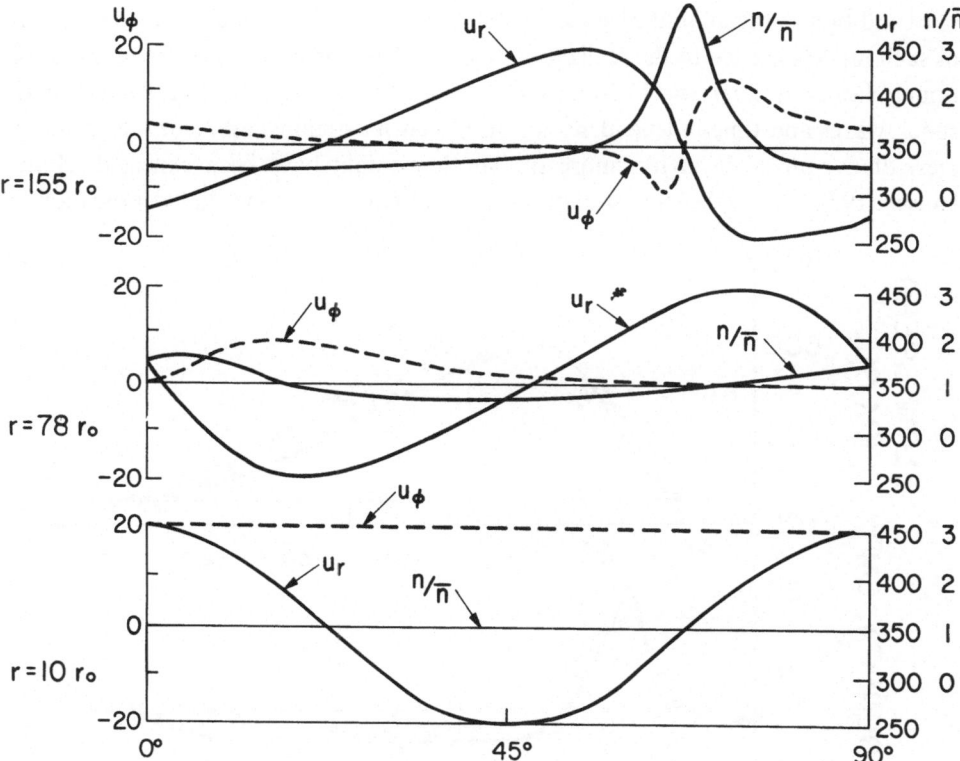

Fig. 2.   The radial velocity component u_r, azimuthal component u_ϕ, and density n, as functions of solar longitude predicted by a model of a corotating stream structure at three different heliocentric distances (Goldstein, 1971).

(or 10 solar radii in the lowest frame of the Figure) as a sinusoidal variation of the radial expansion velocity u_r with solar longitude ϕ; other solar wind parameters, such as the density and the azimuthal velocity component u_ϕ are taken to be constant (independent of ϕ) at this 'inner boundary' of the model. The resulting variations of u_r, n, and u_ϕ with longitude are shown at two larger heliocentric distances in the upper two frames of Figure 2 (note that the time t for a stationary observer at given r would run from right to left in this display). The sinusoidal velocity wave introduced at $r = 1/20$ AU is seen to evolve to a fast rise, slow decay form at larger r; this is merely the effect of nonlinear steepening of the original, large amplitude perturbation. The expected compression and rarefaction arise from this steepening process; at $r = 3/4$ AU (the largest r to which the model was carried), high densities occur during the short interval of rising speeds, while below average densities occur during the longer interval of falling speeds. This density variation, closely resembling those actually observed in the solar wind, is *entirely* due to the interplanetary evolution of the disturbance. The sharp variation of u_ϕ within the compression region results from the nonradial pressure forces associated with the spiral configuration (or nonspherical nature) of the plasma stream.

In addition to the quantitative results illustrated in Figures 1 and 2, the development of such models has led to the realization that the theoretical idealizations of the traditional classes of solar wind disturbances – namely transient, spherically-symmetric shock waves and time-independent, localized plasma streams, are, from the point of view of dynamical behavior, more similar than might have been expected (Hundhausen, 1973). A simple expansion of the fluid equations for nonradial solar wind flow

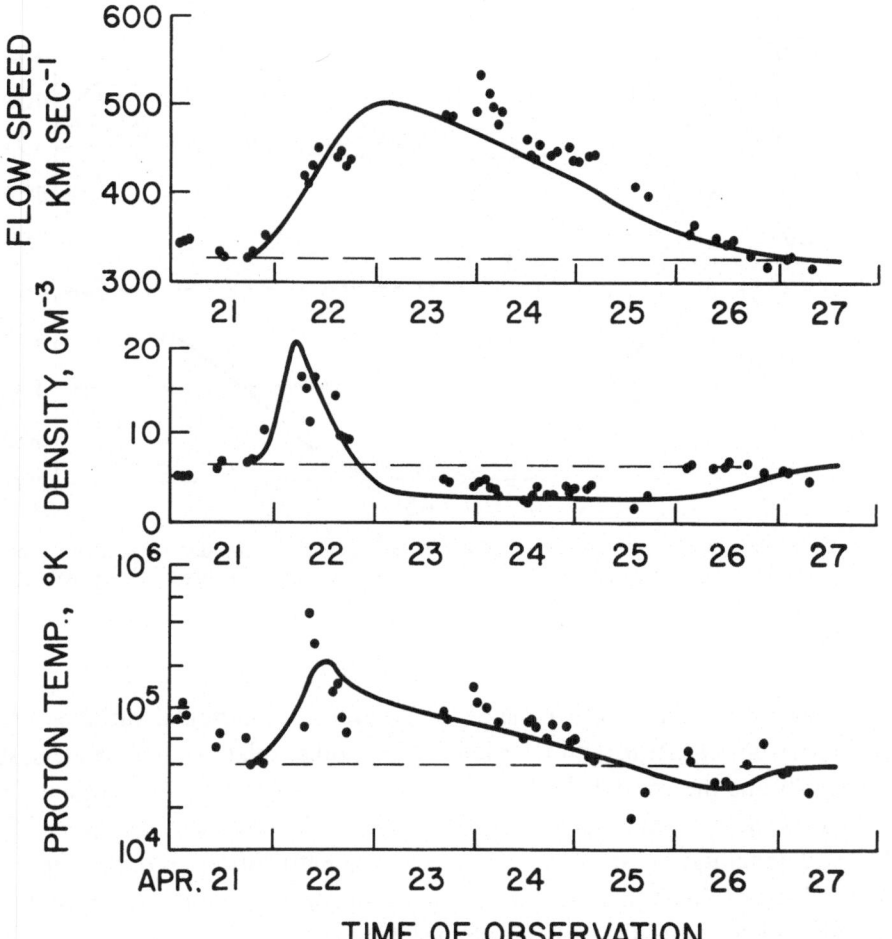

Fig. 3. A comparison of observed and predicted solar wind properties (as a function of time at 1 AU) in a high-speed plasma stream (Hundhausen, 1973).

reveals differences with respect to the simpler case of purely radial flow only in terms of the order u_t/u_r, where u_t is the nonradial velocity component. The ratio u_t/u_r is observed to be small near 1 AU, can be shown to reach a maximum value somewhere near 1 AU (Siscoe, 1970), and is found in the specific model illustrated in Figure 2 to be less than ~ 0.06 even in the stream compression region. It is thus reasonable (for those willing to accept $\sim 10\%$ accuracy) to consider all large solar wind speed inhomogeneities, including the traditional transient shocks, long-lived streams, and the possible intermediate cases, as members of a single dynamical class. Implicit in the speed differences intrinsic to such solar wind disturbances is a density structure related to the interplanetary evolution of the disturbance.

Our discussion thus far has been based largely upon theoretical models, and some evidence that these models give an acceptable description of actual solar wind structures might be comforting. Figure 3 compares the observed variations of the solar wind speed, number density, and temperature in a high-speed stream with predictions of a stream model (Hundhausen, 1973); this model contains two free parameters that

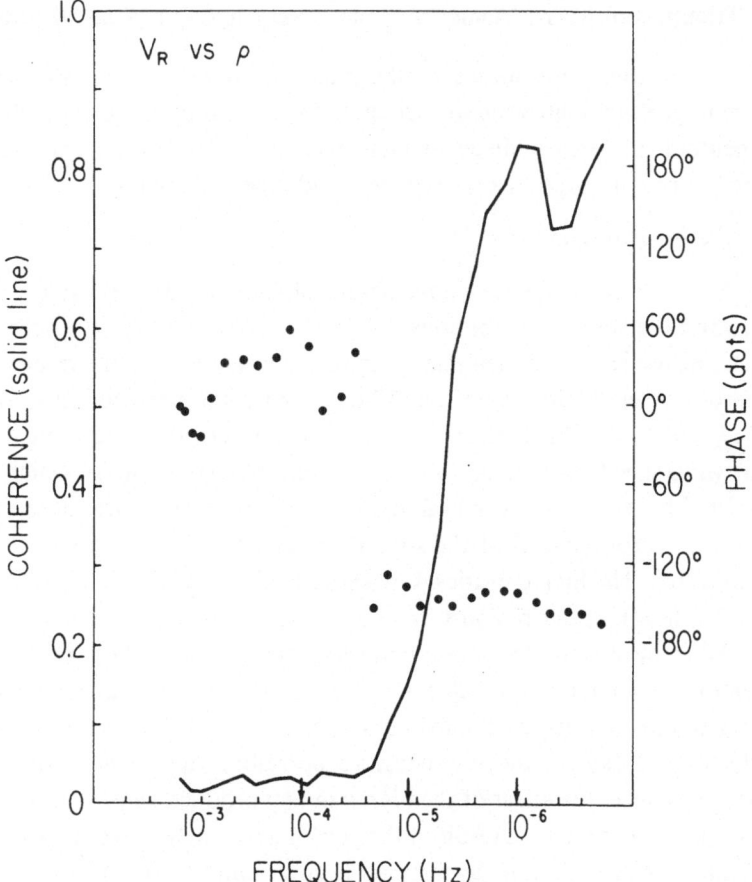

Fig. 4.   A cross spectrum between the radial expansion speed u_r and solar wind density n (Goldstein and Siscoe, 1972).

characterize a 'perturbation' of ambient conditions near the sun and that could be adjusted in such a comparison. The resulting agreement is far from perfect (there are, for example, density variations on a shorter time scale than the model can, or was intended, to account for), but is reasonable. In particular the gross features of the compression-rarefaction structure are well represented by the model (again without assumption of any density modulation at the inner boundary). A similar, broad agreement between such models and a larger sample of stream observations has been found by Gosling *et al.* (1972) and Pizzo *et al.* (1973). A final piece of evidence for the basic validity of this view of solar wind density speed structure stems from the spectral analysis of large quantities of solar wind data. Figure 4 (Goldstein and Siscoe, 1972) shows the results of taking cross spectra of solar wind density and speed fluctuations. Not only do these two parameters vary in a coherent manner for periods longer than ~ 1 day (or for the large-scale structure under discussion here), but the density variations lead the speed variations by $\sim 150°$, a result consistent with the stream models mentioned above.

3. The Transport of Mass, Momentum, and Energy in Solar Wind Disturbances

One of the interesting implications of the relationship between density and speed variations in large-scale solar wind disturbances emerges from the consideration of the mass, momentum, and energy fluxes in such structures. For example, the solar wind carries energy through a unit area on a suncentered sphere at the rate

$$f \approx \tfrac{1}{2}\varrho\, u^3 \; (\text{erg cm}^{-2}\, \text{s}^{-1})$$

where ϱ is the density, the nearly radial nature of solar wind flow has been used to write $u_r \approx u$, and the small contributions due to the internal energy (in both particles and magnetic fields) have been neglected. Figure 5 shows three-hour averages of solar wind properties observed during a single 27 day solar rotation period in 1967 (Montgomery *et al.*, 1972). A large high-speed stream was observed to begin on July 11, in which solar wind speeds as high as 700 km s^{-1} were observed, and in which elevated speeds persisted for six days. Very high densities, about seven times those prevailing for the five days before arrival of the stream, occurred while the speed rose toward its maximum value. The high densities were short lived, and a long interval of densities distinctly below normal occurred on July 12 to 14. Thus this stream clearly and strongly displays the compression rarefaction pattern discussed above. The energy flux density f is seen to reach a maximum value of ~ 1.0 erg cm^{-2} s^{-1} at a time between those at which maximum density and maximum speed are attained. Large values of f persist only for ~ 1 day, as the low densities prevailing during most of the speed elevation largely cancel the effect of the high u in any product, such as f, of these two parameters. This is, of course, a result of the correlated, nearly out-of-phase nature of the density and speed variations in large-scale solar wind disturbances, and can be generalized as follows:

(1) The maximum value assumed by f is much smaller than might be estimated

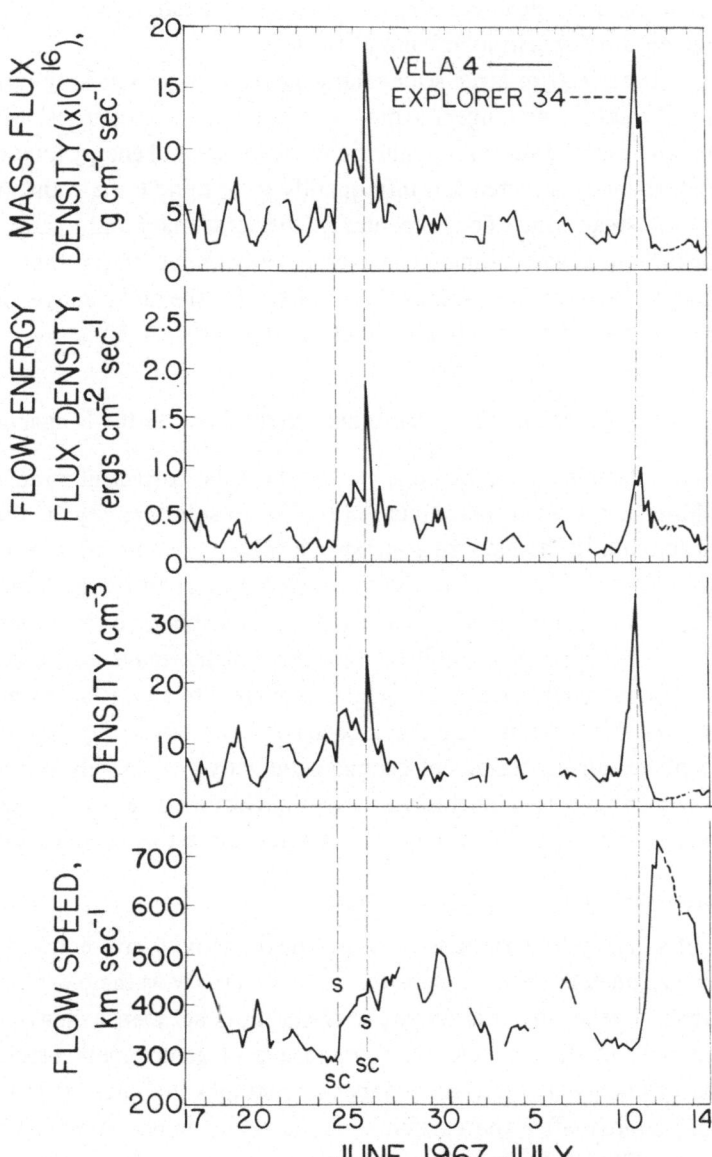

Fig. 5.   Three-hour averages of solar wind properties observed by Vela 4 and Explorer 34 spacecraft during a 27 day solar rotation period in mid-1967 (Montgomery *et al.*, 1972).

from the maximum density ϱ_{max} and maximum speed u_{max} in the disturbance; i.e.,

$$f_{max} \ll \tfrac{1}{2}\varrho_{max}u_{max}^{3}$$

(for the paricular example above, f_{max} is about 10% of $\tfrac{1}{2}\varrho_{max}u_{max}^{3}$).

(2) Highly elevated values of f prevail only near the leading edge of the disturbance,

and are well above normal of ambient values for a small fraction of the interval over which the speed remains elevated. Similar conclusions could be drawn regarding the mass and momentum fluxes in solar wind disturbances.

In a study of the effects of large-scale solar wind disturbances on long-term (i.e., the scale of a solar rotation or longer) averages of solar wind properties, Montgomery *et al.* (1972) found that the short intervals of elevated mass and energy fluxes associated with such disturbances occurred too infrequently to dominate the transport of these quantities by the solar wind. To the contrary, the total mass and energy associated with these disturbances was estimated to average only about 10% of that carried past 1 AU by the solar wind during a solar rotation. Even for the most 'active' or disturbed solar rotation included in this study, this fraction rose only to ~30%.

4. Implications Regarding the Solar Wind Energy Input to the Magnetosphere

Early studies of solar-terrestrial relations viewed the 'solar corpuscular radiation', then generally held to be a special solar emission related to solar activity, as the source of geomagnetic activity. In the modern context of a solar wind emitted continually from the Sun, variations in solar wind properties would appear to be the logical heirs to this role. Although recent emphasis on internal magnetospheric instabilities may seem to relegate the solar wind to an indirect influence on the magnetosphere, the solar wind must still be invoked to stress the magnetosphere into an unstable state. Thus an important relationship remains, and it can be argued that the energy dissipated within the magnetosphere during periods of geomagnetic activity is largely supplied by the solar wind. In this light, the conclusions drawn in Section 3 regarding energy transport in large-scale solar wind disturbances have some interesting ramifications.

4.1. VARIATIONS RELATED TO THE SOLAR CYCLE

The subject of solar cycle variations in geomagnetic activity and hence, by implication, in the interplanetary agent responsible for that activity, is the oldest in the realm of solar-terrestrial relations; the apparent connection between a newly discovered ~10 yr periodicity in the frequency and amplitude of geomagnetic storms and the sunspot cycle was advocated in three independent studies (Sabine, 1852; Wolf, 1852; Gautier, 1852) shortly after the existence of the latter gained scientific acceptance (see Meadows, 1970). The availability of *in situ* solar wind observations since 1962 raises the interesting possibility of identifying the interplanetary variations central to this connection. Curiously, attempts at this identification have proven inconclusive, and the very detection of any clear solar-cycle variation in solar wind properties remains controversial.

Figure 6 (Montgomery *et al.*, 1972) shows 27 day averages of the solar wind number density, flow speed, mass flux dnsity ϱu, and kinetic energy flux density $\frac{1}{2}\varrho u^3$ observed by Vela 3 and 4 spacecraft from July 1965 to October 1969. This display covers the rising portion of the present sunspot cycle (see the top frame of the Figure) with observations from a similar, and thus easily compared, set of spacecraft-borne instru-

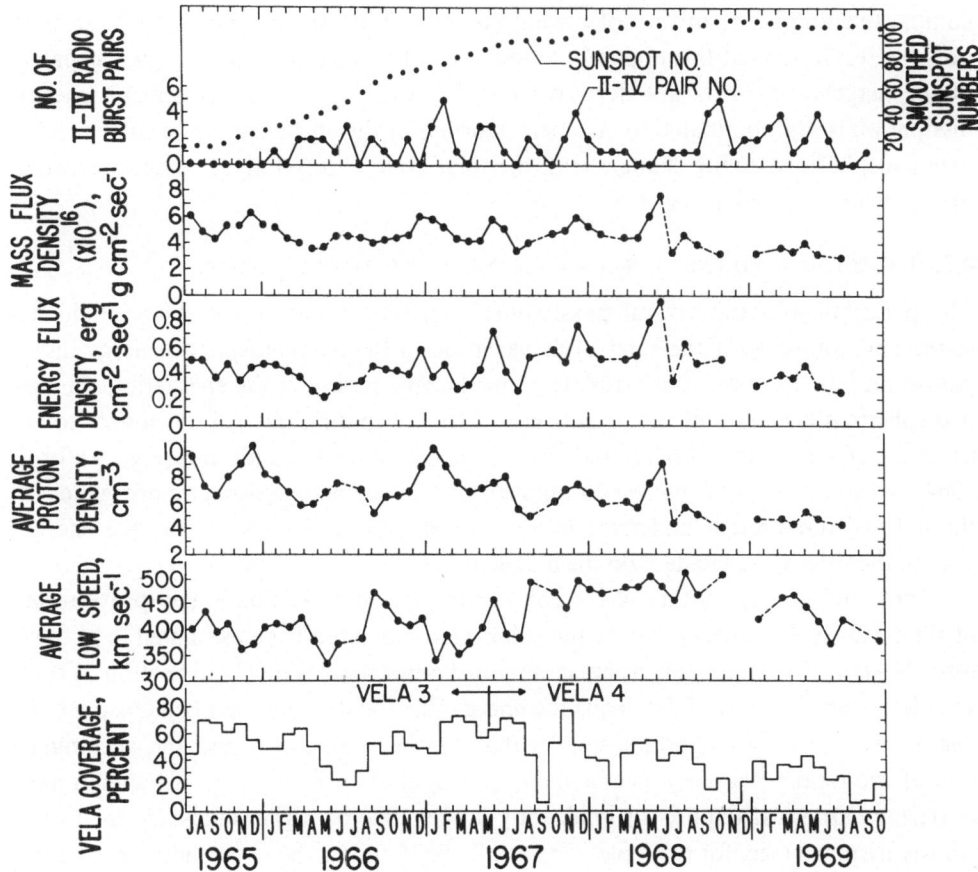

Fig. 6. 27 day averages of solar wind and solar characteristics observed between July 1965 and October 1969 (Montgomery *et al.*, 1972).

ments, and is well suited for the study of possible solar cycle variations. It is immediately apparent that, if such variations were present during this period, they were not large. Gosling *et al.* (1971) drew a similar inference regarding the solar wind speed from a broader set of solar wind observations, while Hirshberg (1973) has argued that the ~ 100 km s^{-1} change in solar wind speed (and a simultaneous 1 γ change in the interplanetary magnetic field strength) between 1967 and 1968 is a solar cycle variation. The downward trend in density throughout this entire period might also be related to the solar cycle (Egidi *et al.*, 1970).

It is also clear from Figure 6 that any variations in the solar wind energy flux density, $\frac{1}{2}\varrho\,u^3$, during this period were not large. This may be regarded as puzzling in view of the large variations in the level of geomagnetic activity associated with the solar cycle. In particular, the frequency of geomagnetic storms changes sufficiently during this cycle that, *if* energy dissipation within the magnetosphere were dominated by storms, a variation by one or two orders of magnitude in this dissipaion rate would be expected. If such an effect is present, it is clear that it is not produced by changes of comparable amplitude in the solar wind energy flux on the magnetosphere. Several possible resolutions of this apparent paradox can be suggested: the energy dissipated in the magnetosphere is not derived from the solar wind; magnetospheric energy dissipation is not dominated by magnetic storms; or the efficiency of the solar wind – magnetospheric coupling changes with the solar sycle in response to some solar wind property not reflected in Figure 6.

4.2. The solar wind energy flux incident on the magnetosphere

The plausibility of the several possibilities suggested immediately above could be tested and compared if the solar cycle variation in the magnetospheric energy dissipation rate were known. Unfortunately, our understanding of the energetics of magnetospheric processes remains primitive, and the dissipation rate is known only to an accuracy of $\sim$ an order of magnitude for geomagnetic storm conditions (e.g., Axford, 1964). It would seem of interest to pursue the topic of magnetospheric energetics in the light of our present understanding the properties, and in particular the energy flux, of the solar wind incident on the magnetosphere.

In any such study, another of the conclusions drawn in Section 3 may prove to be of significance. The energy flux in the solar wind has often been estimated (e.g., Axford, 1964) as $f = \frac{1}{2} \varrho_{max} u_{max}^3$, whereas we now know (conclusion (1) in Section 3) that such levels are never attained. Thus the energy flux incident on the magnetosphere is one or two orders of magnitude smaller than has been assumed, and the accumulation of the total energy in a geomagnetic storm, about 10^{23} erg (e.g., Chapman and Bartels, 1950), must require a significantly longer time or higher efficiency then previously assumed (see, for example, Obayashi, 1967), unless the total energy is smaller than previously estimated.

Acknowledgement

The National Center for Atmospheric Research is sponsored by the National Science Foundation.

References

Axford, W. I.: 1964, *Planetary Space Sci.* **12**, 45.
Chapman, S. and Bartels, J.: 1950, *Geomagnetism, Vol. II*, Clarendon Press, Oxford.
Egidi, A., Formisano, V., Palmiotto, F., and Sarenceno, P.: 1970, *J. Geophys. Res.* **75**, 6999.
Gautier, A.: 1852, *Arch. Sci.* **21**, 194.
Goldstein, B.: 1971, preprint.
Goldstein, B. and Siscoe, G. L.: 1972, in C. P. Sonett, P. J. Coleman, and J. M. Wilcox (eds.), *Solar Wind*, NASA SP-308, Washington.
Gosling, J. T., Hundhausen, A. J., Pizzo, V., and Asbridge, J. R.: 1972, *J. Geophys. Res.* **77**, 5442.
Gosling, J. T., Hansen, R. T., and Bame, S. J.: 1971, *J. Geophys. Res.* **76**, 1811.
Hirshberg, J.: 1973, *Astrophys. Space Sci.* **20**, 473.
Hundhausen, A. J.: 1972, *Coronal Expansion and Solar Wind*, Springer-Verlag, New York.
Hundhausen, A. J.: 1973, *J. Geophys. Res.* **78**, 1528.
Hundhausen, A. J. and Gentry, R. A.: 1969, *J. Geophys. Res.* **74**, 2908.
Meadows, A. J.: 1970, *Early Solar Physics*, Pergamon Press, Oxford.
Montgomery, M. D., Bame, S. J., and Hundhausen, A. J.: 1972, *J. Geophys. Res.* **77**, 5432.

Obayashi, T.: 1967, in J. W. King and W. S. Newman (eds.), *Solar-Terrestrial Physics*, Academic Press, London.
Parker, E. N.: 1963, *Interplanetary Dynamical Processes*, Interscience, New York.
Pizzo, V., Gosling, J. T., and Hundhausen, A. J.: 1973, *J. Geophys. Res.* **78**, to be published.
Sabine, E.: 1852, *Phil. Trans.* **142**, 103.
Sarahbai, V.: 1963, *J. Geophys. Res.* **68**, 1555.
Siscoe, G. L.: 1970, *Solar Phys.* **13**, 490.
Wolf, R.: 1852, *Compt. Rend.* **35**, 364.

THEORY OF THE JOVIAN RADIATION BELTS

F. M. NEUBAUER

Institut für Geophysik und Meteorologie, Technische Universität, Braunschweig, F.R.G.

1. Introduction

The present picture of the Jovian radiation belts and magnetosphere is based on radioastronomical observations, i.e., the observations of decametric and decimetric emissions and theoretical concepts, most of which were developed for the terrestrial magnetosphere. Since the terrestrial magnetosphere and radiation belts are far from being understood, a number of surprises can be expected from space missions to Jupiter on Pioneer 10 and 11 and Mariner-Jupiter-Saturn 1977.

We shall first sumarize a few parts of the general picture of the magnetosphere relevant to the theory of the radiation belts. The magnetic field can be considered to be dipolar with an equatorial surface field of about 10 G and a tilt of $\approx 10°$, where the orientation is opposite to the terrestrial one. Much less conclusive evidence is available for the higher multipole contributions. Extrapolating the solar wind to Jupiter's orbit an average speed of 400 km s^{-1} an average proton concentration of 0.2 cm^{-3} and a magnetic field of 1 γ are expected. The pressure balance equation yields a magnetic field of $\approx 12.5 \gamma$ just inside the subsolar point at a distance of ≈ 50 Jupiter radii R$_J$. Simple scaling of the terrestrial magnetosphere leads to polar caps of 9° diameter. According to Brice and Ioannidis (1970) the magnetospheric plasma is corotating nearly out to the boundary of the magnetosphere, which is expected to be turbulent and much less clearly defined than the terrestrial magnetopause because of the velocity shear between the magnetospheric speed of ≈ 600 km s^{-1} and the shocked solar wind.

The lower boundary of the magnetosphere is given by the Jovian ionosphere (Hunten, 1969) on top of the Jovian atmosphere, which is mainly composed of H and He. Photoelectrons from the ionosphere are the starting point for the model of the magnetospheric plasma distribution by Ioannidis and Brice (1971). This model is characterized by an electron concentration $N_e \approx N_p$ of 0.3 cm^{-3} for small L, a maximum of 100 cm^{-3} at $L \approx 8$ and a subsequent decrease of L^{-4} with a width of 1 R$_J$ perpendicular to the equatorial plane. L characterizes a field line by its equatorial distance from Jupiter measured in planetary radii R$_J$.

The neutral gas density from Jupiter's exosphere vanishes rapidly because of the strong gravitational field and low atmospheric temperature. However, the possibly existing thin atmospheres of the Galilean satellites may contribute significant amounts of neutral gas along the lines suggested by McDonough and Brice (1973) for the Saturnian satellite Titan.

Decimetric emissions first observed in 1959 are generally believed to be due to synchrotron radiation from trapped relativistic electrons in the inner magnetosphere.

B. M. McCormac (ed.), Magnetospheric Physics, 85-92. All Rights Reserved.

Sections 2 and 3 give some details of charged particle dynamics and the present theoretical picture of the Jovian radiation belts, respectively.

2. Charged Particle Motion in the Average Magnetic and Electric Fields of Jupiter

In this section we shall briefly discuss the characteristics of motion of charged particles in the average magnetic and electric field of Jupiter. An aligned, centered dipole magnetic field of $B_{eq} = 10$ G needs only to be considered to clarify the important differences compared with the terrestrial radiation belt particles. Table I (Mead and

TABLE I

Dynamical characteristics of charged particles in the Jovian magnetosphere (essentially from Mead and Hess, 1973)

	$L = 1$	Amalthea (V)	Io (I)	Europa (II)	Ganymede (III)	Callisto (IV)
L	1.00	2.55	5.95	9.47	15.10	26.56
B, G	10.0	0.6048	0.0475	0.0118	0.0029	0.00053
R_{co}, GeV	5300	815	150	59	23.2	7.5
			Protons			
E, MeV	410	29.8	2.37	0.589	0.145	0.027
r_c, km	3.2	13.1	46.9	94.1	190	442
τ_b, min	0.016	0.122	0.991	3.16	10.1	41.6
τ_d, days	0.35	1.63	8.65	21.8	55.6	172
			Electrons			
E, MeV	22.1	5.07	1.128	0.418	0.129	0.026
r_c, km	0.075	0.31	1.09	2.20	4.42	10.3
τ_b, min	0.012	0.030	0.074	0.13	0.30	1.02
τ_d, days	10.8	17.3	27.7	39.9	69.6	181

Hess, 1973) shows the cyclotron radii r_c, bounce periods τ_b, gradient drift periods τ_d of equatorial electrons and protons of energy E for $L = 1$ and the L values of the inner satellites for a constant value of $\mu = 50$ MeV G^{-1}, the first adiabatic invariant. The value of μ chosen is typical for theories of radial diffusion in the Jovian magnetosphere.

It is clear from the Table, that the drift periods τ_d are generally much larger than $T_J = 0.42$ days, the period of revolution or equivalently the period of the drift by the corotational electric field $\Omega R_J \times LB$, where $\Omega = 2\pi/T_J$. This is in sharp contrast to the terrestrial belts, where $\tau_d \ll 1$ day is typically true for particles at intermediate and high energies. The bounce periods τ_d are increased by the ratio of planetary radii for given E and L at Jupiter compared with the Earth. The cyclotron radii are all smaller than the radii of the Galilean satellites, which is important for the sweeping effects of the satellites discussed later. Note, that for a given L r_c varies like $\mu^{1/2}$, τ_b like $\mu^{-1/2}$ and

τ_d like μ^{-1} in the nonrelativistic case. In the ultra relativistic case, which is important particularly for the electrons, these quantities vary like $\mu^{1/2}$, not at all, and $\mu^{-1/2}$, respectively.

For the CRAND and SPAND sources of radiation belt particles the vertical cutoff rigidity R_c of cosmic ray particles is important. For a singly charged particle rigidity is defined as pc, where p is the momentum of the particle. Störmer theory for a dipole shows that only particles of rigidity greater than $R_c = 5300$ GeV $\cos^4 \lambda$ can vertically reach Jupiter's dense atmosphere at zenomagnetic latitude λ. Table I also gives the equatorial cutoff rigidities $R_{co} \approx E_{co}$ at the location of the Galilean satellites, although at the outer satellites the dipole approximation somewhat underestimates R_{co}.

3. Models of the Jovian Radiation Belts

In this section we shall review the physical mechanisms operating in the Jovian radiation belts, i.e. sources, deviations from adiabatic behavior like radial diffusion violating the third adiabatic invariant and pitch angle diffusion, and the loss processes according to present theoretical understanding.

Source mechanisms considered so far are injection from the solar wind, neutron decay, and satellite-magnetosphere interaction. The loss mechanisms which have been considered are scattering by thermal electrons, atmospheric losses, pitch angle diffusion into the loss cone, the sweeping effect of the inner satellites Amalthea through Callisto, and synchrotron radiation losses.

The bulk of radiation belts is expected to be due to injection from the solar wind in current models. The particles, i.e., essentially protons and electrons then diffuse inward conserving their first and second adiabatic invariants. The injection rate is generally taken as a free parameter, since the injection mechanism is only poorly understood even for the terrestrial magnetopause. One possibility, the transport of intact mass elements of plasma containing an equal number of protons and electrons across the boundary, would lead to equal injection rates for protons and electrons. There are certainly other possibilities. A lesser uncertainty exists for the μ distribution of injected particles. Electric fields due to magnetic pumping or fluctuating convection fields then cause inward diffusion of injected particles. The average time τ_D it takes to diffuse from the boundary to L is shown in Figure 1 for $\mu = 50$ MeV G^{-1} and protons and electrons under the assumption that a random convection field $E_{conv} \approx$ ≈ 7 kV R_J^{-1} causes the diffusion. This is one quarter in magnetitude of the electric field due to the average solar wind and is based on the assumption that the convective field 3 kV R_j^{-1} of Brice and Ioannidis (1970) is somewhat enhanced by the turbulence at the boundary. We have used the approximate relativistic expression

$$\tau_D \approx T_J (\pi B_{eq} R_J / 4 E_{conv} T_J)^2 (L^{-2} - 50^{-2})^2,$$

where T_J the rotational period of Jupiter.

Inward diffusion from the outer boundary, which is characterized by a faster growth of the perpendicular component of momentum than the parallel one, and the

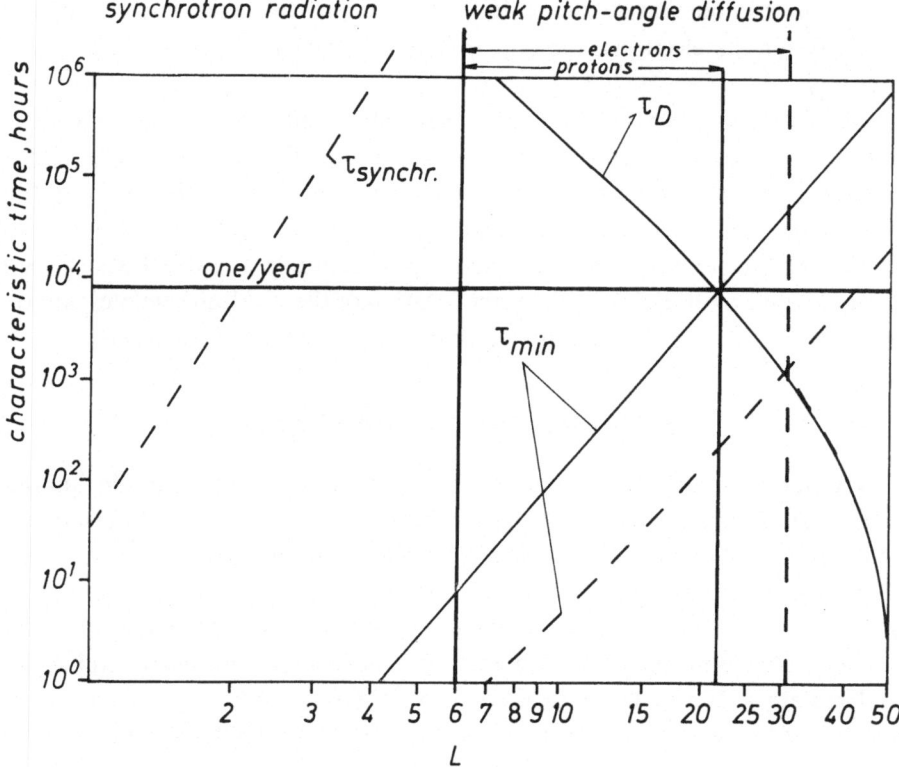

Fig. 1.   Synchrotron life time $\tau_{\text{synchr.}}$ for electrons and radial diffusion time τ_{D} and minimum pitch angle diffusion life time τ_{min} for electrons (dashed) and protons (solid) of $\mu = 50$ MeV G^{-1}. The extreme boundaries for the weak pitch angle diffusion region are also shown.

loss cone would lead to an increasing anisotropy in the sense $(\partial f/\partial\alpha) > 0$, if it were not for pitch angle diffusion by whistler mode turbulence for the electrons and ion cyclotron turbulence for the protons. Diffusion in the pitch angle α causes a steady flux of particles towards the loss cone, where they are lost. Pitch angle diffusion by self-sustained wave turbulence of the type just mentioned requires the existence of particles above the minimum resonant energies $E_{\text{crit}} A^{-2} (1+A)^{-1}$ and $E_{\text{crit}} A^{-1} (1+A)^{-2}$ for protons and electrons, respectively, where $E_{\text{crit}} = B^2/8\pi N$ is the magnetic energy per particle and A the anisotrophy factor. A varies from 1/7 to 1/12 in the region of interest. It turns out that electrons only will be subject to pitch angle diffusion by self-sustained whistler mode turbulence for a Ioannidis and Brice (1971) plasma distribution and $\mu = 50$ MeV G^{-1}. Protons will become unstable if μ is increased or N is increased. Let us assume for a moment that the minimum resonant energies are actually exceeded in the region of interest. The minimum lifetime τ_{min} for this type of loss mechanism has been discussed in detail by Kennel (1969) and is given by

$$\tau_{\text{min}} \approx 6 R_{\text{J}} L^4/v,$$

where v is the particle velocity. It is also plotted in Figure 1 as a function of L for

$\mu = 50$ MeV G^{-1}. It turns out to be much larger than τ_D above $L \approx 22$ for protons and $L \approx 30$ for electrons.

In other words, above these L values, radial diffusion and pitch angle diffusion caused by it act so fast that there is no time to form an appreciable loss cone by emptying the particles from the loss cone into the atmosphere. Therefore the particle distribution functions are expected to be essentially isotropic with only a small $(\partial f/\partial \alpha) > 0$. This region is called the strong pitch angle diffusion region (Kennel, 1969).

The loss cone becomes marked in the weak pitch angle diffusion region with $\tau_D \gg \tau_{min}$ below $L \approx 22$ and 30. In this region the concept of an upper limit on the particle flux first developed by Kennel and Petschek (1966) becomes applicable. This concept has been applied to Jupiter by Brice (1972) and Thorne and Coroniti (1972). Using $\mu = 100$ MeV G^{-1} Brice (1972) arrived at $\beta \approx 0.01$ for the ratio of proton kinetic energy density over magnetic field energy density in the weak diffusion region. For a given relative width of the velocity distribution function β depends on the typical μ like $\mu^{1/2}$ in the classical case. Neubauer (1972), taking into account relativistic effects, derived upper limits for given spectral indices and anisotropies of the proton distribution function but without assuming a particular mechanism for the formation of the belts like radial diffusion.

The upper limit on the flux in the weak pitch angle diffusion region would have some interesting consequencies. If the limit is really reached, e.g., by a sufficiently large injection rate there will be particle precipitation along the field lines leading to a second 'auroral' zone around $\lambda \approx \pm 75°$. For variations of the injection rate above the critical injection rate leading to the upper limit the precipitation rate will vary but not the upper limit flux. The weak diffusion region acts as a filter region and provides a boundary condition for the treatment of the inner radiation belts in which pitch angle diffusion is not important any more because $(B^2/8\pi N)$ increases very rapidly as L decreases below about $L \approx 7$.

There is a difficulty with the diffusion models of the inner belts of Jupiter with an injection source at the 'magnetopause' and terrestrial diffusion mechanisms, however. Unfolding of measured intensity distributions of decimetric radio emissions at 10.4 cm and 21 cm has led Luthey (1972) to derive synchrotron radiation lifetimes of less than 1 yr out to $L \approx 4$. As Figure 1 shows, diffusion by fluctuating convection electric fields is much too slow by orders of magnitude to realize the necessary injection time, i.e., diffusion time into the inner belts. The same is true for magnetic pumping. Another difficulty is associated with the sweeping effect of the satellites. Mead (1972) and Mead and Hess (1973) have computed the mean life-time of protons and electrons on L shells, which cut through the satellites Amalthea through Callisto. Mead and Hess (1973) assumed no distortion of the dipolar magnetic field of Jupiter by the satellites and complete absorption of a particle after impact on the satellite. According to their results, there should be no electrons inside $L = 5.95$ (the orbit of the satellite Io) if the diffusion coefficients due to magnetic pumping or fluctuations of potential electric fields are simply scaled from Earth to Jupiter. The satellites Amalthea, Io, and Europa are most efficient in sweeping up particles.

Since strong electron belts are observed out to $L \approx 4$, one or more assumptions in the model developed so far must be abandoned. Several suggestions have been made. Brice (1972), Brice and McDonough (1972), and Coroniti (1973) have suggested an ionospheric dynamo as a source of electric field fluctuations yielding a much faster diffusion coefficient $D = kL^3$ at low L values in contrast to the L^{10} and L^6 dependences for magnetic pumping and potential electric field diffusion, respectively. Their diffusion coefficients are sufficiently large at low L values to allow the particles to pass by Io's L shell. Their theory predicts daily variations in flux and asymmetry of the belts. Although plausible, their theory has to be further substantiated theoretically and observationally in view of the appreciable difficulties inherent in theories of ionospheric winds produced by tides, gravity waves, etc. Jacques and Davis (1972) derive a diffusion coefficient varying like $L^2(L-1)$ for the ionospheric winds. In an inverse approach, Birmingham et al. (1973) use the observations of decimetric emissions to determine the constants k and m in the expression $D = kL^m$ for the diffusion coefficient and μ_1 for the initial value of μ. As did Jacques and Davis (1972), these authors also take into account synchrotron radiation losses. Their best fit value for m is $m = 2.6 \pm \pm 0.5$, for $\mu_1 = 590 \pm 300$ MeV G^{-1}. The value of k corresponds to a value smaller than the one deduced by Brice and McDonough (1972). The value of μ_1 is large compared with values chosen by other authors. The diffusion coefficients in the latter three papers are sufficient to diffuse the particles even by Io at sufficient rates. Detailed calculations on Jovian diffusion radiation belts have also been performed by Stansberry (1973) for electrons and protons. Other possibilities to overcome the difficulty mentioned above are energy sources inside the orbit of Io and a strong distortion of the field configuration by the satellites such that particles aiming at one of the satellites are not absorbed. In the case of Io a strong interaction with the magnetosphere is indicated by the radio astronomical Io-effect. The exact nature of this interaction depends on the internal conductivity profile of the satellite (e.g., Lewis, 1973).

The paper by Mead and Hess (1973) on the sweeping effect is the only paper on the influence of the satellites on the trapped particle distribution.

Apart from this work Io has only been investigated with the intention to explain decametric emissions. Goldreich and Lynden-Bell (1969) treated in considerable detail the current system generated by the motional electric field of Io through the Jovian ionosphere. This work was refined by Gurnett (1972), Shawhan et al. (1972), and Hubbard (1972) to include the very important plasma sheath effects. In Hubbard's model voltages of more than 10^5 kV develop in the plasma sheath around Io. This voltage could accelerate electrons and protons to energies of the same order. It is interesting to note, that an electron energy of 5×10^5 eV at $L = 5.9$ would lead to 4.8 MeV at $L = 1.8$, an energy which could produce the observed synchrotron radiation in view of the uncertainties of the unfolding problem. 5×10^5 V is a typical voltage across the Debye sheath of Io (Hubbard, 1972). It is furthermore interesting that only a fraction of roughly 2×10^{-4} of the total photoelectron flux from Io according to Hubbard's (1972) model together with a diffusion coefficient $D : L^3$ would be sufficient as a source of Jupiter's relativistic electron belts. This could therefore be an important and perhaps the major source of radiation belt electrons. To assess the significance of this source of energetic electrons and protons, it is necessary to compute the number of accelerated particles missing Jupiter's ionosphere. The contribution of Io to the diffusion coefficient requires the calculation of the dc and ac disturbances created by the satellite. Both important and perhaps decisive problems have not been solved in detail yet. The same is true for the problem of the influence of the intense decametric radiation fields on the radiation belt particles.

Finally we briefly discuss the neutron decay source of radiation belt particles. Only the CRAND source has been considered in some detail. As the high cutoff rigidity at low latitudes mentioned in Section 2 shows, the main source of CRAND neutrons should be the polar atmospheres. Thomas and Doherty (1972) have used observations of the proton population at $L=1.5$ in the terrestrial inner belts to calculate CRAND proton fluxes around $L \approx 18$ between ≈ 100 and ≈ 800 MeV. Although the resulting fluxes with electron scattering as the only loss mechanism lead to fluxes much lower than the estimated fluxes of diffusion models, their energy places them far above diffusion energies of at most a few MeV for solar wind injected particles at this L value. Solar cosmic rays of low energies could also diffuse into the Jovian magnetosphere to contribute to the high energy tail of the particle distribution. SPAND can be considered to be negligible by the same reasons as in the terrestrial case. The solar neutron decay source strength can be shown to be smaller than the terrestrial one by 6×10^{-6} for 50 MeV and 1.1×10^{-2} for 500 MeV solar neutrons. Fluxes of Jovian solar neutron decay protons and electrons have not been calculated however.

Inspection of Table I shows that the cutoff rigidities at the orbits of the Galilean satellites, in particular Ganymede and Callisto, are low enough to allow CRAND sources or even SPAND sources of marked intensity in the intermediate and outer magnetosphere of Jupiter inspite of the small size of the satellites. Although no conclusive evidence is available regarding atmospheres on the Galilean satellites, there are reasons to believe that thin atmospheres could exist. The albedo neutron source would then partly be the atmosphere and the solid body of the satellites.

4. Future Problems

There are only a few quantitative theoretical papers on the Jovian radiation belts. The most important problems for future theoretical work for Jupiter seem to be the following:

(a) How do Io and the other satellites contribute to the source, loss, and transport mechanisms in the Jovian radiation belts?

(b) A more precise treatment of particles generated by the decay of neutrons from the planet and its outer Galilean satellites is warranted.

(c) How can magnetic and electric field fluctuations observed in the Jovian magnetosphere be used to obtain information on the belts and the magnetosphere (Liemohn, 1972)?

(d) What can be said theoretically about the propagation and source mechanisms of waves in the lower atmosphere and ionosphere? Since dynamo electric fields play an important role in diffusion theories, the formal solution of the tidal problem should be attempted.

(e) A more refined model of the plasma distribution with a more rigorous treatment of the outer convective magnetosphere is possible and necessary. A neutral H distribution possibly produced by the satellites should be investigated.

A more speculative question is, how significant are those substorm type activities in the Jovian magnetosphere?

The results from the four fly-by's at Jupiter until 1980 will redirect theoretical research on problems of the Jovian radiation belts.

References

Birmingham, T., Hess, W., Northrop, T., Baxter, R., and Lojko, M.: 1973, *J. Geophys. Res.* submitted.
Brice, N. M.: 1972, Proc. of Jupiter Radiation Belt Workshop, 283.
Brice, N. M. and Ioannidis, G. A.: 1970, *Icarus* 13, 173.
Brice, N. M. and McDonough, T. R.: 1972, Cornell Univ. Report.
Coroniti, F. V.: 1973, Paper, American Geophys. Union.
Goldreich, P. and Lynden-Bell, D.: 1969, *Astrophys. J.* 156, 59.
Gurnett, D. A.: 1972, *Astrophys. J.* 175, 525.
Hubbard, R. F.: 1972, Univ. of Iowa Report.
Hunten, D. M.: 1969, *J. Atmospheric Sci.* 26, 826.
Ioannidis, G. A. and Brice, N. M.: 1971, *Icarus* 14, 360.
Jacques, S. A. and Davis, Jr., L.: 1972, Calif. Inst. of Technol. Report.
Kennel, C. F. and Petschek, H. E.: 1966, *J. Geophys. Res.* 71, 1.
Kennel, C. F.: 1969, *Rev. Geophys. Space Phys.* 7, 379.
Lewis, J. S.: 1973, *Space Sci. Rev.* 14, 401.
Liemohn, H.: 1972, Proc. of Jupiter Radiation Belt Workshop, 381.
Luthey, J. L.: 1972, Proc. of Jupiter Radiation Belt Workshop, 47.
McDonough, T. R. and Brice, N. M.: 1973, *Nature* 242, 513.
Mead, G. D.: 1972, Proc. of Jupiter Radiation Belt Workshop, 271.
Mead, G. D. and Hess, W. N.: 1973, *J. Geophys. Res.* 78, 2793.
Neubauer, F. M.: 1972, Proc. of Jupiter Radiation Belt Workshop, 405.
Shawhan, S. D., Hubbard, R. F., Joyce, G., and Gurnett, D. A.: 1972, University of Iowa Report.
Stansberry, K. G.: 1973, Aerospace Rep. No. ATR-73(9990)-3.
Thomas, J. and Doherty, W. R.: 1972, Proc. of Jupiter Radiation Belt Workshop, 315.
Thorne, R. M. and Coroniti, F. V.: 1972, Proc. of Jupiter Radiation Belt Workshop, 363.

PITCH ANGLE DISTRIBUTIONS OF ENERGETIC ELECTRONS
IN THE EQUATORIAL REGIONS OF
THE OUTER MAGNETOSPHERE – OGO-5 OBSERVATIONS

HARRY I. WEST, Jr. and RICHARD M. BUCK

University of California, Lawrence Livermore Laboratory, Livermore, Calif., U.S.A.

1. Introduction

Pitch angle distributions (PAD) of energetic electrons relate to a wide variety of magnetospheric processes. Wave-particle interactions (turbulence, instabilities) act on a time scale comparable to the particle gyro period; betatron acceleration is important on time scales of the bounce period to minutes; and drift-shell effects are important on time scales of the azimuthal drift period. A study of pitch angle effects can provide important information on the static and dynamic aspects of field topology. Our interest in this paper is primarily with the drift shell effects.

Prior to OGO-5 it was well known that in the inner belt and in much of the outer belt the normal loss cone distribution (peaked at 90° to **B**) prevails. ATS experimenters had observed the butterfly PAD (flux depression at ∼90° to **B**) in the nighttime magnetosphere, and Serlimitsos (1966) and Haskell (1969) had observed this distribution in the more distant nighttime magnetosphere; the experimenters recognized the importance of drift-shell splitting in causing the butterfly PAD. Despite the attractiveness of pitch angle data very few really detailed measurements were available prior to OGO-5.

Two experiments on OGO-5 are providing a fairly detailed picture of the magnetosphere: the UCLA scintillation counter experiment (six counters in fixed directions relative to the satellite) of Farley and Kivelson, and our LLL experiment. The UCLA experimenters have concentrated their efforts on wave particle interactions and betatron acceleration and we (West *et al.*, 1972, 1973a, b; Buck *et al.*, 1973) have concentrated on drift-shell effects and field topology. In this paper we present a brief but unified view of our studies to date.

2. Instrumentation and Spacecraft

The LLL experiment consisted of a seven-channel magnetic electron spectrometer and a seven-channel proton range-energy telescope. Our concern here is only with the electron results. Pertinent channel details are given in Table I. OGO-5 was stabilized, hence it was necessary to scan the experiment relative to the spacecraft for obtaining pitch angle information. The scan was $3° \, s^{-1}$ through excursions of $\pm 115°$. UCLA vector magnetometer data were used in determining the PAD.

OGO-5 was launched March 4, 1968, with its apogee of 24 R_E geocentric on the

TABLE I

Electron and proton channel characteristics

	E_1	E_2	E_3	E_4	E_5
Energy	79 ± 23 keV	158 ± 36	266 ± 36	479 ± 52	822 ± 185
Geometry	0.18 cm² keV sr	0.277	0.390	0.605	4.43

	E_6	E_7	P_1		P_2
Energy	1530 ± 260	2820 ± 270	100–150 keV		230–57
Geometry	8.57	3.88	2.06×10^{-3} cm² sr		1.3×10^{-2}

morning side of the Earth. Its orbital inclination was initially 31° and its period 62 h. Many inbound passes during 1968 were close to the magnetic equator when OGO-5 was in the outer magnetosphere; it is data from these passes that we report here.

3. Results

3.1. SURVEY

In this section we wish to provide an overall view of PAD's of energetic electrons throughout the equatorial regions of the magnetosphere. We do not dwell on subtle pitch angle effects, but rather, present the data in terms of three broad categories: normal, butterfly, and isotropic. 'Isotropic' means 'isotropic over the range of pitch angles scanned'; we did not always observe at small pitch angles, so that at times we had to use judgement in this assignment (however, often for important results, the UCLA scintillation counter experiment was able to extend our coverage).

For the overview, we chose to concentrate on data from our lowest-energy electron channel E_1 (79 keV). Figure 1 shows the pitch angle changes that occurred on a number of inbound orbits in 1968 when OGO-5 was close to the geomagnetic equator or in the plasma sheet. Many of the effects are energy dependent as becomes evident later. Note the coordinates of Figure 1 are geocentric solar magnetospheric in terms of azimuth only. To reduce some of the presentation problems the radial plots are in terms of true radial diatance in R_E; the projection of R on XY plane is usually so close to R that differences are inconsequential. Since electrons drift eastward around the earth, our presentation proceeds likewise. Our discussion in this section is brief with the explanation coming later.

We start in the prenoon magnetosphere. Almost without exception the normal PAD is observed (at $\sim 1000 \pm 3$ LT) from the magnetopause inwards.

In the early afternoon magnetosphere, we encounter the butterfly PAD at extended distances. Note for Day 28, 1969 we find a slight butterfly effect near the magneto-pause ($j_\perp/j_\parallel \sim 0.8$); further momentary evidence is found as we approach the Earth. As we proceed in azimuth towards dusk we find the butterfly PAD prevailing closer to the Earth; also the depression in $j_\perp$ becomes greater.

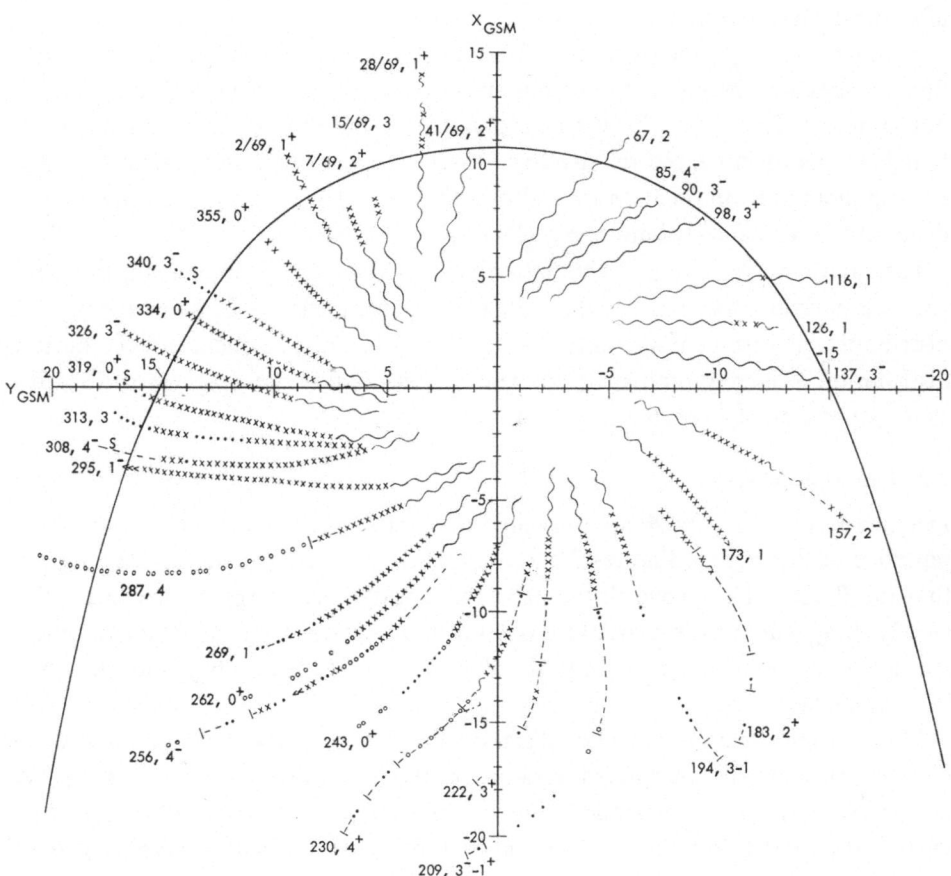

Fig. 1. Pitch angle distribution survey for ~ 79 keV electrons in 1968 and early 1969. Each inbound pass is annotated with the day of year and average K_p. Symbols are used to indicate the PAD form and are: wavy line, normal; period, isotropic and less than 25 electrons cm^{-2} sr^{-1} keV^{-1} s^{-1}; dash, isotropic and flux greater than 25; <, butterfly less than 25; and x, butterfly greater than 25. The symbol o means short isolated gusts; the cross bar, an abrupt flux increase; and s, solar particles.

In the regions past dusk ($\gtrsim 9$ R$_E$) we usually find large depletions of $j_\perp$ ($j_\perp/j_\parallel$ often <0.01). Major magnetic storms occurred on Days 305 and 306. Days 308 and 313 show some of the after effects; the dots and dashes indicate isotropic fluxes of solar electrons but note the butterfly PADs still prevail despite the level of magnetic activity. The disturbed day (Day 287), is interesting; isolated gusts of particles are observed early in the orbit followed by an abrupt increase at 11 R$_E$.

Plus or minus a few hours from midnight we show a number of orbits with abrupt flux increases. Many of these are positively identified as substorm expansion phases by UCLA workers (Aubry *et al.*, 1972; Russell *et al.*, 1971). For example, note the flux increases on Day 230 at 19.2, 17.5, 14.5, and 9.2 R$_E$; the first three substorms appear as flux islands. Note that closer to the Earth the butterfly PAD reemerges shortly after substorm expansion.

Well past midnight the signature of the substorm is not so obvious. Although rapid flux changes are observed, the obvious substorm signatures of Days 222 and 230 are not apparent. Past midnight when we observe the butterfly PAD at $\gtrsim 9\,R_E$ we find that $j_\perp$ is greatly enhanced relative to $j_\parallel$ when compared with fluxes observed in the corresponding region premidnight. Also in these extended regions, we often find evidence of the normal distribution, e.g. (Day 157).

Past dawn we rapidly lose the butterfly PAD. Day 137 is interesting in that E_1 shows a normal PAD whereas the higher-energy channels E_2–E_6 show the butterfly distribution at various points into $7.4\,R_E$. Past ~ 0800 LT, evidence of the butterfly distribution is largely limited to momentary effects near the magnetopause and to slight depression of $j_\perp$ when nearer the Earth.

3.2. PRENOON RESULTS

Detailed results for Day 90 are presented in West *et al.* (1973a) and are quite representative of this region. Figures 2 and 3 present some similar data for Day 85. Note that the PAD's at extended distances show considerable energy dependence, which largely disappears in the heart of the outer belt. It is interesting to note that, in general, strong energy dependence shows in the slot (West *et al.*, 1973a) but is mostly absent in the inner belt.

Many of the PAD's at extended distances in Figure 3 show skirts on the distributions which might be interpreted as components of isotropy. To provide the proper perspective, one should examine contours of constant equatorial B such as those measured by Fairfield (1968) and reproduced in Figure 4. Roederer (1967) has developed the framework for studying the azimuthal drift motion of particles in a distorted magnetosphere. Particles mirroring at the equator, drift at constant B as in Figure 4 (conservation of the first adiabatic invariant). Particles with small equatorial pitch angle drift so as to keep the distance from mirror point to mirror point about constant (conservation of the second invariant). Hence, the Day 85 PAD's that we observe at $9\,R_E$ consist of equatorially mirroring particles that came from $\sim 7\,R_E$ at midnight and of particles at small pitch angles that may have been fairly deep in the magnetotail when near midnight. We later find that if the latter particles crossed the equator at much greater than $\sim 9\,R_E$ at midnight there is a good probability that they became isotropic as a result of the strong tail fields that normally exist there. In addition to these drift-shell effects, wave particle interactions in the region from midnight to 0900 LT probably contributed to the shape of the PAD's.

3.3. AFTERNOON MAGNETOSPHERE

Consider what happens to the PAD's as electrons drift through the distorted noon magnetosphere. Starting at 0900 LT we note that equatorial mirroring particles in the pseudo-trapping regions (i.e., beyond the constant B contour that maps from the noon magnetopause to midnight of Figure 4: $\gtrsim 9\,R_E$ at 0900 LT) will drift to the

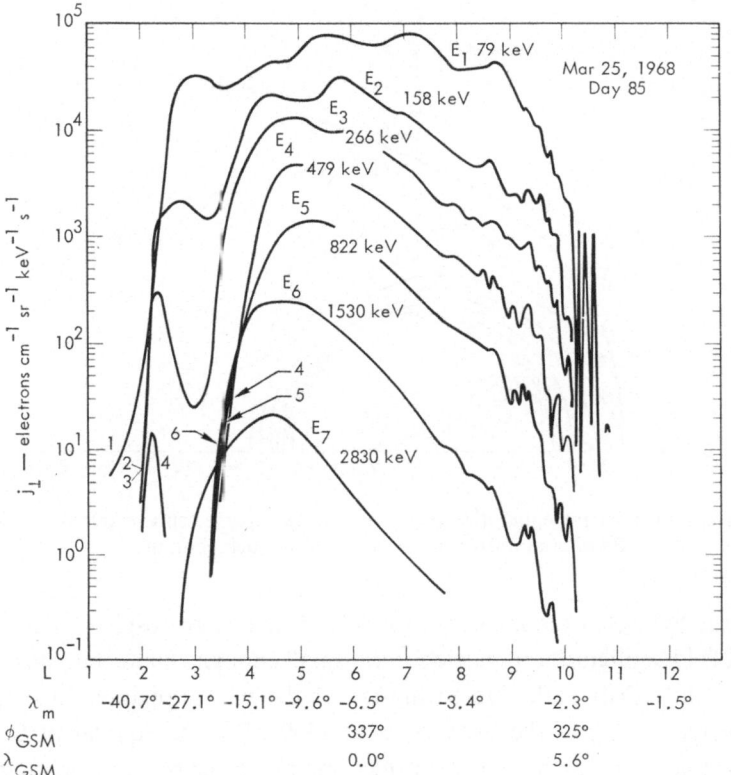

Fig. 2. Radial flux profile on the morning side of the Earth.

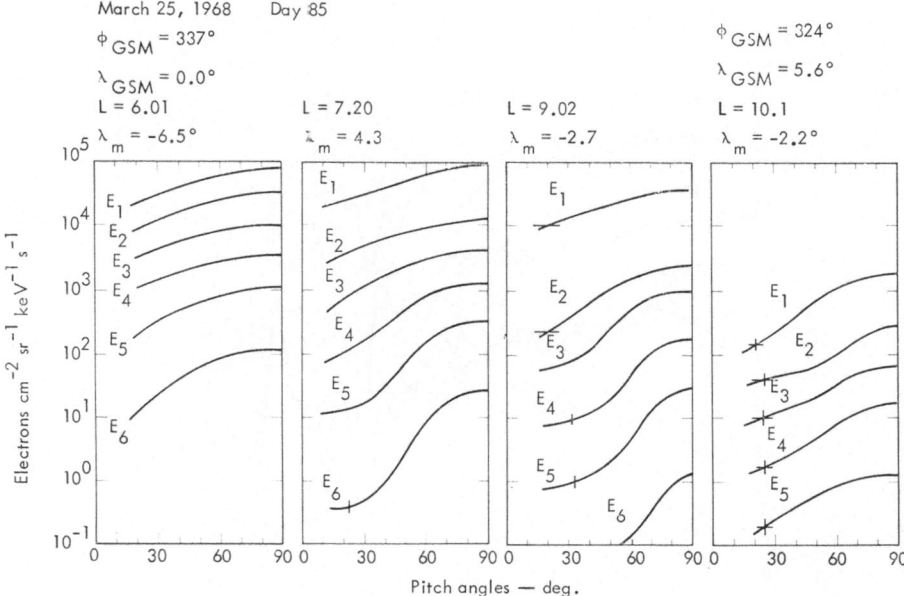

Fig. 3. Pitch angle distributions obtained on the morning side of the Earth (see Figure 2).

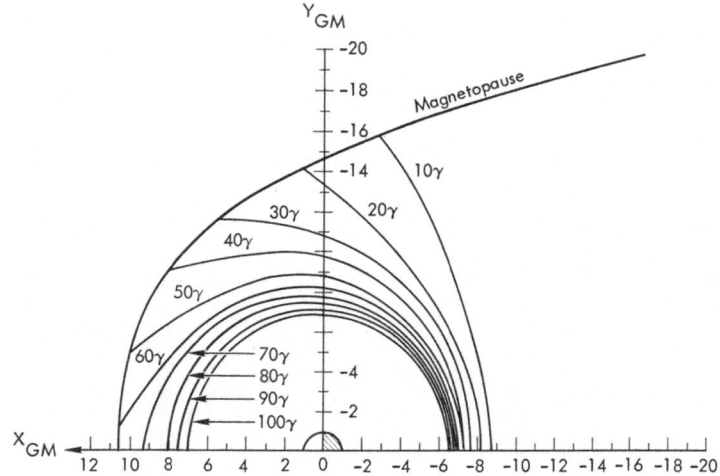

Fig. 4.   Contours of constant equatorial *B* for an average magnetosphere (after Fairfield, 1968).
Equatorial mirroring particles follow such contours.

magnetopasue to be changed in pitch angle or leave the magnetosphere. Because of the compressed field configuration, particles with small enough equatorial pitch angle can drift so that their drift paths are earthward of the magnetopause. In a symmetric magnetosphere, we expect the drift shells at 1500 LT to be equivalent to those at 0900 LT. Past noon we might expect a depletion in $j_\perp$ in the pseudo-trapping regions; this is indeed what we observe as shown by the January 7, 1969 data in Figures 5 and 6

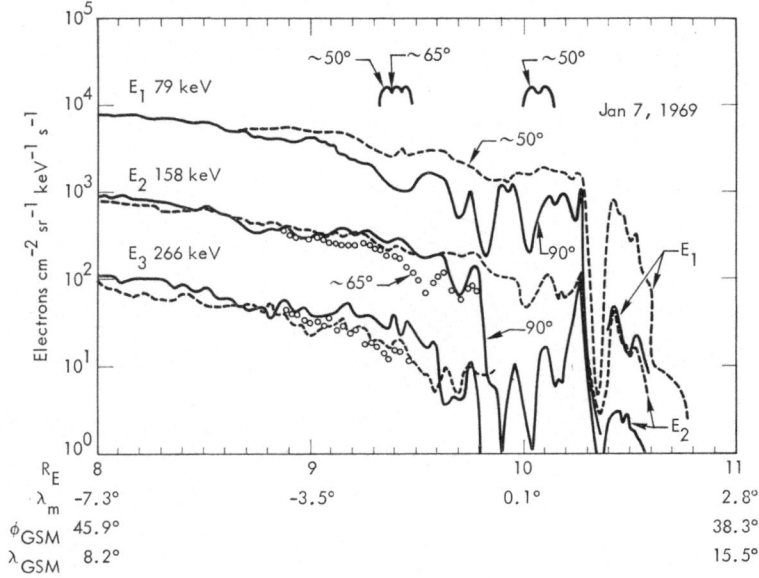

Fig. 5.   Early afternoon data showing effects due to the magnetopause shadowing of drift paths.

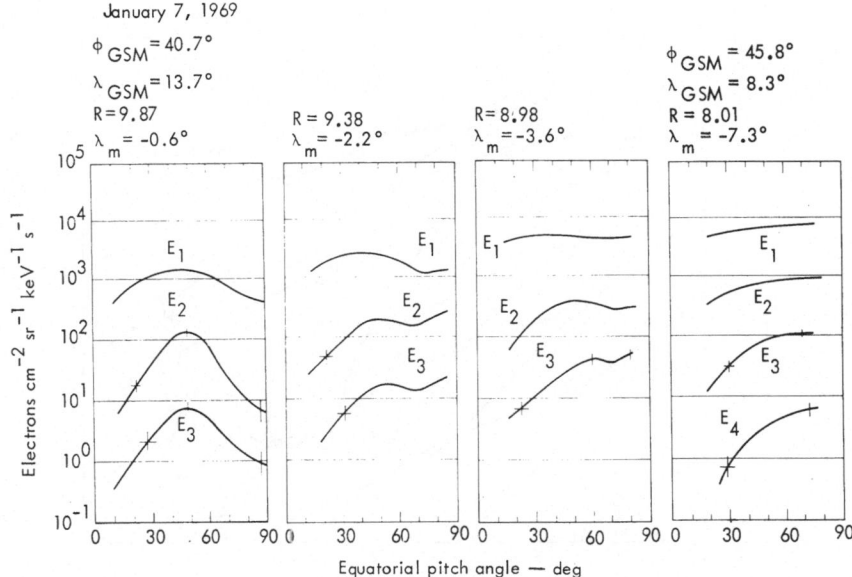

Fig. 6. Pitch angle distributions for Figure 5 (after West *et al.*, 1973a).

(West *et al.*, 1973a). This magnetopause shadowing effect also explains the appearance of the butterfly PAD's in the afternoon regions of our survey in Figure 1.

We should mention another aspect of the picture. Shabansky (1971) shows how near equatorially mirroring particles, as they drift eastward, may branch north or south through minimum B regions in the field configuration near noon so as to stay in the magnetosphere. We have considered this problem from the viewpoint of PAD's and have shown qualitatively (Buck and West, 1973) how the PAD's of Figure 6 (panels 2 and 3 showing minima at $\sim 65°$) might have resulted from such drift motion. We still, however, consider the role of minimum B regions in the high latitude dayside field topology to be unclear.

3.4. PRE-MIDNIGHT MAGNETOSPHERE

As electrons drift into the nighttime magnetosphere they encounter magnetic fields that are increasingly tail-like. Following Roederer (1967) one notes that particles drifting at constant B (90° pitch angles) move closer to the Earth. Particles with large second invariant (small equatorial pitch angles) have equatorial crossings farther from the Earth and their equatorial pitch angles reduced relative to that occurring earlier in their drift path (note in these considerations, that as long as adiabatic motion is occurring, Liouvilles theorem holds and $j(\theta_{eq}) = $ constant with $\theta_{eq} = \sin^{-1}\sqrt{(B_{eq}/B_m)}$.

The quiet-time data of Day 295 exemplifies these effects. Radial profiles of $j_{\perp}$ and $j_{\parallel}$ are shown in Figure 7 and corresponding PAD's in Figure 8. The data at 5 to 9 R_E reflect the magnetic field configuration changes the electrons encounter in their drift from the day to night side of the Earth. To appreciate this, one needs to examine the negative radial flux gradients and PAD's earlier in the drift path (e.g., Day 85, Figures

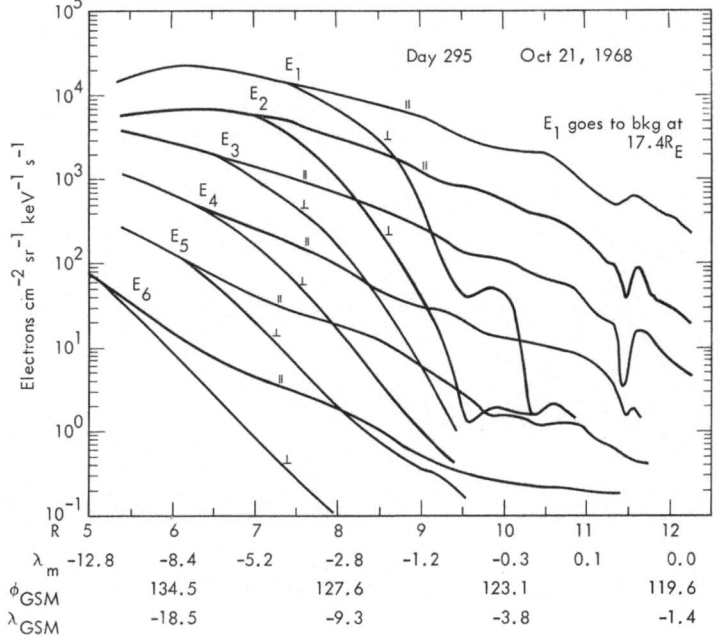

Fig. 7. Radial profile of $j_\perp$ and $j_\parallel$ for a quiet day inbound pass. $K_p = 0^+$. Note that $j_\parallel$ is the peak flux at small pitch angles (10° to 30°).

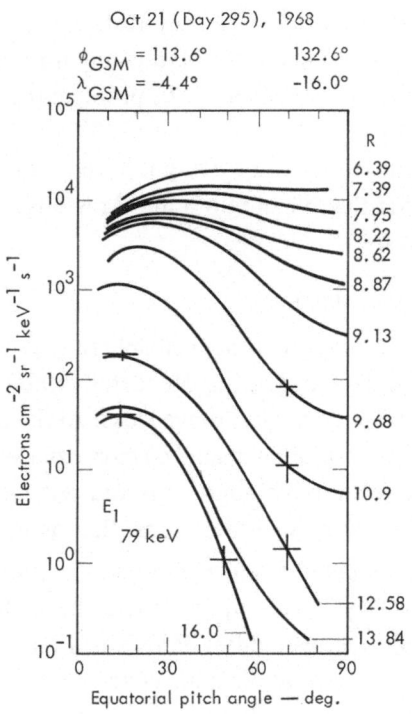

Fig. 8. Pitch angle distributions for Figure 7.

2 and 3). The data beyond ~ 9 R_E in Figure 7 are due to changes both in field configuration and magnetopause shadowing.

3.5. Midnight and substorms

As seen in the survey (Figure 1) the butterfly PAD is found in the dusk magnetosphere even during disturbed periods. Near midnight ($\sim 2300 \pm 2$ LT) the picture changes; the effects of substorms and tail-like fields causes marked changes in the particle fluxes and PAD's. OGO-5 experimenters have published a wide variety of observations in this region. We (West *et al.*, 1973b; Buck *et al.*, 1973) participated in a nine-paper study of substorms on Day 228 (August 15, 1965). The substorm with expansion phase at 0714 UT is especially interesting and the PAD results are fairly general. Prior to the growth phase, the field configuration was only moderately tail-like and the butterfly PAD was observed. At 0640 UT the field at OGO-5 (at ~ 9 R_E) started becoming more tail-like and by 0655 UT the PAD's were isotropic. The particle fluxes then decreased several orders of magnitude until ~ 0714 UT. With the onset of expansion and the appearance of a more dipole-like field configuration, fresh electrons showing the butterfly PAD drifted into the region of observation; the higher energy electrons changed first in time. It is interesting to note that betratron effects (Kivelson *et al.*, 1973) and enhanced ELF (Scarf *et al.*, 1973) were observed during substorm recovery; both observations relate to the PAD changes that occurred.

In Figure 9 we show substorm data for Day 222 (August 9, 1968). Here we note expansions occurring at 17.8, 15.2, 12.6, and 9.8 R_E (see Russell *et al.*, 1971, for discussion of the field data). Simultaneously with the appearance of more dipole-like fields we note the rapid increase in both electron and proton fluxes. The bottom panel marked 'scan angle' shows the pitch angles of the particles being detected at a given instance but is too compressed for that information. However, due to the orientation of the experiment the outer envelope of 'scan angle' is the field inclination and its complement and provides a descriptive picture of the field configuration: narrow when tail-like, wide when dipole-like. Note, with the possible qualification of the occasional limited range of look angles, the lack of modulation of the electron data means isotropy and the presence of modulation means the butterfly PAD. An especially interesting period starts ~ 2100 UT; we see the emergence of the butterfly PAD for E_2 followed by changes in E_1 at ~ 2115 UT. The delay suggests that the drift time of the electrons from an undisturbed source is important.

At 2212 UT note the abrupt loss of modulation, indicating isotropy, as the field became more tail-like. This is the sort of thing we observed earlier in the growth phase of the 0714 UT August 15 substorm. The re-emergence of the butterfly PAD after 0130 UT in the fourth substorm of August 9 parallels the expansion phase of the 0714 UT August 15 substorm.

We consider the transition to isotropy in the growth phase as an indicator of the extent the field has become tail-like, not so much at the position of OGO-5 but farther

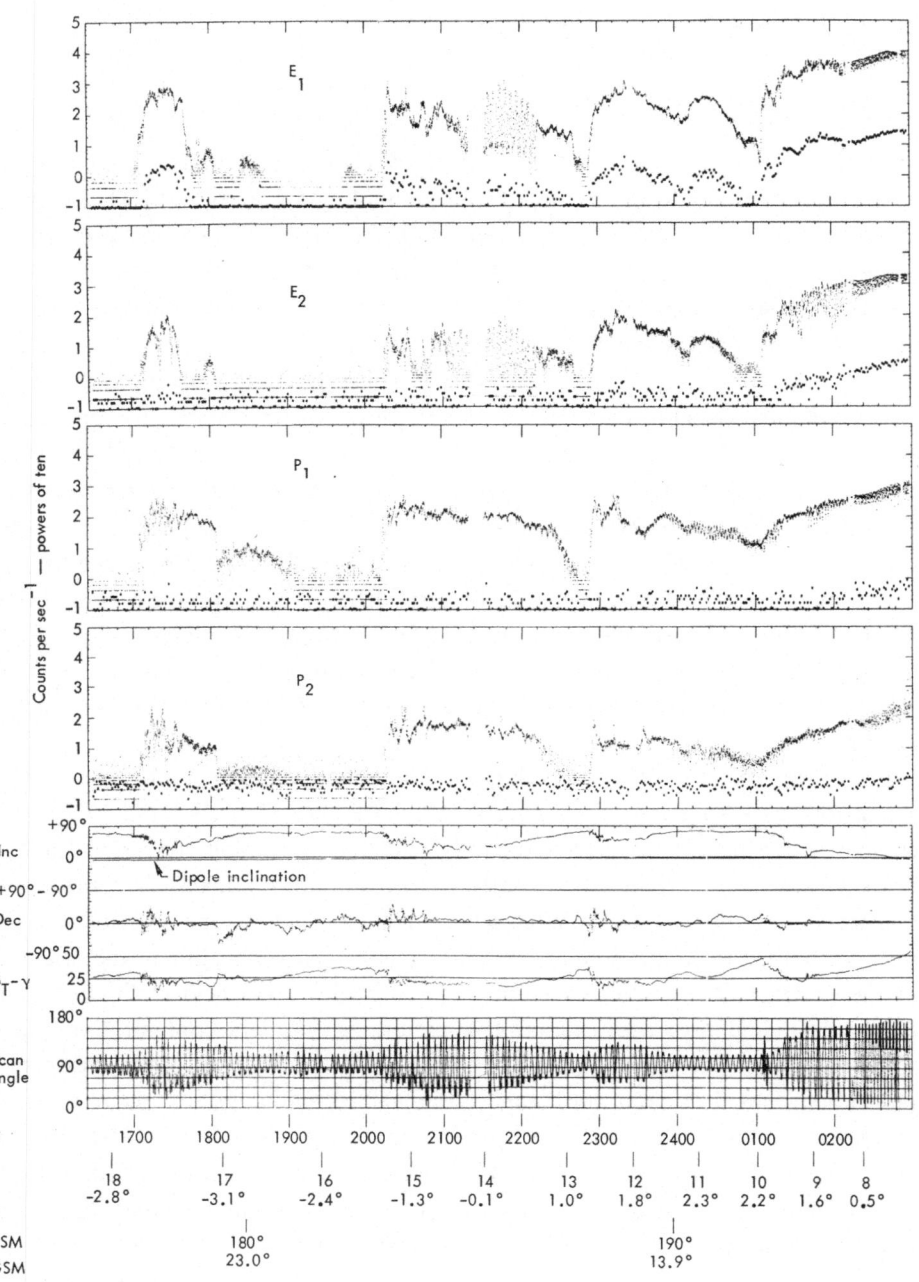

Fig. 9.   Inbound data obtained near midnight on Day 222 (August 9, 1968). The data are plotted at all pitch angles reached by the experiment as it scanned. Large modulation for E_1 and E_2 means the butterfly PAD and the absence of modulation, isotropy.

down the tail where the field crosses the neutral sheet. Presumably when the gyro-radii of the particles are comparable to the field curvature across the sheet or are of the scale of the neutral-sheet thickness, adiabatic guiding-center mostion is no longer possible (Speiser, 1971). Speiser (1967), Shabansky (1971), Sonnerup (1971), and Eastwood (1972) have discussed how particles can leave their guiding-center motion and become caught in a mode in which they oscillate about the field reversals they encounter on either side of the neutral sheet. Assuming this happens, electrons cannot stay in this mode for long since any finite B_z will quickly turn them towards the Earth; some precipitate but many mirror and return to the interaction region. Along with these effects, magnetic noise in the neutral sheet is most certainly important in randomizing the electron motions.

We have made several observations past midnight during quiet periods that tie into this picture. OGO-5 was inbound from the north lobe of the tail. Pitch angle isotropy was first noted followed by an abrupt transition to the butterfly PAD as OGO-5 neared the Earth. A semi-quantitative analysis of the field configuration indicates that isotropy was being observed on field lines which appeared to pass through the neutral sheet. A more careful study of these and other OGO-5 data may allow us to make stronger statements about the field configuration in the neutral-sheet region.

3.6. POST MIDNIGHT

In terms of interpreting our PAD's, this is the least understood region of the outer magnetosphere. Clearly electrons showing the butterfly PAD can drift through midnight during substorm expansion. However, the almost complete dropout in $j_\perp$ seen near dusk (>9 R_E) is never seen near dawn. Turbulent interactions of some sort have enhanced $j_\perp$ relative to $j_\parallel$ and the isotropic or normal PAD is often observed. At times, we have observed what we would describe as plasma-sheet oscillations (in the 10 min range) that may fit into the theories of Siscoe (1969) and McKensie (1971). In any respect, at extended distances the butterfly PAD appears to be lost as the result of randomizing processes. Nearer the Earth the changes in magnetic field configuration with increasing azimuth reverse the effects of drift shellsplitting so that much past dawn only the normal PAD is observed. This brings us full circle in the drift of the electrons around the Earth and the description of the pitch angle effects that occur.

Acknowledgements

We have profited greatly through our contacts with the UCLA OGO-5 experimenters Drs M. G. Kivelson, C. T. Russell, R. L. McPherron, and T. A. Farley. We thank Drs P. J. Coleman and C. T. Russell for providing us with OGO-5 magnetometer data. Dr J. R. Walton of LLL developed some of the computer codes used. The work was done under the auspices of the U.S. Atomic Energy Commission. Early aspects of the work were funded in part by NASA through P. O. S-7001-G. The first author was aided by a NSF Travel Grant GA-39862.

References

Aubry, M. P., Kivelson, M. G., McPherron, R. L., Russell, C. T., and Colburn, D. S.: 1972, *J. Geophys. Res.* **77**, 5487.

Buck, R. M. and West, H. I., Jr.: 1973, Paper, Amer. Geophys. Union, Wash. D.C.

Buck, R. M., West, H. I., Jr., and D'Arcy, R. G., Jr.: 1973, *J. Geophys. Res.* **78**, 3103.

Eastwood, J. W.: 1972, *Planetary Space Sci.* **20**, 1555.

Fairfield, D. H.: 1968, *J. Geophys. Res.* **73**, 7329.

Haskell, G. P.: 1969, *J. Geophys. Res.* **74**, 1740.

Kivelson, M. G., Aubry, M. P., and Farley, T. A.: 1973, *J. Geophys. Res.* **78**, 3079.

McKensie, J. F.: 1971, *J. Geophys. Res.* **76**, 2958.

Roederer, J. G.: 1967, *J. Geophys. Res.* **72**, 981.

Roederer, J. G.: 1969, *Rev. Geophys.* **77**, 77.

Russell, C. T., McPherron, R. L., and Colman, P. J. Jr.: 1971, *J. Geophys. Res.* **76**, 1823.

Scarf, F. L., Fredricks, R. W., Kennell, C. F., and Coroniti, F. V.: 1973, *J. Geophys. Res.* **78**, 3119.

Serlimitsos, P.: 1966, *J. Geophys. Res.* **71**, 61.

Shabansky, V. P.: 1971, *Space Sci. Rev.* **12**, 299.

Siscoe, G. L.: 1969, *J. Geophys. Res.* **74**, 6482.

Sonnerup, B. U. Ö.: 1971, *J. Geophys. Res.* **76**, 8211.

Speiser, T. W.: 1967, *J. Geophys. Res.* **72**, 3919.

Speiser, T. W.: 1971, *Radio Sci.* **6**, 1971.

West, H. I., Jr., Buck, R. M., and Walton, J. R.: 1972, *Nature Phys. Sci.* **240**, 6.

West, H. I., Jr., Buck, R. M., and Walton, J. R.: 1973a, *J. Geophys. Res.* **78**, 1064.

West, H. I., Jr., Buck, R. M., and Walton, J. R.: 1973b, *J. Geophys. Res.* **78**, 3093.

THE PITCH ANGLE DISTRIBUTION OF > 40 keV ELECTRONS

T. A. FRITZ*

Space Environment Laboratory NOAA/ERL Boulder, Colo., U.S.A.

1. Introduction

The subject of wave-particle interactions inside the magnetosphere has become an increasingly active area of research in recent years. Of the many interactions and instabilities that have been discussed in the literature those dealing with gyro-resonant interactions have received considerable attention. A recent review by Gendrin (1972) on these types of wave-particle interactions summarizes the present state of the theories and points out that "uncertainties about the anisotropy of the particle distribution functions are a heavy handicap for obtaining good comparisons between theory and experiment." The present paper presents conclusions drawn about the pitch angle distribution of > 40 keV electrons in the outer radiation zone which should add some insight into the behavior of this often studied but not well understood component of the magnetospheric particle population.

2. Observations

Data from three satellites operational in 1963 which carried thin-windowed 213-type G. M. tubes are shown in Figure 1 at two IN Lat (Λ) for a two week period in June of that year. Injun 3 was a low altitude, magnetically aligned satellite which measured > 40 keV electrons in two constant pitch angle intervals: $\alpha = 90° \pm 13°$ and $\alpha = 0° \pm 43°$. Alouette 1 was a low altitude, spinning satellite which measured > 40 keV electrons in a limited and variable pitch angle range. Explorer 14 provided a spin averaged integral measurement of the > 40 keV electron intensity near the equatorial plane. Characteristics of these detectors have been discussed by O'Brien *et al.* (1963), Mc Diarmid and Burrows (1964), Frank *et al.* (1964), and Fritz (1967). The measurements for the particle flux from Injun 3 and Alouette 1 are for the locally mirroring > 40 keV electrons at the position of the satellite. The position of the satellite is coded in the B values (in gauss) of the various observations. A curve has been drawn through the data points belonging to the most often sampled B interval for the Injun 3 data. The same curve was then normalized to the Alouette 1 measurements.

Fritz (1968, 1970) used the ratio of the responses of the two Injun 3 detectors, one oriented perpendicular and the other parallel to the local **B** vector, to define an angular distribution parameter, ϕ. It was shown that the value of ϕ was remarkably constant for a large range of latitude even though there were large variations of the

* Work done while spending a year in residence at the Max-Planck-Institut für Aeronomie, Institut für Stratosphärenphysik, Lindau/Harz, Federal Republic of Germany.

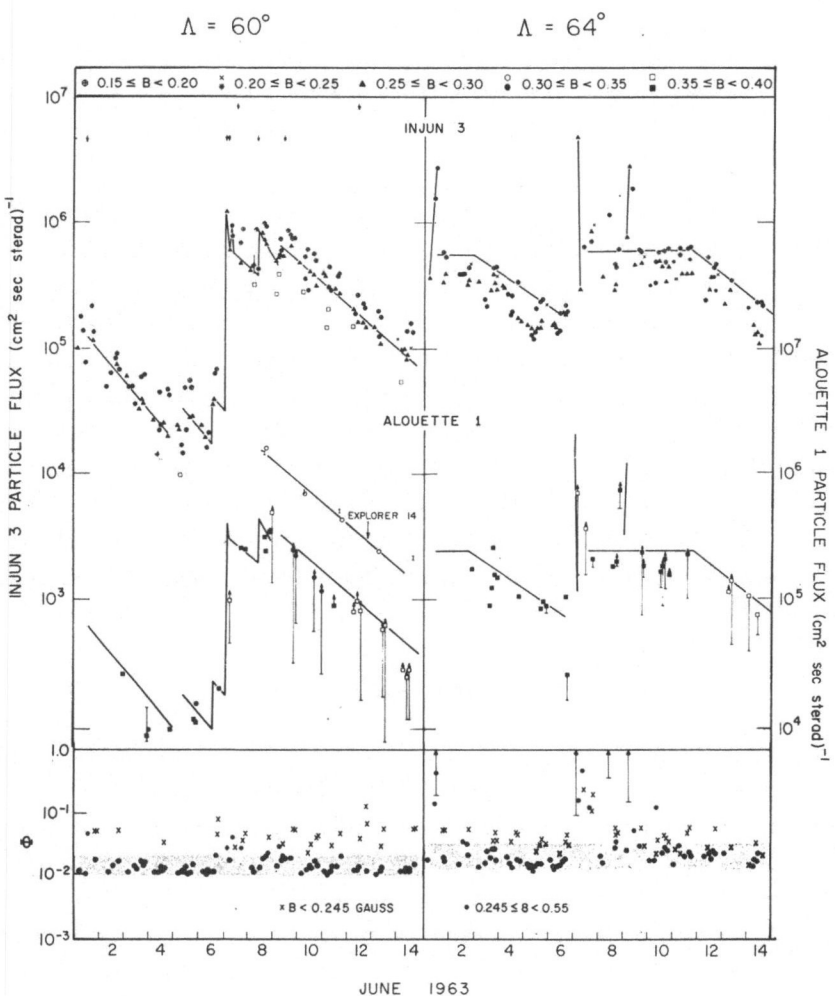

Fig. 1. Variations of the intensity of > 40 keV electrons measured by three satellites at two IN Lat in June 1963. Injun 3 was a low altitude satellite which measured $j_\perp (E_e > 40$ keV) at the position of the satellite (coded in the B values in gauss). Alouette 1 was a low altitude spinning satellite which measured $j_{\alpha'} (E_e > 40$ keV) over a limited range of local pitch angles, α' which on occasions permitted the determination of $j_\perp (E_e > 40$ keV) at 1000 km altitude. Explorer 14 provided a spin-averaged, integral measurement of the flux of > 40 keV electrons near the equatorial plane with measurements on the inbound (I) and outbound (o) portions of the orbit indicated. The pitch angle distribution parameter $\phi = j_{\alpha'=0°\pm43°} (E_e > 40$ keV$)/j_{\alpha'=90°\pm13°} (E_e > 40$ keV$)$ is also shown.

electron intensities over this same latitude range. The time history of the value of ϕ is presented in Figure 1 for the two IN Lat shown. Fritz (1968) also observed that the position of the high latitude boundary (intensity cutoff) varied by $\Delta\Lambda \approx 7°$ during a small substorm on a magnetically quiet day while at the same time the position where the ϕ parameter increased initially from 10^{-2} toward 1 (isotropy) remained unchanged at a position of $\Lambda \approx 65°$. Since it appears that the complex region associated with the high latitude boundary region is confined to $\Lambda > 65°$ during small and moderate substorm activity, the present study was confined to latitudes below $\Lambda = 65°$. Intensity variations below 65° are then associated with the large geomagnetic storms and are somewhat independent of the complex high latitude boundary region phenomena.

From Figure 1 and from an examination of 10 months of the Injun 3 data and selected simultaneous intervals with Alouette 1 the following conclusions are drawn:

(a) The long-term decay of the >40 keV electron fluxes occurs at the same rate at low altitudes and near the equatorial plane. When these fluxes begin a long-term decrease they decay with a time constant τ of 4 ± 1 day where a form of $e^{-t/\tau}$ has been assumed.

(b) There appears to be short-term intensity enhancements (injection of new electrons) which occur during magnetically active periods.

(c) The decay of these 'newly injected' intensities occurs at a rate faster than the resolving time of the orbits of the satellites (<1 h). This decay brings the intensities back to a 'limiting stable flux' condition.

(d) The 'limiting stable flux' value is a constant from event to event at a given latitude and exhibits a dependence on B (see Table I).

(e) When the value of B is between 0.245 and 0.55 G the ϕ parameter is amazingly constant at a value of $\phi = 2 \pm 1 \times 10^{-2}$ (shaded area).

(f) If $B < 0.245$ G, the value of ϕ increases.

(g) The value of ϕ is not dependent on the value of $j_\perp$ ($E_e > 40$ keV) except during the short-term enhancements of $j_\perp$ ($E_e > 40$ keV) above the 'limiting stable flux' noted in (b) and (c) above. During these brief periods the value of ϕ approaches 1 (isotropy). The conclusions noted in (a) and (b) above are not new and have been discussed by many authors in the past. The remainder of the conclusions need to be examined in terms of their implications on the electron pitch angle distribution and pitch angle diffusion.

3. Pitch Angle Diffusion

Kennel (1969) has developed a formulation of steady state pitch angle diffusion in which particles injected on a given velocity space-energy shell on a given tube of force diffuse in pitch angle until they reach the loss cone, so that precipitation balances injection. The loss cone angle, α_0, was given by

$$\sin^2\alpha_0 = Beq/B(100)$$

where $B(100)$ is the Earth's field strength at 100 km. Using a pitch angle diffusion coefficient of the form

$$D(\alpha) = (\Delta\alpha)^2/2t_B = D_0 \sin^q \alpha \approx D_0 \alpha^q$$

where t_B is the particle bounce period and α is the particles equatorial pitch angle, Kennel was able to derive an expression for the pitch angle distribution $h(\alpha)$ in and near the loss cone:

$$h(\alpha) = \frac{1}{D_0}\left\{(D_0 T_B/\alpha_0^2)^{1/2}\,\alpha^{-q/2}\,\frac{I_{q/(2-q)}\left[\dfrac{2}{2-q}\,(\alpha^{2-q}/D_0 T_B)^{1/2}\right]}{I_{2/(2-q)}\left[\dfrac{2}{2-q}\,(\alpha_0^{2-q}/D_0 T_B)^{1/2}\right]}\right\}$$

where the Bessel functions of order p and imaginary argument are denoted by I_p and $T_B = 1/4\,t_B$.

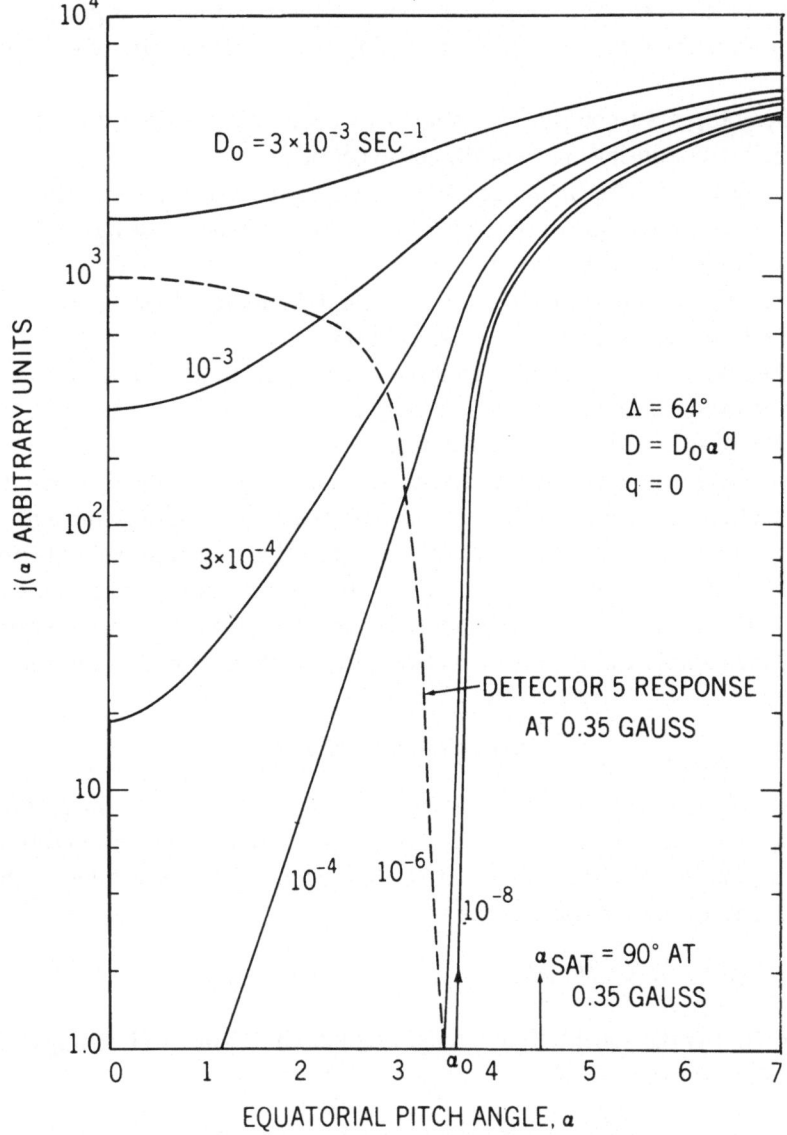

Fig. 2. Solutions for the pitch angle distribution near the loss cone from a steady state pitch angle diffusion formulation of Kennel (1969). The experimentally determined response function for the Injun 3 detector 5 transformed to an equatorial pitch angle function as it would appear at a satellite altitude corresponding to 0.35 G.

When this form of the pitch angle distribution is applied to measurements of 40 keV electrons at a given latitude, the values of T_B and α_0 are fixed and only the two adjustable parameters of the diffusion coefficient, D_0 and q, remain to be specified. In Figure 2 the pitch angle distribution for $q=0$ at $\Lambda=64°$ is presented. Note the exponential variation of the distribution function inside the loss cone in the weak diffusion limit which seems to be a standard feature of calculations of this sort (Theodoridis and Paolini, 1967; Kennel, 1969; Gendrin, 1972). The experimentally determined response function of the Injun 3 detector oriented parallel to the local **B** vector (detector 5) has been transformed to the equatorial pitch angle distribution using the magnetic moment invariance relation. This response function is also presented in Figure 2 as it would be if the satellite were at an altitude corresponding to $B=0.35$ G. The response function of the Injun 3 detector 5 is conical and rotationally symmetric around the local **B** vector as is the pitch angle distribution. It is therefore a straightforward calculation to numerically integrate the pitch angle distribution over the response function of detector 5. The resultant value represents the flux measured by detector 5 for a given pitch angle distribution which can then be normalized with the intensity of particles which have a local pitch angle of 90° determined from the same distribution function. In this manner a value of the pitch angle distribution parameter ϕ is determined. By varying the two adjustable parameters of $h(\alpha)$, various values of ϕ can be calculated. These are presented in Figure 3.

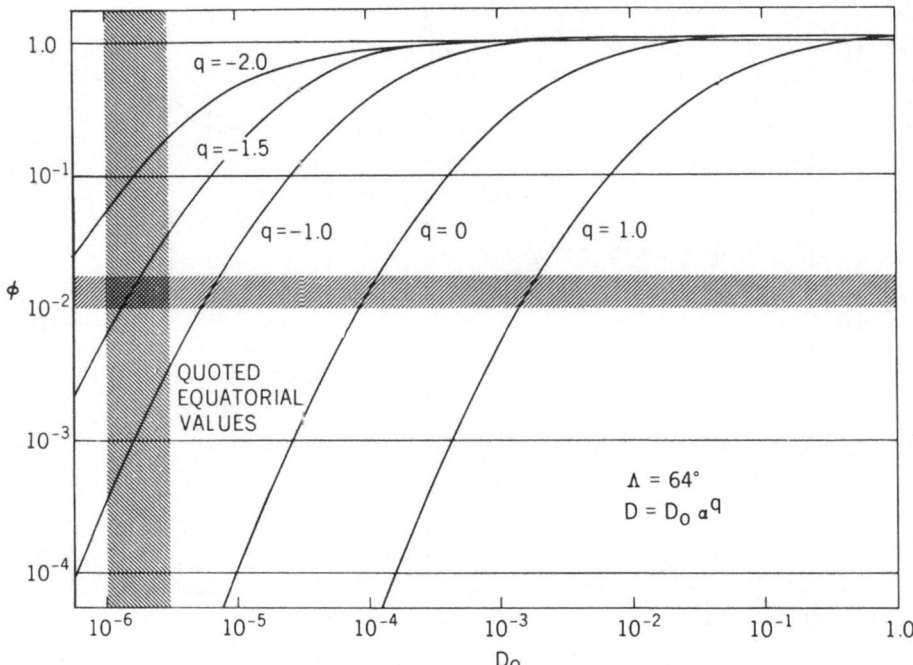

Fig. 3. Calculations of the pitch angle distribution parameter ϕ for various values of D_0 and q in the form of $h(\alpha)$.

Since the purpose of the calculation is to reproduce a value of $\phi = 2 \pm 1 \times 10^{-2}$ by varying the two adjustable parameters, there is no unique solution. A second condition is added to the problem if conclusion (a) in Section 2 is used to determine the value of D_0. With $\tau = 4$ days $D_0 = 2.8 \times 10^{-6} \, \mathrm{s}^{-1}$ and the companion value of q required to produce the proper value of ϕ is observed in Figure 3 to be $q \approx -1.5$. This form of the pitch angle distribution is presented in Figure 4 and can successfully

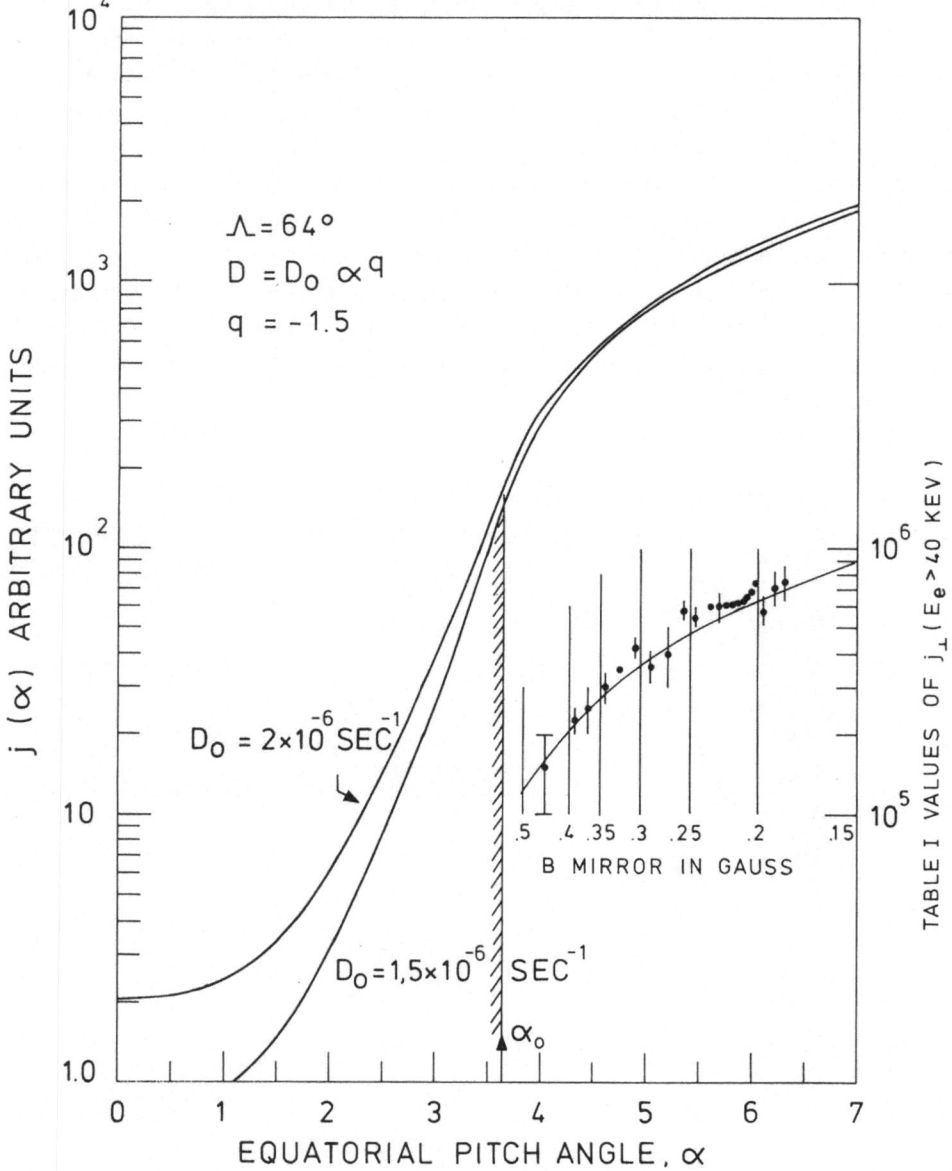

Fig. 4.   Pitch angle distribution required to explain the observed behavior of $> 40 \, \mathrm{keV}$ electrons. Points are the values of the 'limiting stable flux' presented in Table I.

explain conclusions (a) and (e) noted in Section 2. With this form now determined it is possible to examine the implications of the B dependences noted in conclusions (d), (e), and (f).

In Table I a list of the observed values of the 'limiting stable flux' at $\varLambda = 64°$ is presented for a number of magnetically disturbed periods in 1963. During the periods indicated the value of $j_\perp (E_e > 40$ keV) was maintained at or near the value given in the

TABLE I

Observed values of the 'limiting stable flux' at $\varLambda = 64°$

Time interval of the observation in 1963	Range of B-values in G corresponding to the observation	Observed 'limiting stable' value of $j_\perp (E_e > 40$ keV) Electrons cm^{-2} s^{-1} sr^{-1}
January 13–19	0.20 to 0.25	6.0×10^5
January 30–31	0.40 to 0.50	$1.5 \pm 0.5 \times 10^5$
February 11–16	0.20 to 0.25	$5.5 \pm 0.5 \times 10^5$
March 8–13	0.35 to 0.40	$2.0 - 2.5 \times 10^5$
March 8–13	0.30 to 0.35	$4.0 - 4.5 \times 10^5$
April 4–5	0.30 to 0.35	3.5×10^5
April 14–19	0.25 to 0.30	$5.8 \pm 0.6 \times 10^5$
May 3–9	0.20 tp 0.25	6.2×10^5
May 3–9	0.30 to 0.35	$3.0 \pm 0.4 \times 10^5$
May 10–15	0.20 to 0.25	6.1×10^5
May 28–June 3	0.25 to 0.30	$3.6 \pm 0.5 \times 10^5$
June 7–12	0.15 to 0.20	$5.8 \pm 0.8 \times 10^5$
June 7–12	0.25 to 0.30	$4.0 \pm 1.0 \times 10^5$
June 7–12	0.35 to 0.40	$2.5 \pm 0.5 \times 10^5$
June 17–21	0.15 to 0.20	$7.5 \pm 1.1 \times 10^5$
June 17–21	0.20 to 0.25	$6.0 \pm 0.8 \times 10^5$
June 26–30	0.20 to 0.25	6.9×10^5
July 5–11	0.20 to 0.25	6.6×10^5
July 21–26	0.20 to 0.25	7.4×10^5
July 29–August 10	0.15 to 0.20	$7.1 \pm 1.0 \times 10^5$
July 29–August 10	0.20 to 0.25	6.2×10^5
August 24–Sept. 4	0.20 to 0.25	6.5×10^5
September 8–13	0.20 to 0.25	6.3×10^5

Table for the specified range in B. Since the values of $j_\perp (E_e > 40$ keV) correspond to electrons mirroring at the position of the satellite, the corresponding ranges in B can be transformed to an equivalent equatorial pitch angle. The intervals in equatorial pitch angle corresponding to the various intervals of $\mathbf{B}$ given in Table I are shown on Figure 4. Taking the liberty of plotting the minimum value of the flux given in Table I for a specific range of B values near the smaller corresponding pitch angles and similiarly the largest values of $j_\perp (E_e > 40$ keV) in Table I near the larger corresponding pitch angles, the distribution of points shown in Figure 4 is produced. A curve normalized to these points corresponding to the form of the pitch angle distribution determined above is drawn through the points. This curve appears to provide a very good fit to the plotted points.

With the distribution function presented in Figure 4 it is possible to calculate the expected value of the parameter ϕ at various satellite altitudes. The variation in altitude has the effect of changing the range of pitch angles over which the response function of detector 5 (Figure 2) integrates. The results of this calculation are presented in Figure 5 along with the range in B (shaded area) in which ϕ is observed to maintain the value of $\phi = 2 \pm 1 \times 10^{-2}$. The calculation agrees very well with conclu-

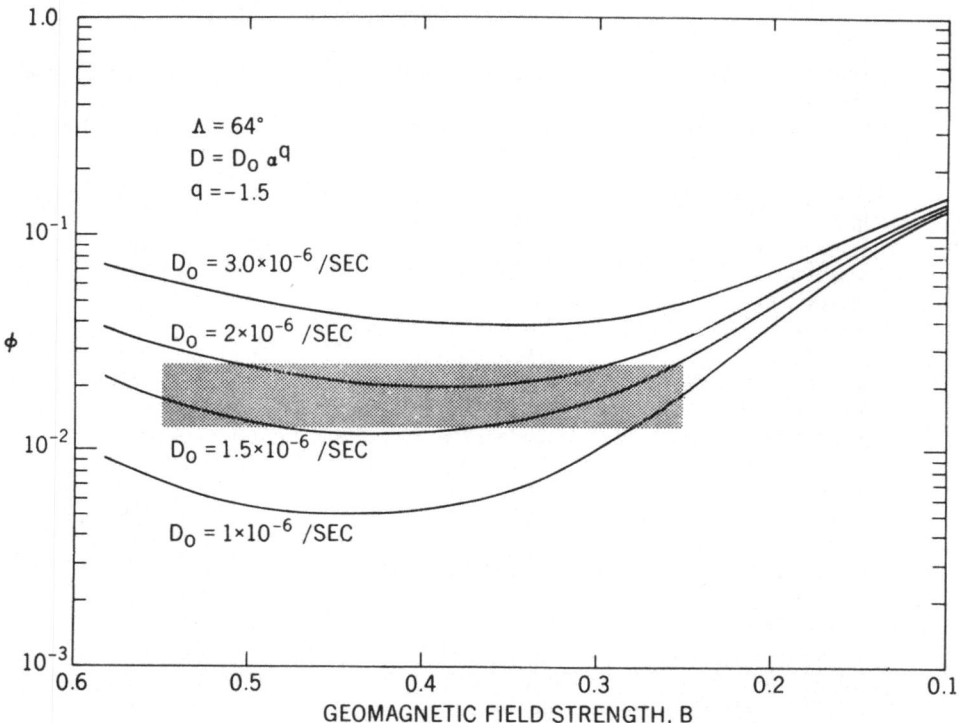

Fig. 5. Calculation of the pitch angle distribution parameter ϕ as a function of B.

sion (e). The increase noted in conclusion (f) for values of ϕ at $B < 0.245$ G is also predicted. Since the minimum value of B sampled by Injun 3 (apogee = 2785 km) is approximately 0.18 G the magnitude of the increase in the value of ϕ for low B values (points denoted by 'X' in Figure 1) is also predicted by this distribution function.

4. Conclusions

4.1. PITCH ANGLE DISTRIBUTION

Observations of the behavior of fluxes of > 40 keV electrons have been fitted to a generalized pitch angle diffusion calculation of Kennel (1969). The resultant steady state pitch angle distribution is in agreement with all of the observed characteristics of

>40 keV electrons discussed in this paper. No specific process to drive the diffusion process has been invoked.

No value of ϕ much less than 10^{-2} has been observed in the Injun 3 data although Fritz (1967) has shown that values of $\phi < 10^{-4}$ could have been readily observed before any limitations of the detectors were reached. Therefore the pitch angle distribution presented in Figure 4 must be regarded as demonstrating the greatest degree of aniso­tropy which > 40 keV electrons achieve.

The behavior of > 40 keV electrons for $\Lambda < 65°$ characterized in this paper can be regarded as a three step process: (a) injection, (b) rapid decay, and (c) slow decay. During periods of geomagnetic disturbances, electrons are (a) injected on to low latitude (altitude) field lines. At low altitudes this injection appears in the form of an intensity increase and an isotropic pitch angle distribution. On a rapid time scale (<1 h) both the intensity and pitch angle distribution (b) decay to their respective values characterized by the 'limiting stable flux' and pitch angle distribution presented in Figure 4. The subsequent decay (c) of the > 40 keV electrons preserves the form of the pitch angle distribution while reducing the intensity with a time constant $\tau \approx 4 \pm 1$ day.

4.2. Implications

Kennel (1969) has shown that a particle lifetime τ^* can be determined from the diffusion coefficient by calculating the time it takes a particle to diffuse from some small pitch angle α^* to $\alpha = 0$.

$$\tau^* \approx \frac{\alpha^{*2}}{D(\alpha)} \approx \frac{\alpha^{*(2-q)}}{D_0} \approx \frac{\sin^{2-q} \alpha^*}{D_0}.$$

Substituting the values of $q = -1.5$ and $D_0 = 2 \times 10^{-6}$ s^{-1} a lifetime can be calculated for the particle to diffuse from some radial distance R on a given L shell to the atmosphere. Normalizing these values with the decay constant at the equatorial plane, it is found that a result discussed by McDiarmid and Burrows (1966) is reproduced. This is presented in Figure 6. They used the observation noted in conclusion (a) of Section 2 and argued that the mirror point velocities must increase down the field line at least as fast as the magnetic field strength B increases (i.e., inversely as the cross sectional area of a tube of force), since otherwise a 'pile up' of electrons would occur at low altitudes. With this method they were able to place the upper limit to the mean electron lifetime given in Figure 6. The curve labeled 'Fritz' is the result of the present work and represents an actual calculated lifetime rather than an upper limit. As noted by McDiarmid and Burrows the lifetime of particles observed by low altitude satellites is quite short compared to their longitudinal drift times.

4.3. Qualifications

Although the vast majority of data corresponds to the picture presented, not every measurement of > 40 keV electron intensities at a given latitude can be explained in this manner. An example of this is the phenomenon that is commonly described as a

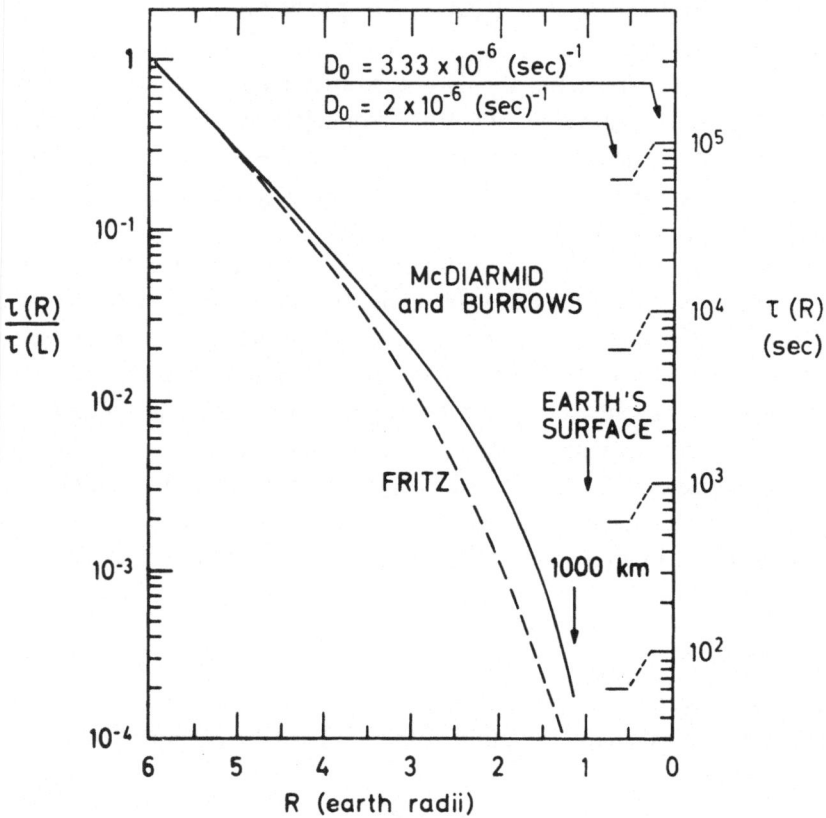

Fig. 6. $\tau(R)/\tau(L)$: the 'lifetime' of a > 40 keV electron to diffuse from a radial distance R to $R = 1$ divided by the 'lifetime' of an electron at the equator on the $L = 6$ shell ($\Lambda = 65.9°$). Curve labeled 'McDiarmid and Burrows' (1966) is an upper limit calculation. Curve labeled 'Fritz' represents the results of the present calculation. Right scale represents values of $\tau(R)$ for two different equatorial lifetimes $\tau = 1/D_0$.

'boundary collapse'. This usually occurs during the initial phases of strong geomagnetic disturbances and has the characteristic at low altitudes that the high latitude boundary moves to much lower latitudes (lower than 64° for example). On a plot such as Figure 1, the intensity appears to decrease by two orders of magnitude or more but then returns to a value near the 'limiting stable flux' value on the next pass or two of the satellite.

The region of the high latitude boundary is usually in a state of strong pitch angle diffusion and exhibits complex behavior that cannot be explained by the application of the steady state theory used here.

Acknowledgements

This work was begun while the author was a post-doctoral fellow at the National Research Council of Canada, Ottawa, and was brought essentially to its present form while the author was a NRC/NAS post-doctoral resident research associate at Goddard Space Flight Center, Greenbelt, Md. Appreciation is expressed to Dr I. B. McDiarmid for many discussions and for the use of the Alouette 1 data, to Dr Donald J. Williams for many discussions, to Dr James A. Van Allen for use of the Injun 3 and Explorer 14 data, and to Dr L. Rossberg for reviving the author's interest through many discussions in seeing this work published.

The author wishes to express his appreciation to Professor G. Pfotzer and Dr Donald J. Williams for the special arrangements which have made the present year in residence at MPI/Lindau a possibility.

References

Frank, L. A., Van Allen, J. A., and Hills, H. K.: 1964, *J. Geophys. Res.* **69**, 2171.
Fritz, T. A.: 1967, U. of Iowa Res. Rept. 67–72.
Fritz, T. A.: 1968, *J. Geophys. Res.* **73**, 7245.
Fritz, T. A.: 1970, *J. Geophys. Res.* **75**, 5387.
Gendrin, R.: 1972, in E. R. Dyer (ed.), *Solar-Terrestrial Physics/1970: Part III*, D. Reidel Publishing Company, Dordrecht-Holland, p. 236.
Kennel, C. F.: 1969, *Rev. Geophys.* **7**, 379.
McDiarmid, I. B. and Burrows, J. R.: 1964, *Can. J. Phys.* **42**, 616.
McDiarmid, I. B. and Burrows, J. R.: 1966, *Can. J. Phys.* **44**, 669.
O'Brien, B. J., Laughlin, C. D., and Gurnett, D. A.: 1963, *J. Geophys. Res.* **69**, 1.
Theodoridis, G. C. and Paolini, F. R.: 1967, *Ann. Geophys.* **23**, 375.

PITCH ANGLE DIFFUSION OF RELATIVISTIC ELECTRONS
IN THE PLASMASPHERE

GEORGE A. KUCK

ADC (XPQD), Ent AFB, Colorado Springs, Colo. 80916, U.S.A.

Abstract. Whistler mode electromagnetic waves, which are highly variable in both space and time, cause pitch angle diffusion of electrons. The effective diffusion coefficient can be radically different from the instantaneous diffusion coefficient since electrons both bounce and drift average the waves. Analysis of the time evolution of the 1.2 and 2.1 MeV electron pitch angle distributions following the October 28 and November 1, 1962, Russian high altitude nuclear tests allow the effective diffusion coefficients to be obtained for $1.75 \leqslant L \leqslant 2.5$. The evolution is consistent with the assumption of scattering by broad band white noise near the principal cyclotron harmonic. The data for both electron energies can be explained by a diffusion coefficient related to a single power law frequency spectrum of the form $B_{wave} = Af^{-n}$, $1.4 \leqslant n \leqslant 1.7$ above 300 Hz. Below 300 Hz, the power rises as a function of frequency.

1. Introduction

Pitch angle diffusion by electromagnetic waves is an important loss mechanism for geomagnetically trapped particles. Experimental coefficients have been derived by examining the structure of the pitch angle distribution (Theodoridis and Paolini, 1967), the buildup of intensities in the loss cone as a function of longitude (Berg and Søraas 1972; Imhof, 1968), and the angular distribution and intensity time profile in the loss cone (Whalen and McDiarmid, 1973). Numerous theoretical calculations have also been made (Dungey, 1963; Roberts, 1966; Lyons and Thorne, 1972; Lyons *et al.*, 1972). In this paper, the effective diffusion coefficients are derived by analyzing the time behavior of $E \geqslant 1.2$ and 2.1 MeV electrons following the artificial injections on October 28 and November 1, 1962. The pitch angle distributions are presented and interpreted in terms of pitch angle diffusion caused by broad band white electromagnetic noise (Roberts, 1968).

The results presented in the paper were obtained from a scintillation spectrometer flown on USAF satellite 1962BK. Preliminary results and the instruments description have been reported elsewhere (Katz *et al.*, 1964; Giaconni *et al.*, 1964). These results suffered from important satellite and instrument related problems. Discussion of how these problems were resolved may be found in Kuck (1973).

The theoretical framework used in analyzing the data was that of pure pitch angle scattering by broad band white noise (Roberts, 1968; Roederer, 1970). The diffusion equation used was:

$$\frac{\partial j}{\partial t} = \frac{1}{\mu \tau_B} \frac{\partial}{\partial \mu} \left[D_{\mu\mu} \mu \tau_B \frac{\partial j}{\partial \mu} \right] + Q - S \qquad (1)$$

where

j = unidirectional flux,
B = electron bounce period,

B. M. McCormac (ed.), Magnetospheric Physics, 117–127. All Rights Reserved.

μ = equatorial pitch angle cosine,

t = time,

$D_{\mu\mu}$ = bounce averaged equatorial pitch angle diffusion coefficient,

Q = source term, and

S = sink term.

The source term can be set to zero because the injection of relativistic electrons by natural processes is negligible when compared to the artificial injection. The sink term may be included in the boundary conditions. Equation (1) may then be solved by standard central difference techniques (Richtmyer and Morton, 1967). The diffusion coefficient, $D_{\mu\mu}$, was calculated using the theoretical framework presented by Roberts (1968, 1969). The more sophisticated treatment of Lyons et al. (1972) could not be used because the VLF wave fields are variable in space and time (Russell et al., 1970, 1972; Gurnett, 1966; Thorne et al., 1973). Without simultaneous particle wave measurements, there is no real justification for making the more sophisticated calculation. The diffusion coefficient must be drift-as well as bounce-averaged and the drift averaging could not be done. Furthermore, the drift-averaged diffusion coefficient depends upon the electron density at different longitudes which varies by at least a factor of 2 over the drift orbit of the electron (Bewersdorff and Sagalyn, 1972). Although the electron density was measured on the same satellite (Cormier et al., 1965), the longitudinal variation of the electron density was not determined. In view of these uncertainties, it was decided to use the simple derivation of the diffusion coefficient given by Roberts (1969). The local diffusion coefficient used was:

$$D_{\mu_1\mu_1} = \left(\frac{e}{mc}\right)^2 (1 - \mu_l^2)\left[1 + \frac{V_\parallel}{V_g(\omega_0)}\right]^{-1} B_x(f_0) \tag{2}$$

where

$D_{\mu_1\mu_l}$ = local diffusion coefficient,

e = electron charge,

m = relativistic electron mass,

c = velocity of light,

μ_l = local pitch angle cosine,

$V_\parallel$ = parallel electron velocity,

$V_g(\omega_0)$ = whistler mode group velocity, and

$B_x(f_0)$ = power spectral density function of one component of the disturbance field.

The electron density was assumed to be uniform over the electron bounce and drift orbit. The density was scaled as L^{-4} with the normalization of 4000 electrons cm^{-3} at $L = 1.75$ (Cormier et al., 1965). The diffusion coefficient was bounce-averaged along the electron's spiral orbit and the resulting diffusion coefficient was used in the numerical solution of Equation (1). The initial condition used was the experimentally determined pitch angle distribution at $t = 10^4$ s after the burst in order to ensure that the electrons were uniformly spread in longitude (Davidson and Henricks, 1971).

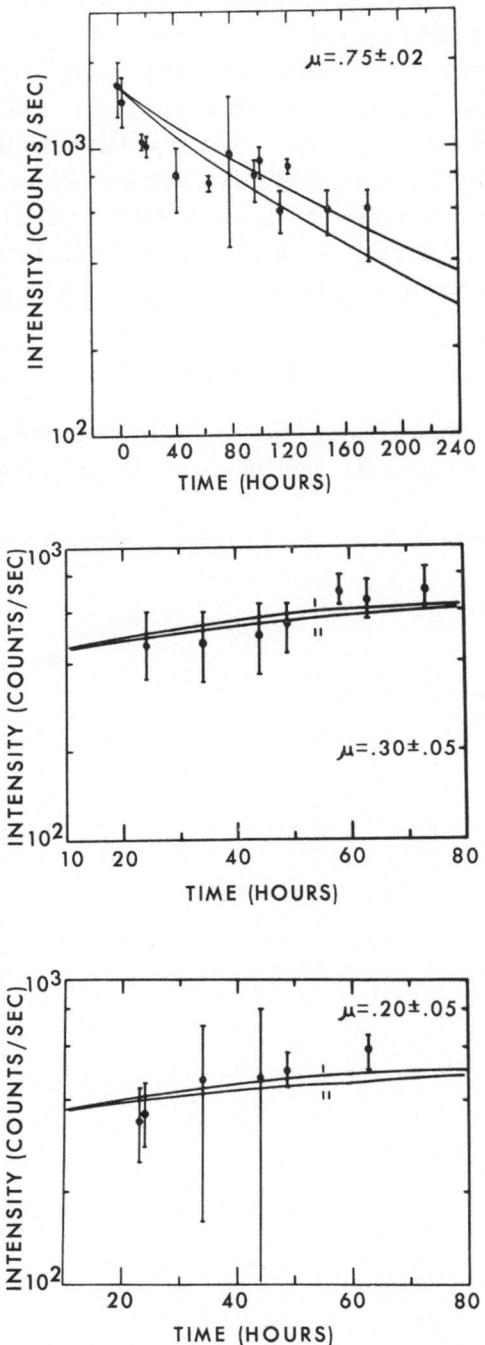

Fig. 1.   Unidirectional intensity on $L = 1.85$ for three different μ values. The curves labeled I are for $B_{\mathrm{wave}} = 6 \times 10^{-10} \, \gamma^2 \, \mathrm{Hz}^{-1}$ while those labeled II are for $B_{\mathrm{wave}} = 8 \times 10^{-10} \, \gamma^2 \, \mathrm{Hz}^{-1}$; (a) $\mu = 0.75$; (b) $\mu = 0.30$; (c) $\mu = 0.2$.

For high geomagnetic latitudes, the flux decreased with time while at low geomagnetic latitudes, the flux increased and then decreased. Figure 1 shows the time behavior
of the flux for three different values of the equatorial cosine. The lines give the predicted flux, assuming the diffusion is caused by broadband white noise for B_{wave} of
$6 \times 10^{-10} \gamma^2 \, \mathrm{Hz}^{-1}$ and $8 \times 10^{-10} \gamma^2 \, \mathrm{Hz}^{-1}$. The fit to this and other data could be improved by varying the diffusion coefficient as a function of time. However, since there
is little information available concerning how the wave fields in the plasmasphere vary
as a function of magnetic activity, this analysis used the simplest assumption that the
waves in the plasmasphere had a constant amplitude during the period of observation.

2. Results

The time behavior of the electrons injected by the November 1, 1962, test was
examined from approximately 3 h until 80 h after the burst. Figure 2 shows that the

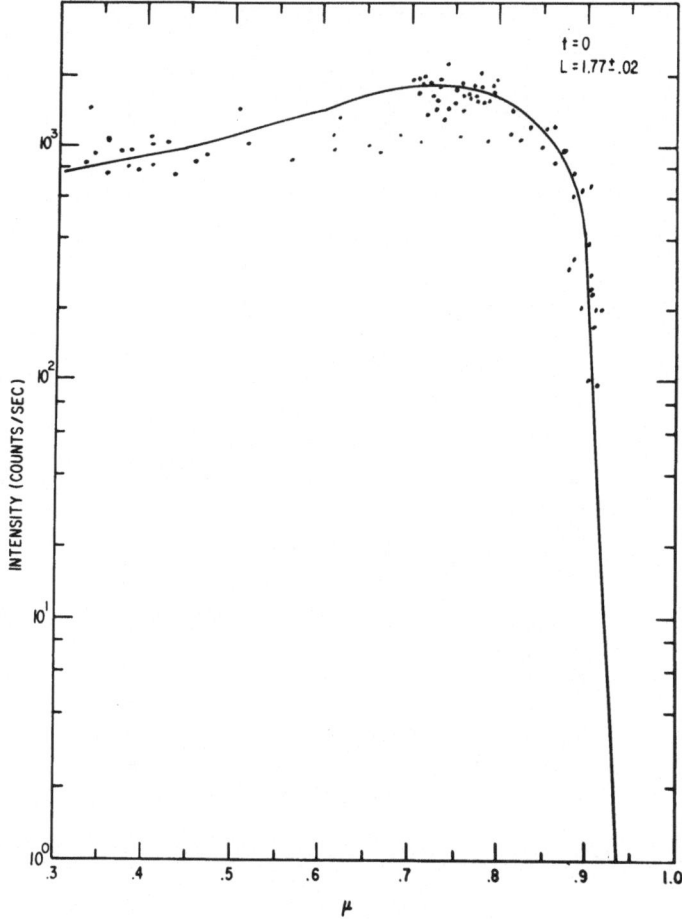

Fig. 2.   Equatorial pitch angle distribution of $E \geqslant 1.2$ MeV electrons injected on the $L = 1.77 \pm 0.02$
shell. Distribution is for electrons 3 h after the November 1, 1962, burst.

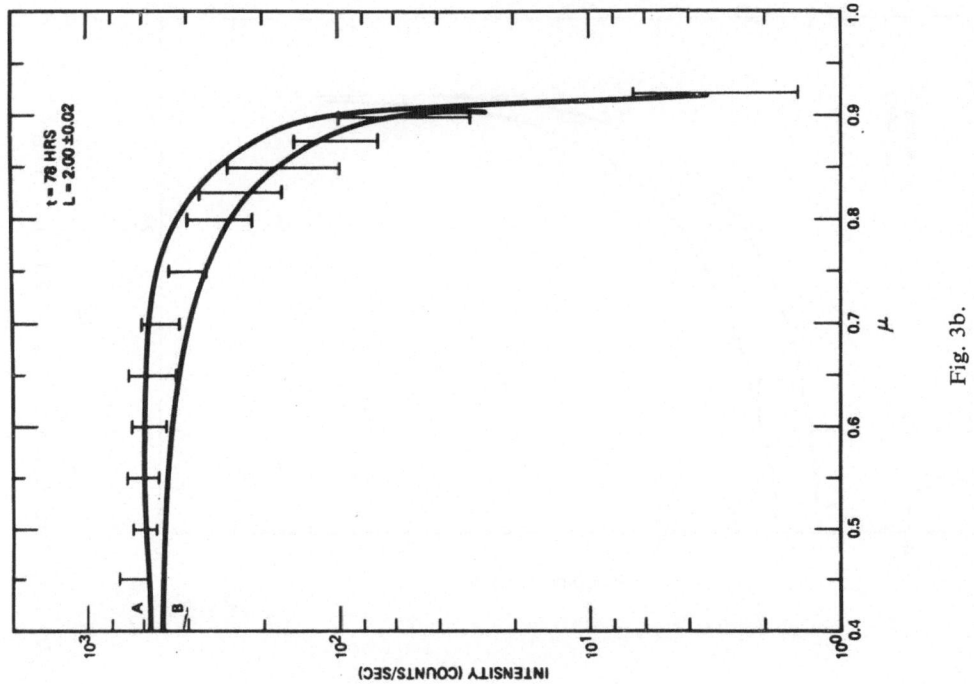

Fig. 3b.

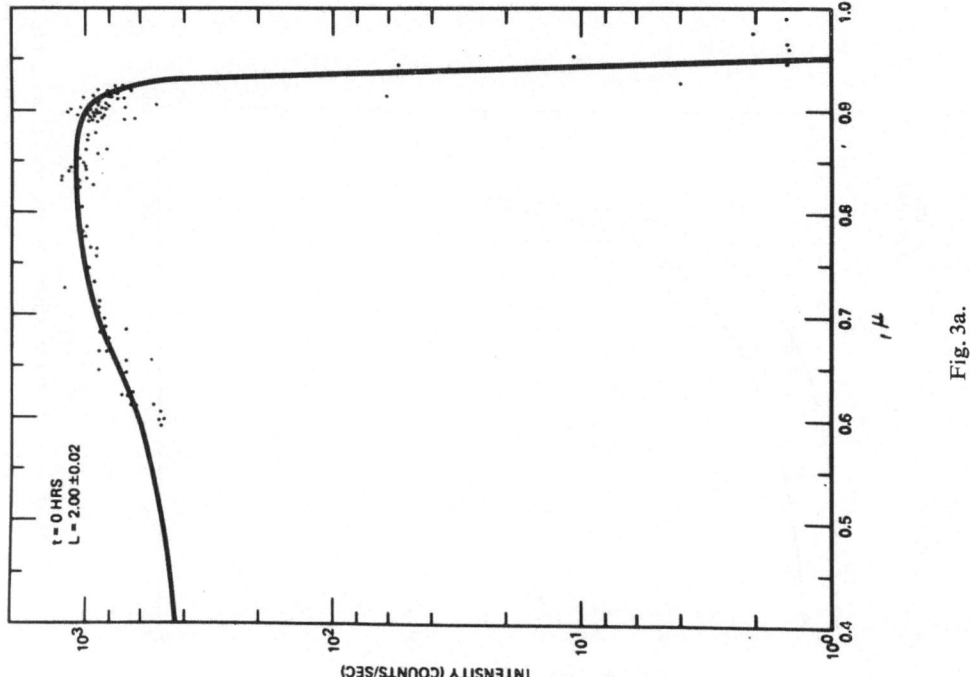

Fig. 3a.

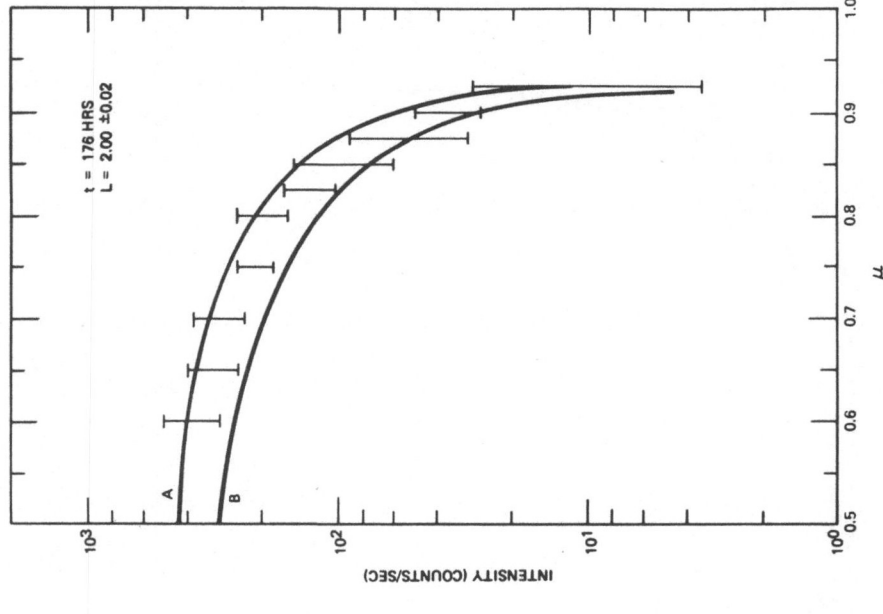

Fig. 3d.

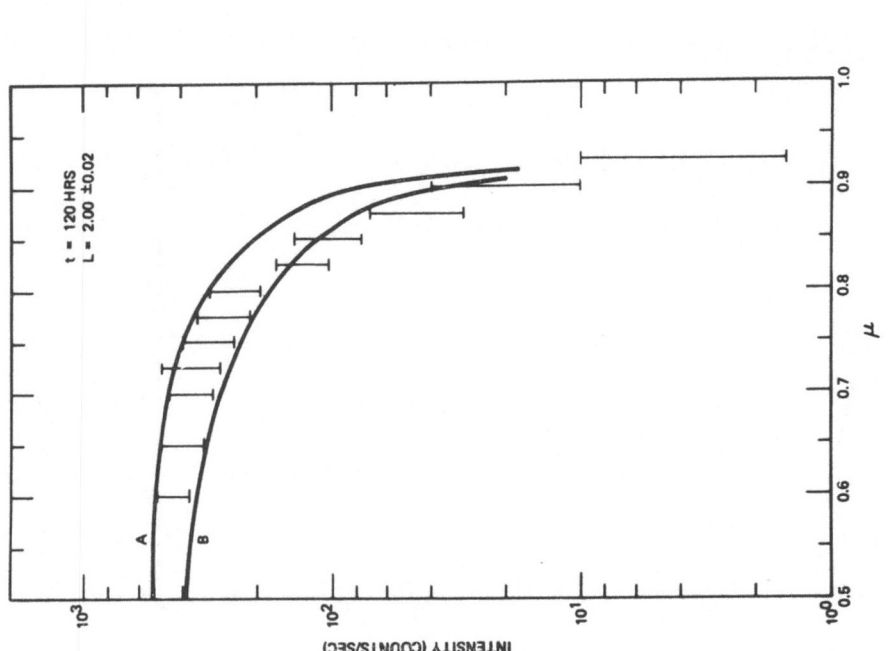

Fig. 3c.

Fig. 3a–d.   Pitch angle distributions for $E \geqslant 1.2$ MeV electrons for $L = 2.0 \pm 0.02$.   Curve A: $B_{\text{wave}} = 9.0 \times 10^{-10} \, \gamma^2 \, \text{Hz}^{-1}$.
Curve B: $B_{\text{wave}} = 2.0 \times 10^{-9} \, \gamma^2 \, \text{Hz}^{-1}$.

injection did occur along the entire field line. The flux in the peak was three times higher than the flux near the geomagnetic equator. There have been no other pitch angle distributions published for this event (Walt, 1971). Other published results on this test were obtained from omnidirectional detectors for which this variation would not have been experimentally observable. The time evolution could be explained by wave fields of the strength $(6 \text{ to } 4) \times 10^{-10} \, \gamma^2 \, \text{Hz}^{-1}$. In terms of wave amplitudes, this is $(2.5 \pm 1.0) \times 10^{-2} \, m\gamma \, \text{Hz}^{-1/2}$. Waves of this strength above 1 kHz have been observed on satellites (Koons and McPherson, 1972; Gurnett, 1966).

The October 28, 1962, Russian high altitude nuclear test injected electrons on $L \geqslant 1.80$ (Willard and Kenney, 1964). The 1.2 and 2.1 MeV electrons were observed by this satellite above $L = 3$. However, extrapolation of the particle fluxes to low μ values on L shells above $L = 2.5$ gave too great an uncertainty in the wave field determinations. This extrapolation was necessary because of the satellite orbit. Figures 3a through 3d present the experimental time evolution of the $E \geqslant 1.2$ MeV

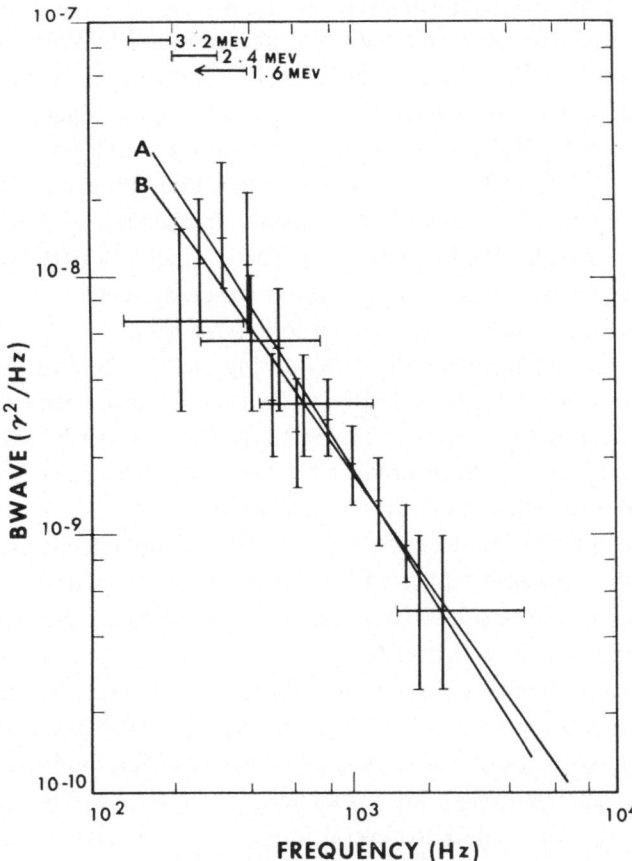

Fig. 4.   The measured wave fields best fitting the experimental data for $t \geqslant 80$ h vs. resonant wave frequency at $\mu = 0.7$. Curve A represents a least squares fit omitting the 2.1 MeV data points for $L \geqslant 2.3$; Curve B fits all data points. Horizontal lines labeled 1.6 MeV, 2.4 MeV, and 3.2 MeV are plotted at the frequencies for which West (1966) saw a distinct break in the power law lifetimes.

electron distributions. The solid lines present the theoretical evolution for different field strengths. The errors associated with the $E \geqslant 2.1$ MeV electrons were larger than those presented in this Figure due to lower particle fluxes. Errors associated with the $E \geqslant 4$ MeV electrons were such that these higher energy particles could not be examined.

An electron of one particular energy resonates with waves of different frequencies as it is pitch angle scattered. On the L shells examined, the frequency change is approximately a factor of 2 as the electron moves from $\mu = 0.5$ to $\mu = 0.9$. Figure 4 presents the wave intensities plotted against the resonant frequency at $\mu = 0.7$ for the different experimentally determined wave fields for *both* the $E \geqslant 1.2$ and $\geqslant 2.1$ MeV electrons. The horizontal error bars are associated with the change in frequency as the electron covers the range from $0.5 \leqslant \mu \leqslant 0.9$. The straight line labeled A was a least squares power law fit to the data points except for the 2.1 MeV points at $L = 2.3$ and 2.4. The frequency dependence in this case is $f^{-1.65}$. Using all the data points the frequency dependence is $f^{-1.41}$.

The break in the curve at between 200 and 300 Hz is probably real. The f^{-n} behavior of the wave fields does not extend to zero frequency. This was also evident in the previously reported data by West (1966). The data obtained by West were fitted using a power law from $t \approx 10^4$ to 3×10^5 s after the burst. When the decay coefficients for the peak of the injected distributions were obtained for the higher energy electrons at $E \approx 2.4$ and $E \approx 3.2$ MeV, West found a sharp drop in the power law coefficient between $L = 2.1$ and $L = 2.3$. Physically, this implies that the waves causing the diffusion were less intense. In terms of the resonant frequency, this decrease occurred between 100 and 300 Hz. The lower energy particles which resonate with waves above 300 Hz showed approximately a constant power law decay coefficient.

The wave fields necessary to explain the long term behavior of the electron pitch angle distributions underestimate the decay during the first 40 h for both the 1.2 and 2.1 MeV electrons on all L shells. The wave fields could have been enhanced for the first two days following the October 28, 1962 test. Dst did vary by a factor of 2 over this time period. If the wave fields produced by the equatorial ring current vary as Dst, then the derived wave fields should vary by a factor of 2.

The variation in the diffusion coefficient with time was obtained from a plot of wave field strengths vs. resonant frequency for times less than 48 h. This power spectrum is presented as Figure 5. The solid line is the $f^{-1.65}$ result from Figure 4 increased by a factor of 2. Note that the shape of the power spectrum is approximately the same. The change is due to the change in the normalization of the spectrum which is approximately the same as the change in Dst. Thus the most probable cause of the increased diffusion is the change in Dst. Measurements of the wave fields indicate that the wave strengths in the magnetosphere are a function of magnetic activity (Kaiser, 1972).

The wave values are subject to several important uncertainties which can systematically change the absolute wave values. For example, the electron density variation along the drift path and bounce path can cause a factor of 2 variation in the diffusion coefficient within the theoretical framework used in this report. Current theories do not take the variation of electron density along the drift path and bounce

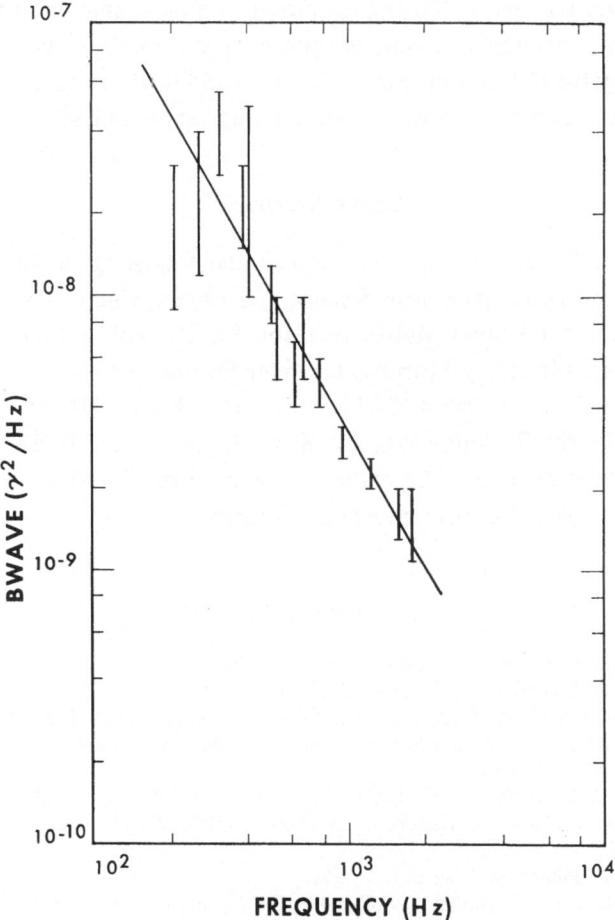

Fig. 5.   The measured wave fields best fitting the experimental data for $t \leqslant 80$ h vs. the resonant wave frequency at $\mu = 0.7$. The solid line represents curve A of Figure 4. Normalization has been changed by a factor of 2.

path into account (Lyons and Thorne, 1972; Lyons *et al.*, 1972). Fortunately, the errors introduced by these different effects are systematic errors. Thus, all the wave field values determined from the electron diffusion coefficients are systematically lower than the actual wave field values.

3. Conclusions

An estimate of the power spectrum of whistler mode waves causing pitch angle diffusion of relativistic electrons may be obtained from electron pitch angle distributions. The behavior of the electrons injected by the Russian high altitude nuclear tests of October 28 and November 1, 1972, allowed one to make this determination. The decay of the electrons was consistent with a power spectrum of the for $B_{\text{wave}} = A f^{-n}$ with

$n = 1.65$ above 300 Hz. Below 300 Hz, the power rises as a function of frequency. This power spectrum is not the instantaneous power spectrum. It is the 'effective' power spectrum seen by the electron over many bounce- and drift- averaged paths. The data also indicate that this effective power spectrum does vary with Dst.

Acknowledgement

I wish to gratefully acknowledge the help of the following people: Dr Riccardo Giaconni and coworkers (American Science and Engineering of Cambridge, Mass); Mr. Ludwig Katz, Mr Robert McInerny, Capt. Virgil Webb (AFCRL); Lt Col William A. Whitaker, Mr Harry Murphy, Dr Peter Eiseman (AFWL); Dr Lawrence L. Lyons (NOAA); Dr Harry West (LLL); Dr Charles Blank (DNA); Dr Christopher P. Leavitt, Dr Harjit S. Ahluwalia, Dr Roy Thomas and Mr Thomas Summers (University of New Mexico). The research was performed under Program Element 62601F, AFWL Project 8809/01 and AFCRL Project 8600/06/03.

References

Berg, L. E. and Søraas, F.: 1972, *J. Geophys. Res.* **77**, 6708.
Bewersdorff, A. B. and Sagalyn, R. C.: 1972, *J. Geophys. Res.* **77**, 4732.
Cormier, R. J., Ulwick, J. C., Klobuchear, J. A., Pfister, W., and Keneshea, T. J.: 1965, in S. L. Valley (ed.), *Handbook of Geophysics and Space Environments*, Air Force Cambridge Research Laboratories, Bedford, Massachusetts.
Davidson, G. T. and Hendricks, R. W., Jr.: 1971, in J. Claudis, G. T. Davidson, and L. L. Newkirk (eds.), *The Trapped Radiation Handbook*, Section 7, DNA 2524H, Defense Nuclear Agency, Washington, D.C.
Dungey, J. W.: 1963, *Planetary Space Sci.* **11**, 591.
Giaconni, R., Paolini, F. R., Hadley, W. C., and Talbot, R., Jr.: 1964, *A Research Program to Investigate Artificially Injected Trapped Radiation*, AFCRL-64-917, Air Force Cambridge Research Laboratories, Bedford, Massachusetts, 30 October 1964.
Gurnett, D. A.: 1966, *J. Geophys. Res.* **71**, 5599.
Imhof, W. L.: 1968, *J. Geophys. Res.* **73**, 4167.
Kaiser, T. R.: 1972, in B. M. McCormac (ed.), *Earth's Magnetospheric Processes*, D. Reidel Publishing Company, Dordrecht, Holland, p. 340.
Katz, L., Smart, D., Paolini, F. R., Giacconni, R., and Talbot, R., Jr.: 1964, in *Space Research IV*, North-Holland Publishing Company, Amsterdam, Holland, p. 646.
Koons, H. C. and McPherson, D. A.: 1972, *J. Geophys. Res.* **77**, 3475.
Kuck, G. A.: 1973, Ph.D. Dissertation, Univ. of New Mexico.
Lyons, L. R. and Thorne, R. M.: 1972, *J. Geophys. Res.* **77**, 5608.
Lyons, L. R., Thorne, R. M., and Kennel, C. F.: 1972, *J. Geophys. Res.* **77**, 3455.
Richtmyer, R. D. and Morton, K. W.: 1967, *Difference Methods for Initial-Value Problems* (2nd ed.), Interscience Publishers, New York.
Roberts, C. S.: 1966, in B. M. McCormac (ed.), *Radiation Trapped in the Earth's Magnetic Field*, D. Reidel Publishing Company, Dordrecht, Holland, p. 403.
Roberts, C. S.: 1968, in B. M. McCormac (ed.), *Earth's Particles and Fields*, Reinhold Book Co., New York, p. 319.
Roberts, C. S.: 1969, *Rev. Geophys.* **7**, 305.
Roederer, J. G.: 1970, *Dynamics of Geomagnetically Trapped Radiation*, Springer-Verlag, New York.
Russell, C. T., Holtzer, R. E., and Smith, E. J.: 1970, *J. Geophys. Res.* **75**, 755.
Russell, C. T., McPherron, R. L., and Coleman, P. J., Jr.: 1972, *Space Sci. Rev.* **12**, 810.
Theodoridis, G. C. and Paolini, F. R.: 1967, *Ann. Geophys.* **23**, 375.

Thorne, R. M., Smith, E. J., Burton, R. K., and Holzer, R. W.: 1973, *J. Geophys. Res.* **78**, 1581.

Walt, M.: 1971, in J. Cladis, G. T. Davidson, and L. L. Newkirk (eds.), *The Trapped Radiation Handbook*, Section 6, DNA 2524H, Defense Nuclear Agency, Washington, D.C.

West, H. I.: 1966, in B. M. McCormac (ed.), *Radiation Trapped in the Earth's Magnetic Field*, D. Reidel Publishing Company, Dordrecht, Holland, p. 634.

Whalen, B. A. and McDiarmid, I. B.: 1973, *J. Geophys. Res.* **78**, 1608.

Willard, H. R. and Kenney, J. F.: 1964, *Geophysical Effects of High-Altitude Nuclear Explosions*, Boeing Scientific Research Laboratories Report D1-82-0372, Seattle, Washington.

L DEPENDENT PEAKS IN THE ENERGY SPECTRA OF ELECTRONS PRECIPITATING FROM THE INNER BELT

W. L. IMHOF, E. E. GAINES and J. B. REAGAN

Lockheed Palo Alto Research Laboratory, Lockheed Missiles and Space Company, 3251 Hanover Street Palo Alto, Calif. 94304, U.S.A.

1. Introduction

At low satellite altitudes, peaks have recently been reported to occur often in the energy spectra of electrons precipitating from the outer edge of the inner radiation belt (Imhof *et al.*, 1973). The energies of the peaks in the L shell interval ~ 1.4 to ~ 1.8 at $h_{min} < 0$ often decrease very rapidly with increasing L value. This paper complements the previous study by presenting some new data on the distributions in energy of the peaks at a given L value, on the variations in the slope of the energy vs. L value, and on the local times of occurrence. The observations are interpreted in terms of the possible resonant interaction of electrons with VLF waves of a nearly constant frequency over the L range of concern.

In the observations discussed here the precipitating electrons are quasi-trapped, having B, L points with a minimum longitude trace altitude, h_{min}, of less than 0 km. The data were taken with plastic scintillator-photomultiplier type sensors, each with its own 256 channel digital pulser height analyzer. Most of the measurements presented in this paper were performed with one of two spectrometers placed on the gravity-gradient stabilized polar-orbiting satellite 1971-089A which was launched into orbit on October 17, 1971. One of the counters was placed at 90° with respect to the zenith whereas the other was placed at 20° to the zenith. Each of the spectrometers measures the intensities and energy spectra of electrons over the range of energies from 130 keV to 2.8 MeV with a channel width of about 11 keV. Some substantiating data taken with a similar type spectrometer placed on the spinning satellite 1972–076B are also presented.

2. Observations

The energy spectra observed in the outer portions of the inner belt are frequently very smooth or very low in intensity for $h_{min} < 0$ km. However, on certain passes pronounced peaks are observed. These peaks often display a rapid change in energy with L (as illustrated in Figure 1a) for data taken in every other 10 s interval during a pass of the satellite through the western edge of the South Atlantic Anomaly.

The variations with L of the central energies of the peaks are shown in Figure 1b for one pass of each of the two satellites. These peaks occurred at times differing by 31 min. Since it has been found empirically that the L dependence of the energies can generally be fitted rather well to the expression $E \propto L^{-n}$, the points are plotted on log-log paper. Best fit lines are drawn through the points for each of the two satellite

B. M. McCormac (ed.), Magnetospheric Physics, 129–133. All Rights Reserved.

passes. The peaks shown in Figure 1b are representative examples, but significant variations in both the energies and in the slopes of the energy vs. L curves frequently occur. Although the data can generally be fitted well to the form $E \propto L^{-n}$, the energy range over which this fit occurs varies considerably and sometimes the slope changes significantly within a pass.

Data from over 300 satellite passes have recently been processed to study the morphology of the peaked spectra. The distribution in n value for the best fit curves and the distribution in peak energy at $L = 1.65$ are shown in Figure 2. The spread in central energies has a FWHM (full width-at-half maximum) of approximately 150 keV whereas the FWHM of the peaks on a given pass is usually less that 100 keV.

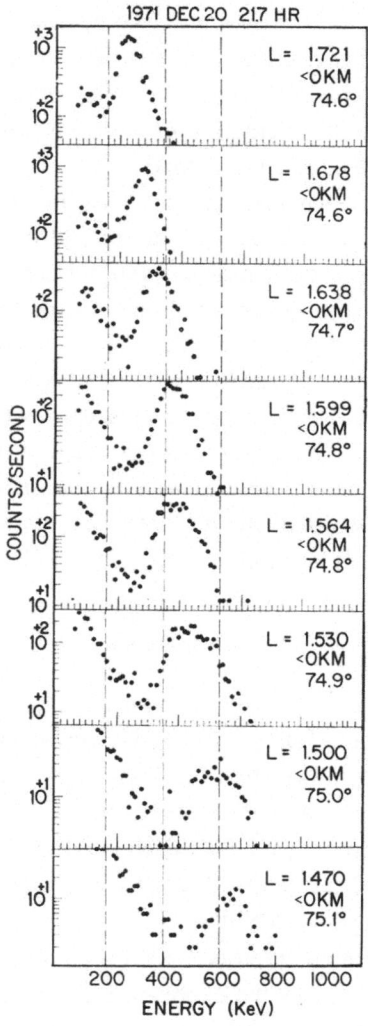

Fig. 1a.

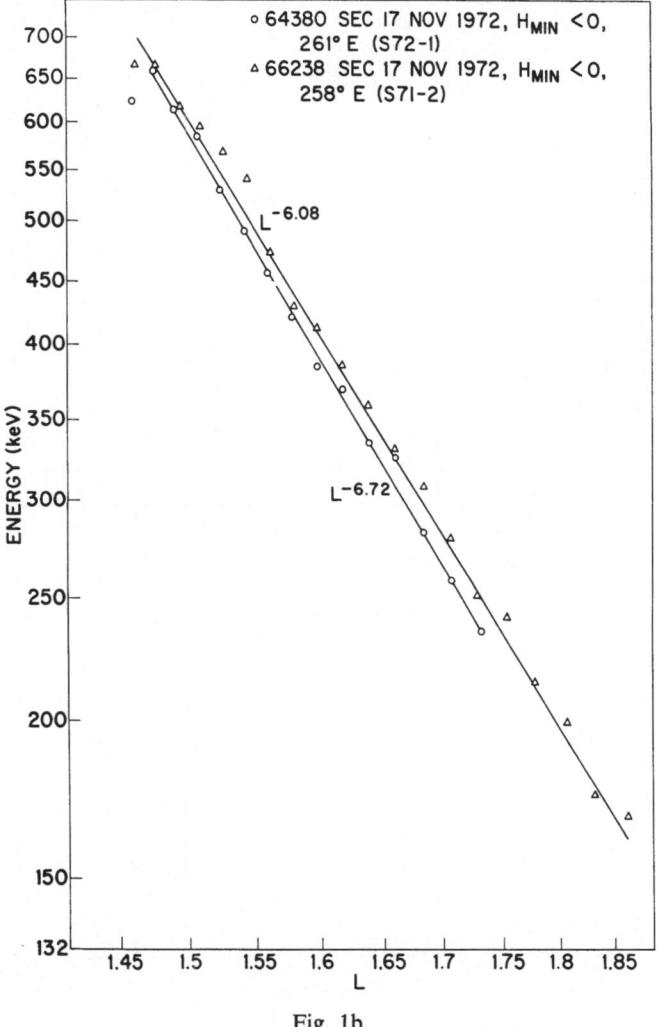

Fig. 1b.

Fig. 1. (a) The energy spectra measured during alternate successive 10 s intervals on a given pass of the satellite. The time at $L = 1.65$ is given at the top of the column. Each spectrum is labeled by its central L value, the value of h_{min}, and the pitch angle of observation. (b) Central energies of the peaks plotted as a function L for one pass of each of the two satellites, S71-2 (1971-098A) and S72-1 (1972-076B) at nearby times. Best fit straight lines, of the form $E \propto L^{-n}$, are drawn through the points for each of the passes.

Below the electron energy abscissa is another scale giving the wave frequency for first order cyclotron resonance at the equator, as discussed below.

From data taken with the 1971-089A satellite over a period of more than 1 yr an investigation was made of the probability of occurrence of peaked spectra as a function of MLT. Since the precipitating electrons measured are quasi-trapped there is not a one-to-one correspondence between the time of observation and the time of precipitation of the electrons to h_{min} values below 100 km. The local time measurements are thus performed with a time resolution ranging from ~ 2 h for data taken

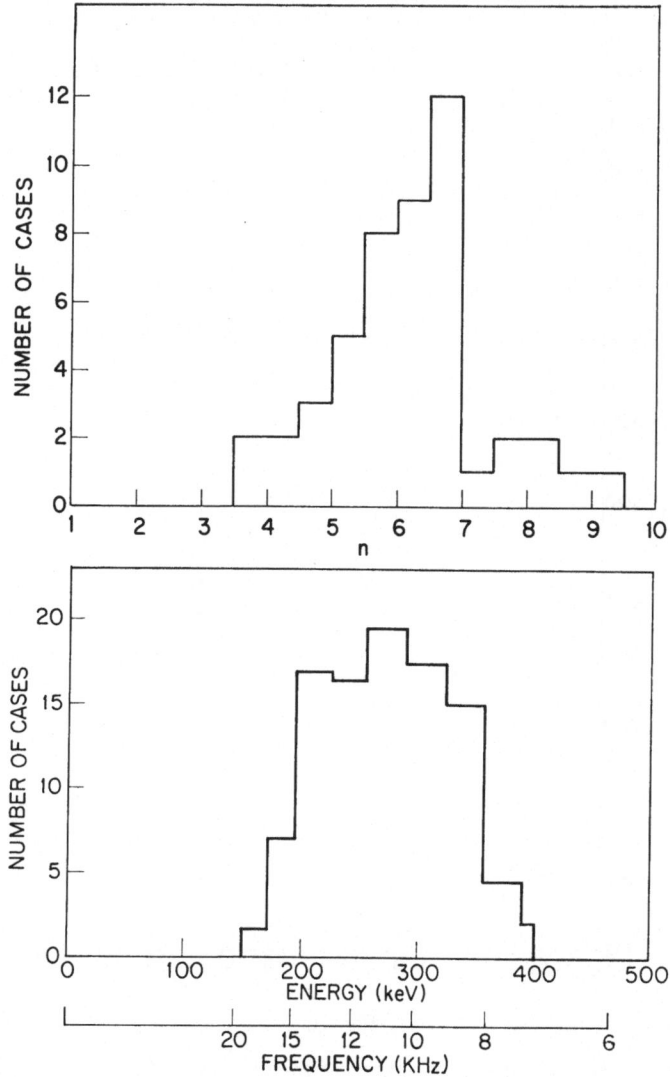

Fig. 2. The distribution of n values, where $E \propto L^{-n}$, and the distribution of central energies at $L = 1.65$ for electrons observed at $h_{\min} < 0$ km. In the lower figure there is an additional scale giving the wave frequency for first order cyclotron resonance at the equator.

just east of the South Atlantic Anomaly to ~ 16 h for points taken just west of the anomaly. To minimize any possible geographical effects data were processed in 3 longitude intervals; 245° E to 295° E, 175° E to 205° E and 53° E to 73° E. From these analyses it is concluded that the precipitation of peaked spectra occurs preferentially in the early dawn hours. A preliminary study of the dependences of the peaked spectra upon geomagnetic activity has revealed no obvious variations with the K_p index.

　　The data acquired in this experiment indicate energy selectivity in the process(es) for precipitating electrons from the outer edge of the inner radiation belt. Electron pitch angle diffusion has been related to cyclotron resonant interactions with wide-band whistler mode noise (Cornwall, 1964; Kennel and Petschek, 1966; Roberts, 1966). The simple gyroresonance concept has also been invoked to account for discrete VLF emissions (Helliwell, 1967; Brice, 1964). In the data presented here the central energies of the peaks vary with *L* as expected for cyclotron resonance with waves of a constant frequency (Imhof *et al.*, 1973). The required wave frequencies are generally in the range of ~ 8 to ~ 15 kHz. In many cases the observed peaks could result from a broad wave band with a fairly sharp upper cutoff frequency independent of *L*. Since the higher energy portion of the peak is often less steep than the low energy side and since the population of trapped electrons falls off with increasing energy the wave band could often extend down in frequency with either a broad or no lower frequency cutoff. An evaluation of the validity of this possible precipitation mechanism is hampered by the lack of available ELF and VLF measurements at the latitudes of interest. Neither the frequency nor local time distributions are well documented in the literature. Even the possible contribution from man-made signals, as suggested by C. Russell during the discussion period, is difficult to evaluate at present.

Acknowledgements

The continued support of the Defense Nuclear Agency through the Office of Naval Research (Contract Number N00014-70-C-0203) was essential to the development of the S71-2 satellite payload and to the performance of this experiment. Much of data analysis presented here was supported by the Lockheed Independent Research Program.

References

Brice, N: 1964, *J. Geophys. Res.* **69**, 4515.
Cornwall, J. M.: 1964, *J. Geophys. Res.* **69**, 1251.
Helliwell, R. A.: 1967, *J. Geophys. Res.* **72**, 4773.
Imhof, W. L., Gaines, E. E., and Reagan, J. B.: 1973, *J. Geophys. Res.* **78**, 4568.
Kennel, C. F. and Petschek, H. E.: 1966, *J. Geophys. Res.* **71**, 1.
Roberts, C. S.: 1966, in B. M. McCormac (ed.), *Radiation Trapped in the Earth's Magnetic Field*, D. Reidel Publishing Co. Dordrecht, Holland, p. 403.

MORPHOLOGY OF ENERGETIC O$^+$ IN THE MAGNETOSPHERE

E. G. SHELLEY, R. G. JOHNSON and R. D. SHARP

Lockheed Palo Alto Research Laboratory, Palo Alto, Calif. 94304, U.S.A.

This report is a very brief review and summary of our present knowledge of the morphology of the energetic O$^+$ fluxes which have been observed in the magnetosphere during magnetically disturbed periods (Shelley *et al.*, 1972b; Sharp *et al.*, 1973b).

Most measurements of energetic plasma in the magnetosphere do not distinguish between ionic species. In the energy range of tens of eV to tens of keV, typically either the energy-per-unit-charge distribution or the momentum-per-unit-charge distribution of all ions combined is measured. It has generally been assumed that the major ionic constituent of the energetic plasma in the magnetosphere is H$^+$ and that other components are of interest primarily as indicators of plasma origin. On two recent polar-orbiting satellites (OV1-18 and STP 71-2), the Lockheed group flew energetic ion mass spectrometers consisting of a series of velocity selectors and energy analyzers. These spectrometers were capable of separating the plasma into its ionic components in the mass region between 1 and 32 AMU per-unit-charge and the energy region between 0.7 and 12 keV per-unit-charge. The instrumentation is described in more detail in Shelley *et al.* (1972b). It was discovered that during geomagnetic storms the precipitating O$^+$ fluxes frequently were comparable to or exceeded the H$^+$ fluxes in the energy range of the spectrometers. This is demonstrated in Figure 1 which shows the relative number fluxes of both H$^+$ and O$^+$ measured on six consecutive nightside auroral zone passes during the December 16 to 18, 1971, magnetic storm. These measurements were made at approximately 800 km altitude and at a fixed zenith angle of 55°. There are several general characteristics of the O$^+$ fluxes which are demonstrated in Figure 1. As mentioned above, the peak O$^+$ fluxes are frequently comparable to or exceed the accompanying H$^+$ fluxes in the measured energy range. The O$^+$ fluxes are observed over a wide latitudinal range, $2 \leqslant L \leqslant 9$. The O$^+$ fluxes frequently extend to lower L values by about one-half an L unit than do the H$^+$ fluxes in the same energy range. Although both the O$^+$ and H$^+$ energy spectra are quite variable (Shelley *et al.*, 1972a), the spectra are sometimes similar in the regions where both are observed. The similarity of the spectra suggests that the two ion species have both experienced the same acceleration and transport processes.

On the basis of the presently analyzed data, it is found that the O$^+$ fluxes are comparable to the H$^+$ fluxes only during magnetic storms. Though we have not as yet performed a detailed quantitative comparison of the O$^+$ fluxes with magnetic activity indices, Figure 2 shows that at least during this one storm the peak O$^+$ fluxes qualitatively followed the K_p index. The data in the lower curve are proportional to the peak O$^+$ fluxes observed on selected passes throughout the period of the storm. In the regions where the data are indicated by down-pointing arrows, the fluxes are not necessarily below the spectrometer sensitivity level, but counts in the O$^+$ channels

B. M. McCormac (ed.), Magnetospheric Physics, 135–139. All Rights Reserved.
Copyright © 1974 by D. Reidel Publishing Company, Dordrecht-Holland.

acquired during a single 0.032 s accumulation interval are not statistically significant. Some averaging would be required to establish flux levels.

The data presented in Figures 1 and 2 were acquired in the region around 0300. In order to examine the local time dependence of these O^+ fluxes, we recently performed a coarse survey of eleven major storms during a 12 m period during which time the

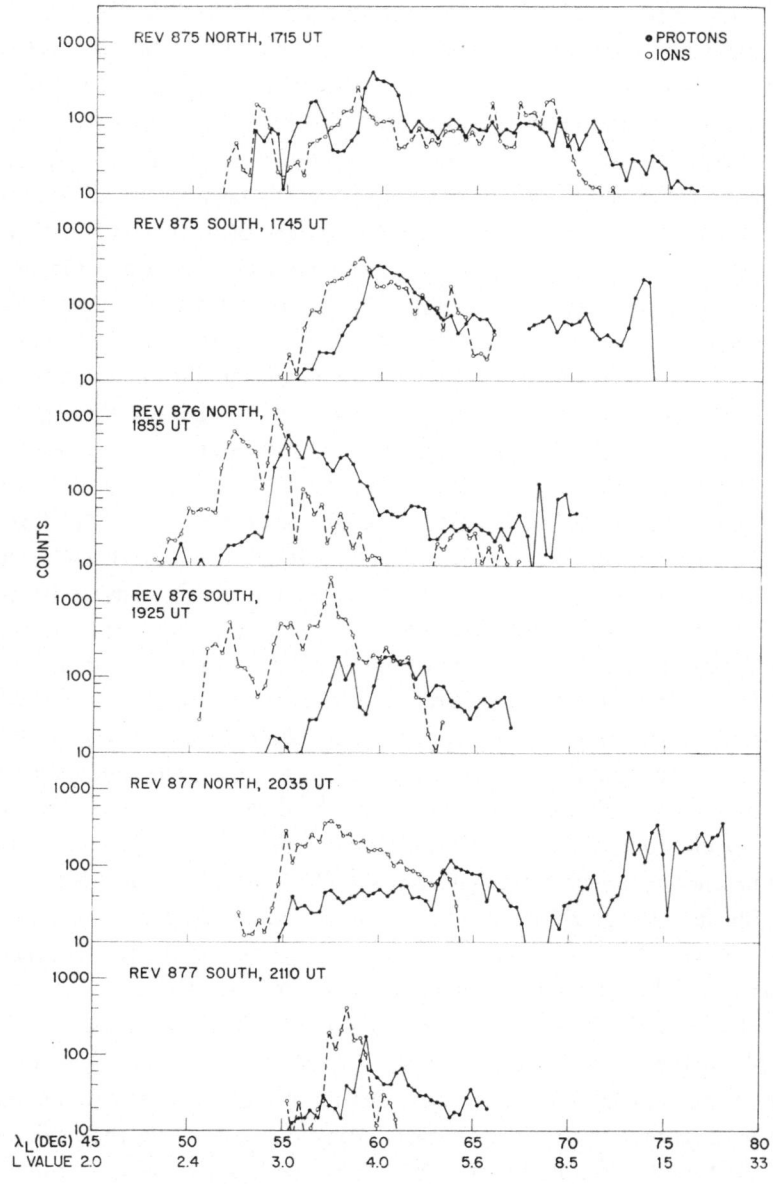

Fig. 1.   Relative number fluxes of H^+ ions (solid curve and closed circles) and $^{16}O^+$ ions (dashed curve and open circles) on six consecutive passes through the nightside auroral zone during the Dec. 16–18, 1971, magnetic storms. The 'counts' are proportional to the total number fluxes in the energy-per-unit-charge range of 0.7 keV to 12 keV (taken from Shelley *et al.*, 1972b).

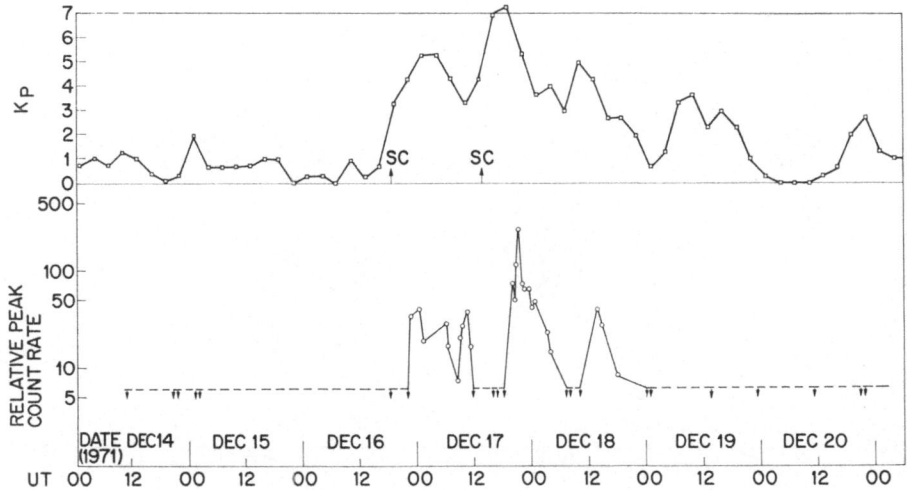

Fig. 2. The upper panel shows the K_p range for the one-week period including the Dec. 16–18, 1971, magnetic storm. In the lower panel, the relative count rate is proportional to the peak O⁺ flux on selected passes throughout the storm (taken from Shelley *et al.*, 1972b).

satellite orbit precessed through all local times (Johnson *et al.*, 1972). The survey consisted of selecting passes near the beginning, middle and end of each of the storms. The peak O⁺ flux was determined for each pass as well as the IN Lat and MLT of the peak flux and the upper and lower latitude limits of the observable O⁺ fluxes above the sensitivity threshold of approximately 2×10^5 ions cm⁻² s⁻¹ sr⁻¹. In Figure 3 we show the range of observable O⁺ flux for each pass by the solid lines. The dot on each line indicates the position of maximum flux intensity. This survey has clearly shown that the O⁺ fluxes are observable at all local times. There appears to be an asymmetry in the latitudinal extent of the precipitation region with the region extending to lower latitudes on the morning side. It must be remembered, however, that each storm is observed in a rather limited local time sector and that a relatively small number of storms contributes to the data in each local time sector. Thus the apparent asymmetry may result from variations between storms. The peak fluxes are generally found to lie below the energetic electron trapping boundary (defined by > 130 keV electrons) and the high-latitude limit is generally near the trapping boundary (Johnson *et al.*, 1972).

The MLT dependence of the peak directional O⁺ fluxes is shown in Figure 4. The solid horizontal bars indicate the median values in each 3 h sector of MLT. There is a clear maximum in the late evening to dawn sector and a minimum in the noon region with a reduction in the median flux of approximately one order of magnitude. The dashed line indicates a flux level which for an isotropic flux and an average energy of 4 keV would correspond to a precipitated energy flux of 0.1 erg cm⁻² s⁻¹. Thus, one sees that the peak flux in the morning sector frequently corresponds to several tenths of an erg cm⁻² s⁻¹ and is thus a nonnegligible energy input to the atmosphere. These oxygen fluxes should produce luminosity that is readily observable with present

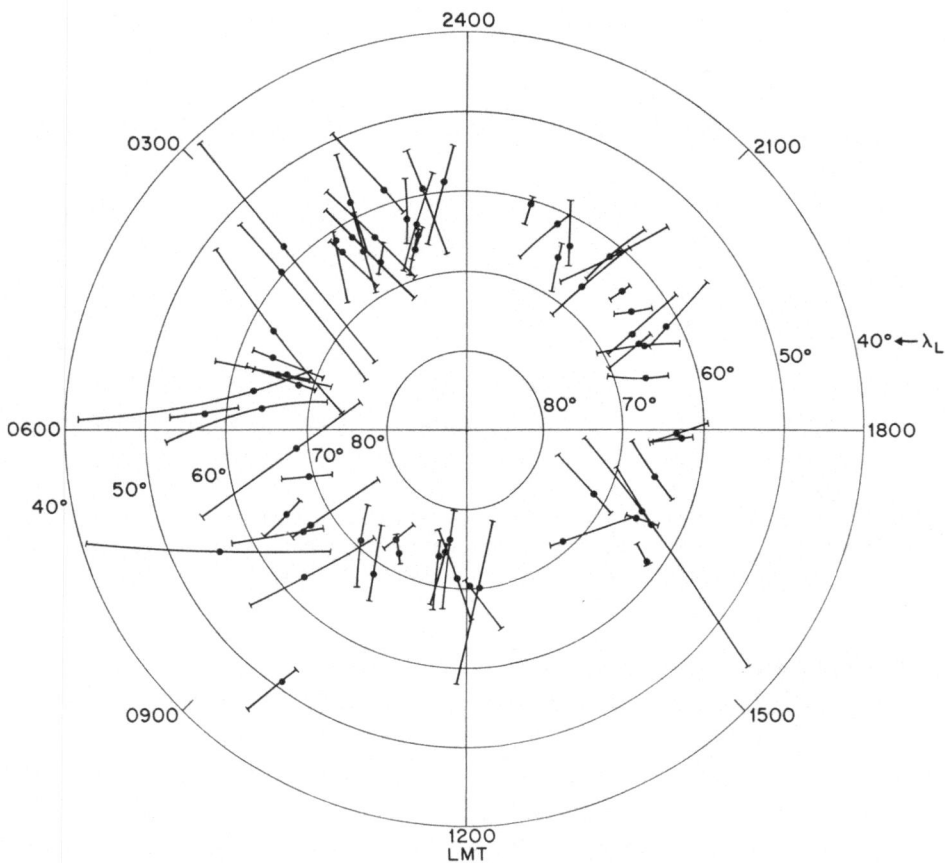

Fig. 3.   Polar plot in IN Lat and MLT of regions where the O⁺ fluxes have been observed. The solid
lines along the orbital paths indicate ranges where the O⁺ fluxes were above the spectrometer sensitivity
threshold and the solid circles indicate the location of the peak flux observed on each pass.

ground-based optical techniques; however, since oxygen is normally present in the
atmosphere and is excited by all types of precipitating energetic particles, the optical
emissions cannot be simply utilized to identify the primary ion species.

The source of the O^+ fluxes is almost certainly in the ionosphere. If charge components of oxygen other than O^+ are present, their fluxes are small compared with the
observed fluxes of O^+. The primary charge component in the solar wind is $^{16}O^{6+}$
(Bame *et al.*, 1968, 1970), while O^+ is the dominant component of the cold plasma in
the ionosphere in the high latitude regions up to greater than 1000 km altitude
(Taylor, 1973). Cladis (1973a, b) has shown that ion cyclotron waves generated by the
proton ring current could accelerate this cold O^+. Assuming that the O^+ is indeed of
ionospheric origin, one can then infer that at least part of the accompanying H^+ may
also be of ionospheric origin. Thus, a significant part of the stormtime ring current
might consist of both H^+ and O^+ originating in the ionosphere. The importance of
the O^+ in the magnetospheric energy balance depends on whether it is primarily a

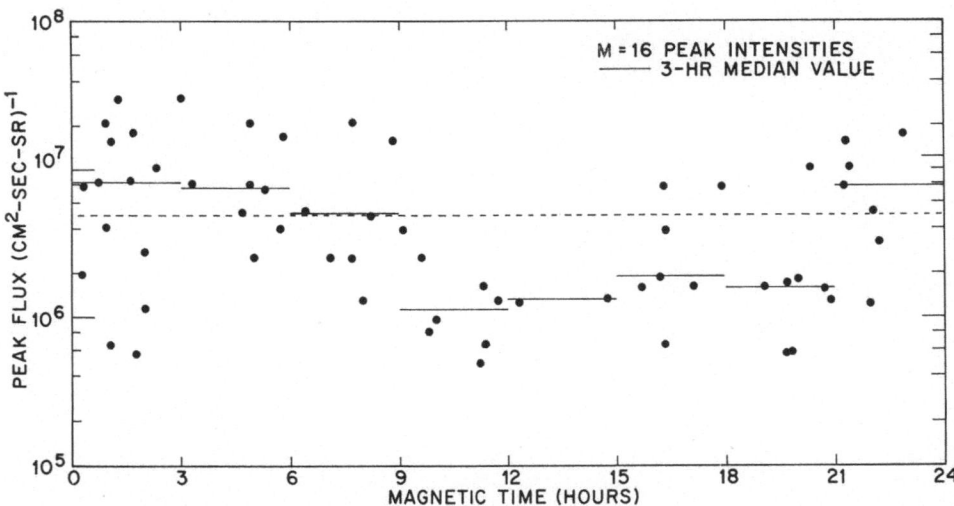

Fig. 4.   Peak number flux of O$^+$ as a function of MLT. The dashed horizontal line indicates a level corresponding to an energy flux of approximately 0.1 erg cm^{-2} s^{-1}. The shorter solid lines indicate the median flux values in the corresponding 3 h MLT sectors.

low-altitude phenomenon or if it also contributes significantly to the plasma energy density in the equatorial region. As yet there have been no measurements of this type except at relatively low altitudes.

Acknowledgements

The satellite experiment was supported by the Defense Nuclear Agency and the Office of Naval Research. This analysis has been supported by the Air Force Office of Scientific Research/AFSC, United States Air Force, under Contract F44620-73-C-0050, and the Lockheed Independent Research Program.

References

Bame, S. J., Hundhausen, A. J., Asbridge, J. R., and Strong, I. B.: 1968, *Phys. Rev. Letters* **20**, 393.
Bame, S. J., Asbridge, J. R., Hundhausen, A. J., and Montgomery, M. D.: 1970, *J. Geophys. Res.* **75**, 6360.
Cladis, J. B.: 1973a, *Radio Sci.* **8**, 1029.
Cladis, J. B.: 1973b, *J. Geophys. Res.*, **78**, 8129.
Johnson, R. G., Sharp, R. D., and Shelley, E. G.: 1972, *Trans. Amer. Geophys. Union* **53**, 1092.
Sharp, R. D., Johnson, R. G., and Shelley, E. G.: 1973a, *Trans. Amer. Geophys. Union* **54**, 432.
Sharp, R. D., Johnson, R. G., Shelley, E. G., and Harris, K. K.: 1973b, *J. Geophys. Res.*, submitted.
Shelley, E. G., Johnson, R. G., Sharp, R. D.: 1972a, *Trans. Amer. Geophys. Union* **53**, 1092.
Shelley, E. G., Johnson, R. G., and Sharp, R. D.: 1972b, *J. Geophys. Res.* **77**, 6104.
Taylor, H. A., Jr.: 1973, *J. Geophys. Res.* **78**, 315.

PART III

PLASMA CONVECTION AND ELECTRIC FIELDS

SUBSTORM INJECTION BOUNDARIES

CARL E. McILWAIN

Physics Department, University of California, San Diego, La Jolla, Calif. 92037, U.S.A.

Abstract. An improved magnetospheric electric field model is used to compute the initial locations of particles injected by several substorms. Trajectories are traced from the time of their encounter with the ATS-5 satellite backwards to the onset time given by ground-based magnetometers. A spiral shaped inner boundary of injection is found which is quite similar to that found by a statistical analysis. This injection boundary is shown to move in an energy dependent fashion which can explain the soft energy spectra observed at the inner edge of the electrons plasma sheet.

1. Background

A static electric field model, labeled E3, was constructed (McIlwain, 1972) which helped provide an understanding of many of the overall characteristics and behavior of the plasma encountered in synchronous orbit (DeForest and McIlwain, 1971).

The electric and magnetic fields can be thought of as an optical system which images particle trajectories from the injection region onto the satellite's path. The inverse problem of looking at energy-time spectrograms (the images) and deducing the electric field (the optical system) and the injection boundaries (the objects) cannot be uniquely solved using data from a single satellite. Information from other satellites and from ground-based observations is thus used in the search for solutions.

2. Existence of Injection

The details of particle injection are too complex to be determined by data available at the present time. The sudden appearance of hot plasma in the region inside the previously existing hot plasma in the magnetotail (the plasma sheet) is easy to observe, however, so that the existence of injection during the onset of each magnetospheric substorm is widely accepted.

The ATS-5 data show considerable evidence for the presence of simple inward motion and compressive heating. Other aspects of the data indicate that inward convection is far from the whole story. One important example is that in the midnight region, both positive and negative particles of all energies are sometimes observed to arrive almost simultaneously. Further, when this occurs during an eclipse, the satellite potential quickly changes from less than 50 V to over 2000 V or more (see DeForest, 1972) indicating the disappearance of the cold ionospheric plasma put into the region during the period following the previous substorm. The ATS-5 data, taken together with the observation that the initial bright auroral breakup light is on the same inner set of field lines (see Mende *et al.* (1972) as modified by the improved magnetic field model of Suigura *et al.* (1971)), thus strongly suggest the existence of *in situ* heating such as by parallel electric fields. The auroral data indicate that this heating spreads to

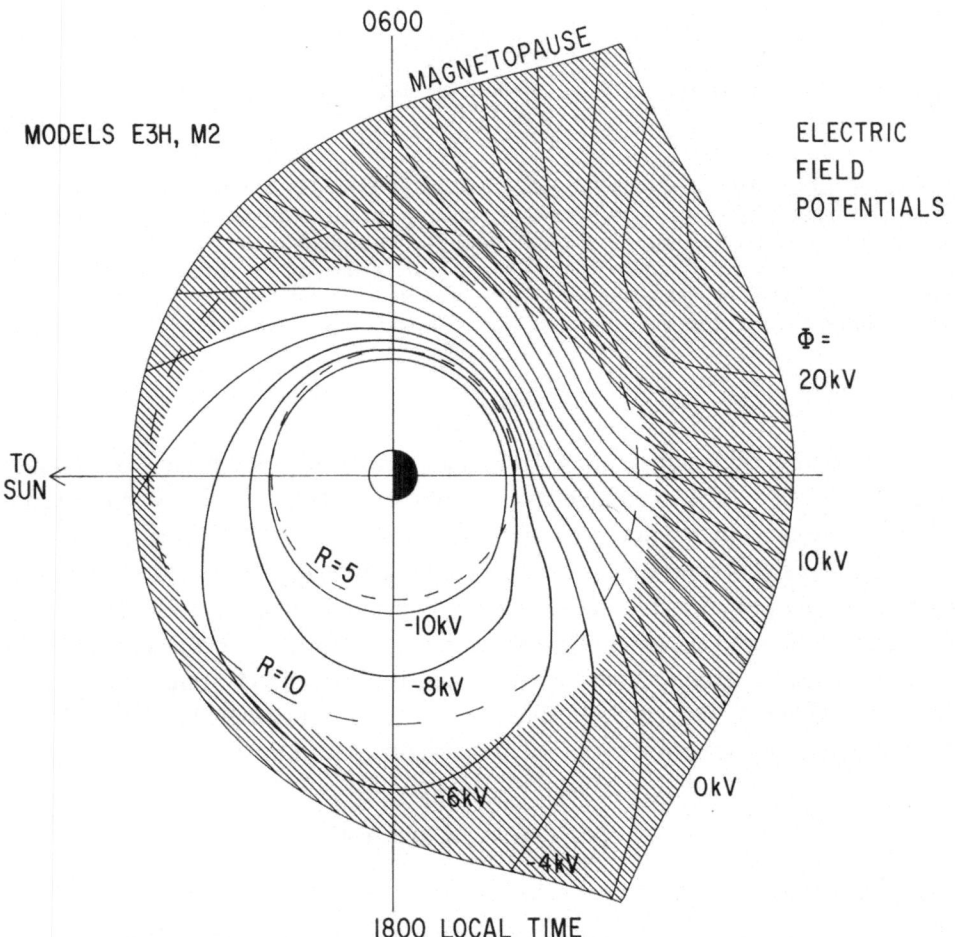

Fig. 1.   A plot of contours of constant potentials given by Model E3H mapped onto the equatorial plane by Model M2. The hatched area denotes regions where there is no reason to expect the model to be useful of accuracy.

higher lines and thus may be important in the entire plasma sheet region as well as near its inner edge.

Rather than attack the questions of how the plasma was heated and where it comes from, the present paper seeks only to determine the location of the inner boundary of the fresh plasma *after* the first 5 min or so of rapid heating. It is assumed that there is no further heating of the plasma other than that from the convective flow driven by the model electric field.

3. Electric Field Model E3H

Since the construction of model E3 (McIlwain, 1972) a number of attempts have been

TABLE I

Model E3H 8/5/73

j	A_{1j}	A_{2j}	A_{3j}	A_{4j}	A_{5j}	A_{6j}	φ_j	C_j
1	6.49	3.26	2.30	1.12	−0.24	0.17	4	2
2	1.37	0.83	0.38	0.48	−0.17	0.13	6	2
3	1.73	0.84	0.36	0.50	−0.26	0.08	8	2
4	0.69	−0.13	−0.31	0.16	−0.28	−0.03	10	2
5	0.90	−0.49	−0.53	−0.09	−0.41	−0.17	12	2
6	0.06	−1.04	−0.88	−0.37	−0.46	−0.26	14	2
7	−0.96	−1.78	−1.11	−0.64	−0.62	−0.38	16	2
8	−2.30	−1.78	−1.28	−0.53	−0.60	−0.32	18	2
9	−3.20	−2.13	−1.75	−0.89	−0.83	−0.43	20	2
10	−0.78	−1.09	−0.06	−0.63	0	−0.17	21	1
11	1.00	−1.10	−0.29	−0.49	−0.27	−0.19	22	1
12	0.59	0.07	−0.04	0.03	−0.16	−0.05	22.5	0.5
13	1.20	0.28	0.17	0.13	−0.20	−0.05	23	0.5
14	2.24	0.77	0.57	0.29	−0.29	−0.08	23.5	0.5
15	2.64	1.13	0.94	0.40	−0.23	−0.06	0	0.5
16	3.47	1.67	1.38	0.53	−0.22	−0.06	0.5	0.5
17	2.48	1.27	1.05	0.38	−0.12	−0.03	1	0.5
18	2.00	1.04	0.84	0.30	−0.08	−0.02	1.5	0.5
19	5.67	2.92	2.28	0.76	−0.13	−0.02	2	1
20	2.23	1.00	0.72	0.30	−0.03	0.06	3	1
i	1	2	3	4	5	6		
B_i	0	40	100	180	280	400		
d_i	30	50	70	90	110	130		

made to improve on it. Model E3A was one in which an effort was made to include a westward flow in the outer dusk sector (as indicated by some ionospheric measurements). While the ATS-5 data do indicate this kind of flow occurs during the initial minutes after a substorm onset, it does not seem compatible with a persistent westward flow in this region.

The most recent model, E3H, includes modifications which predict the time and space dependence of the inner edge of the plasma sheet much more accurately than earlier models.

Figure 1 shows the mapping of model E3H onto the magnetic equatorial plane using the previously defined magnetic field model M2. Table I gives the coefficients A_{ij} for this field. The values of B_i, d_i, φ_j and C_j remain the same as were given for model E3. As before, the field is time independent, but one type of variation which might be useful is to effectively change the scale size by a factor of f, using the conversion $B_i' = B_i/f^3$, $a_i' = f^6 a_i$ and $A_{ij}' = A_{ij}/f$. This conversion could be used to obtain better correspondence to K_p values outside the range of 1 to 2, but only $f = 1.0$ is used in the present paper.

4. Plasma Sheet Boundaries

The ATS-5 satellite is in a geosynchronous orbit and thus encounters the inner edge of

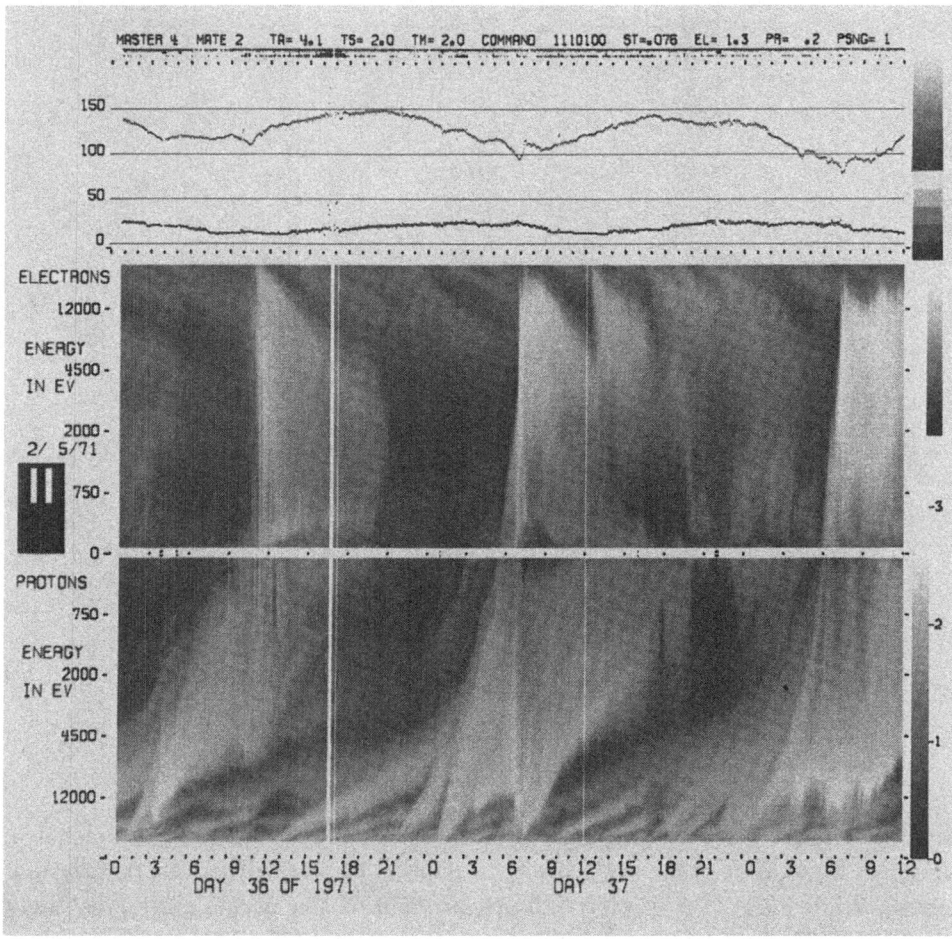

Fig. 2.   A spectrogram of ATS-5 data covering a period of 60 h showing the daily encounter with
the plasma sheet.

the plasma sheet once each day unless the K_p value stays near zero. Frank (1968) and
Vasyliunas (1968) have shown that the electron spectrum becomes 'softer' at the inner
edge of the plasma sheet. ATS-5 data show this same dependence: the upper energy
limit of substantial electron fluxes rises from less than 50 eV to greater than 3000 eV as
the satellite runs into the plasma sheet. Spectrograms (such as Figures 2 and 4) also
show that the plasma sheet protons are encountered earlier than the electrons and thus
represent a further extension of the plasma sheet toward the earth.

As before (DeForest and McIlwain, 1971; McIlwain, 1972), the ATS-5 data are
interpreted by assuming that the plasma sheet is reconstituted after each magnetospheric substorm and that the freshly inserted plasma proceeds to disperse in energy
dependent clouds driven by the residual electric field and by the magnetic gradient and
curvature drifts. The intensity often rises rapidly as each energy particle is encountered,

showing that the inner boundary of the plasma sheet is well defined, but is energy dependent. It seems natural to assume that at some earlier time, presumably the substorm onset time, all particles have a common inner boundary.

Each new substorm obliterates part of the pre-existing distributions, but an important portion of the earlier boundaries will have moved far from the injection region and remain relatively undisturbed. Other than during rapidly rising levels of activity, four or more residues of previous plasma sheet boundaries can be found coexisting with the one formed during the most recent substorm.

5. Backtracking

Near the middle of Figure 2, can be seen the ATS-5 encounter of the electron portion

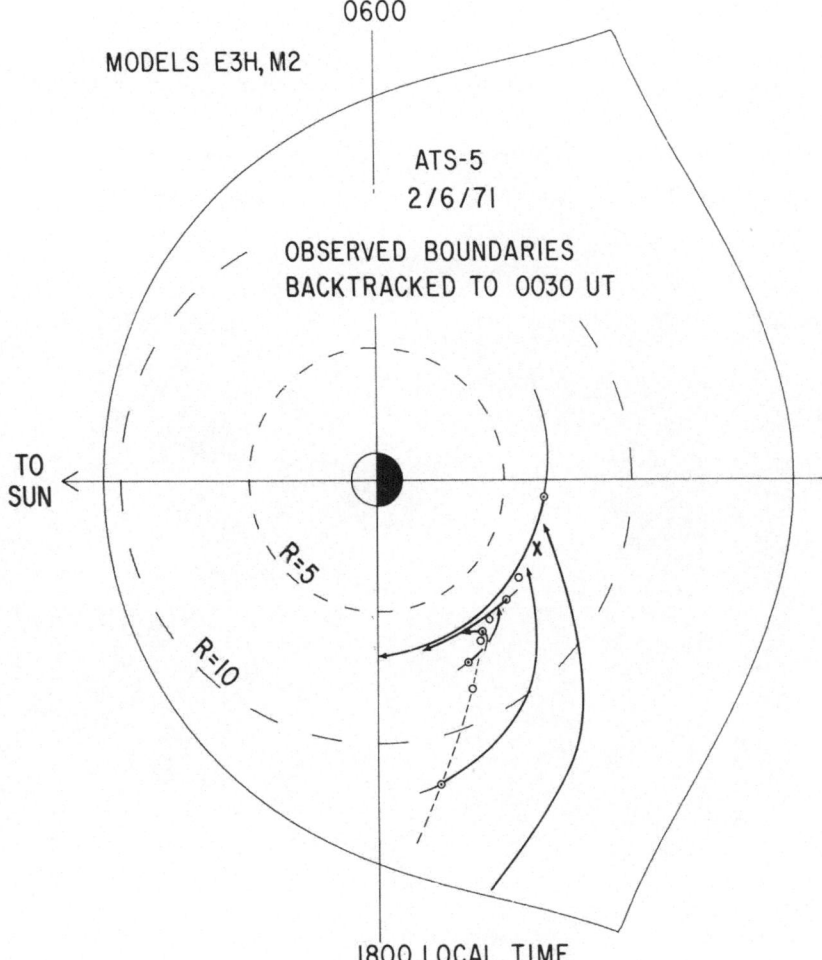

Fig. 3. Trajectories computed backwards from their encounter with ATS-5 to the onset time of the associated substorm.

of the plasma sheet at about 0530 UT on day 37 of 1971 (February 6, 1971). During the 5 h preceding the electron encounter, protons with progressively decreasing minimum energies were observed, starting with the highest energy being measured (50 keV) and reaching the lowest energy (50 eV) just before low energy electrons began appearing. Extrapolating the proton encounters back to infinite energy indicates that the most recent re-constitution of the plasma sheet was during a substorm beginning at 0030 ± 5 UT. This is confirmed by an examination of ground magnetometer records. In particular, Leirvogur, with an IN Lat of 66.7, was near local midnight and recorded a single event with a rise of $+110 \gamma$ at 0027 UT and a sharp decrease of 370γ at 0035 UT in the horizontal field component.

The energy being encountered at each hour between 0100 and 0600 UT has been read from an expanded version of Figure 2 and used to determine the trajectories of these particles according to the field models E3H and M2. Figure 3 shows the result,

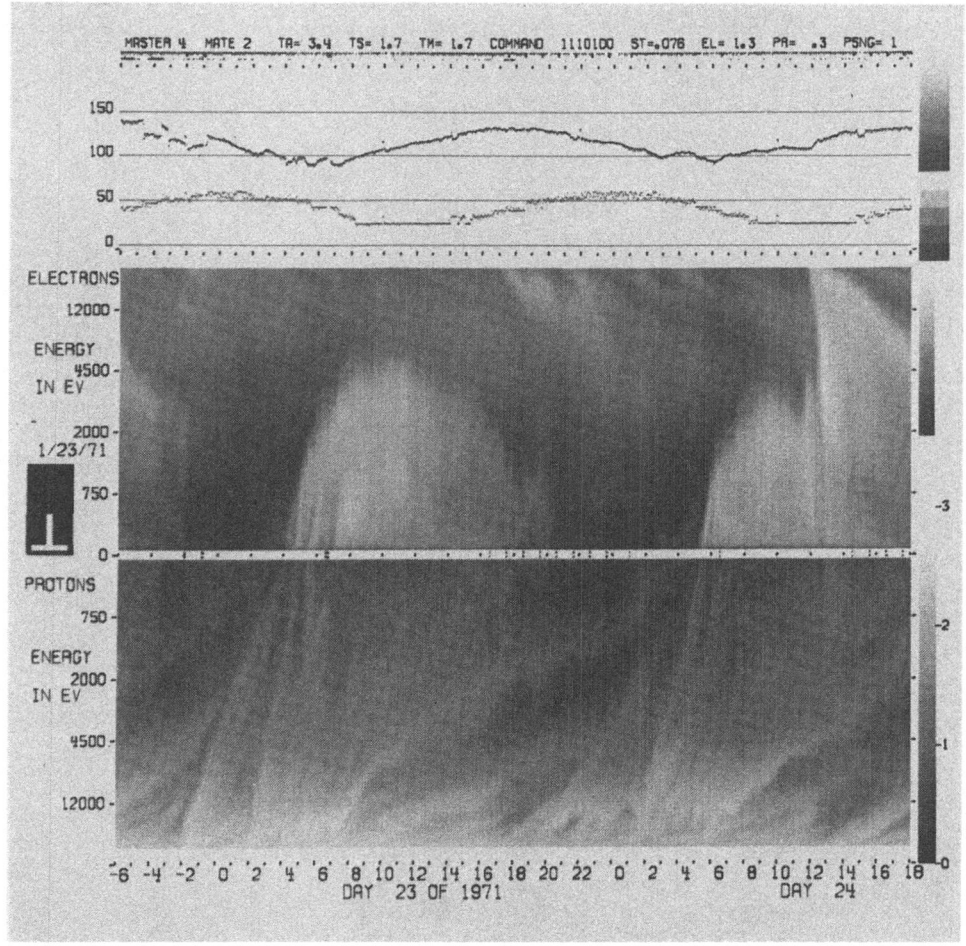

Fig. 4.   A spectrogram of ATS-5 data covering a period of 48 h.

where the arrow-heads indicate the satellite encounter, and the circled dots indicate the computed locations at 0030 UT. The cross indicates the encounter with the inner boundary of plasma sheet electrons which is presumably the same as the outer boundary of cold plasma.

The right hand side of Figure 4 shows another example at a similar level of activity ($K_p = 1$ to $2+$). The backtracked trajectories are shown in Figure 5. The left hand side of Figure 4 and the trajectories in Figure 6 show how the initial plasma sheet boundary is still spiral shaped at higher levels of activity ($K_p = 2+$ to 3) but is found at smaller radial distances.

A statistical study of the change of scale with K_p has been performed by Carpenter (1967). He has found that the post midnight-plasma pause locations could be represented by

$$R_p = 5.7\text{–}0.47\,K_p. \tag{1}$$

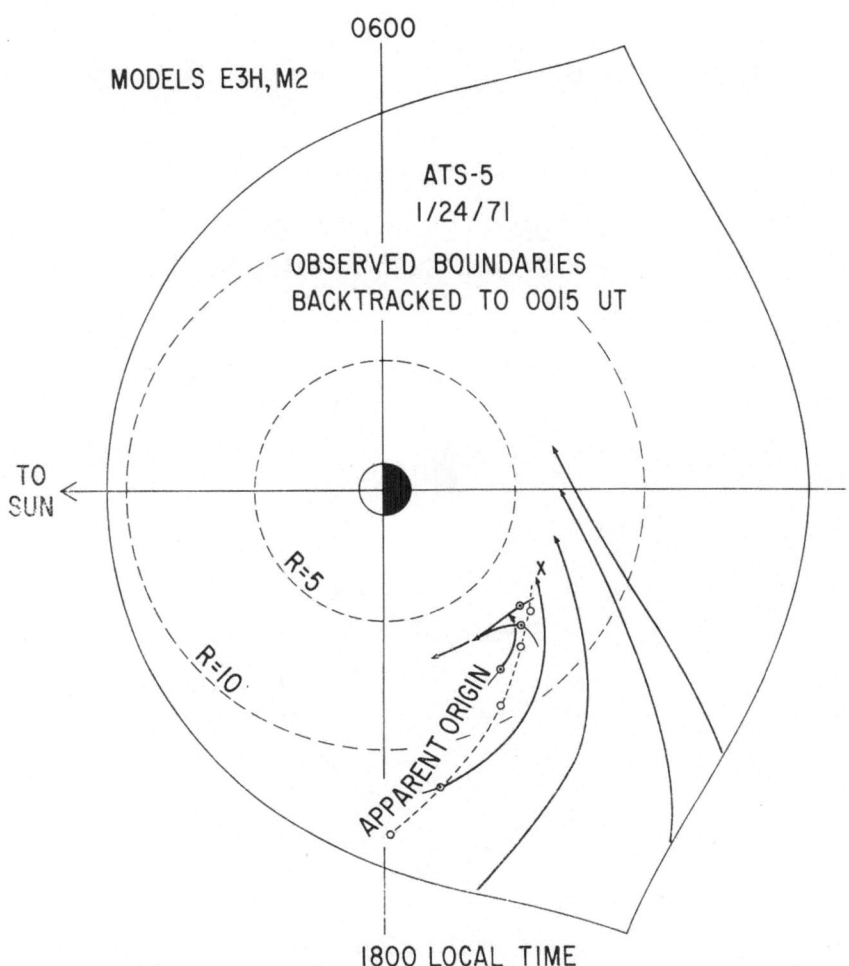

Fig. 5. Similar to Figure 3.

A statistical study of ATS-5 encounters with the cold plasma boundary has been performed (Mauk and McIlwain, 1973) which shows the local time of encounter at $R = 6.7$ to be approximately

$$\varphi = 25.5 - 1.5 \, K_p \tag{2}$$

for $K_p = 0+$ to $5+$. Combining this equation with Carpenter's radial distance scaling gives the cold plasma boundary to be

$$R_b = (122 - 10 \, K_p)/(\varphi - 7.3) \tag{3}$$

for $\varphi = 18$ to 24 h. Figure 7 shows the boundaries given by this equation for K_p values of 2 and 5.

Taken together then, these statistical studies can be seen to give a spiral boundary

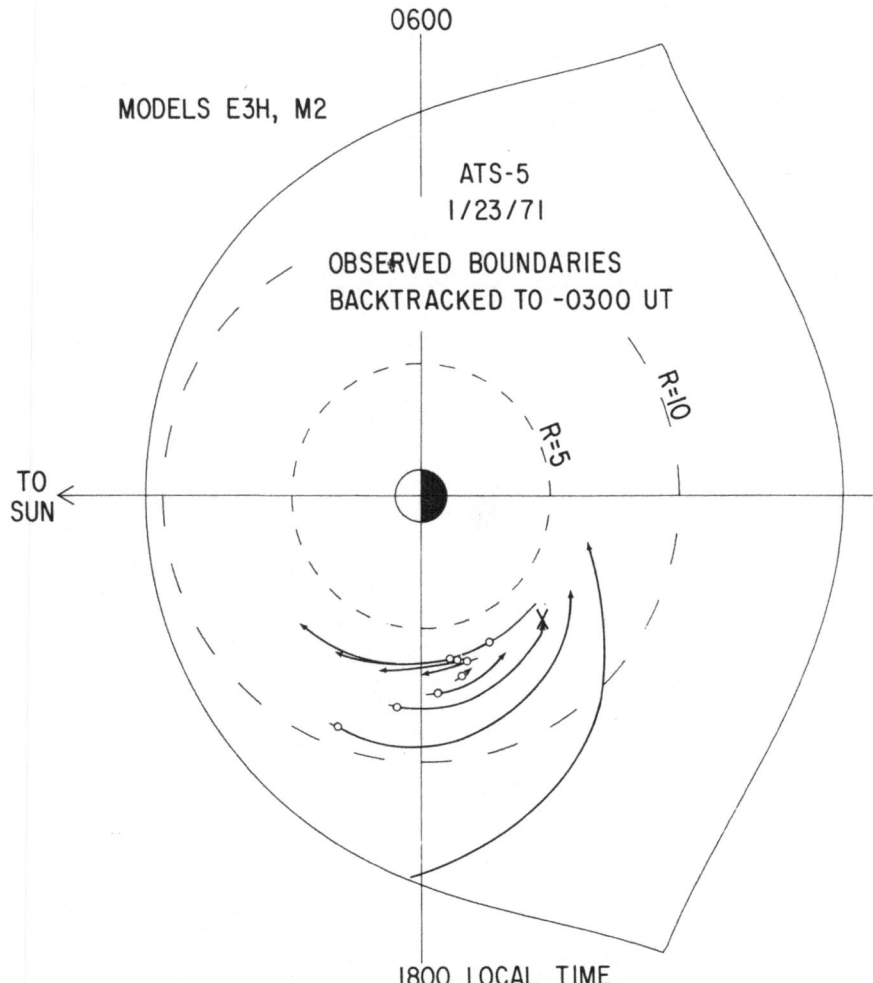

Fig. 6.   Similar to Figure 3 but for a slightly higher level of activity.

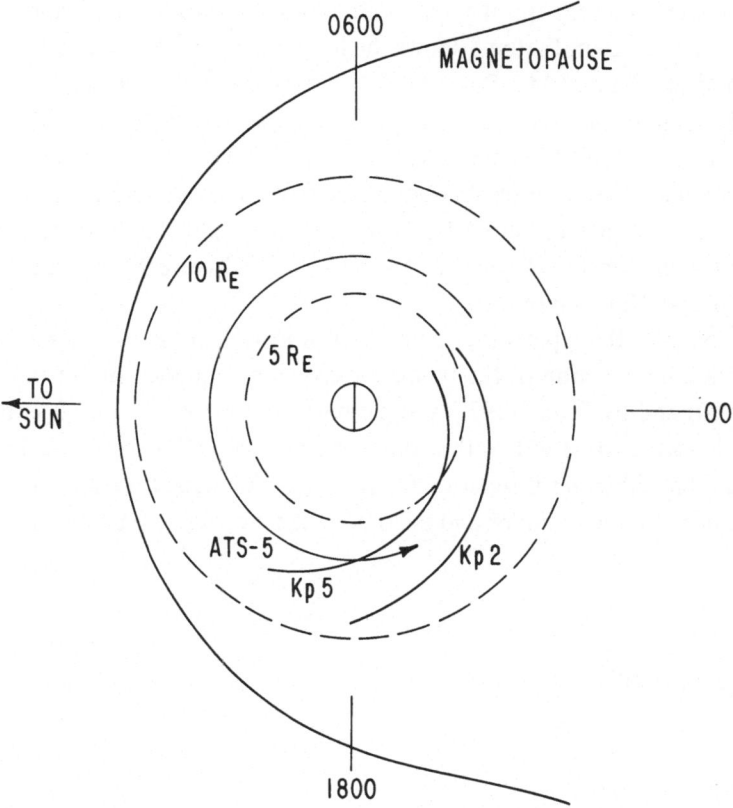

Fig. 7.   A plot of Equation (3) for K_p equal 2 and 5.

in the dusk to midnight region which is quite similar to the boundaries given by the circles in Figures 3, 5, and 6. The electric field model has been tailored to permit this agreement however, so it is important to obtain independent confirmation.

6. Boundaries Versus Time

The synchronous orbit of ATS-5 is advantageous in that it lingers in the transition region between the plasma sheet and the plasmasphere. There are two aspects of the orbit, however, which complicate interpretation: (1) it encounters the plasma boundaries at shallow angles, and (2) the drift velocities of many of the particles are very similar in direction and magnitude to the spacecraft velocity. In other words, ATS-5 tends to move parallel to the spatial structures, parts of which tend to corotate with the satellite.

Fortunately, satellites with near equatorial eccentric orbits, such as some of those in the OGO and IMP series, make rapid radial cuts through the spatial structure. To compare the present results with data obtained from such satellites, it is useful to compute the boundaries of the different particles at specific times after a substorm.

As an explicit example, particles with magnetic moments of 40, 20, 0, -20, -40 eV

γ^{-1} were all started on the spiral given by Equation (3) with $K_p = 1.5$, and with a local time extent of 16 to 24 h. Figure 8 shows the loci of the particles after 2 h of motion in the field models E3H and M2. Some of the more energetic electrons ($\mu = -40$) can be seen racing around past the dawn meridian but the ones remaining in the dusk to midnight sector have not deviated radically from the injection spiral which is indicated here by the circles. The cold plasma boundary is given by the middle line ($\mu = 0$) and can be seen to have moved into 5 R_E in the post-midnight region in agreement with the Carpenter relationship (Equation (1)) for $K_p = 1.5$. Figure 9 shows the boundaries after an additional 2 h of motion.

If 2 h is taken as the typical time since a substorm, then the $\mu = 0$ trace in Figure 8 should predict the location of the boundary crossings obtained by Vasyliunas (1968, see Figure 20) and by Frank (1971) using data from the OGO 1 and 3 satellites. The prediction is found to be well within the scatter of the OGO crossings. Presumably, the fit would be still better if separate predictions were made for each OGO crossing, taking into account the K_p index and the time of the most recent substorm.

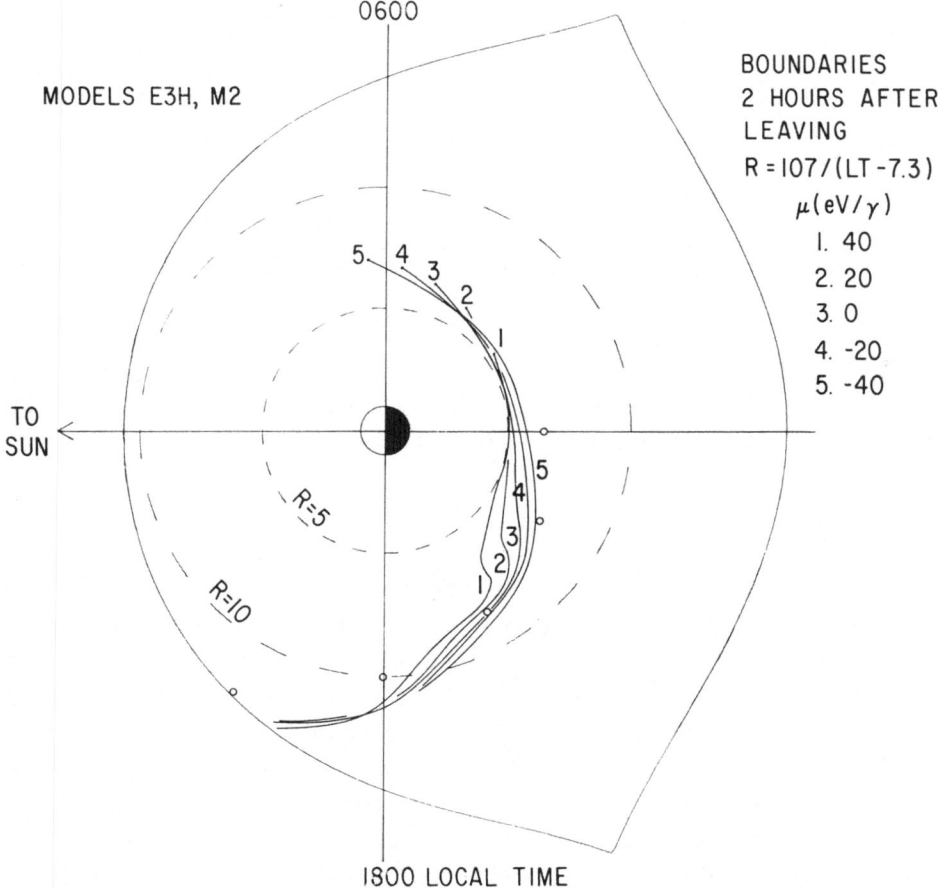

Fig. 8.   Computed locations of particles with five different magnetic moment 2 h after leaving a model boundary (Equation (3) with $K_P = 1.5$).

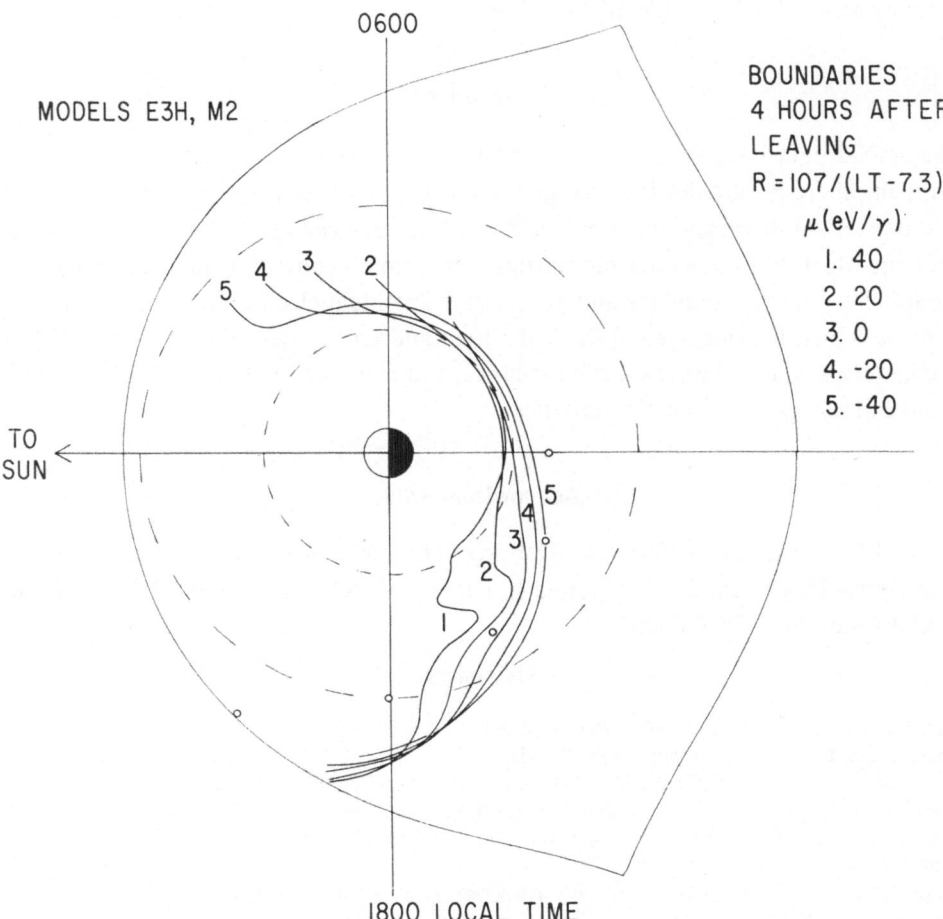

Fig. 9. Similar to Figure 8. Four hours after injection.

The agreement in shape is in large part controlled by the location of the initial boundary, that is, by Equation (3). The splitting of the boundaries of the particles with differing magnetic moments, however, is also strongly controlled by the electric and magnetic field models. Various explanations (Kavanagh *et al.*, 1968; Kennel, 1969) have been offered to explain boundary splitting since the initial observations of the phenomenon obtained by the OGO 1 and 3 spacecraft (Vasyliunas, 1968; Frank, 1968; Schield and Frank, 1970). The patterns shown in Figures 8 and 9 provide at least a qualitative fit to the earlier observations. In the future, quantitative comparisons should lead the way to further improvements in the models.

The boundary splitting in the present models is due to the energy (and charge) dependence of the drift velocity component which is proportional to the gradient of the magnetic field. To obtain the observed amount of splitting, the electric field component of the drift velocity must be quite small. In the region around 9 R_E between

18 and 21 h LT, it now seems likely that the electric field is typically smaller than that given by Model E3H (0.1 mV m^{-1} and less).

7. Summary

The ATS-5 data have been studied using the hypothesis that large transient electric fields inject fresh particles into a region outside a sharp boundary during the initial part of each substorm and that these fields are then replaced by small time independent fields in which both new and old particles proceed to convect around the Earth. A model of the initial boundary and an improved residual electric field model are constructed which provide a good fit to the time and energy dependence of the ATS-5 encounters with the plasma sheet boundaries, and at least a qualitative fit to the OGO 1 and 3 observations of these boundaries.

Acknowledgments

I would like to thank Dr Sherman DeForest and Dr V. Vasyliunas for their helpful comments. This research is supported in part by NASA Contract NAS 5-10364 and NASA Grant NGL 05-005-007.

References

Carpenter, D. L.: 1967, *J. Geophys. Res.* **72**, 2969.
DeForest, S. E.: 1972, *J. Geophys. Res.* **77**, 651.
DeForest, S. E. and McIlwain, C. E.: 1971, *J. Geophys. Res.* **76**, 3587.
Frank, L. A.: 1968, in R. L. Carovillano, J. F. McClay, and H. R. Radoski, (eds.), *Physics of the Magnetosphere*, D. Reidel Publishing Company, Dordrecht-Holland, p. 271.
Frank, L. A.: 1971, *J. Geophys. Res.* **76**, 2265.
Kavanagh, L. D., Jr., Freeman, J. W., Jr., and Chen, A. J.: 1968, *J. Geophys. Res.* **73**, 5511.
Kennel, C. F.: 1969, *Rev. Geophys.* **7**, 379.
Mauk, B. and McIlwain, C. E.: 1973, *J. Geophys. Res.*, submitted.
McIlwain, C. E.: 1972, in B. M. McCormac (ed.), *Earth's Magnetospheric Processes*, D. Reidel Publishing Company, Dordrecht-Holland, p. 268.
Mende, S. B., Sharp, R. D., Shelley, E. G., Haerendel, G., and Hones, E. W.: 1972, *J. Geophys. Res.* **77**, 4682.
Schield, M. A. and Frank, L. A.: 1970, *J. Geophys. Res.* **75**, 5401.
Suigura, M., Ledley, B. G., Skillman, T. L., and Heppner, J. P.: 1971, *J. Geophys. Res.* **76**, 7552.
Vasyliunas, V. M.: 1968, *J. Geophys. Res.* **73**, 2839.

FIELD ALIGNED CURRENTS IN THE MAGNETOSPHERE

JAMES C. ARMSTRONG

Johns Hopkins University, Applied Physics Laboratory, Silver Spring, Md., U.S.A.

Abstract. This paper deals with field aligned currents in general. Specific topics are: a brief review of the well established features; new results from the triaxial magnetometer experiment on the TRIAD satellite; comparison of the results from a variety of studies; relation of field aligned currents to other areas of research; and comments on future studies.

1. Introduction

This paper focuses primarily on one aspect of the coupling between the magnetosphere and the ionosphere, i.e., the field aligned currents which flow between these regions of space. These are often referred to as Birkeland currents in recognition of his initial suggestion for their existence (Birkeland, 1908). Birkeland's hypothesis was considered dubious on physical grounds for many years, but today it is well known that field aligned currents exist and furthermore play a central role in magnetosphere-ionosphere coupling at auroral latitudes.

Here we review very briefly the well-established facts concerning field aligned currents, present some new three axis magnetometer data from the 800 km altitude polar orbiting TRIAD satellite, discuss points of agreement and disagreement among the various measurements, relate the field aligned currents to other areas of research, and conclude with comments on future studies needed for further progress.

2. Brief Review of Previous Work

The first space observations supporting the existence of field aligned currents were made by Zmuda and co-workers (Zmuda *et al.*, 1966, 1967) using a polar orbiting satellite at 1100 km altitude. They detected transverse magnetic disturbances at auroral latitudes which Cummings and Dessler (1967) interpreted as being due to field aligned currents. The data had several limitations, including the restriction to one magnetometer axis and (in most cases) insufficient information to determine attitude when the magnetometer was used to acquire transverse magnetic disturbance data. Nevertheless, it was possible (Zmuda *et al.*, 1970) to interpret the magnetic disturbances in terms of field aligned sheet currents and define the morphology of these currents at all local times and under a variety of magnetic conditions. The significant points are reiterated here: (a) The region of occurrence was statistically coincident with the visual auroral oval as determined by Feldstein (1966). (b) The currents were detected on $>90\%$ of the satellite passes through the auroral oval, mainly in the form of east-west oriented sheets, with current densities ranging from $\sim 2 \times 10^{-7}$ A m^{-2} (the limit of sensitivity) to $\sim 9 \times 10^{-5}$ A m^{-2}; north-south sheet thickness ranged from ~ 4 km (the approximate spatial resolution limit) to >100 km. (c) The dynamics of this region as a function of magnetic activity was similar to that of the auroral oval. (d) Average maximum current density increased with increasing magnetic activity as measured by K_p. (e) Average maximum current densities were larger on the noon and midnight meridians than on the morning and afternoon meridians. From (e) it was concluded that the likely source for the currents on the dayside was the direct impingement of the solar wind on the magnetosphere. For completeness it should be noted that the concept of distinct current sheets, while useful, is potentially misleading. A more accurate concept is that of continuous current flow into and out of the ionosphere within a distinct latitude interval (which on the average is the auroral oval), with rather abrupt changes in current density separating regions of approximately constant current density along a fixed longitude. An example of modelling field aligned and ionospheric current sheets to fit simultaneous satellite and ground data has been presented by Armstrong and Zmuda (1970).

Since the initial observations a large and growing body of data on field aligned currents at auroral latitudes has been published in the literature. Space limitations prevent even listing all these; a short list relevant to subsequent discussion in this paper and useful for general reference is as follows: for ground based observations of effects due to currents similar to the type initially proposed by Birkeland, in addition to Birkeland's original work – Akasofu and Meng (1969), Bonnevier et al. (1970), and Kisabeth and Rostoker (1971); for rocket observations of sheet currents at ionospheric altitudes – Cloutier et al. (1970), Choy et al. (1971), and Whalen and McDiarmid (1972); for low altitude satellite observations in addition to those mentioned – Theile and Praetorius (1973), Berko et al. (1973), and Armstrong and Zmuda (1973a); for observations in the magnetotail – Aubry et al. (1972) and Fairfield (1973); for observations in the dayside polar cusp – Fairfield and Ness (1972); and for possible observations of these currents at synchronous altitudes – Cummings et al. (1968) and DeForest and McIlwain (1971). Theoretical studies of field aligned currents also comprise a large body of literature. A selected list is: for mechanisms to generate sheets of current – Kern (1962), Boström (1964), Atkinson (1970), Vasyliunas (1970), and Taylor and Perkins (1971); for models of field aligned currents during substorms – Atkinson (1967), Akasofu (1970), and Boström (1974); for other phenomena where field aligned currents play a critical role – Swift (1965), Kindel and Kennel (1971), and Block (1972) for discussions of current-driven plasma instabilities which might lead to parallel electric fields; Block and Fälthammar (1968) for possible effects on the electron density in the ionosphere; and Armstrong and Zmuda (1973b) for heating of the ionosphere via the ionospheric closing segment of the field aligned currents.

3. New Results from the TRIAD Satellite

The Navy/APL TRIAD satellite was launched on September 2, 1972, and carried, among other systems, a triaxial fluxgate magnetometer – 13 bit analog to digital converter system for making high resolution magnetic field measurements. The satellite orbit, magnetometer experiment, magnetometer orientation, etc., have been described in detail by Armstrong and Zmuda (1973a), and two examples of the data presented and discussed in a preliminary way. For present purposes the orbit can be considered to be circular polar at an altitude of 800 km. The experiment returns magnetometer samples at a rate of 2.25 samples axis^{-1} s^{-1} to a resolut on of 12γ. Two stations, one at APL in Howard County, Maryland, and one in College, Alaska, are presently collecting data from the experiment in real time when the satellite is in view. There is no onboard storage of data, and data passes are limited to $\sim$15 min maximum, rather shorter than desired since complete traversals of the auroral oval and polar cap cannot be acquired. Latitudinal coverage at College is $\sim$45$°$ to 85$°$. Satellite attitude is well stabilized (to better than 3$°$) in all three axes, but variations in attitude of less than 1$°$ from the beginning to the end of many data passes cause difficulty in defining a 'baseline' level consistent with the experiment resolution and relative accuracy. These small attitude variations cause a drift in ΔB ($= B_{\text{measured}} - B_{\text{theoretical}}$) away from zero even when no field aligned currents are present, since $B_{\text{theoretical}}$ for each sensor output is calculated on the basis of nominal satellite attitude. Consequently we must estimate a true baseline for the data even after subtracting out the main field component, in much the same manner as do those who work with ground based magnetometers. When over Alaska (where all the data presented here were taken), nominal satellite attitude and magnetometer orientation on the satellite are such that the sensors point approximately in the magnetic east-west, magnetic north-south, and parallel to the main field directions.

The TRIAD results represent the first total vector field measurements of the magnetic perturbations across the complete field aligned current system at a given local time, permit repeated sampling of the auroral oval under a variety of interplanetary and magnetic activity conditions, and will eventually allow sampling the oval at all local times. Furthermore simultaneous correlative data are actively being taken during satellite overpasses. The TRIAD results have already been used to demonstrate (Armstrong and Zmuda, 1973a) that the total perturbation field vector at auroral latitudes is transverse to the main field to within experimental sensitivity, thus confirming in detail that the magnetic perturbations result from field aligned currents; furthermore it was shown that the total perturbation field for the representative examples of the data presented was primarily in the magnetic east-west direction, consistent with expectations from magnetic east-west oriented sheets of field aligned current. It should be noted here that while there are dramatic exceptions to the typical magnetic east-west orientation of the perturbation vector, there are *no* exceptions to the absence of perturbations parallel to the main field within the presently available body of data.

Two final points must be mentioned before proceeding with the presentation of new data: (a) The local time intervals of the data sample available at this writing are $\sim$0600 to 1300 h and $\sim$1800 to 0100 h, thus conclusions presented here apply to only these local times. (b) Conclusions based on TRIAD results should be regarded as preliminary to some degree, although they are expected to withstand more detailed future study using whatever correlative measurements are available.

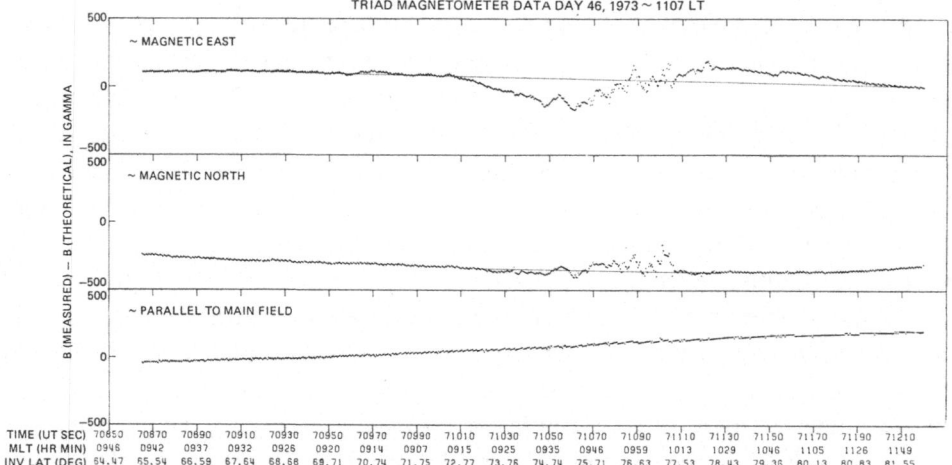

Fig. 1. TRIAD dayside data showing magnetometer output vs. time for all three axes after subtracting out the main field contribution based on the IGRF 1965.0 model. The smooth trace is a straight line which fits the data prior to encountering the disturbance field region and is not necessarily the best estimate for a baseline level. Note the absence of a disturbance field parallel to the main field. The highly structured disturbance field and apparent lack of baseline level shift is characteristic of dayside passes.

Figure 1 shows a very typical example of the dayside data and illustrates a pattern of magnetic fluctuations observed frequently for dayside passes. The word 'pattern' has its usual meaning and attendant set of limitations when used to synthesize complex magnetospheric phenomena. As is usually the case, little effort is required to find exceptions to the patterns described below.

The discernible features of the dayside pattern are: (a) highly structured magnetic fluctuations in both the magnetic east-west and north-south direction, indicating passage through rather thin but very intense sheets of field aligned currents. The most likely variation from this feature is to have the structure appear predominantly in the magnetic east-west direction. Predominance of magnetic north-south fluctuations almost never occurs. (b) No apparent shift in the magnetometer baseline level is passing through the dayside current region, implying equal amounts of current into and out of the ionosphere when summed algebraically over the satellite trajectory (but note the cautionary remarks on the baseline given above). (c) A tendency to have an overall outline like one or the other of those shown schematically in Figure 2 (a), with the small scale structure superimposed.

In general, the largest field aligned current densities are observed on the dayside and correspond to the small scale (but large amplitude) fluctuations evident in Figure 1. Current densities as large as 5×10^{-5} A m^{-2} have been observed. The actual current desnsities might exceed this value at times since there are examples in the data where the sampling rate of the experiment appears to be too low to follow the fluctuations completely. Also no period of exceptionally high magnetic activity (such as that studied by Armstrong and Zmuda, 1970) has occurred since the TRIAD launch; the maximum current density can be expected to occur during such a period.

Figures 3 and 4 show two passes typical of the nightside data available to date. Again, a 'pattern' in the same sense as above is evident. For the nightside pattern the discernible features are: (a) A large decrease in the westward component at the poleward boundary of the field aligned current region, corresponding to a sheet of

EAST-WEST ΔB "OUTLINES"

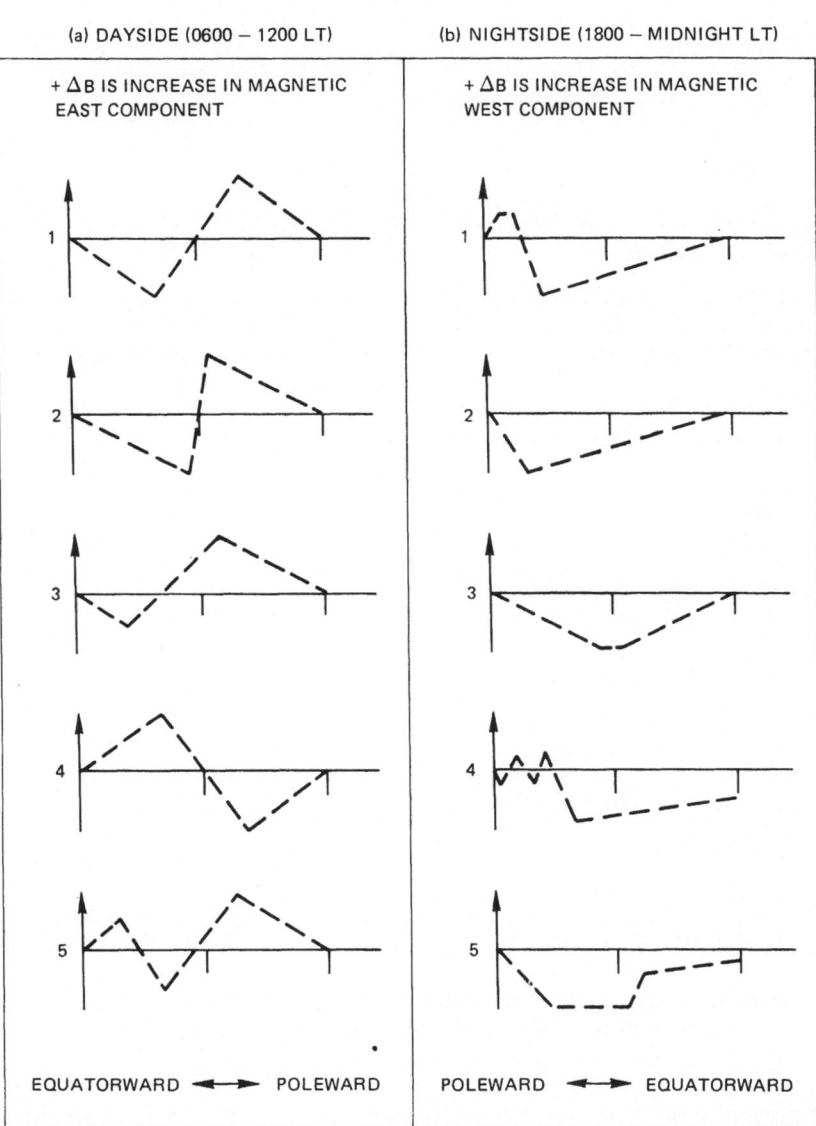

Fig. 2. 'Outlines' of the east-west magnetic disturbance field on the dayside and nightside observed by the TRIAD satellite. Small scale structure within the overall boundaries is not shown. Illustrations 4 and 5 on the nightside indicate the possibility that more current flows away from the Earth at the poleward boundary than flows toward the Earth at lower latitudes for some of the presently available data

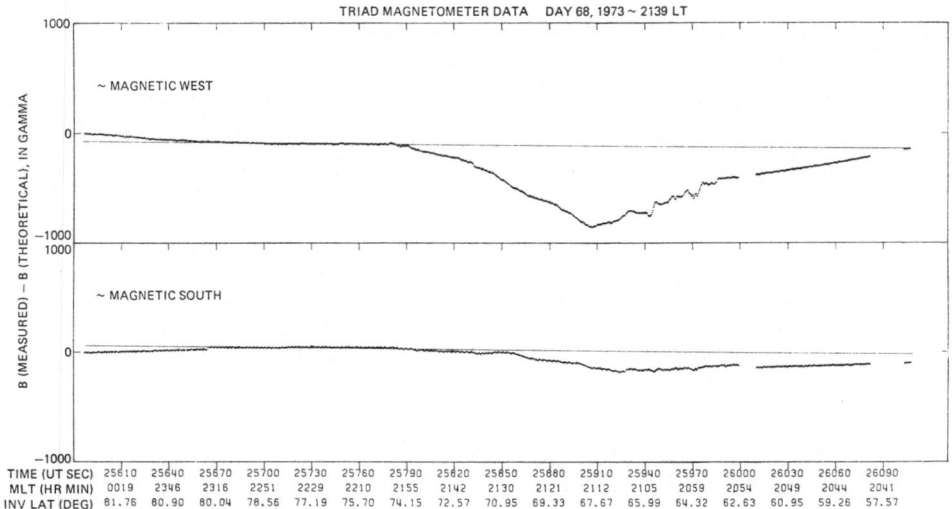

Fig. 3. TRIAD nightside data. The large negative decrease in the magnetic west component at the poleward boundary of the disturbance field region corresponds to a current out of the ionosphere, which appears to close to some extent at lower latitudes. The component parallel to the main field is less than 12 γ and is not shown.

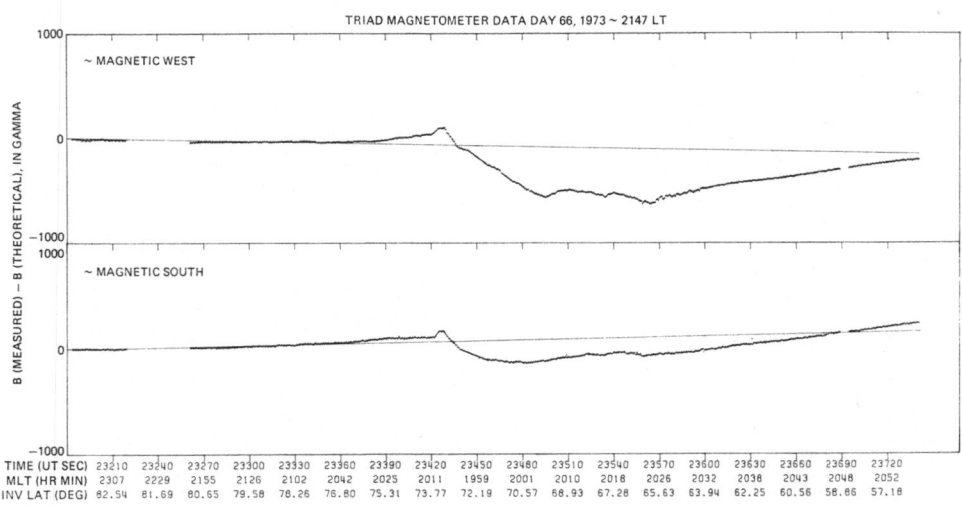

Fig. 4. Similar to the data in Figure 3, except that the large high latitude current out of the iono-sphere is preceded by a much smaller current into the ionosphere.

current flowing *away* from the Earth. In many cases ($\sim 50\%$) this is preceded by a much smaller increase in the westward component (see Figure 3), corresponding to a smaller current flowing toward the earth. (b) A gradual recovery of the westward com-ponent to its level prior to encountering the poleward current sheet. In many cases this recovery is not completed before the satellite passes out of view (see Figure 2b).

There are cases where it appears that this recovery in only partial. This is one area which is critically dependent on the baseline level determination, and a word of caution is in order here since baseline determination for the TRIAD data is not exact. (c) Small scale structure within the poleward and equatorward boundaries, but much less so than for dayside data. (d) Roughly equal probability for an increase or a decrease in the magnetic north component, with amplitude considerably less than the westward disturbance component, indicating the various current sheets can be tilted slightly from magnetic east-west in either direction.

To summarize, the current sheet flowing away from the earth at the poleward boundary of the current region for the local times being considered appears to be a permanent feature; it seems to close to some extent (in some cases, completely) at the same local time interval but at a lower latitude. The structure within this region probably corresponds to features in the plasma sheet particle population and or structure (especially conductivity variations) within the ionosphere. Some might relate to particles producing visual aurora.

Variations in the pre-midnight disturbance field outline are shown schematically in Figure 2b.

4. Comparison of Various Results

The field aligned currents detected at altitudes of ~ 1000 km have been shown to extend down to the ionosphere and upward into the outer portions of the magnetosphere (see the reference list above). The results of these observations have been used to construct the field aligned current distribution within the magnetosphere shown in Figure 5. There is no doubt that the various results mentioned previously are related in some general way, but there are interesting dissimilarities among them. For example, Cloutier et al. (1970) and Park and Cloutier (1971) describe a two-sheet field aligned current system associated with a quiet auroral arc which has a north-south spatial extent of < 50 km. While they discuss a possible uncertainty in the relative location of the two sheets, extensive arguments are presented that the most probable north-south dimension of the entire field aligned current system existing at the time of their rocket flight is < 50 km. The TRIAD satellite has sampled the same local time (~ 2000) repeatedly, under a variety of magnetic conditions, and never observed a total field aligned current region with north-south spatial extent less than ~ 400 km. Nor has any example occurred which could be interpreted entirely in terms of two sheets of equal and opposite field aligned current. Thus we must conclude that Cloutier et al. and Park and Cloutier either made their observation under a very unusual set of conditions or that they detected only a very small portion of the overall field aligned current distribution existing at the time of their rocket flight, corresponding perhaps to one of the minor variations within the overall region of disturbance shown in Figures 3 and 4.

In the rocket flight of Choy et al. (1971) only field aligned currents flowing out of (electrons into) the ionosphere were detected. The satellite experiments see roughly equal amounts of current into and out of the ionosphere on a given traversal of the auroral oval (with occasional exceptions, as discussed above); the possibilities for resolving this discrepancy seem to be: (a) the rocket did not penetrate the return portion of the overall field aligned current system, (b) not all the current-carrying particles were measured by the particle detectors; in particular measurements of electrons below 25 eV were not obtained (the rocket magnetometer results were considered inconclusive and requiring further study). There are numerous differences as well

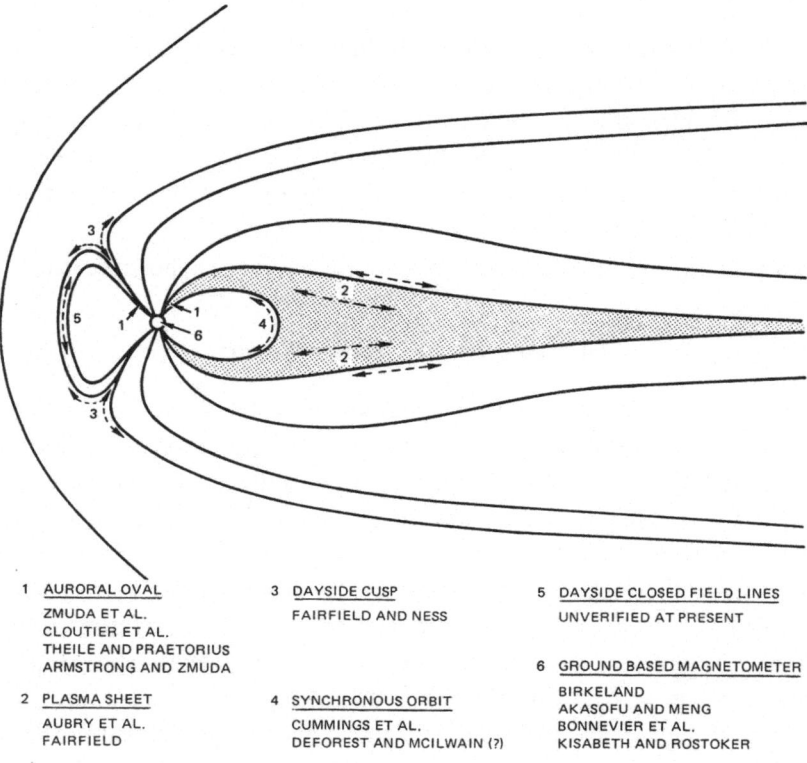

1	AURORAL OVAL	3	DAYSIDE CUSP	5	DAYSIDE CLOSED FIELD LINES
	ZMUDA ET AL.		FAIRFIELD AND NESS		UNVERIFIED AT PRESENT
	CLOUTIER ET AL.				
	THEILE AND PRAETORIUS				
	ARMSTRONG AND ZMUDA			6	GROUND BASED MAGNETOMETER
					BIRKELAND
2	PLASMA SHEET	4	SYNCHRONOUS ORBIT		AKASOFU AND MENG
	AUBRY ET AL.		CUMMINGS ET AL.		BONNEVIER ET AL.
	FAIRFIELD		DEFOREST AND MCILWAIN (?)		KISABETH AND ROSTOKER

Fig. 5. A summary showing regions of the magnetosphere where field aligned currents have been observed. The list of observers is intended as representative and does not imply that these are the only observations of field aligned currents. Item 5 is suggested to help account for the latitudinal extent of the transverse magnetic disturbance field region on the dayside.

between the results of Choy *et al.* and Park and Cloutier, including relative placement of the auroral arc and the ionospheric electrojet as well as placement of these features with respect to the most intense field aligned current sheet. However a detailed intercomparison is probably not warranted since observing conditions could have been grossly different for the two rocket flights.

The polar cusp field aligned current measurements of Fairfield and Ness (1972) appear to agree in magnitude with the low altitude satellite results (after accounting for field line divergence) but map into a low altitude width of <10 km, much too narrow to account for the overall field aligned current region detected by, for instance, the TRIAD satellite on the dayside. The orbit of their satellite (IMP-5) for the period studied in relation to field aligned currents is such that it passes initially into the magnetosphere at an intermediate latitude, stays within the magnetosphere for some time, then passes out of the magnetosphere and through the polar cusp at a higher latitude. They discuss only the data within the cusp in terms of field aligned currents, where the large change in field declination but small change in total field magnitude clearly corresponds to effects due to sheets of field aligned currents. However the data from initial penetration to well within the magnetosphere (i.e., on closed dayside field lines) also show large changes in declination, and while interpretation in terms of field aligned currents is complicated by simultaneous changes in total field magnitude, this does not rule out the existence of field aligned currents on these field lines. In fact, we suggest that field aligned currents do flow continuously on these field lines and account for a large portion of the current region observed at low altitudes. The currents might also flow frequently or continuously on field lines poleward of the cusp, but the magnetometer experiment of Fairfield and Ness lacked the range to thoroughly explore this region.

Finally we consider the magnetotail measurements of Aubry *et al.* (1972) and Fairfield (1973), after pointing out that Vasyliunas (1970) appears to have first interpreted the magnetotail measurements of Behannon (1969) in terms of field aligned currents on the outer boundary of the plasma sheet, determined that the currents flow toward the Earth on the dawn side of the tail and away from the Earth on the dusk side, and showed that this current configuration was consistent with his mathematical model of magnetospheric convection (in three sentences). It has been shown by both Aubry *et al.* and Fairfield that their results are consistent in several respects with the low altitude results of Zmuda *et al.* (1970). A major discrepancy is that they detect these currents only during the expansion phase of substorms, whereas the low altitude satellites detect the currents at a frequency approaching 100% (based on a preliminary assessment of the TRIAD results). Aubry *et al.* propose a very plausible reason for this, namely that the steady state plasma sheet boundary is diffuse and the satellite velocity is small normal to this boundary, so no sharp magnetic signature is apparent. Also, during the interval required to cross the quiet time plasma sheet boundary there is a high probability of a plasma sheet expansion. Fairfield shows that statistically these currents flow toward the Earth on the dawn side and away from the Earth on the dusk side and in both cases map onto the poleward boundary of the auroral oval. A preliminary assessment of the TRIAD data shows complete agreement for a flow away from the Earth at the poleward boundary of the field aligned current region in the 2000 – midnight local time interval, except that this is frequently preceded by a much smaller current flow in the opposite direction just poleward of the primary sheet current. Figures 3 and 4 are representative examples. Data for postmidnight local times are not available as yet. Another discrepancy (at this time unexplained) is that the low altitude satellites appear to see the closing portion (or at least a major part of it) of this current at the same longitude but several degrees lower in latitude. Such a return current has never been observed by the high altitude satellites. One reason might be that it is much more diffuse than the poleward current sheep. We can speculate that the low latitude closing portion coincides with the *inner* edge of the plasma sheet but do not have the space to explore this possibility here. Also for special cases, it might relate to the sheets of field aligned current observed at synchronous altitudes by Cummings *et al.* (1968).

In summary we can make the following observations regarding field aligned current measurements. Rocket studies are useful for detailed correlations of field aligned currents with visual aurora and in particular with particle spectra and pitch angle distribution, but they are not likely to sample the entire field aligned current distribution existing at a given local time. Furthermore, an enormous number of rockets would be required to build up a statistically significant body of data at all local times and under a variety of magnetic conditions. The high altitude satellites (IMP's and OGO's) demonstrate the existence of field aligned currents in the outer portions of the magnetosphere and show how they relate to the plasma sheet and dayside cusp. However only the most intense currents (for example, those on the outer edge of the plasma sheet during substorm expansion (Fairfield, 1973)) can be modelled, and no case-by-case correlations with aurora, ionospheric electric fields, etc., are possible. Satellites at synchronous orbit encounter field aligned currents only rarely and even then under unusual circumstances. The polar orbiting satellites appear to give the most comprehensive results. Difficulties with previous experiments of this type are merely technological and could easily be overcome on suitably funded and motivated satellite missions. Detailed comparisons between the features in the satellite data and simultaneously measured ionospheric features such as auroral arcs, electric fields, etc., can be carried out with a relative accuracy of ~ 2 km, comparable to the rocket results. The most fundamental limitations on the (single) polar orbiting satellite technique are the inability to separate time and space variations, and the necessity for waiting ~ 100 min between samples of the same region of space.

5. Relation to Other Areas of Magnetospheric Research

Field aligned currents appear to permeate the plasma sheet to some extent and couple to currents flowing within the ionosphere at essentially all times. Furthermore, being a permanent feature of the dayside magnetosphere, they must play some fundamental role in the solar wind – interplanetary magnetic field – magnetospheric interaction, which to our knowledge has never been adequately treated theoretically. Important, large scale magnetospheric phenomena such as convection and particle energy spectra in the plasma sheet and distant tail region are affected by field aligned currents. Boström (1974) discusses the effects on convection. Vasyliunas (1972) has considered some fundamental implications (via Liouville's theorem) of the particle energy spectra in the magnetotail based on the collisionless nature of this region of space. We point out there hat when the hot magnetotail plasma flows into the ionosphere in the form of a current, the return current offers the possibility of replacing the hot plasma with a cold component of ionospheric origin. In effect this introduces 'collisions' into the particle dynamics and resulting particle energy spectra of the magnetotail plasma.

The magnetospheric substorm is another area where field aligned currents are likely to play a fundamental role. One popular substorm model, originally due to Atkinson (1967), has the substorm expansion phase beginning with a disruption of the cross-tail current and redirection of it via field aligned currents into the ionosphere, where it feeds the associated westward electrojet.

Aurora are known to be intimately related to field aligned currents since, statistically, they both occur within the auroral oval. One question of interest is whether or not auroral electrons ($1 \leqslant E_e \leqslant 10$ keV) form one segment of a three dimensional current system; the answer has important implications for mechanisms concerned with the production of auroral electrons.

Finally we mention effects on the ionosphere. It has been suggested that field aligned currents can affect the composition and the density of the ionosphere if they persist for times of order several hours. It is easy to show that, if the field aligned sheet currents close in the ionosphere via Pedersen currents, a 'typical' current system would provide more energy input into the nightside ionosphere than would be deposited by the particles producing a 'typical' auroral arc. Furthermore closure of the more complex field aligned current systems via ionospheric Pedersen currents would affect the electric field structure in the auroral ionosphere, in addition to the structure produced there by conductivity variations. This can be visualized as a series of magnetospheric 'batteries' driving varying amounts of current through various parts of the resistive ionosphere (for the 'batteries' one could mention the plasma clouds of DeForest and McIlwain (1971) which correspond to volumes of excess particle energy density and which would relax to a background value by discharge to the ionosphere via field aligned currents). The critical role which field aligned currents might play in formation of parallel electric fields (via anomalous resistivity and/or formation of electrostatic double layers) was mentioned previously in Section 1 above.

6. Future Research

At least initially, future efforts in field aligned current research should be directed toward sorting out the detailed, statistically significant relationship between the currents and other magnetospheric-ionospheric observables such as visual aurora, ionospheric electric fields, the auroral electrojet, and perhaps variations in interplanetary quantities. Establishment of firm relationships will prevent unnecessary quibbling in future research efforts where field aligned currents play a role; furthermore, they can provide a basis for additional insight into magnetospheric phenomena.

Conclusive studies on the general, overall structure of field aligned currents during non-active periods are needed so that phenomena such as substorms can be studied as deviations from these basic patterns. The ultimate goal might be to acquire sufficient understanding to allow development into a tool for monitoring magnetospherci conditions from the relatively inexpensive low altitude polar orbit.

Acknowledgment

This work was supported by the Naval Ordnance System Command under contract number N00017-72-C-4401 with the Johns Hopkins University, and by the Office of Naval Research through Task Problem Z620 under the above contract number.

References

Akasofu, S.-I. and Meng C.-I.: 1969, *J. Geophys. Res.* **74**, 293.
Akasofu, S.-I.: 1970, in B. M. McCormac (ed.), *Particles and Fields in the Magnetosphere* D. Reidel Publishing Company, Dordrecht-Holland, p. 34.
Armstrong, J. C. and Zmuda, A. J.: 1970, *J. Geophys. Res.* **75**, 7122.
Armstrong, J. C. and Zmuda, A. J.: 1973a, *J. Geophys. Res.* to be published.
Armstrong, J. C. and Zmuda, A. J.: 1973b, *Radio Sci.* **8**, 401.
Atkinson, G.: 1967, *J. Geophys. Res.* **72**, 5373.
Atkinson, G.: 1970, *J. Geophys. Res.* **75**, 4746.
Aubry, M. P., Kivelson, M. G., McPherron, R. L., Russell, C. T., and Colburn, D. S.: 1972, *J. Geophys. Res.* **77**, 5487.
Behannon, K. W.: 1969, GSFC Preprint X-616-69-146.
Berko, F. W., Hoffman, R. A., Burton, R. K., and Holzer, R. E.: 1973, GSFC Preprint X-646-73-45.
Birkeland, K.: 1908, in *The Norwegian Aurora Polaris Expedition 1902-1903*, H. Aschehoug, Christiania, Norway, Vol. 1, Sec. 1.
Block, L. P. and Fälthammar, C.-G.: 1968, *J. Geophys. Res.* **73**, 4807.
Block, L. P.: 1972, *Cosmic Electrodyn.* **3**, 349.
Bonnevier, B., Boström, R., and Rostoker, G.: 1970, *J. Geophys. Res.* **75**, 107.
Boström, R.: 1964, *J. Geophys. Res.* **69**, 4983.
Boström, R.: 1974, this volume, p. 45.
Choy, L. W., Arnoldy, R. L., Potter, W., Kintner, P., and Cahill, L. J., Jr.: 1971, *J. Geophys. Res.* **76**, 8279.
Cloutier, P. A., Anderson, H. R., Parks, R. J., Vondrak, R. R., Spiger, R. J., and Sandel, B. R.: 1970, *J. Geophys. Res.* **75**, 2595.
Cummings, W. D. and Dessler, A. J.: 1967, *J. Geophys. Res.* **72**, 1007.
Cummings, W. D., Barfield, J. N., and Coleman, P. J., Jr.: 1968, *J. Geophys. Res.* **73**, 6687.
DeForest, S. E. and McIlwain, C. E.: 1971, *J. Geophys. Res.* **76**, 3587.
Fairfield, D. H. and Ness, N. F.: 1972, *J. Geophys. Res.* **77**, 611.
Fairfield, D. H.: 1973, *J. Geophys. Res.* **78**, 1553.
Feldstein, Y. I.: 1966, *Planetary Space Sci.* **14**, 121.
Kern, J. W.: 1962, *J. Geophys. Res.* **67**, 2649.
Kindel, J. M. and Kennel, C. F.: 1971, *J. Geophys. Res.* **76**, 3055.
Kisabeth, J. L. and Rostoker, G.: 1971, *J. Geophys. Res.* **76**, 6815.
Park, R. J. and Cloutier, P. A.: 1971, *J. Geophys. Res.* **76**, 7714.
Swift, D. W.: 1965, *J. Geophys. Res.* **70**, 3061.
Taylor, H. E. and Perkins, F. W.: 1971, *J. Geophys. Res.* **76**, 272; Correction, 1971, *J. Geophys. Res.* **76**, 6210.
Theile, B. and Praetorius, H. M.: 1973, *Planetary Space Sci.* **21**, 179.
Vasyliunas, V. M.: 1970, in B. M. McCormac (ed.), *Particles and Fields in the Magnetosphere*, D. Reidel Publishing Company, Dordrecht-Holland, p. 60.
Vasyliunas, V. M.: 1972, Paper, Solar-Terrestrial Relations Conference, Calgary, Canada.
Whalen, B. A. and McDiarmid, I. B.: 1972, *J. Geophys. Res.* **77**, 191.
Zmuda, A. J., Martin, J. H., and Heuring, F. T.: 1966, *J. Geophys. Res.* **71**, 5033.
Zmuda, A. J., Heuring, F. T., and Martin, J. H.: 1967, *J. Geophys. Res.* **72**, 1115.
Zmuda, A. J., Armstrong, J. C., and Heuring, F. T.: 1970, *J. Geophys. Res.* **75**, 4757.

CALCULATIONS OF MAGNETOSPHERIC
ELECTRIC FIELDS

R. A. WOLF

Department of Space Physics and Astronomy, Rice University, Houston, Tex., U.S.A.

1. Introduction

This paper will describe efforts to compute theoretically the electric field and plasma distributions in the magnetosphere. In the last few years, such efforts have been made by Block (1966), Iwasaki and Nishida (1967), Karlson (1971), Swift (1971), Vasyliunas (1970, 1972), Wolf (1970), and Jaggi and Wolf (1973). The physical basis of all this work is pretty much the same; the differences lie in the details of the approximations.

In discussing the theory of magnetospheric convection, it is convenient to divide the ionosphere and magnetosphere into two groups of magnetic flux tubes, as shown in Figure 1. One group of flux tubes, connected to the polar cap ionosphere, crosses the equatorial plane within the magnetopause boundary layer (dotted region) or far out in the tail (cross hatched region), or connects directly to solar wind field lines. Flux tubes of the other group cross the ionosphere equatorward of the polar cap and do not extend to great distances from the Earth. In Figure 1, these two groups of flux tubes are separated by a surface I, whose intersections with the ionosphere and equatorial

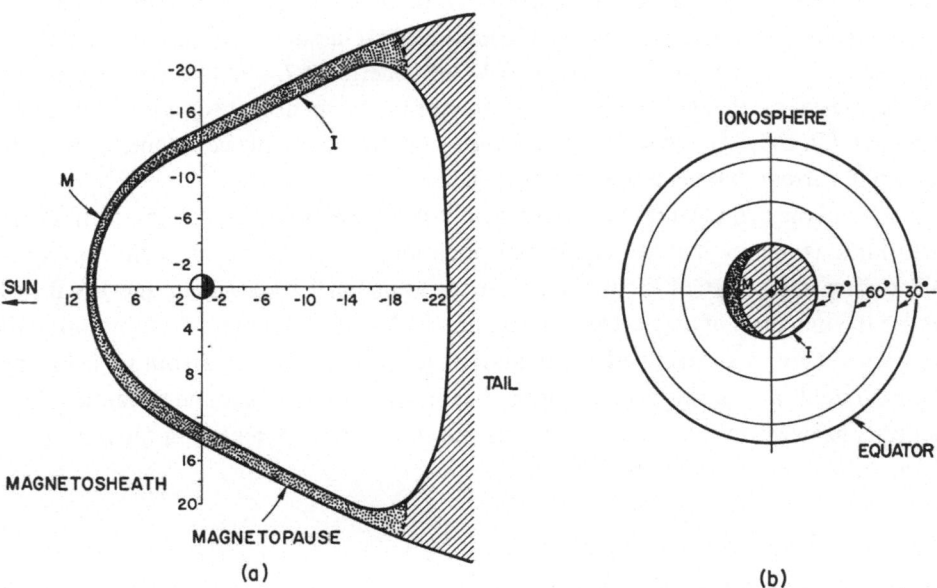

Fig. 1. Corresponding regions in the ionosphere and equatorial plane. Diagram (a) shows the magnetospheric equatorial plane, diagram (b) the ionosphere viewed from high above the north magnetic pole. The surface I separates polarcap field lines from lower L values; its intersections with the ionosphere and magnetospheric equatorial plane are shown.

B. M. McCormac (ed.), Magnetospheric Physics, 167–177. All Rights Reserved.
Copyright © 1974 by D. Reidel Publishing Company, Dordrecht-Holland.

plane are indicated. Different physical processes dominate the two regions. Motions of polar cap field lines are governed mainly by their intimate contact with the solar wind, due to merging or some viscous interaction; ionospheric conductivity plays a lesser role. On the other hand, motion of lower L shell field lines is strongly affected by ionospheric conductivity, pressure gradients in the magnetospheric plasma, and probably dynamo winds. Other papers in this volume discuss theoretical models of the polar cap field lines. This one will deal instead with some physical processes governing the convection of lower L value field lines; the rate of convection across the polar cap will be taken as a boundary condition.

We present some new model calculations, with a model particle population that simulates the typical quiet time ring current. A possible observational test of the model is discussed. At the end of the paper is a summary of the status of these theoretical electric field calculations: physical processes included in the calculations and those not included; which observed general features of magnetospheric convection the models fit satisfactorily and where there seem to be substantial conflicts.

2. Calculations With Quiet Time Ring Current

We present here some computer calculations designed to model the quiet time ring current, i.e., the plasma sheet ions at $L \lesssim 10$.

We assume that at time $t=0$, a sheet of ions is positioned far out in the tail, at a geocentric distance of about 16 R_E at local midnight. The sheet is taken to have 6.3×10^{20} ions per weber of magnetic flux, corresponding to a number density of 1.2 cm^{-3} at $L=7$ for an isotropic distribution. The particles are assumed to have magnetic moment 200 eV γ^{-1}, which corresponds to 17 keV energy at $L=7$. The parameters used for Figures 6–11 of Jaggi and Wolf (1973) were the same, except for a lower magnetic moment (50 eV γ^{-1}), which was designed to represent of the component the storm time ring current that penetrates to low L values.

To avoid the complexities of drift shell splitting, we assume that the ions drift across field lines as if they had 90° equatorial pitch angles and mirrored in the equatorial plane. The inner edge of the ion sheet moves as its particles gradient and $\mathbf{E} \times \mathbf{B}$ drift under the influence of an electrostatic field $\mathbf{E} = -\nabla V$. The potential distribution in the magnetosphere is determined from that in the ionosphere by assuming field lines equipotentials and taking account of the Earth's rotation. We assume the atmospheric neutrals have zero velocity, and compute the electrostatic potential in the ionosphere from the following equations

$$\mathbf{j}_h = \boldsymbol{\sigma} \cdot (-\nabla_h V) \tag{1}$$

$$\nabla_h \cdot \mathbf{j}_h = j_{\parallel} \tag{2}$$

where $\mathbf{j}_h$ is the height integrated horizontal ionospheric current, and $\boldsymbol{\sigma}$ is the height integrated conductivity tensor. Our model conductivity is assumed independent of time, but is otherwise fairly sophisticated, including Hall and Pedersen conductivities, day-night asymmetry, dayside solar zenith angle effect, auroral zone enhancement, and dip angle effect (Jaggi and Wolf, 1973). The quantity $j_{\parallel}$ is the density of Birkeland (field aligned) currents generated by plasma pressure gradients in the magnetosphere – in the present case at the Alfvén layer, the inner edge of the ion sheet. (Pressure gradients in the electrons are neglected since the pressure of plasma sheet ions exceeds that of the electrons by ~ a factor of 5.) For a detailed discussion of the generation of Birkeland currents at such an Alfvén layer, see Jaggi and Wolf (1973) or Wolf and Jaggi (1973). Equations (1) and (2) are solved subject to two boundary conditions. First, the latitudinal current is set equal to zero at 21° Lat. (an approximation to the symmetry condition that no currents flow across the equator). Second, the potential

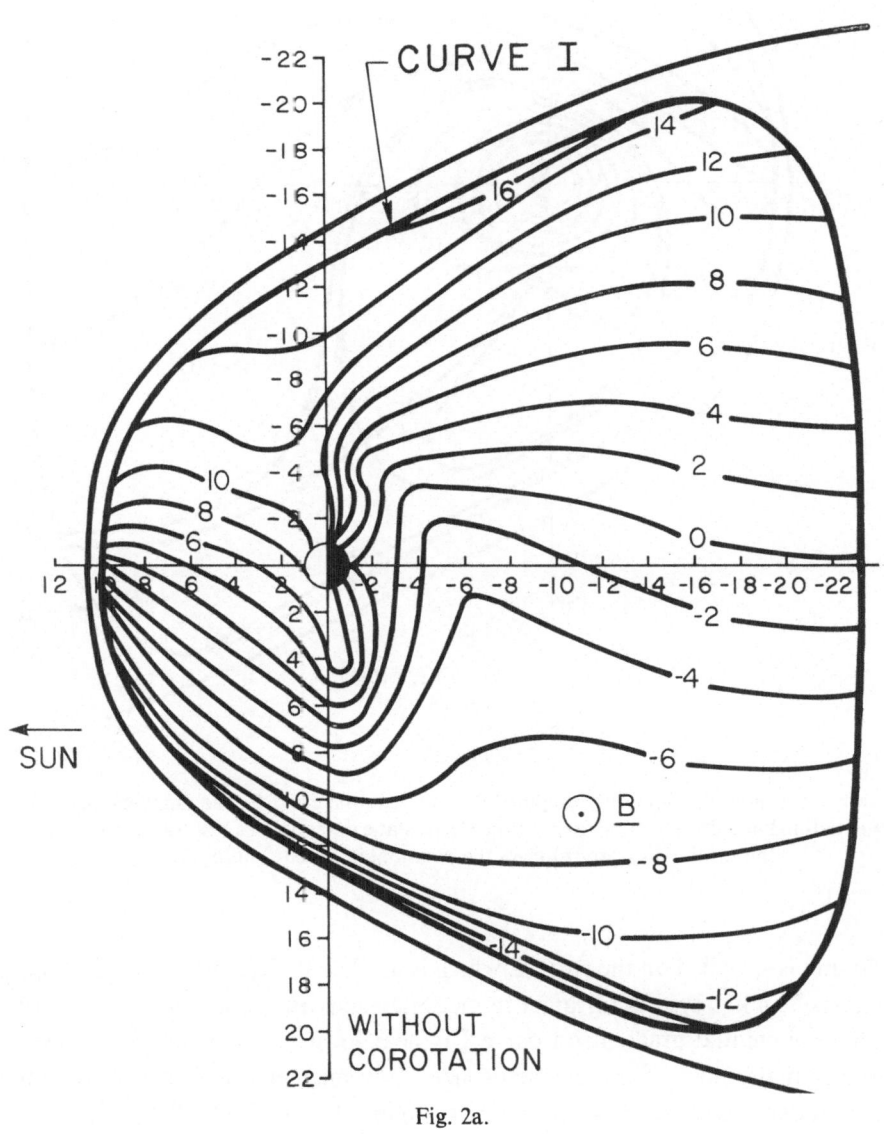

Fig. 2a.

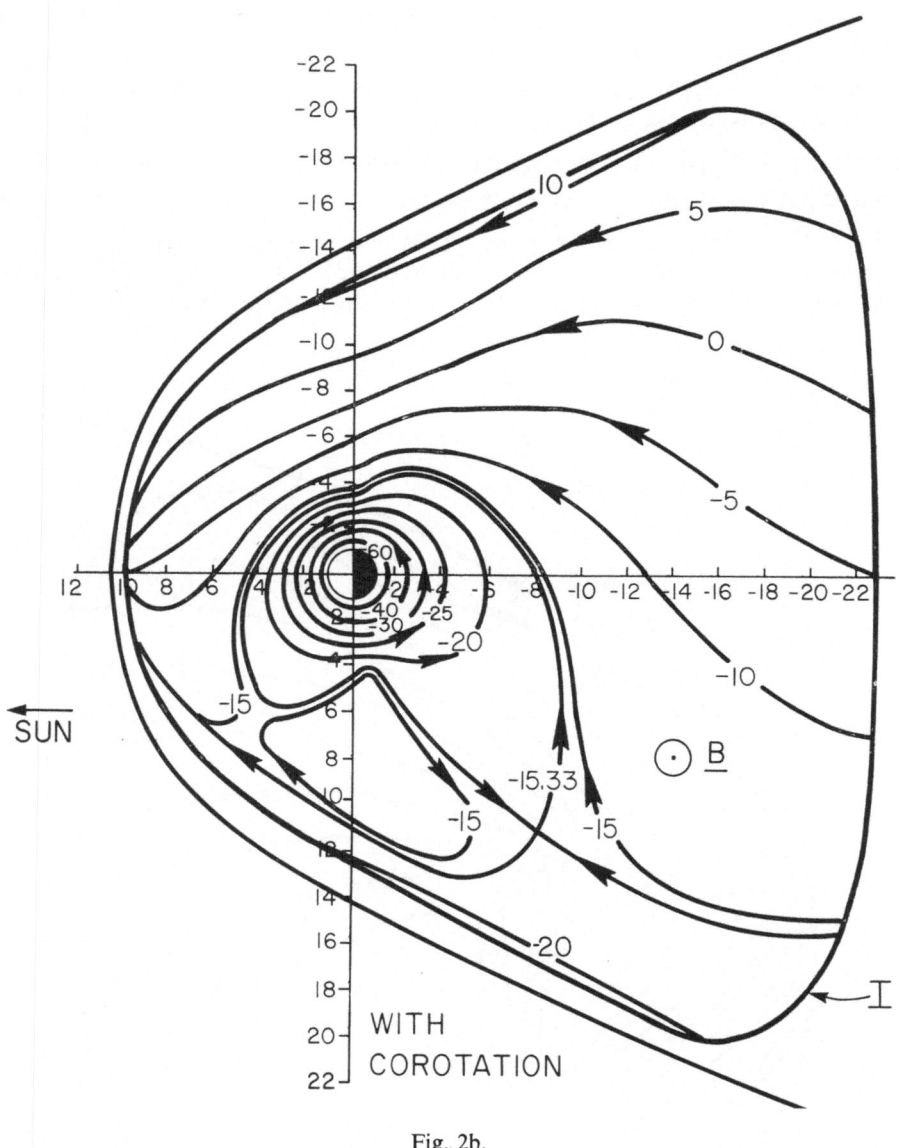

Fig. 2b.

Figs. 2a-b. Potential distributions computed for the case where $j_{\parallel} = 0$. The numbers on the curves are potentials in kilovolts. Diagram (a) gives the electrostatic potential in a rest frame that rotates with the Earth. Diagram (b) gives the potential in a non-rotating frame.

distribution is specified on the curve I, which is at 77° Lat. The total dawn-dusk potential drop is 33.4 kV, typical for quiet to moderate activity (Bohse and Aggson, 1973). The largest potential gradient on curve I is near local noon, which corresponds to merging near the nose of the magnetosphere. The magnetic field model, an untilted dipole that does not change in time, is described in detail by Wolf (1970).

Figure 2 shows a potential distribution computed for the case where $j_{\parallel} = 0$. Figure 3 shows the potential distribution computed including Birkeland current from the Alfvén layer. The configuration corresponds to 4 h after the ion sheet started in from the tail. Comparison of Figures 2a and 3 indicates that Birkeland currents from the Alfvén layer have reduced the nightside convection field, earthward of the layer, to a small fraction of its original value; dayside electric fields earthward of the layer have been reduced somewhat.

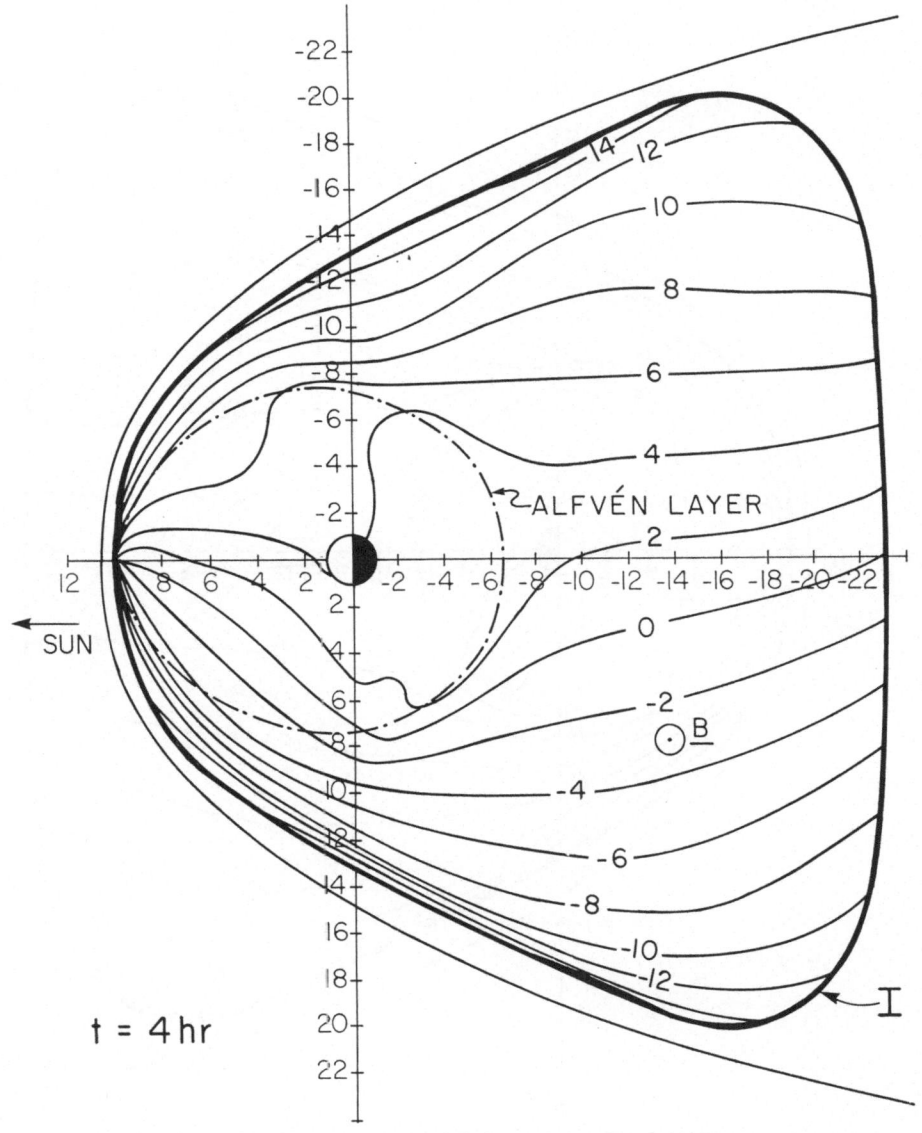

Fig. 3.   Potential distribution 4 h after the ion sheet started in from the tail, in the rest frame of the rotating Earth. The numbers on the curves are potentials in kilovolts.

Figure 4 shows the potential distribution 3 h later, 7 h after the sheet started in from the tail. Note that the inner edge penetrates to about 6.8 R_E at dusk, 6.2 R_E at midnight, not far from the normal observed position of the inner boundary of the plasma sheet ions. Note also that the convection electric field is almost zero earthward of the Alfvén layer. This general shielding effect has been pointed out by Karlson (1971), Swift (1971), Vasyliunas (1972), and others.

By the time a steady state is reached, the low L values are shielded from the high

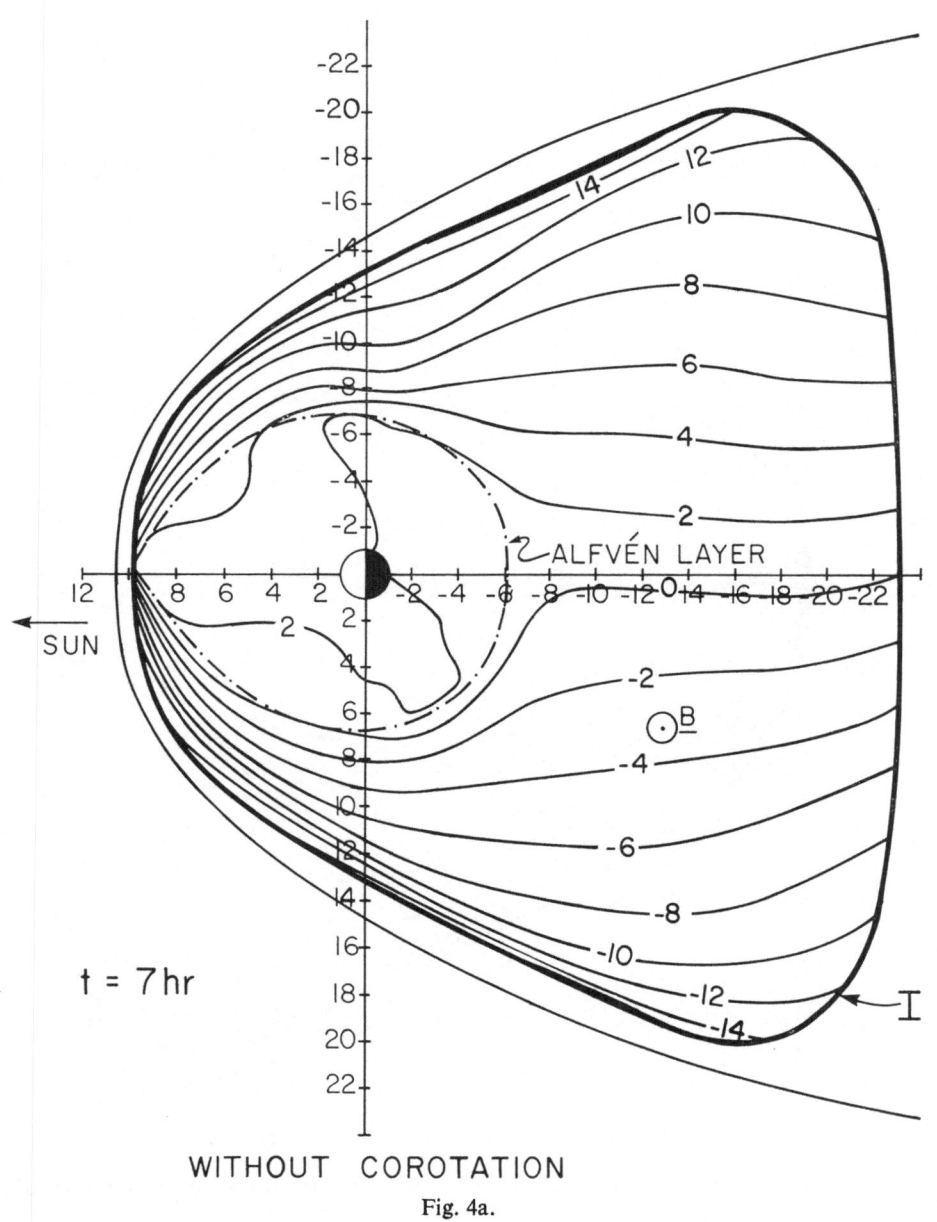

WITHOUT   COROTATION

Fig. 4a.

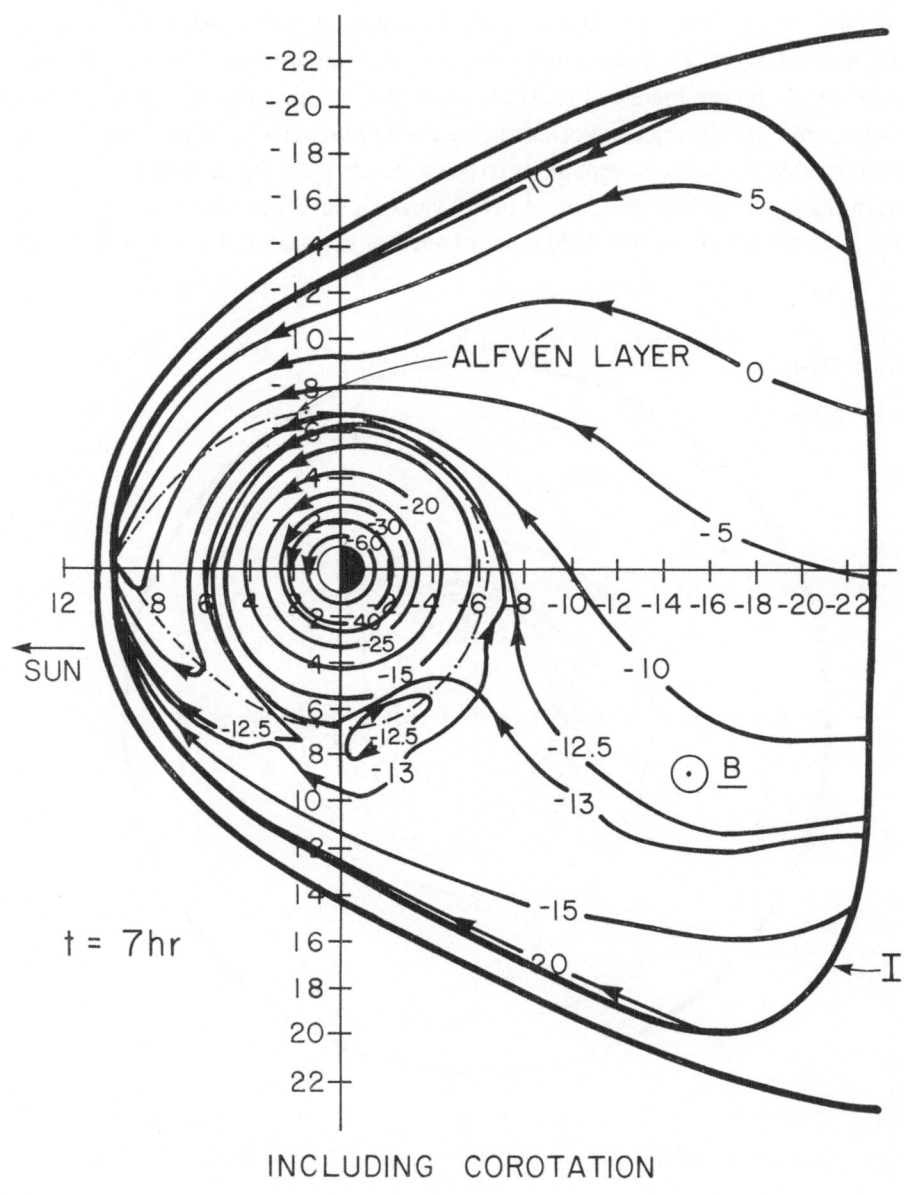

INCLUDING COROTATION

Fig. 4b.

Figs. 4a-b. Potential distribution 7 h after the ion sheet started in from the tail. Diagram (a) shows the potential in the rest frame of the rotating Earth; diagram (b) gives the potential in a non-rotating frame.

latitude electric field provided that

$$\eta e \gg \sigma_p, \tag{3}$$

where η is the number of plasma sheet or ring current particles per unit magnetic flux, e is the charge of the electron, and σ_p is the height integrated Pedersen conductivity (Vasyliunas, 1972; Wolf and Jaggi, 1973). The inequality is satisfied in the present quiet time models, which involve nighttime eastward and westward auroral electrojets of about 10^5 A. Increasing the auroral zone conductivity an order of magnitude to obtain substorm-like electrojets involves raising the conductivity to values comparable to ηe; so these shielding effects seem likely to be inoperative during sustorms.

But in the present case, shielding of low L values is rather complete as illustrated in Figure 5. There is little electric field equatorward of the dash-dot curve, which marks

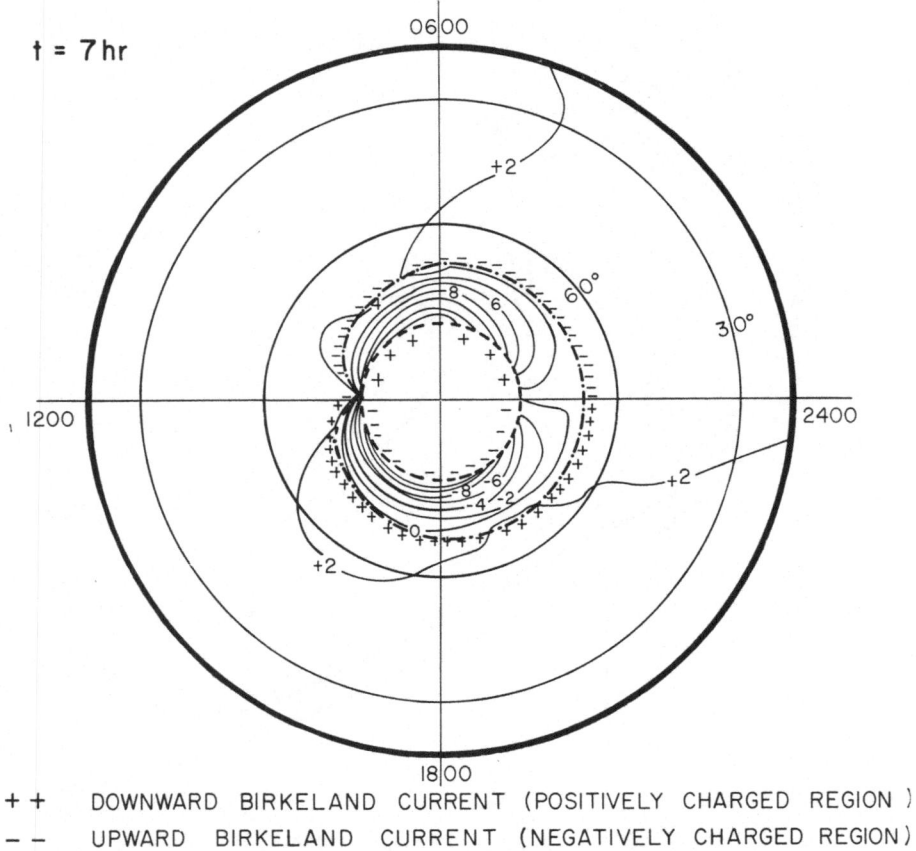

+ + +   DOWNWARD BIRKELAND CURRENT (POSITIVELY CHARGED REGION)
- - -   UPWARD   BIRKELAND   CURRENT (NEGATIVELY CHARGED REGION)

Fig. 5.   Potential distribution 7 h after the ion sheet started in from the tail, shown in the ionosphere.
The potential is in the rest frame of the rotating Earth.

the ionospheric end of the Alfvén layer. Note also that, in the auroral zone, the electric field is poleward before midnight, equatorward after. Far north of the Alfvén layer there is a substantial westward field near midnight. This disappears as you move nearer the Alfvén layer. The layer acts like a good conductor, shielding the region equatorward of it from any electric field and thus not allowing a significant east-west field in the region immediately poleward of it.

The plus (minus) signs represent the polarization charges resulting from Birkeland currents down into (up from) the ionosphere. The main Birkeland currents driving the convection and causing the dawn-dusk field in the ionosphere are indicated poleward of curve I, down into the ionosphere near dawn, up and out near dusk; they are presumably strongest near the electric field reversals seen by Gurnett and Frank (1973) and Heppner (1972). The Birkeland currents from the Alfvén layer have the opposite polarity. Their strength is that required to cancel the electric field equatorward of the layer. The general configuration of these currents has been discussed by Schield *et al.* (1969) and Vasyliunas (1970, 1972). Birkeland currents from the inner edge of the nightside Alfvén layer may be small due to small conductivity at low latitude at night. The currents are bigger on the dayside, about 10^6 A. This is required to shield the low latitudes from the high latitude electric field. The quiet time ring current has to be asymmetric to the extent of about 10^6 A, requiring significant distortion of the drift shells. Note that in Figure 4 the inner edge of the ring current touches the magnetopause at local noon. Of course, this model has only one particle species – it neglects electrons and some protons, and these other particles may do part of the shielding and contribute part of the asymmetry of the ring current.

3. Predicted Asymmetry of the Quiet Time Ring Current

We have no multispecies model calculations to present, but we can make a general, though approximate, prediction concerning the tendency for the ring current to be pulled sunward on the dayside, leaving fewer particles, and less energy density in particles, near noon than near dawn or dusk. For the case of an isotropic distribution

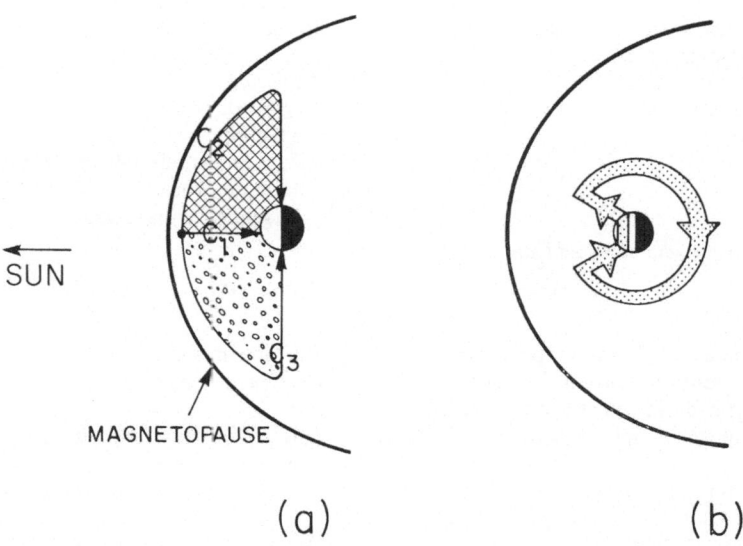

(a)   (b)

Fig. 6.   (a) Integration contours in the magnetic equatorial plane; (b) quiet time asymmetric ring current configuration.

in the equatorial plane, the approximate relation is as follows:

$$0.6 \int_{c_2} \beta_e \frac{r_e}{B_e} \, d \left(\frac{B_e^2}{2\mu_0} + \frac{p_e}{2} \right) + 0.6 \int_{c_3} \beta_e \frac{r_e}{B_e} \, d \left(\frac{B_e^2}{2\mu_0} + \frac{p_e}{2} \right) =$$

$$= 1.2 \int_{c_1} \beta_e \frac{r_e}{B_e} \, d \left(\frac{B_e^2}{2\mu_0} + \frac{p_e}{2} \right) + (2.7 \times 10^6 \ \text{A}) \left(\frac{\Phi_0}{50 \ \text{kV}} \right) \left(\frac{\sigma_p}{10 \ \text{mhos}} \right), \qquad (4)$$

where the integration paths C_1, C_2, and C_3 in the magnetic equatorial plane are shown in Figure 6; r_e is the geocentric distance in the equatorial plane; β_e is the ratio of particle pressure to magnetic pressure; p_e is the particle pressure; B_e is the magnetic field magnitude in the equatorial plane; Φ_0 is the potential across the dayside polar cap, and σ_p is the dayside ionospheric Pedersen conductivity. For each path, the range in $\frac{1}{2} B_e^2/\mu_0 + \frac{1}{2} p_e$ is the same. The relation essentially says the average β_e is greater near dawn and dusk than near noon. The physical basis of the equation is that the drift currents past dawn and dusk exceed the drift current past noon by the amount of the Birkeland currents required to shield low ionospheric latitudes from the high latitude electric field, i.e., by the amount of ring current that has to be diverted through the

TABLE I

Status of convection models

A. Theoretical status

Element	Included?
Merging or viscous interaction.	Included as boundary condition.
Ionospheric conductivity.	Included. Complicated model, though still time independent.
Currents generated by pressure gradients in magnetospheric plasma.	Included, but using only one component ion sheet.
Parallel electric fields.	Neglected.
Time variations in **B**.	Neglected.
Neutral winds.	Neglected.
Precipitation.	Effects on particle pressure gradients neglected. Effects on auroral conductivity included in large-scale average sense.

B. Consistency with observed features

Feature	Agreement
~ 50 kV potential drop across polar cap.	Built into model.
Eastward, westward auroral electrojets.	Consistent (almost any convection theory is.)
Position of nightside quiettime ring current.	Agrees.
Injection of storm time ring current to $L \approx 3.5$.	One component ionsheet calculations are in adequate agreement.
Tendency for reduced electric field inside plasmasphere.	Agrees (to the extent that observers agree).
Size of plasmasphere.	Agrees (to the extent that observers agree).
Shape of plasmasphere.	Mediocre agreement. No clean interpretation of bulge.

ionosphere. The relation is corrected for currents into and out of the magnetopause.

The author is aware of no comprehensive observational study of the quiet time ring current on the dayside, analogous to that of Frank and Owens' (1970) study of the evening sector. However, Heppner *et al.* (1967) have observed substantial inflation of the field near dawn.

Note that the predicted asymmetry between noon and dawn or dusk is large only if a substantial part of the total 50 kV drop across the polar cap exists on the sunlit part of the ionosphere. Aside from this, the most crucial (and not clearly valid) assumption involved in Equation (4) is the assumption that field lines can carry the million-amp Birkeland currents keeping potential drops along field lines much smaller than, say, 25 kV. If the field lines are unable to carry such heavy currents, the ring current must pull itself into a complete symmetric ring, the Alfvén layers must become almost circular, and the last term of Equation (4) must be deleted.

Table I summarizes the status of our convection calculations, from both theoretical and observational points of view.

Acknowledgments

The author is grateful to R. K. Jaggi for running the computer programs and to A. J. Dessler, T. W. Hill and H. B. Garrett for helpful comments. The work was supported in part by the Atmospheric Sciences Section of the National Science Foundation, under Grant GA-27206. The computations were performed on the Univac 1108 machine at NASA-L. B. Johnson Space Center.

References

Block, L. P.: 1966, *J. Geophys. Res.* **71**, 855.
Bohse, J. R. and Aggson, T. L.: 1973, *EOS, Trans. Amer. Geophys. Union* **54**, 417.
Frank, L. A. and Owens, H. D.: 1970, *J. Geophys. Res.* **75**, 1269.
Gurnett, D. A. and Frank, L. A.: 1973, *J. Geophys. Res.* **78**, 145.
Heppner, J. P.: 1972, *J. Geophys. Res.* **77**, 4877.
Heppner, J. P., Sugiura, M., Skillman, T. L., Ledley, B. G., and Campbell, M.: 1967, *J. Geophys. Res.* **72**, 5417.
Iwasaki, N. and Nishida, A.: 1967, *Rept. Ionosph. Space Res. Japan*, **21**, 17.
Jaggi, R. K. and Wolf, R. A.: 1973, *J. Geophys. Res.* **78**, 2852.
Karlson, E. T.: 1971, *Cosmic Electrodyn.* **1**, 474.
Schield, M. A., Freeman, J. W., Jr., and Dessler, A. J.: 1969, *J. Geophys. Res.* **74**, 247.
Swift, D. W.: 1971, *J. Geophys. Res.* **76**, 2276.
Vasyliunas, V. M.: 1970, in B. M. McCormac (ed.), *Particles and Fields in the Magnetosphere*, D. Reidel Publishing Company, Dordrecht-Holland, p. 60.
Vasyliunas, V. M.: 1972, in B. M. McCormac (ed.), *Earth's Magnetospheric Processes*, D. Reidel Publishing Company, Dordrecht-Holland, p. 29.
Wolf, R. A.: 1970, *J. Geophys. Res.* **75**, 4677.
Wolf, R. A. and Jaggi, R. K.: 1973, *Comm. Astrophys. Space Phys.* **5**, 99.

THE PLASMA SHEET IN THE EVENING SECTOR

J. R. BURROWS

Division of Physics, National Research Council, Ottawa K1A OR6, Canada

Abstract. Some factors are considered which contribute to asymmetry between the premidnight and postmidnight sectors of the plasma both in the equatorial plane and at low altitudes. In particular the greater probability of occurrence of extended auroral arcs in the premidnight sector is noted. The formation of these quiet arcs, their relationship to 'inverted V' structures and their effect on energetic particle trapping are considered. It is concluded that the energetic electron 'trapping boundary' is not necessarily the limit of closed field lines but instead is caused by wave turbulence in the plasma sheet at the most equatorward of the extended arcs. Various implications of this conclusion for plasma sheet structure are discussed.

1. Introduction

The existence of the plasma sheet and its time dependent configuration is primarily a consequence of the entry of solar wind plasma into the magnetosphere, its subsequent heating in non-adiabatic processes and its convection within the magnetosphere. Here emphasis will be placed on phenomena occurring in the evening side of the plasma sheet as observed at low altitudes although it is recognized that the majority of the particles and energy resides near the equatorial plane.

Although it is more usual to think of the plasma sheet as a nightside phenomenon associated with the tail, one should recognize the possibility of it extending to all local times since, on the average, magnetospheric convection occurs at all local times centred on the auroral oval. 'The plasma sheet' is specifically defined here to be that region of the magnetosphere where a hot plasma is contained by the geomagnetic field on closed field lines and where a major part of that plasma's motion is due to convection. Note that by this definition energetic electrons and protons, with motions dominated by gradient curvature drift may pass into or out of the plasma sheet at different times in their history. The definition excludes magnetosheath plasma occurring on geomagnetic field lines directly connected to the magnetosheath field in the cleft or via the high latitude magnetotail. It also excludes, perhaps somewhat arbitrarily, hot plasma near the inner edge of the plasma sheet which is no longer convecting due to changes in the convection electric field as a function of time. This inner edge transition region has particular properties discussed by Vasyliunas (1972) and Frank (1971).

The following subjects are treated below. In the second section, some qualitative differences between the evening and morning sides of the plasma sheet are noted. In the third and fourth sections, some observations near the equatorial plane and at low altitudes are briefly reviewed. Recent observations from the ISIS-2 satellite are presented in the fifth section followed by discussion and a summary.

2. Asymmetries About the Noon-Midnight Plane

The need to construct simple models and diagrams of the magnetosphere frequently

B. M. McCormac (ed.), Magnetospheric Physics, 179–197. All Rights Reserved.
Copyright © 1974 by D. Reidel Publishing Company, Dordrecht-Holland.

leads one to resort to noon-midnight cross sections of the magnetosphere or projections onto the polar cap that are essentially symmetrical about the noon midnight axis. Instead, let us consider factors contributing to a basic asymmetry about the noon-midnight plane.

(1) The co-rotation electric field imposes an asymmetry which is most effectively coupled from the ionosphere to the magnetosphere within the plasmasphere but also exists at higher latitudes with diminished magnitude and effectiveness of coupling (Fälthammar, 1972).

(2) The trans-polar electric field directed from morning to evening (Cauffman and Gurnett, 1972; Heppner, 1972; Gurnett and Frank, 1973) supports the inference of a variable two cell convection pattern as represented schematically in Figure 1. Note that one would expect an electric field component directed from dawn to dusk in both the noon and midnight sectors near the electric field reversal boundary between sunward and antisunward convection. Although plasma drift velocities derived from Ba releases should show this component, a summary of observations (Haerendel, 1972)

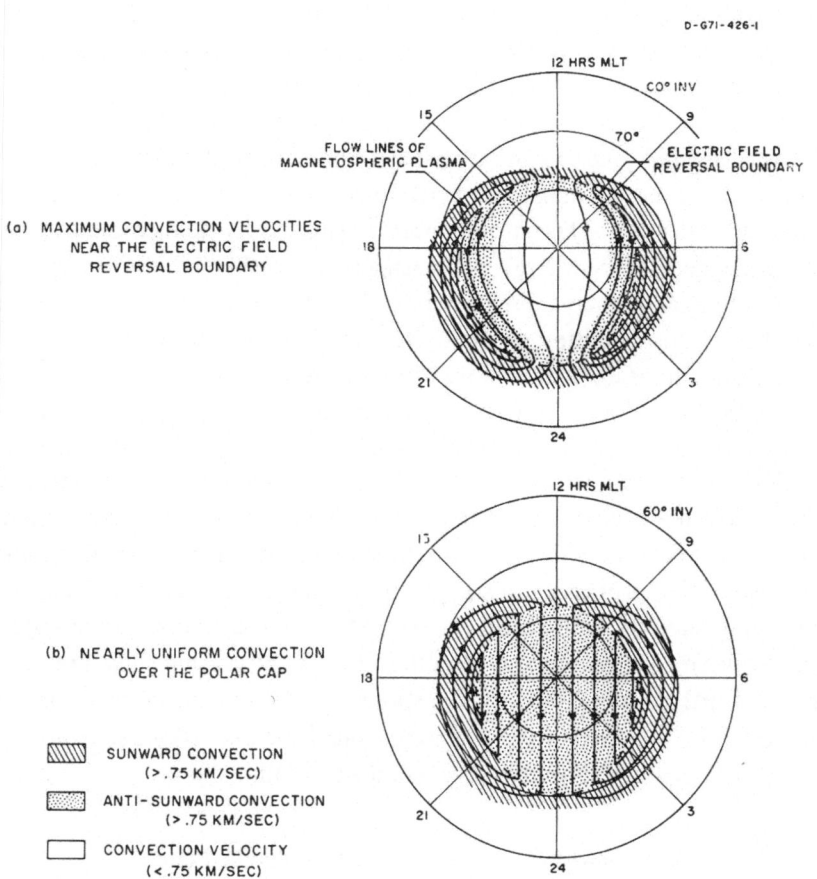

Fig. 1.   Two types of average convection patterns drawn schematically in MLT – IN Lat coordinates from Gurnett (1972).

shows few releases between 70° and 75° IN Lat where such a field would most probably occur. Mozer (1973) has reported the growth of an enhanced westward component of the electric field based on 15 min averages from 19 balloon flights taken during the hour before a negative bay onset at the nearest ground magnetometer.

(3) Gradient and curvature drift motions, eastward for electrons and westward for protons, introduce asymmetries. As a result, the sunward convection of non-thermal electrons around the evening auroral oval is opposed by their drift while in the post-midnight sector the two motions are additive. The reverse is true for protons. Then 1 to 10 keV electrons should tend to accumulate along the premidnight oval and to intrude to lower latitudes in the postmidnight sector, i.e., the premidnight Alfvén layer should be more abrupt.

(4) The plasma pressure in the plasma sheet modifies the electric field and the convection pattern (Vasyliunas, 1972) leading to convection asymmetries in the inner (near-earth) part of the plasma sheet which are self-consistent with the asymmetric storm time ring current distribution.

3. The Plasma Sheet Near the Equatorial Plane

A schematic composite of the distribution of the plasma sheet's electron component in the equatorial plane is shown in Figure 2. Note the extension of the plasma sheet far around into the dayside and the asymmetry, with electrons reaching the plasma-pause after midnight but forming a distinct inner edge separated from the plasma-pause in the evening sector.

An opposite asymmetry in the proton component, which penetrates to L shells lower than the plasma sheet electrons in the evening sector results in the asymmetric ring current during the growth phase of magnetospheric storms.

Related asymmetries are found in time-independent electric field distributions modelled to fit the dispersion traces of plasma observed at geosynchronous orbit following substorm injections (McIlwain, 1972). A concentration of westward electric field is located near midnight beyond 7 R_E. In more recent work, McIlwain (1974) infers even weaker field strength in the evening sector inside 10 R_E after the plasma injection event.

Carpenter and Akasofu (1972) have used the equatorward drift of whistler ducts to measure an enhanced westward electric field in the postmidnight sector of the outer plasmasphere ($L \sim 4$), at times close to the enhancement of auroral activity of the quiet arcs on the equatorward horizon ($\sim 66°$ IN Lat) of the ASC at Byrd and the onset of strong negative bay activity on Byrd magnetograms. All of the above observations indicate distinct asymmetries about midnight in the inner part of the plasma sheet.

Farther out in the tail at Vela orbit (18 R_E), little evidence of local time asymmetries in the plasma parameters has been noted except for the greater flux of electrons ($E > 64$ keV) on the morning side (Montgomery, 1968; Walker and Farley, 1972). Vela's magnetopause crossings appear similar at dawn and dusk in respect to the leakage of plasma sheet protons and electrons into the magnetosheath (Hones *et al.*, 1972b) and the penetration of antisunward streaming magnetosheath plasma for a few thousand kilometers inside the magnetopause (Hones *et al.*, 1972a).

Studies of energetic electrons ($E \geqslant 100$ keV) in the plasma sheet near the equatorial plane have shown some local time asymmetries in their pitch angle distributions at large radial distances in the tail (West *et al.*, 1973).

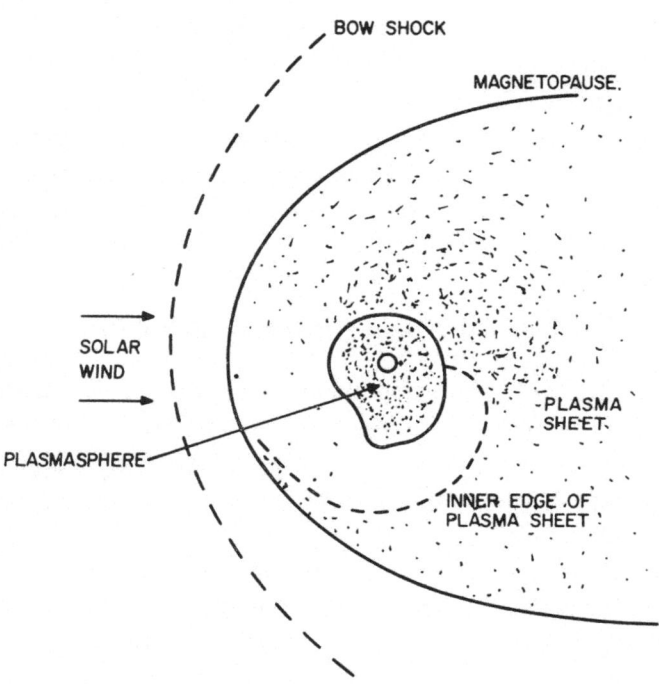

Fig. 2.   A schematic composite diagram of the electron plasma sheet distribution in the equatorial plane by Atkinson (1972) from diagrams by Vasyliunas (1968) and Frank (1971). The density of the dots roughly indicates the hot plasma particle density. Systematic spectral variations within the plasma sheet are not shown.

Isotropic distributions occur more frequently after midnight than before due to pitch angle scattering of the electrons in the equatorial plane as the geomagnetic field becomes more taillike near midnight during the growth phase of substorms. Before midnight, the anisotropic butterfly distributions occur out to 17 R_E and imply at least limited trapping and drift consistent with the predominantly northward Z component of the geomagnetic field observed near the neutral sheet out to 70 R_E (Schindler and Ness, 1972; Behannon, 1970). Erratic proton flows at Vela (Hones *et al.*, 1972a) and a larger variance of the magnetic field vector near the neutral sheet at 60 R_E relative to that in the high latitude magnetotail (Meng and Mihalov, 1972) are evidence for turbulence in the distant plasma sheet.

4. Low Altitude and Ground-Based Observations

The precipitating and low altitude mirroring components of the plasma sheet are measured by polar-orbiting satellites and sounding rockets. Balloons, magnetometers, all-sky cameras and ground-based photometers observe the ionospheric and atmospheric response to the precipitating component, although it does not necessarily follow that all of the auroral oval should be identified with the plasma sheet.

4.1. THE ELECTROJET CURRENT SYSTEMS

These systems in the ionosphere are a more indirect and complex indication of the plasma sheet since modification of the ionospheric conductivity can cause both Hall and Pedersen currents to flow which are not simply related to the convection electric field in the plasma sheet (Rostoker, 1972). However the early model by Sugiura and Heppner (1965) of an eastward electrojet in the evening sector overlapped by a westward electrojet at higher latitudes along the Harang discontinuity appears to be confirmed frequently in the latitude profiles of magnetograms studied by Kisabeth and Rostoker (1971). The slow growth of the westward electrojet prior to the substorm breakup phase appears to result from a changing convection pattern near the outer edge of the plasma sheet. The rapid increase in the strength of the westward electrojet in the evening sector after the substorm breakup appears related to the westward travelling surge and the westward shift of the Harang discontinuity into the evening sector. It should be noted that the westward traveling surge propagates along the quiet arc(s) near the poleward edge of the plasma sheet while the eastward electrojet seems more nearly identified with the inner edge of the plasma sheet in the premidnight sector. Kamide and Fukushima (1971) have suggested that the eastward electrojet is the ionospheric closure of the equatorial asymmetric proton ring current.

The Harang discontinuity in electrojet currents caused by the split in the two cell convective flow is typically asymmetric and displaced to about 2300 h. Heppner's (1972) schematic diagram of the plasma flow, based partly on Ba release experiments, indicates convective continuity near the Harang discontinuity with a more abrupt flow reversal on the evening side.

4.2. PARTICLE PRECIPITATION PATTERNS

Particle precipitation patterns at various local times and the predominance of intense ~ keV electron precipitation events in the premidnight sector were reported by Frank and Ackerson (1972). The observation of 'inverted V' type energy-time profiles was correlated with an auroral arc observed from the ground (Ackerson and Frank, 1972). Gurnett and Frank (1973) have reported the close coincidence of 'inverted V' events with the reversal of the large scale convection electric field near dusk. Closer to midnight the electric field reversal becomes less distinct with large fluctuations near the 'inverted V' structures.

Frank and Gurnett (1971), Ackerson and Frank (1972) and Gurnett (1972) have reported that the 'inverted V' structures occur at or poleward of the 'trapping bound-

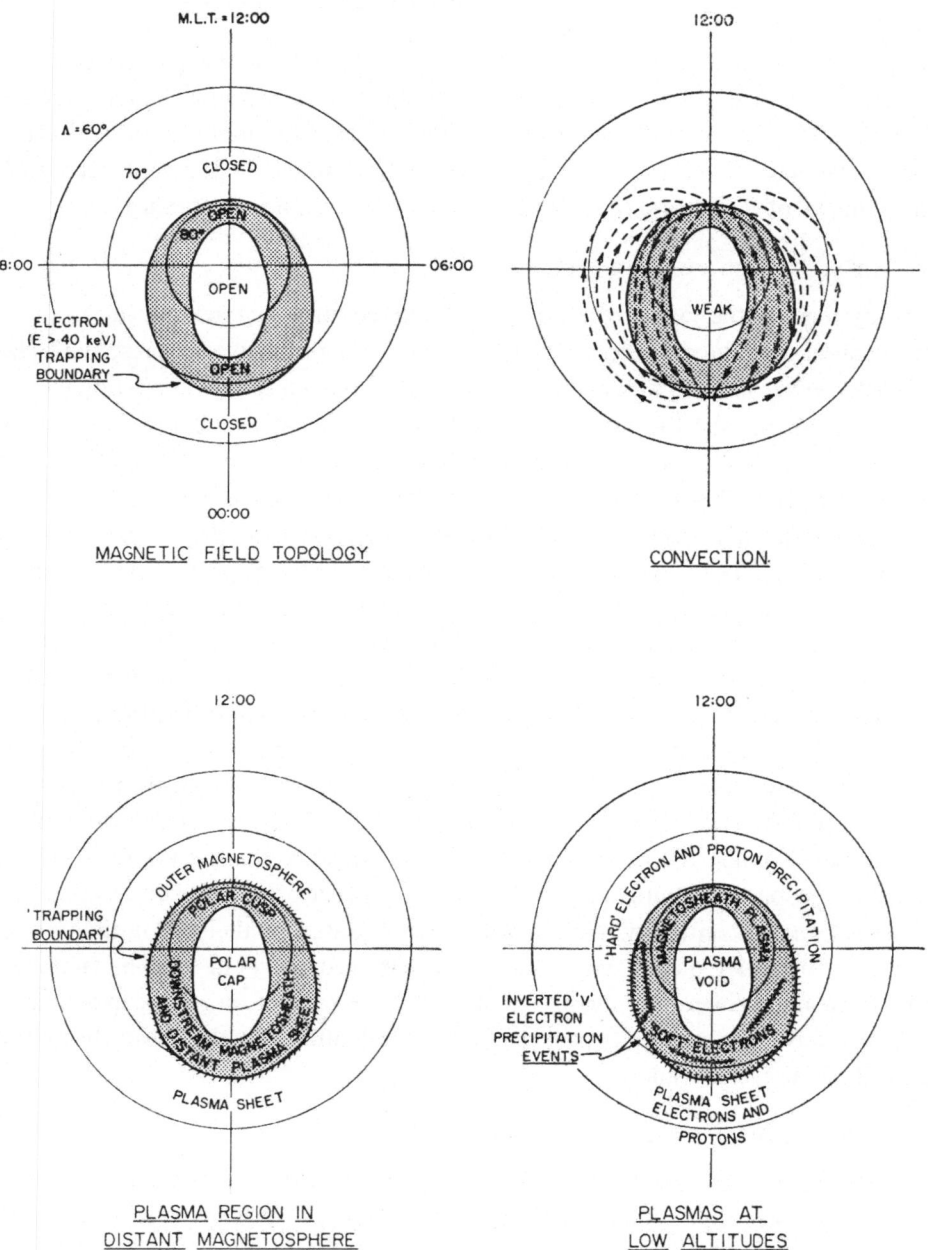

Fig. 3.   Schematic diagrams of magnetospheric features in MLT – IN Lat coordinates which interpret the low altitude signatures of the magnetic field topology, the convection and the plasma region
in the distant magnetosphere and their relationship to plasma observations at low altitudes, from
Frank and Ackerson (1972).

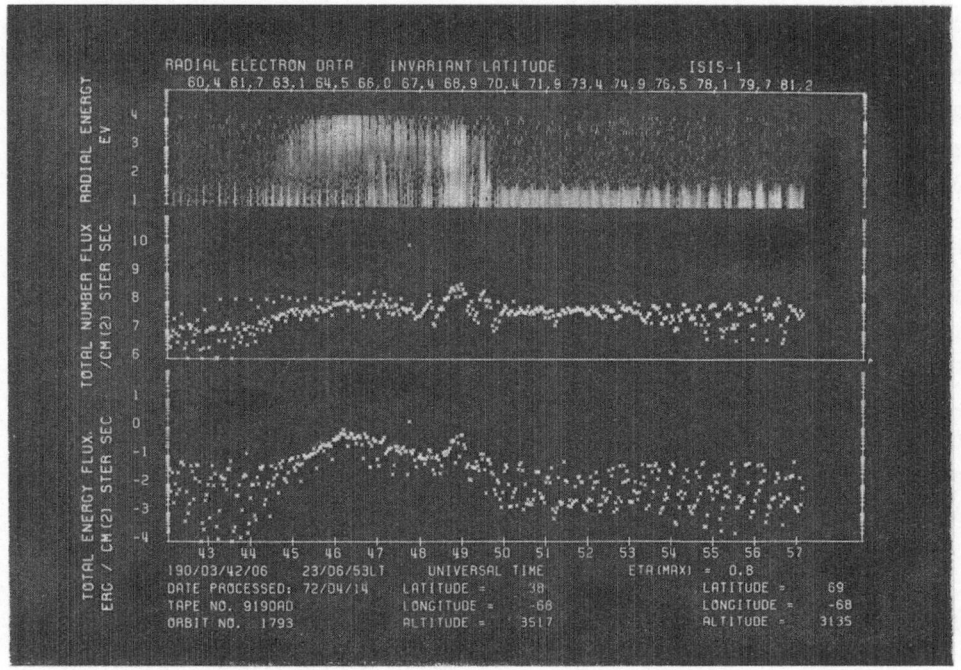

Fig. 4a.

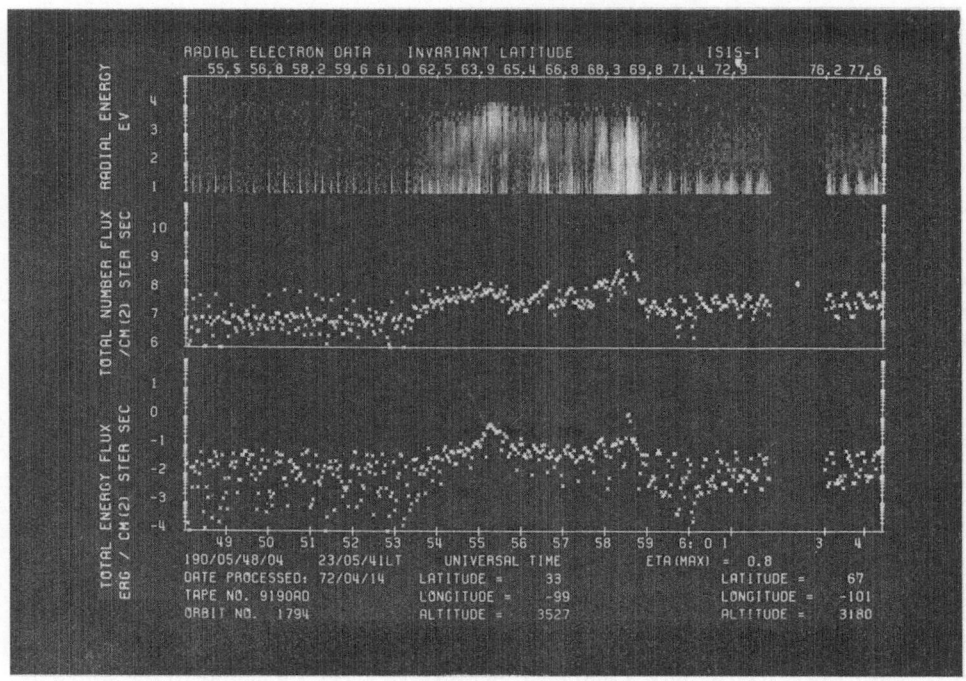

Fig. 4b.

Fig. 4. Observations of the electron component of the plasma sheet by the ISIS-1 soft particle spectrometer on July 9, 1969 at 03:42 and 05:48 UT at an altitude of 3400 km (courtesy of W. J. Heikkila). The corresponding EPD data have been published (Burrows and McDiarmid, 1972, Figures 10, 11 and 12).

ary'. The latter, defined as the latitude at which the 45 keV electron flux mirroring
at the Injun satellite's altitude drops below instrumental threshold, appears to be
essentially the same as the '35 keV background boundary' studied statistically by
McDiarmid and Burrows (1968). All of these authors identified it with the boundary of
closed geomagnetic field. Figure 3 shows this relationship schematically, placing the
'inverted V's' in the open field region. However, there appears to be mounting evidence which contradicts this interpretation. Intense field aligned precipitation ($\sim 10^{11}$
electrons $cm^{-2} s^{-1} sr^{-1} keV^{-1}$ at 500 eV) was observed during a rocket flight through
the northernmost arc of a series of extended arcs in the premidnight (21:30 h LT)
sector while small fluxes of 25 keV electrons, occurring coincident with and poleward
of the arc, were observed to have pitch angle distributions peaked perpendicular to the
magnetic field (Whalen and McDiarmid, 1972). This distribution implies a closed field
region since it can be most plausibly explained as arising from absorption in the
opposite hemisphere's atmosphere.

Evidence from ISIS-1 now suggests that the abrupt intensity decrease called the
'trapping boundary' is caused by turbulence in the plasma sheet. Examples from Burrows and McDiarmid (1972) show that the higher energy electrons (~ 200 keV) were
precipitated by plasma turbulence near the inner edge of the plasma sheet although
lower energy electrons (~ 20 keV) in the plasma sheet were observed poleward of this
with trapped pitch angle distributions. For these ISIS-1 passes, the soft particle
spectrometer simultaneously made electron plasma sheet observations as shown in
Figure 4. The association of the evolving '200 keV trapping boundary' at 65.3° and
64.5° IN Lat with the peak in the electron total energy flux is evident. The exponentially decreasing fluxes of the inner edge region extend from the peak in to $\sim 62°$ IN
Lat while more active acceleration occurs near the poleward edge of the plasma sheet
at $\sim 69°$ IN Lat.

From the data in Section 5 it will be seen that, in a similar manner, turbulence in the
auroral arcs, associated with their typically lower average electron energy, causes precipitation of ~ 45 keV flux, thereby creating the 'trapping boundary' feature at a
lower latitude than the limit of closed field lines.

5. ISIS-2 Observations in the Evening Sector

These new data from the ISIS-2 Energetic Particle Detector (EPD) spectrometers and
the Auroral Scanning Photometer (ASP) provide simultaneous observations of the
auroral oval's large scale morphology and the particle fluxes observed along the
satellite orbit while crossing the auroral oval in the evening sector. The spin stabilized
satellite is in a 1400 km circular polar orbit with a spin period of ~ 20 s. Both the
ASP and EPD detectors have collimated fields of view perpendicular to the spin axis.
The ASP instrumentation and data format are described by Anger et al. (1974) and
Lui and Anger (1973). The EPD instrumentation is described by Venkatarangan and
Burrows (1974). Here, selected channels from its eight channel differential electron
spectrometer (0.15 to 10 keV) are used along with several of the integral threshold

electron detectors (E> 22, >40, >210 keV) to study bright quiet arcs observed during three nights in December 1971.

5.1. DECEMBER 11, 1971

The ASP pictures of the auroral oval are shown in Figure 5. A system of bright arcs extends from ~19 h MLT to midnight, terminating in a bright patchy region. The

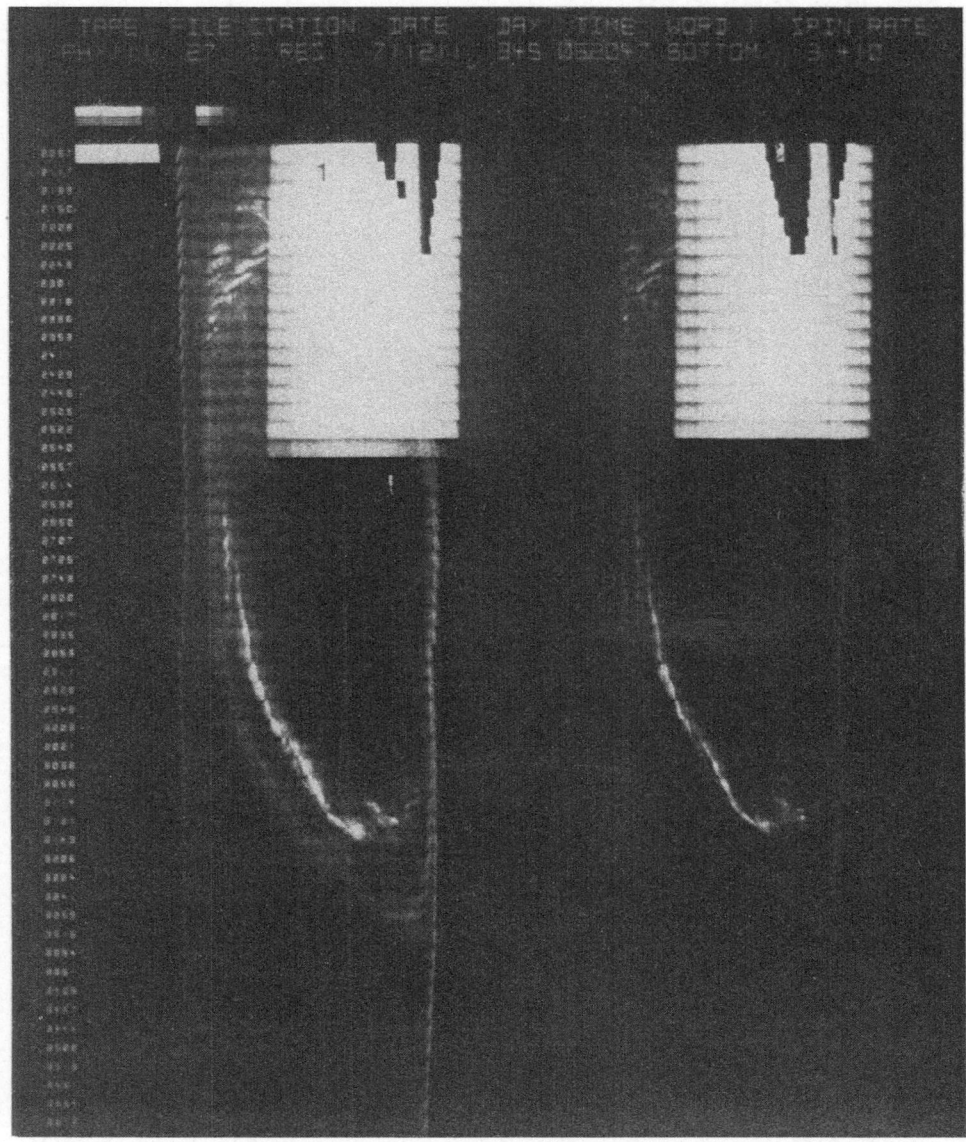

Fig. 5. ASP maps of the auroral oval on Dec. 11, 1971 at 05:20–05:35 UT in two wavelengths (5577 Å at left and 3914 Å at right). Scattered sunlight in the upper right corner obscures the morning part of the dayside oval. The satellite orbit passes down the middle of the two maps intersecting the bright evening arcs at 22.5 h MLT. Much of the morning side of the oval is beyond the horizon.

system's three principal arcs overlap each other with those at later local time occurring to the poleward side. Other arcs are visible in the afternoon sector (upper left) around 14 to 17 h MLT. A continuous diffuse zone extends from the dayside to beyond midnight (Lui *et al.*, 1973). It is most prominent at low latitudes but extends poleward to envelop the bright arc region. The ISIS-2 orbit is along the 93°W meridian on the nightside.

Examination of magnetograms from Canadian stations confirms that it was very quiet until 05:55 UT. The H component fluctuations were less than 20 γ from 00 to 06 h UT at Great Whale River (GWR) (66.8° Geomag. Lat), Ft. Churchill (CH) (68.8°), Baker Lake (BL) (73.9°) Ft. Smith (SM) (68.1°) and Ft. Chipewyan (FTCH) (67.0°) except for a narrow +40 γ excursion at BL at 04:00 UT.

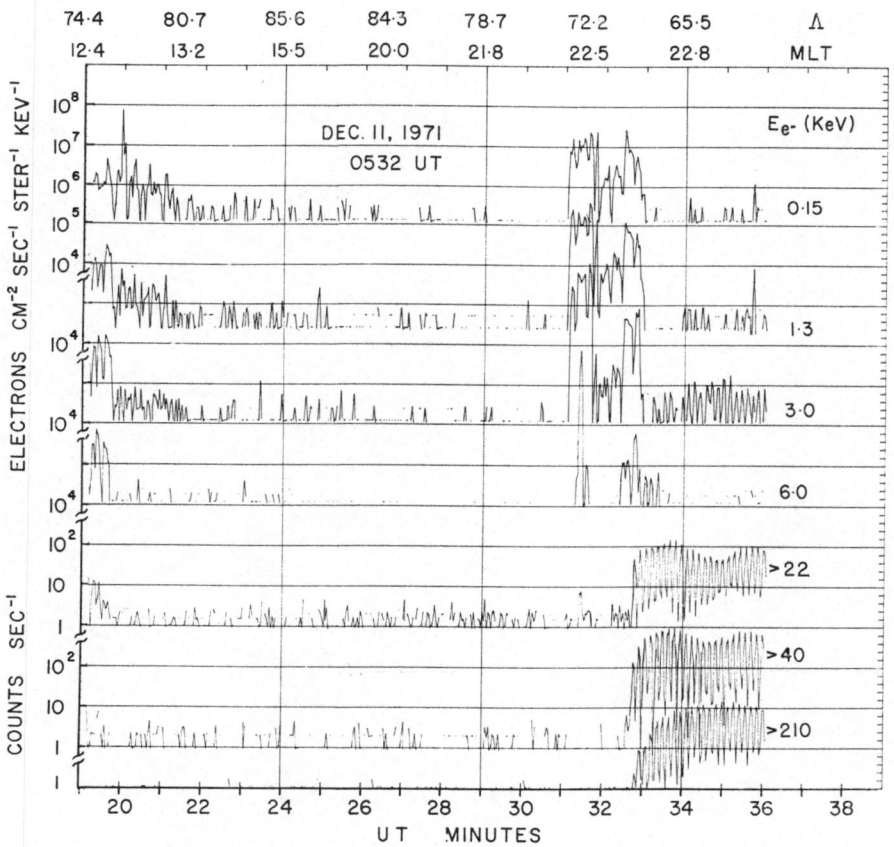

Fig. 6.   The EPD data corresponding to Figure 5 for 7 electron channels, plotted linearly against UT. The energy resolution of the differential channels is $\Delta E/E = 30\%$. Invariant latitude and MLT are marked at the top at 2.5 min intervals. The data are averaged with 2 s resolution. Base lines for the logarithmic intensity scales are displaced as necessary for different channels to avoid averlap. Spin modulation at one-half the spin period due to the pitch angle anisotropy of the flux is most evident after 33 min UT.

The EPD data from four differential and three integral electron channels are shown in Figure 6 as a function of UT. The plasma sheet is observed at 31 of 33 min UT (22.5 h MLT, 73.5° to 68.2° IN Lat). The evident continuity of flux at $E \lesssim 1.3$ keV corresponds to the diffuse zone in Figure 5 with larger fluxes at ~ 3 keV near the low and high latitude edges. The core of 6 keV flux, which corresponds to the principal bright arc, has the E–t characteristics of an 'inverted V'. The 'trapping boundary' o f 40 keV electrons is coincident with the high flux region at the inner edge of the plasma sheet. However, there are fluxes in the plasma sheet poleward of the 'trapping boundary' with pitch angle distributions peaked perpendicular to the field. This is demonstrated in Figure 7 for the bright arc region, where the peak average energy of the 'inverted V' occurs at 31:24 UT. Large flux depressions occur when the atmospheric loss cone is viewed (pitch angles $\sim 170°$). Flux maxima occur at $\sim 90°$ with a minimum

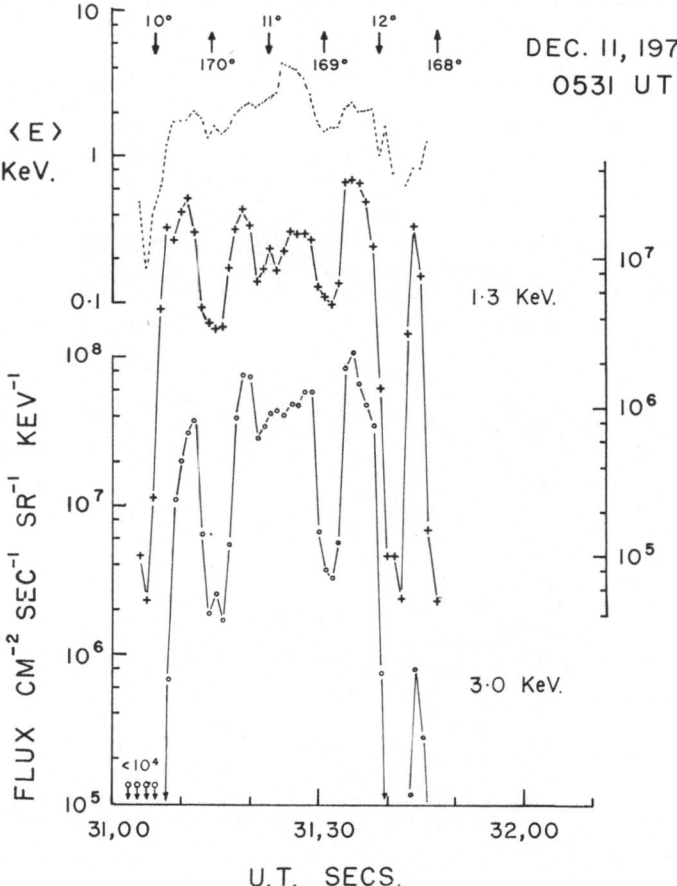

Fig. 7. The 'inverted V' region expanded for 05:31:00 to 0.5:31:45 UT with 1 s resolution. Two differential electron channels and the average energy $\langle E \rangle$ constructed from all eight channels (0.15 to 10 keV) are displayed. Arrows at the top mark the maximum and minimum pitch angles reached during each spin. Pitch angles 0° to 90° correspond to down-coming fluxes.

at 11°. Spatial flux variations obscure pitch angle effects near the edges of the region.

5.2. DECEMBER 12, 1971

The ASP picture for the pass at 04:17 UT is published (Lui and Anger, 1973, Figures 3a and 4a). It shows two bright arcs embedded in a diffuse zone. The shorter arc lies at ∼68°. The diffuse auroral belt extends to earlier and later local times with a well defined equatorward boundary at 65.5° IN Lat. The ISIS-2 orbit follows the 75°W meridian.

Magnetograms indicate that the satellite crossed the arcs just before an expansive phase in the premidnight sector. There is an onset of substorm activity at GWR at 04:26 UT which reaches CH at 05:02 UT. At GWR, $\Delta H < 20\,\gamma$ for the two hours before the pass.

The EPD data are shown in Figure 8. A region of soft structured bursts ($E \gtrsim 1$ KeV) occurs at 77° to 69° IN Lat. The plasma sheet has a sharp poleward edge at 68° and its soft inner edge extends to ∼65.5°. The sharp drop in the >20 and >40 keV flux at

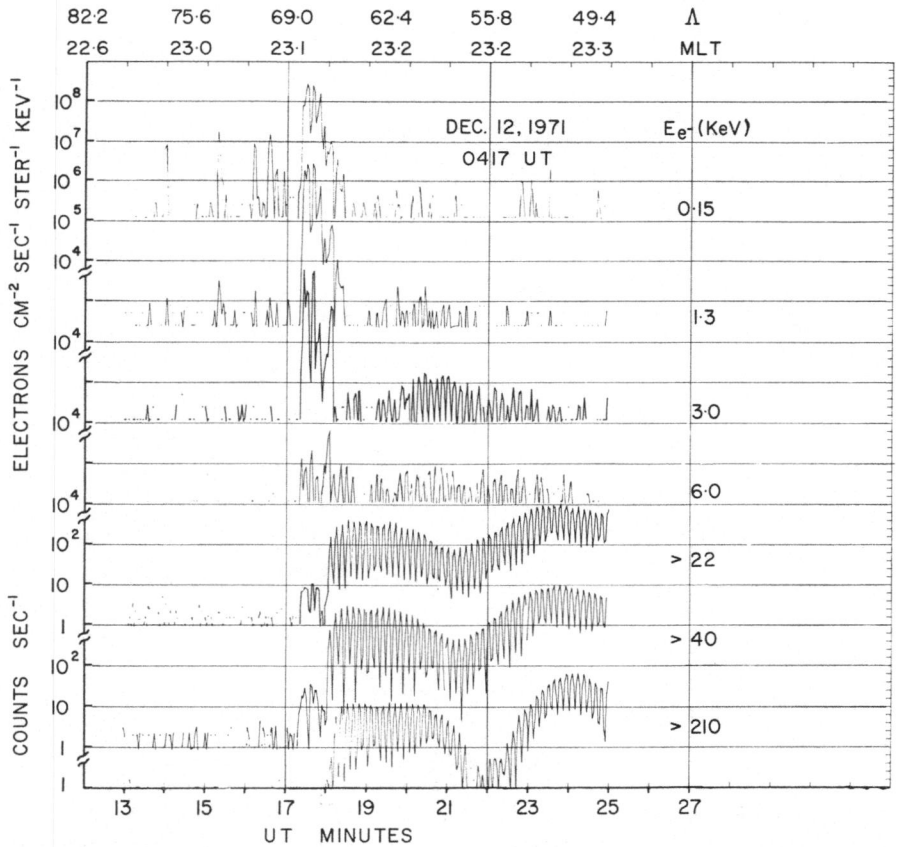

Fig. 8.   The EPD data on December 12, 1971 at 04:13 to 04:25 UT. The format is similar to Figure 6.

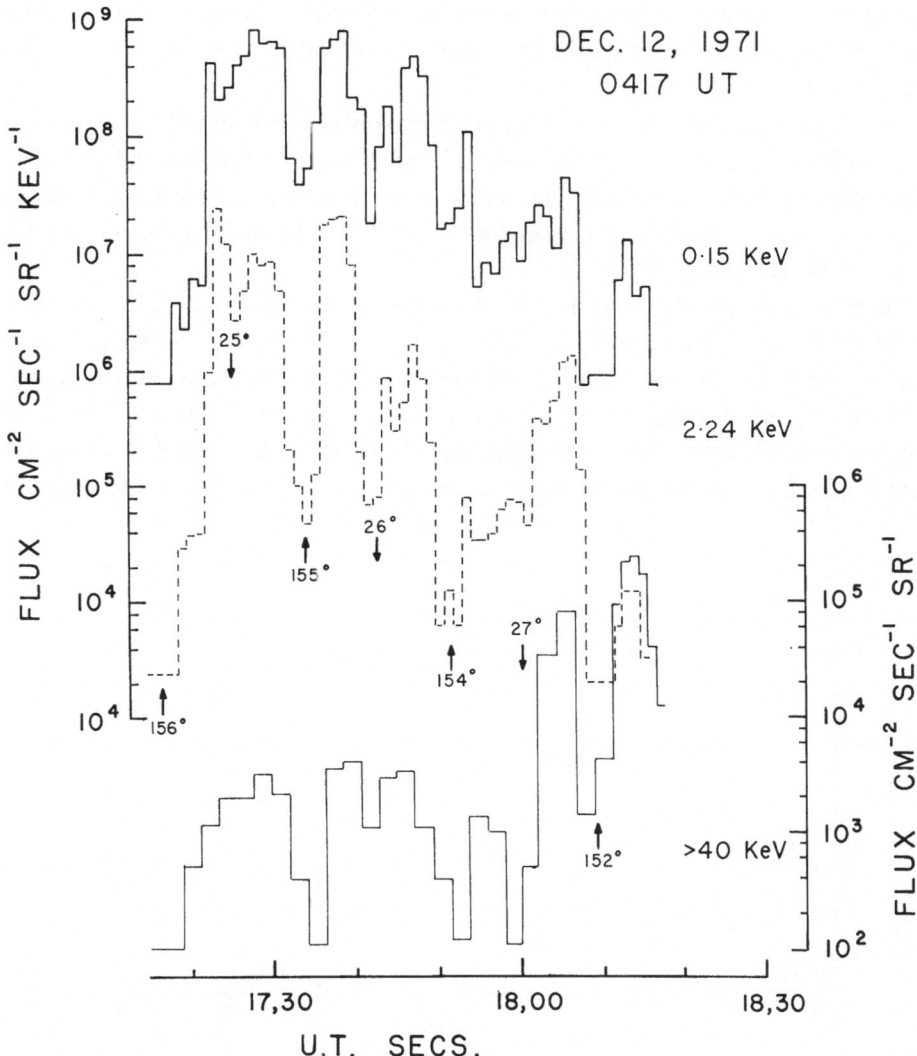

Fig. 9.   The 'inverted V' region at 04:17 UT expanded with 1 s resolution on the differential channels and 2 s resolution on the integral channel. The pitch angle range is somewhat more limited than
that for Figure 7.

66.2° just overlaps the sharp increase in 3.0 keV flux at 66.0°. However, since the
energetic flux extends across the whole plasma sheet, the 'trapping boundary' is ~68°.
The plasma sheet region of intense flux is expanded in Figure 9. Flux minima can be
seen in the down-coming loss cones in spite of the disturbed conditions and the limited
pitch angle range scanned.

5.3. DECEMBER 14, 1971

The ASP picture for the ISIS-2 pass at 07:24 UT has been published and discussed
(Anger *et al.*, 1973). It shows a bright arc extended along the auroral oval in the

evening sector from 17 h MLT with a westward travelling surge propagating from midnight which had not yet reached the satellite orbit plane on the 125° W meridian at 21.5 h MLT.

The magnetogram at CH shows a substorm expansive phase onset ($\Delta H = -90\ \gamma$) an hour earlier at 06:14 UT with the station located near the northern border of the westward electrojet. The northern border jumps northward as the electrojet widens at 06:55 UT and it appears to be centered at about 72° IN Lat during the period 07:10 to 07:30 UT (Rostoker, 1973).

The EPD data are shown in Figure 10. A bursty electron region (E < 400 eV) at 76° to 70° IN Lat is also seen by the SPS experiment. The sharply delineated plasma sheet region (70.3° to 66.8° IN Lat) contains a single 'inverted V' structure. Its peak at 6.0 keV occurs at a slightly lower latitude than the 'trapping boundary' for 22 keV electrons which is marked with a vertical line. The parts marked C and D are expanded in Figure 11, with the arrows marking down-coming and up-going fluxes with the

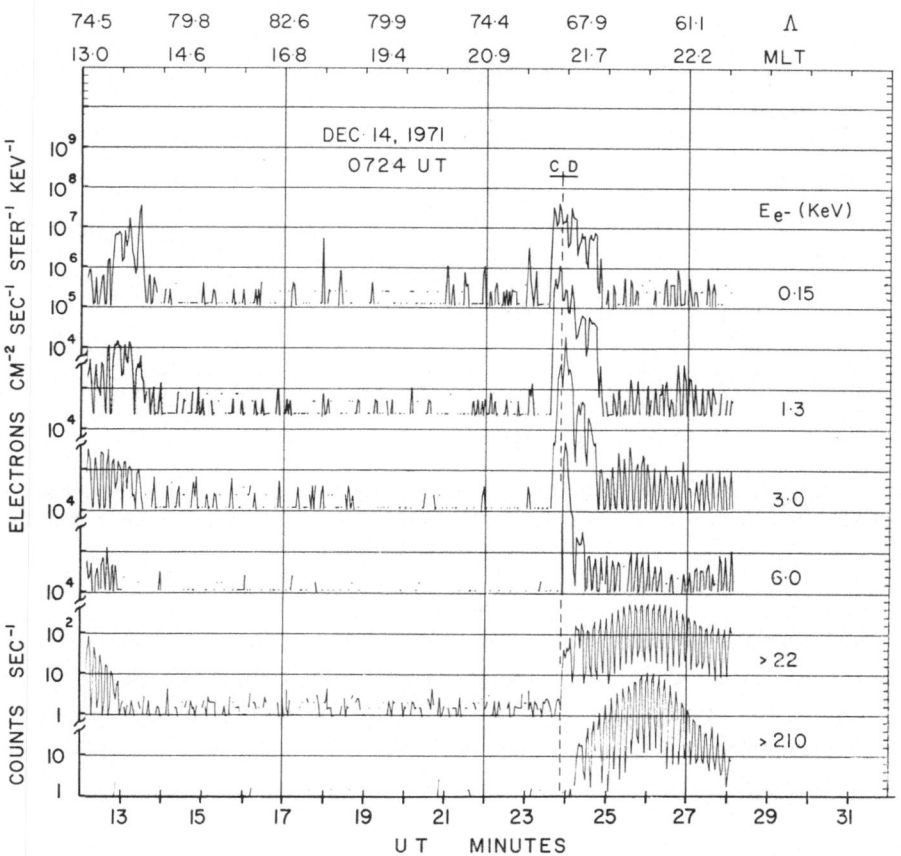

Fig. 10.   The EPD data on Dec. 14, 1971 at 07:12–07:28 UT.
The format is similar to Figure 6.

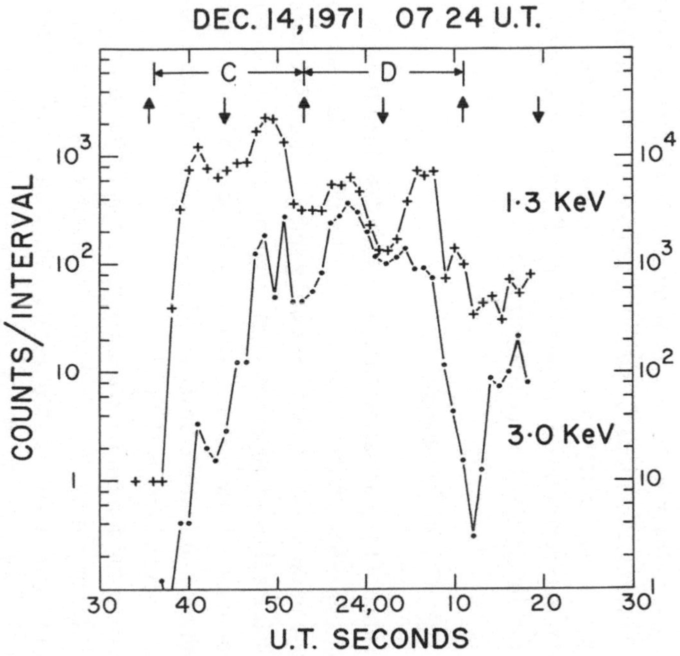

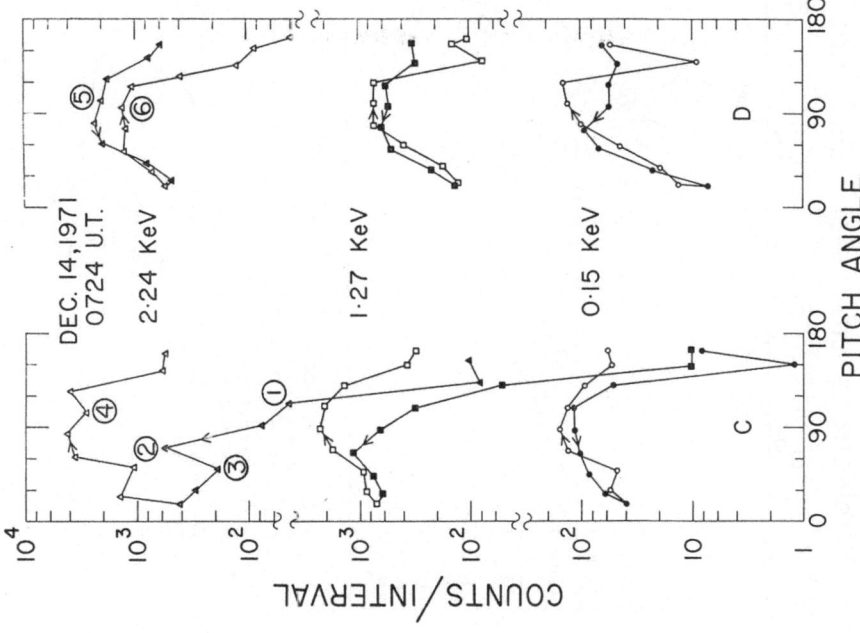

Fig. 11. An expansion of the 'inverted V' region from Figure 10 near 07:24 UT in a format similar to Figure 7. Regions C and D correspond to two adjacent spin periods.

Fig. 12. The pitch angle distribution in regions C and D from Figure 11 for three differential energy channels. Arrows indicate the direction of scan. The points have one second resolution.

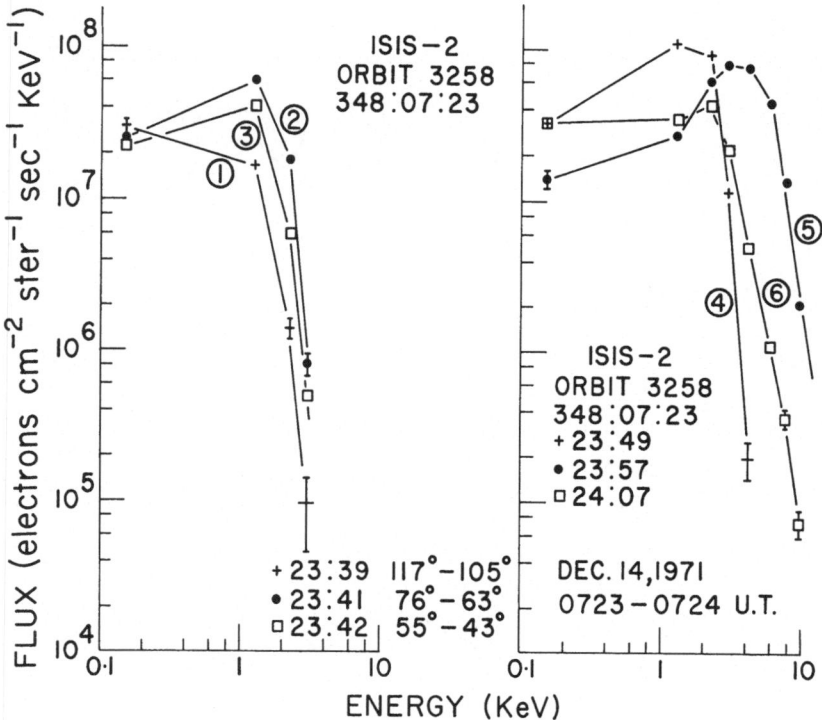

Fig. 13. Electron spectra in the 'inverted V' auroral arc region for the six occasions marked on Figure 12.

minima occurring most clearly in the more broadly distributed 1.3 keV flux. In Figure 12, the count rates from the two spin periods C and D are plotted vs. pitch angle for three energy channels. The minima are apparent in the up and down loss cones. Six spectra are sampled at the numbered points on Figure 12 and displayed in Figure 13. It is clear that the peak of the 'inverted V' occurs in spectrum 5.

6. Discussion

The three evening side crossings of the plasma sheet at low altitude described above were selected by several criteria but other similar passes exist in the ISIS-2 data set. The three are all from quiet days in the period prior to or during small isolated substorms when simultaneous ASP pictures showed the existence of bright auroral arcs extending from early afternoon to midnight. It was also required that the spin axis orientation give good pitch angle coverage and that the 'inverted V' be sufficiently broad to give a complete pitch angle scan in several energy channels.

One notes that all three cases have in common a narrow, sharply delineated plasma sheet which is a characteristic configuration in all ISIS-2 data from early afternoon to midnight on quiet days prior to substorm expansions. In all three cases, residual pitch angle modulation characteristic of trapping in a closed field is found poleward of the steep drop in 40 keV flux. It is not surprising that the modulation is somewhat un-clear in view of the spatial structure and the strong acceleration and pitch angle diffusion processes occurring in the 'inverted V' region.

The spatial relationship between the 40 keV 'trapping boundary' and the 'inverted V' feature is variable with the first case beyond, the second inside and the third at or inside the 'trapping boundary'. The sharp decrease in 40 keV flux is consistently correlated to the most equatorward part of the plasma sheet's isotropic high flux region. It is therefore suggested that wave turbulence within the trapped plasma occurs all along the extended arcs, and that the gradient drifting trapped flux (E ~ 40 keV) encounters this region at a point in the early evening and is precipitated effectively to form a 'trapping boundary' which steepens as it approaches midnight. The typical equatorward motion of these arcs prior to substorm breakup will sometimes produce an evolving 'trapping boundary'. For example, on December 14, the 22 keV flux is approaching isotropy for 68° to 69.5° IN Lat resulting in an eroding boundary; on December 12, 40 keV flux has been and is being substantially eroded for 66° to 68.5° IN Lat; on December 11 the lower latitude turbulent region at 69° is maintaining a stationary erosion boundary.

The plasma sheet configuration following an isolated substorm expansion and following multiple substorm activity has not been treated here. However, the thicker, hotter plasma sheet at those times results in a poleward expansion of the closed field region and frequently a poleward motion of the 'trapping boundary' which suggests that the wave turbulence is weaker or extends less continuously than it did prior to breakup (McDiarmid and Hruska, 1972).

7. Summary

(1) Based on these studies it appears that the 'trapping boundary' should be treated as a lower limit to the limit of closed field lines particularly in the evening sector. In this region it should perhaps more properly be called the 'turbulence boundary'.

(2) It appears plausible to identify arc formation with convection discontinuities and the extended intense evening arc in particular with a convection reversal (Gurnett and Frank, 1973). The tendency of such arcs to terminate near midnight is then con-sistent with a two cell convection pattern. However, it does not necessarily follow that the plasma convecting antisunward is on open field lines particularly in the premid-night hours (20 to 24 h MLT).

(3) One might infer from the overlapping arcs seen on Dec. 11 that the antisunward convecting plasma intrudes into the sunward convecting region in a filamentary way which may be related to the multiple electric field reversals seen in the evening sector (Heppner, 1972).

(4) The electrons accelerated in processes along the arcs probably map to the equa-torial plane well beyond the synchronous orbit and may not be detected at ATS-5. The lower energy electrons will convect sunward in the afternoon sector to form the tongue shown in Figure 2. The higher energy electrons which gradient drift

toward midnight will contribute to the buildup of hotter plasma inflating that region.

(5) The bright arcs in the evening are seldom if ever connected to those in the after-noon except by a fainter diffuse portion of the auroral oval. The relevance of this observation to the plasma convection patterns in the polar cap and plasma sheet needs further study. In the afternoon hours the relative importance of local plasma acceleration vs. convection from later local times is at present undetermined.

(6) Although auroral arcs are seen at all local times in and near the oval, there is a dawn-dusk asymmetry evident in the intensity, average energy and extension along the oval of their causative particle fluxes. They provide evidence of systematic asymmetries in the outer plasma sheet which are fundamental to our understanding of magneto-spheric convection processes.

(7) The relationship of the interplanetary magnetic field direction to the convection over the polar cap and its impact on plasma sheet configuration is an important aspect of magnetospheric convection which has been studied by other authors (Hoffman and Burch, 1973; Heppner, 1972) but has not been developed here.

References

Ackerson, K. L. and Frank, L. A.: 1972, *J. Geophys. Res.* **77**, 1128.

Anger, C. D., Lui, A. T. Y., and Akasofu, S. I.: 1973, *J. Geophys. Res.* **78**, 3020.

Anger, C. D., Sawchuk, W., and Shepherd, G. G.: 1974, this volume, p. 357.

Atkinson, G.: 1972, in Kr. Folkestad, (ed.), *Magnetosphere-Ionosphere Interactions*, Universitets-forlaget, Oslo, p. 203.

Behannon, K. W.: 1970, *J. Geophys. Res.* **75**, 743.

Burrows, J. R. and McDiarmid, I. B.: 1972, Paper, Critical Problems of Magnetospheric Physics, Joint COSPAR, IAGA and URSI Symposium, p. 83.

Carpenter, D. L. and Akasofu, S. I.: 1972, *J. Geophys. Res.* **77**, 6854.

Cauffman, D. P. and Gurnett, D. A.: 1972, *Space Sci. Rev.* **13**, 369.

Fälthammar, C. G.: 1972, in B. M. McCormac (ed.), *Earths Magnetospheric Processes*, D. Reidel Publishing Company, Dordrecht-Holland, p. 21.

Frank, L. A.: 1971, *J. Geophys. Res.* **76**, 2265.

Frank, L. A. and Gurnett, D. A.: 1971, *J. Geophys. Res.* **76**, 6829.

Frank, L. A. and Ackerson, K. L.: 1972, *J. Geophys. Res.* **77**, 4116.

Gurnett, D. A.: 1972, Paper, Critical Problems of Magnetospheric Physics, Joint COSPAR/IAGA/ URSI Symposium, p. 123.

Gurnett, D. A. and Frank, L. A.: 1973, *J. Geophys. Res.* **78**, 145.

Haerendel, G.: 1972, in B. M. McCormac (ed.), *Earth's Magnetospheric Processes*, D. Reidel Pub-lishing Company, Dordrecht-Holland, p. 247.

Heppner, J. P.: 1972, Paper, Critical Problems of Magnetospheric Physics, Joint COSPAR/IAGA/ URSI Symposium, p. 107.

Hoffman, R. A. and Burch, J. L.: 1973, *J. Geophys. Res.* **78**, 2867.

Hones, E. W., Ashbridge, J. R., Bame, S. J., Montgomery, M. D., and Akasofu, S.-I.: 1972a, *J. Geophys. Res.* **77**, 5503.

Hones, E. W., Akasofu, S.-I., Bame, S. J., and Singer, S.: 1972b, *J. Geophys. Res.* **77**, 6688.

Kamide, Y. and Fukushima, N.: 1971, *Rep. Ionos. Res. Space Res. Jap.* **25**, 125.

Kisabeth, J. L. and Rostoker, G.: 1971, *J. Geophys. Res.* **76**, 6815.

Lui, A. T. Y. and Anger, C. D.: 1973, *Planetary Space Sci.* **21**, 799.

Lui, A. T. Y., Perreault, P., Akasofu, S.-I., and Anger, C. D.: 1973, *Planetary Space Sci.* **21**, 857

McDiarmid, I. B. and Burrows, J. R.: 1968, *Can. J. Phys.* **46**, 49.

McDiarmid, I. B. and Hruska, A.: 1972, *J. Geophys. Res.* **77**, 3377.

McIlwain, C. E.: 1972, in B. M. McCormac (ed.), *Earth's Magnetospheric Processes*, D. Reidel Publishing Company, Dordrecht-Holland, p. 272.

McIlwain, C. E.: 1974, this volume, p. 143.

Meng, C. I. and Mihalov, J. D.: 1972, *J. Geophys. Res.* **77**, 1739.

Montgomery, M. D.: 1968, *J. Geophys. Res.* **73**, 871.

Mozer, F. S.: 1973, *J. Geophys. Res.* **78**, 1719.

Rostoker, G.: 1972, *Rev. Geoph. Space Phys.* **19**, 157.

Rostoker, G.: 1973, private communication based on the University of Alberta Magnetometer Chain.

Schindler, K. and Ness, N. F.: 1972, *J. Geophys. Res.* **77**, 91.

Sugiura, M. and Heppner, J. P.: 1965, in W. N. Hess (ed.), *Introduction to Space Science*, Gordon and Breach, New York, p. 45.

Vasyliunas, V. M.: 1968, *J. Geophys. Res.* **73**, 7519.

Vasyliunas, V. M.: 1972, in B. M. McCormac (ed.), *Earth's Magnetospheric Processes*, D. Reidel Publishing Company, Dordrecht-Holland, p. 29.

Venkatarangan, P. and Burrows, J. R.: 1974, *J. Geophys. Res.* submitted.

Walker, R. J. and Farley, T. A.: 1972, *J. Geophys. Res.* **77**, 4650.

West, H. I., Buck, R. M., and Walton, J. R.: 1973, *J. Geophys. Res.* **78**, 1064.

Whalen, B. A. and McDiarmid, I. B.: 1962, *J. Geophys. Res.* **77**, 191.

MOTHER-DAUGHTER ROCKET STUDIES OF LOW ENERGY ELECTRONS NEAR AURORAL ARCS

H. MOESTUE and B. N. MAEHLUM

Norwegian Defence Research Establishment P.O. Box 25, Kjeller, Norway

1. Introduction

Two phenomena in auroral physics have been studied with increasing interest in recent years:

(a) field aligned electron and proton fluxes and their occurrence relative to auroral forms, and

(b) spatial and temporal variations in the fluxes of auroral particles.

Observations of low energy electrons from satellites and rockets have not yet made it possible to determine in detail how the pitch angle and energy distribution of these particles vary through an auroral display.

The assumption that observed particle fluxes are of either temporal or spatial nature has formed the basis for several discussions of the origin of auroral particles and their acceleration (Bryant *et al.*, 1967; Hoffman and Evans, 1968), but rather few observations in which spatial flux changes can be distinguished from temporal ones have been reported in the literature (O'Brien and Reasoner, 1971; Johnstone, 1973; Courtier *et al.*, 1974).

Low energy electron fluxes associated with auroral arcs were measured with a high time resolution and a good pitch angle coverage by two mother-daughter sounding rockets, launched from Andenes in Norway. The experiment was designed to separate spatial structure in the particle precipitation from temporal fluctuations. However, since only three fixed energy channels were used, the observations do not enable us to determine the detailed energy spectrum of the auroral electrons.

In this paper we present observations from the two mother-daughter rockets and discuss how the pitch angle and energy distributions vary across auroral arcs. Preliminary results from this study have already been published by Maehlum and Moestue (1973). Some results from the analysis of spatial and temporal electron flux variations are also given in Section 3 of this paper.

2. Pitch Angle and Energy Distributions

The two rockets (Electra 1 and Electra 2) were launched in a northward direction, transverse to *L* shells from Andenes (IN Lat 66.4°), Norway. Relevant experimental details are listed in Table I. The elevation angle at launch was as low as 69° and 75° for Electra 1 and Electra 2, respectively. The rockets therefore covered a large horizontal distance, the impact ranges being about 400 km.

Both rockets traversed regions of auroral particle precipitation. Electra 1 passed

B. M. McCormac (ed.), Magnetospheric Physics, 199–205. All Rights Reserved.
Copyright © 1974 by D. Reidel Publishing Company, Dordrecht-Holland.

TABLE I

Rocket performance and experimental details

	Electra 1	Electra 2
Launch date and time	Jan 13, '72	Feb. 6, '73
	2104 LT	2057 LT
	21.9 EDT	21.9 EDT
Launch direction		
Azimuth	340°	340°
Elevation	68°	75°
Apogee altitude	192 km	258 km
Impact range	440 km	390 km
Separation		
Time	$T + 85$ s	$T + 85$ s
Altitude	115 km	135 km
Speed	5.1 m s^{-1}	10.6 m s^{-1}
Speed across L		
– shells	3.6 m s^{-1}	8.7 m s^{-1}
Auroral arc(s)	$T + 159$–180 s	$T + 175$–185 s
Crossed during		
flight time	$T + 190$–200 s	
intervals		
Geomagnetic disturbance		
$\Delta X, \Delta Y, \Delta Z$	$\lesssim 10\,\gamma$	$\lesssim 100\,\gamma$
Electron	0.5 keV	0.5 keV
Energy channels,	1.0 keV	1.0 keV
Mother and Daughter	3.0 keV	3.0 keV

over two stable arcs, and Electra 2 flew over one single arc which was moving somewhat equatorward during the flight. During both rocket launchings the auroral arcs, as observed by the ground based and rocket borne photometers, were located near the high latitude boundary of the 40 keV electrons which were monitored by Geiger counters. We therefore believe that the arcs observed were of the 'inverted V' type introduced by Frank and Ackerson (1971).

The electron fluxes observed by the Electra 1 mother payload are shown in Figure 1, together with the positions of the auroral arcs derived from rocket borne photometers (Måseide, 1972), and Figure 2 presents the observations from both the mother and the daughter payloads of the Electra 2 rocket. From the daughter observations only data for pitch angles less than 90° have been used, and a smooth curve has been drawn. The electron fluxes observed by Electra 1 and Electra 2 can be considered as latitude profiles, since the visual auroral forms were fairly stable during both rocket flights. These figures are – to our knowledge – the first complete cross-section through an auroral display ever reported in the literature with high temporal resolution.

The Electra 1 observations demonstrate that equatorward of the aurora all energy channels show field aligned fluxes. The 1 keV electron flux is 10 times more intense

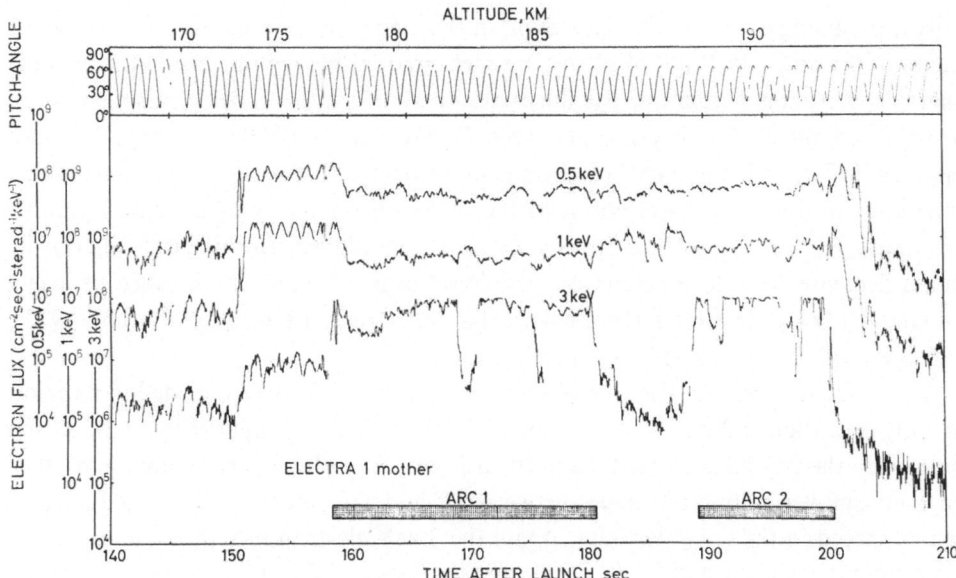

Fig. 1.   Electron fluxes measured by the Electra 1 mother payload.

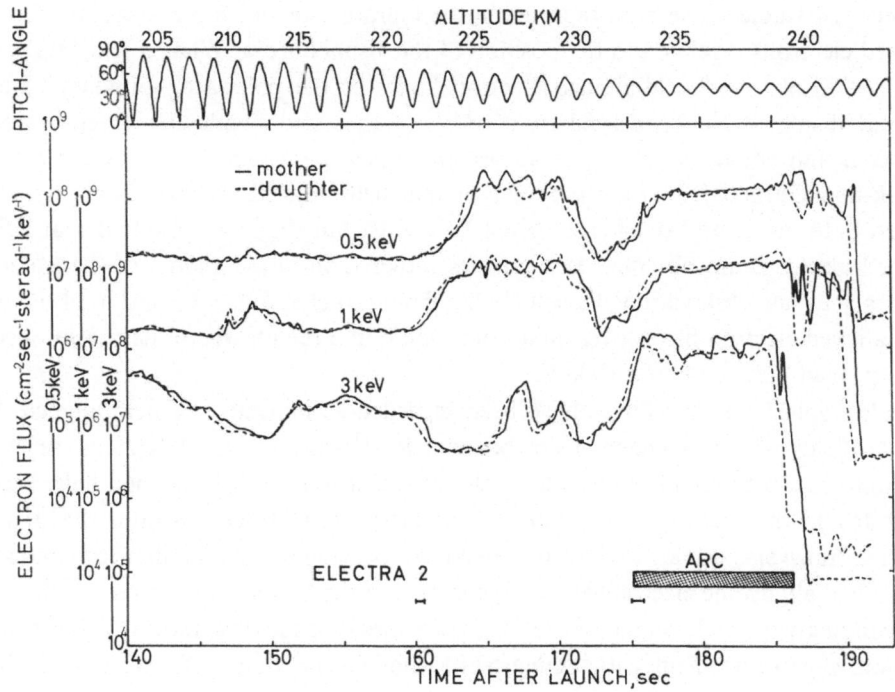

Fig. 2.   Electron fluxes measured by the Electra 2 mother and daughter payloads. Only observations for pitch angles less than 90° have been included in the daughter data and a smooth curve has been drawn. The horizontal bars indicate the expected time difference between mother and daughter observations for a spatial structure.

parallel than normal to the geomagnetic field lines. (The small dips in the curve at the lowest pitch angles are due to saturation in the 1 keV electron counter.) The highest fluxes of 0.5 keV and 1 keV electrons are measured in this region. Inside the arcs the energy spectrum is harder and the fluxes are isotropic (in the upper hemisphere). The 3 keV flux associated with 'Arc 2', between $T+190$ s and $T+200$ s is even higher than appears in Figure 1, the 3 keV electron counter being close to saturation during this time interval. In a narrow region near the poleward boundary of the northernmost arc the 0.5 keV and 1 keV electrons are again field aligned. In this latter region the fluxes decrease by 2 to 3 orders of magnitude over a horizontal distance of a few kilometers. Over the aurora there is an enhanced F region ionization (Tröim, 1973), which we will refer to as the 'Auroral F layer.'

A similar gross pattern with a region of hard, isotropic fluxes embedded in a region of soft, field aligned fluxes was observed also in the Electra 2 flight (Figure 2). At the time when the payloads crossed the poleward boundary of the auroral arc the mother payload sampled only pitch angles between 37° and 53°, but the field aligned pattern can be traced in the daughter data. Again the 3 keV electron flux shows more variability than the 0.5 keV and 1 keV fluxes, and the 'Auroral F layer' is present above the arc (Tröim, 1973). The only way in which the Electra 2 observations possibly are different from the Electra 1 data, is that the southward, soft, field aligned precipitation region is somewhat detached from the harder region inside the arc.

Very few studies have been reported in the literature in which the location of field aligned electrons is related to the position of the visual aurora. Whalen and McDiarmid (1972) observed field-aligned electron fluxes at the poleward boundary of an auroral display which consisted of a series of arcs and bands. However, inside, between, and equatorward of the aurora the fluxes were reported to be isotropic. Frank and Gurnett (1971) reported earlier that field aligned electron fluxes are observed both inside and outside 'inverted V' events, but they too now find that field aligned electrons are observed at the edges rather than in the middle of the events. Contrary to our observations Pazich (1972), from a rocket flight in Alaska, observed field aligned electron fluxes over an auroral band, in a region where the energy spectrum peaked between 8 and 10 keV.

Field aligned electron fluxes have been regarded as indicative of field aligned dc electric fields in the magnetosphere below 1 R_E (Evans et al., 1972), and several mechanisms which could render the pitch angle distribution field aligned have been suggested in the literature. A study by Lennartson (1972) demonstrates that if an external transverse dc electric field is applied to a moving region of enhanced conductivity, field aligned dc electric fields are generated at the boundaries between high and low conductivity. Field aligned fluxes at boundaries of a moving auroral F layer are qualitatively in agreement with Lennartson's model. Our data and also a study by Lundin (1973) show that field aligned electrons are found only in regions where the density of thermal electrons and the precipitating electron flux display sharp gradients in the north-south direction.

Another possible mechanism is the one associated with anomalous resistance caused by ion cyclotron waves (Kindel and Kennel, 1971), whereby parallel electric fields in the topside ionosphere could be sustained for a sufficiently long time to accelerate auroral particles.

Furthermore dc electric fields would arise in conjunction with potential double layers (Block, 1967).

It is difficult to see, however, why the two latter mechanisms should render the electron fluxes field aligned near the boundaries of the region of enhanced particle precipitation rather than near the peak in the fluxes. The key question is how the anisotropic distribution is generated. Are the field aligned distributions generated by the gradients in the fluxes, or do the gradients in the plasma density of the ionospheric *F* layer play an important role? Furthermore, since the auroral particles probably have received their energy from a source other than the one which anisotropize the boundary electrons, one would like to ask whether the boundary electrons originally belonged to the auroral particle population or whether they are 'background' electrons accelerated by a low altitude process.

3. Spatial and Temporal Variations

By comparing the time variations of the electron fluxes observed by the mother and the daughter payloads, we find that most of the variations in the fluxes measured by the Electra 1 payloads are consistent with a stationary precipitation region. The reduction in the fluxes at $T+200$ to 203 s (Figure 1) is identified as a spatial structure, and the fluxes decrease by 2 to 3 orders of magnitude over a horizontal distance of 2 to 5 km. Also purely temporal changes can be seen in these observations, e.g., the drop in the 3 keV electron flux at $T+170$ s and the enhancement at $T+188$ to 190 s, where the flux increases by a factor of 20 in less than a second. However, neither in the Electra 1 data nor in the Electra 2 data do we find such 'spikes', i..e, short-lived temporal flux enhancements, as reported by O'Brien and Reasoner (1971).

During the flight time of Electra 2 the auroral arc was moving slowly equatorward. Assuming that spatial auroral structures are extended in the east-west direction, we find that most of the time differences, as measured by the mother and the daughter payload (Figure 2), between characteristic features in the particle fluxes are consistent with a moving precipitation region, the speed being that of the auroral arc as observed from the ground. However, a cross-correlation analysis between mother and daughter data has revealed a couple of interesting points. Cross-correlation analysis has been performed for the time intervals $T+163$ to 173 s and $T+173$ to 193 s for the 0.5 keV electrons and $T+172$ to 190 s and $T+176$ to 183 s for the 3 keV electrons. The results are shown in Figure 3. The difference between the two curves in the left part of Figure 3 is due to the effect of the strong gradients in the 3 keV electron flux at $T+175$ s and $T+184$ s. The maximum occurs at very nearly the same time difference between mother and daughter observations, about 0.6 s. From this we conclude that the fine structure in the 3 keV electron fluxes moves across the geomagnetic field with the same speed as the bulk of precipitation. Clearly the time difference between flux changes measured

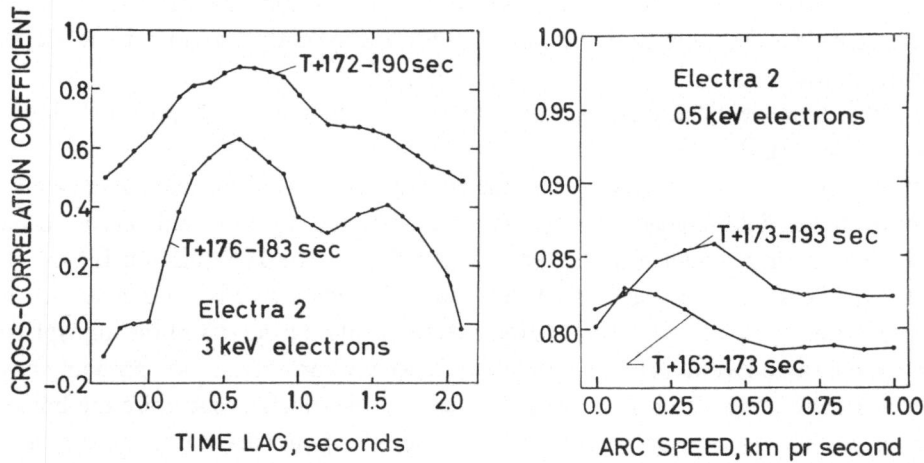

Fig. 3.   Cross correlation of mother and daughter electron flux data from the Electra 2 rocket flight.

by the mother and the daughter payload is dependent on the north-south velocity of the aurora. If we correlate the mother and daughter data in such a way that different speed of the precipitation region is considered, an interesting feature appears in the 0.5 keV electron data. For the time interval $T+163$ to 173 s the cross correlation coefficient maximizes at an auroral arc speed (perpendicular to L shells) of 100 to 200 m s^{-1}, whereas data for the time interval $T+173$ to 193 s give the highest correlation for a velocity of 400 m s^{-1}. In fact this can be readily seen from Figure 2, time differences appearently being smaller at $T+176$ s than at $T+172$ s. This fact may support the suggestion that the soft electrons at the equatorward boundary of an auroral arc are of a different origin than the isotropic and harder electron fluxes over the arc.

Interesting are also the very abrupt flux decreases in the time interval $T+185$ to 192 s. All data are obtained before apogee, so it is certainly not an altitude effect. The time differences observed in this part of the flight indicate that this precipitation boundary moves equatorward perpendicular to L shells with a speed of 300 to 400 m s^{-1}. The gradient is very steep: The 3 keV electron flux decrease by a factor 5000 over a horizontal distance of 0.75 km, 1 keV flux by a factor 300 over 0.51 km, and 0.5 keV electrons by a factor 35 over a horizontal distance of no more than 0.23 km.

4. Summary

Equatorward and poleward of the auroral arcs there is a region of soft, field aligned electron fluxes. Above the auroral arc the electron spectrum is harder and the fluxes are isotropic (in the upper hemisphere).

The 3 keV electron fluxes are more structured than are the fluxes in the two lower eneigy channels. Most of the flux variations in all three energy channels are due to stationary or slowly moving structures in the precipitation regions, which is not surprising since the auroral forms were fairly stable. Flux gradients may be very sharp:

2 to 3 orders of magnitude over a latitude interval of 0.5 to 1.5 km. Also rapid temporal flux changes may occur, even in the case of a pre-breakup aurora, when these observations were made. An increase by a factor of 20 in less than 1 s is observed.

Fine structure in the 3 keV electron flux seems to move along with the gross precipitation pattern, at least right above the aurora.

It appears, based on 0.5 keV electron data, that various regions of precipitation which are associated with an auroral arc may move across L shells with different velocities.

References

Block, L. P.: 1967, *Space Sci. Rev.* **7**, 198.
Bryant, D. A., Colin, H. L., Courtier, G. M., and Johnstone, A. D.: 1967, *Nature* **215**, 45.
Courtier, G. M., Smith, M. J., and Bryant, D. A.: this volume, p. 207.
Evans, D. S., Maehlum, B. N., and Wedde, T.: 1972, *EOS* **53**, 731.
Frank, L. A. and Ackerson, K. L.: 1971, *J. Geophys. Res.* **76**, 3612.
Frank, L. A. and Gurnett, D. A.: 1971, *J. Geophys. Res.* **76**, 6829.
Hoffman, R. A. and Evans, D. S.: 1968, *J. Geophys. Res.* **73**, 6201.
Johnstone, A. D.: 1973, Paper, Symposium on European Sounding Rocket and Related Research at High Latitudes, ESRO, April 1973.
Kindel, J. M. and Kennel, C. F.: 1971, *J. Geophys. Res.* **76**, 3055.
Lennartson, W.: 1972, Royal Institute of Technology, Stockholm, No 72–104.
Lundin, R.: 1973, Paper, Symposium on European Sounding Rocket and Related Research at High Latitudes, ESRO, April 1973.
Maehlum, B. N. and Moestue, H.: 1973, *Planetary Space Sci.* **21**, 1957.
Måseide, K.: 1972, private communication.
O'Brien, B. J. and Reasoner, D. L.: 1971, *J. Geophys. Res.* **76**, 8258.
Pazich, P. M.: 1972, Ph. D. Thesis, Rice University, Texas.
Tröim, J.: 1973, private communication.
Whalen, B. A. and McDiarmid, I. B.: 1972, *J. Geophys. Res.* **77**, 191.

AURORAL ELECTRONS OBSERVED USING A
MOTHER-DAUGHTER ROCKET

G. M. COURTIER, M.J. SMITH and D. A. BRYANT

SRC, Appleton Laboratory, Ditton Park, Slough, SL39JX Bucks. U.K.

1. Introduction

This paper presents results obtained from the Skylark rocket, S77/1, launched from Esrange, Kiruna at 2226:50 U.T. on March 2, 1972. It was launched over a broad auroral band during the onset of a geomagnetic bay. With this experiment it was possible to resolve spatial and temporal intensity variations of precipitating auroral electrons. This was achieved by mounting a separate payload with its own telemetry and ejecting it axially forward from the main payload. It was ejected at an altitude of 93 km with a separation velocity of 24 m s^{-1}. The horizontal separation between the two payloads, measured across magnetic field lines, increased throughout the flight to a maximum of 3 km.

Electron intensities were measured over an energy range of to 1 to 100 keV by channel multipliers with electrostatic analyzers and geiger counters. All detectors on the ejected payload functioned correctly, but a fault on the main payload meant that only the two geiger counters worked. The viewing directions of all the detectors that gave useful data were parallel to the rocket axis, so the range of pitch angles scanned was small.

2. Results

The intensities at five electron energies at the ejected payload are shown in Figure 1. The intensities at 3.8, 5.7 and 9.0 keV were measured by the channel multipliers, and the integral intensities, for >18 and >48 keV, were measured by the geiger counters. Changes in intensity occur sporadically as pulses with duration ~5 s superimposed on a more slowly varying intensity. They occur at all energies, and have a larger amplitude at the higher energies. The pulses seen at the rocket occurred at about the same times as pulsed increases in light intensity recorded by a ground based photometer, which viewed the aurora magnetically beneath the rocket. At about 420 s there is a rapid decrease in intensity caused by the payload re-entering the atmosphere.

To determine whether the pulses are due to temporal variations or spatial structure, the times of the intensity changes of >48 keV electrons at the ejected payload are compared with the times of intensity changes of electrons with approximately the same energy (>51 keV) at the main payload. Cross-correlation of the two data sets, with a varying time shift between them, shows that the changes occurred simultaneously and therefore that the pulses are a temporal phenomenon.

Having established that the event is temporal, the time difference between the intensity changes for the different electron energies may be used to determine the source

B. M. McCormac (ed.), Magnetospheric Physics, 207–271. All Rights Reserved.

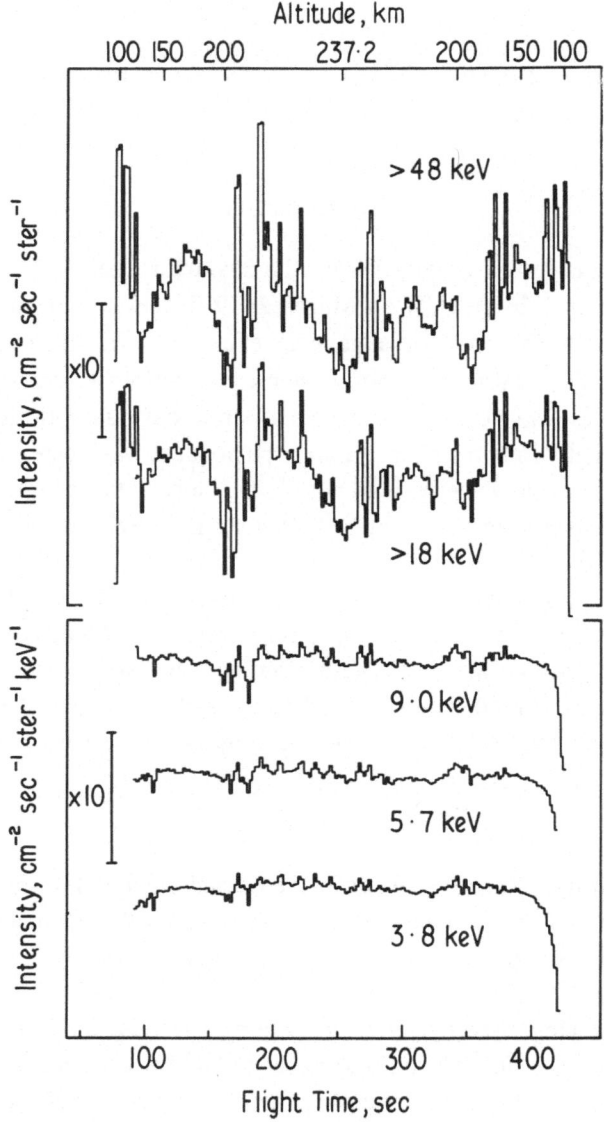

Fig. 1. Electron intensities at the ejected payload of S77/1.

distance, i.e., the distance to the disturbance that caused the pulsed modulations in the data. The intensity at the 5.7 keV energy is used as a reference.

For simultaneous modulation of the electrons at a distance S, the following formula applies

$$\Delta T(5.7, E) = S\left(1/V(5.7) - 1/V(E)\right)$$

where $\Delta T(5.7, E)$ is the time delay, i.e., the difference in the times when the intensity changes occur for the electrons of energy 5.7 keV and the electrons of energy E, $V(5.7)$ is the speed of an electron with energy 5.7 keV and $V(E)$ is the speed of an electron with energy E. E will take the values 3.8, 9.0, 18 and 48 keV, where 18 and 48 keV are the geiger-counter threshold energies. A graph of $\Delta T(5.7, E)$ vs. $(1/V(5.7) - 1/V(E))$ will be a straight line, and the slope of that line will give the source distance.

The flight was divided into 20 s intervals and for each interval where pulses occur, the time delays were found by cross-correlating the intensities at different energies against the intensity at 5.7 keV, with various time shifts introduced.

Each graph showed a linear relationship and hence a source distance was derived for each time interval.

The source distances are plotted in Figure 2 and show a systematic change with time. The 260 to 280 s interval gives a source distance of $51\,000 \pm 6000$ km which is comparable to the distance to the dipole equator (45 000 km for the shell parameter $L = 5.4$ at Esrange). For the intervals 160 to 180 s and 220 to 240 s the source distances are $41\,000 \pm 4000$ km and $90\,000 \pm 7000$ km, respectively. These are the two extremes. The labels on the graph indicate for comparison the locality of the rocket, the distance

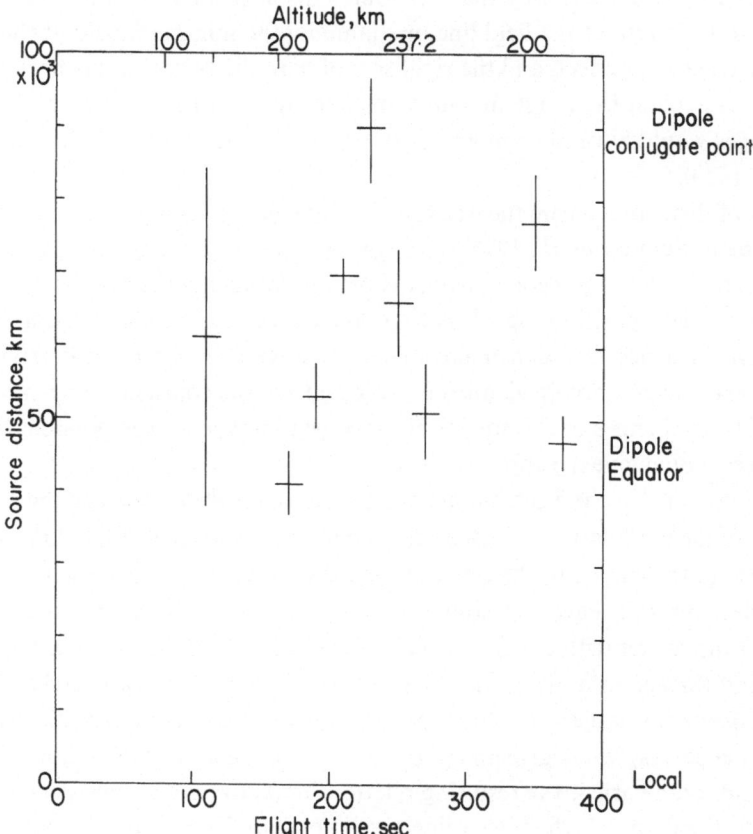

Fig. 2.   Variation in the distance to the source of modulation of electron intensities throughout the flight.

to the dipole equator, and the distance to the dipole conjugate point for $L = 5.4$. Standard errors in the source distance are shown.

3. Discussion

There are two questions which arise:
  (a) What causes the changes in intensity of the precipitated electrons?
  (b) What makes the source distance so variable?

A possible cause of the electron precipitation is pitch angle scattering, since the pitch angle distribution, although nearly isotropic for all intensities, rises slightly towards the larger pitch angles. The near isotropy of the distribution can be explained if the pitch angle scattering angles are comparable to, or larger than, the loss cone half-angle. The intensity changes could then be due to variations in the number of electrons being scattered, while the pitch angle distribution is unaffected. We expect scattering to be most effective where the magnetic field is weakest.

With this constraint, a possible explanation of the varying source distance is that the position of the weakest field changes. The magnetic field distortions would need to be periodic in the same way that the source distance varies throughout the flight. Changes in the length of the field line, maintaining the minimum field at the equator, could give rise to this effect, and the right sort of periodic behavior has been observed. However, to explain the range of source distance implied here, the magnitude of the field changes would have to be much greater than has been observed (McPherron and Coleman, 1970).

A form of distortion where the weakest magnetic field does not occur at the equator has been seen (Sugiura *et al.*, 1971), but again the necessary magnitude of the distortion has not. The types of geomagnetic distortion producing the necessary changes in the minimum field position are likely to produce geomagnetic micropulsations at the ground. No evidence for such micropulsations is seen in the records from ground based magnetograms obtained during the event so we consider these mechanisms unlikely. If the changes in the source distance were not real, it is possible that some other process could give the apparent effect.

Let us consider that the electrons are accelerated somewhere between the rocket and the source of the modulation, which is taken to be at a distance of 50000 km. An energy ΔE is added to the energy of the electrons at a distance x from the rocket. This means that an electron will leave the source of the modulation with one speed and be detected at the rocket with a greater speed. The electron will therefore appear capable of travelling further in a given time than it has. This, in turn, will make the source distance appear too large. For this explanation to be acceptable it must explain the extreme time delays, which give an apparent source distance of 90000 km. No matter how ΔE and x are varied, the resulting relationship between the time delays is inconsistent with the data, which show a linear relationship. The source distance of 90000 km cannot be realized, even by distributing the region of acceleration, and so a model which gives a fixed acceleration at all energies is unacceptable.

Since the theories described here do not satisfy the results, new ideas are being tested. A model in which an acceleration proportional to energy occurs between the rocket and the point of modulation seems, in principle, capable of explaining the results.

Acknowledgments

We are greatly indebted to the ESTEC personnel who were responsible for the design, integration and launching of the S77/1 payload. The work described in this paper was carried out at the Appleton Laboratory of the Science Research Council and is published with the permission of the Director.

References

McPherron, R. L. and Coleman, P. J.: 1970, *J. Geophys. Res.* **75**, 3927.
Sugiura, M., Ledley, B. G., Skillman, T. L., and Heppner, J. P.: 1971, *J. Geophys. Res.* **76**, 7552.

PART IV

WAVE-PARTICLE INTERACTIONS

WAVE PARTICLE INTERACTION – SOME
LABORATORY OBSERVATIONS

C. A. NYACK, P. J. CHRISTIANSEN and G. MARTELLI

Plasma and Space Physics Group, University of Sussex, Falmer, Brighton, U.K.

1. Introduction

Electrostatic waves are both a sign of, and intimately bound up with, significant changes in the particle velocity distribution in natural as well as laboratory plasmas. It has been known from some years of laboratory study that collisionless plasmas exhibit an abundance of electrostatic ($\mathbf{k} \parallel \mathbf{E}$) noise, or 'microinstabilities', in a very wide variety of experiments. The contribution of electrostatic waves to particle acceleration, velocity space diffusion, and bulk properties such as particle diffusion, plasma resistivity and heat transfer are the subject of theoretical (here we include the increasingly important computer simulations) and experimental studies in many laboratories around the world.

The reason for this is clear. In situations in which particle velocity distributions exhibit significant non-maxwellian features, say a thermal distribution with a bump or a spike in the tail (a cool plasma penetrated by a stream of electrons, for example) or a double humped distribution (two interpenetrating plasmas moving at different velocities, or a current of electrons moving with respect to the plasma ions), linear theories show that electrostatic waves can be unstable, via wave-particle interactions, with often large growth rates, of the order of the frequency of the instability. Depending on the plasma parameters, on the details of the velocity distribution and orientation of the external magnetic field, these instabilities cover a wide range of frequencies, from the upper hybrid frequency, plasma frequency and neighborhood of harmonics of the electron cyclotron frequency (plasma and Bernstein waves), down to the ion plasma and lower hybrid frequencies (Buneman or two stream modes) and below (ion sound waves) (Seidl, 1970; Stringer, 1964; Lashmore-Davies and Martin, 1973; Ashour-Abdalla and Cowley, 1974). Since these waves propagate not only in a wide range of orientations with respect to the magnetic field, but also in a wide range of velocities from the thermal electron velocity downwards, the thermal particles, which represent the bulk of the plasma, can be significantly affected. The successful prediction of which wave modes are unstable is by no means the end of the story, and we are faced with the very complicated problem of how the energy is redistributed by subsequent non-linear processes.

The laboratory experiments which we will discuss deal with a small area of this general problem, i.e., with some of the non-linear particle effects that occur in a particularly simple situation: that of a cool plasma penetrated by a fast electron beam. The majority of these experiments involve the Langmuir (or plasma) wave, but the results are applicable to a larger class of phenomena. We hope that they may add a

B. M. McCormac (ed.), Magnetospheric Physics, 215–230. All Rights Reserved.
Copyright © 1974 by D. Reidel Publishing Company, Dordrecht-Holland.

little to the recent increase of interest in electrostatic waves shown by a number of magnetospheric physicists. Seen from a laboratory plasma point of view, the historical emphasis in space research on whistler-particle interactions, in which only a small number of particles in the high energy tail of distribution are affected, seems a little strong (though the thermal electrons may be influenced by non-linear wave decays). The technical problems of detecting and interpreting rapidly fluctuating electric fields and low energy particle fluxes from rockets and satellites are considerable. On the other hand, laboratory experiments have other difficulties, in particular those of scaling and geometrical factors. As will be seen, one has to work at rather high frequencies than in natural plasmas, and use very small probes in order to observe the phenomena of interest within a machine length, and often the effects of finite plasma diameter must be considered in interpreting measurements (although this particular curse may be removed with the recent development of the large volume 'magnetic wall' plasmas at UCLA). However, the laboratory worker does have a measure of control over the object of study, and the ability of reproducing it and of resolving it spatially, while the space experimenter, despite intrinsically better time and local space resolution, must often rely on good sense, or good luck, or both, in his observations, and is at present faced with the difficulty of including non-local effects in interpreting his data.

2. Linear Theory

The linear theory of velocity space microinstabilities, using the Vlasov equation approach (Landau, 1946) is treated in most modern texts. The experiments which we will describe deal with a stationary plasma penetrated by a stream of electrons (Figure 1a), so we first examine the predictions of a one dimensional calculation of this kind.

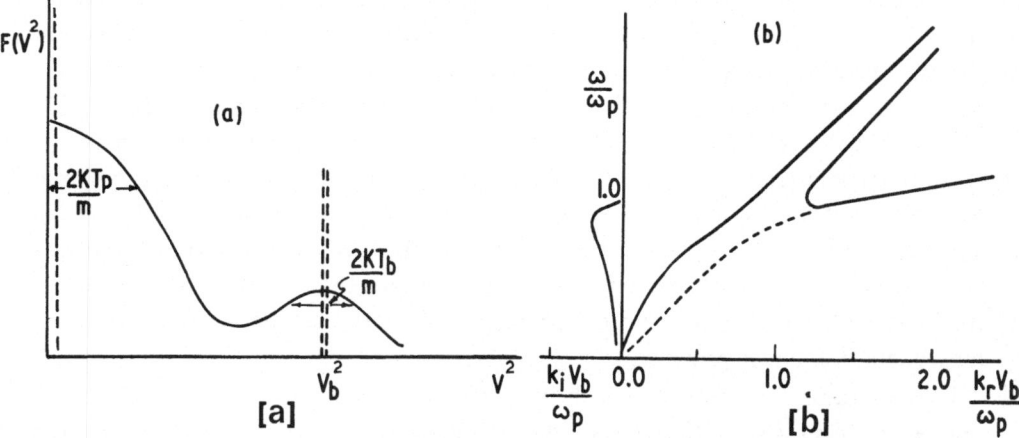

Fig. 1.   (a) Bump in the tail velocity distribution. Dashed lines illustrate the cold beam, cold plasma limit. (b) Real and imaginary dispersion for beam plasma coupling in a cylindrical system. The dashed line is the unstable branch whose spatial growth rate Im(k) is shown. The effect of finite geometry on the plasma branch can be seen at low k_r. Parameters are $\eta = 0.01$, $\omega_{ce}/\omega_p = 2.0$, $\omega_p/V_p = 2.0$.

Intuitively, we would expect waves to grow in regions of velocity space where there is a significant excess of electrons traveling faster than the wave, in the neighborhood of its phase velocity, i.e., where $(\partial f/\partial v)_{\omega/k} > 0$. The 'bump in the tail' theory predicts (see e.g., Krall and Trivelpiece, 1973) that, for quasistatic $(\mathbf{k} \parallel \mathbf{E})$ perturbations, of the form $e^{(i\omega t - kz)}$, where ω is complex to allow for growth or damping, the maximum growth rate occurs at frequencies close to the plasma frequency, the imaginary part of ω being given by

$$\omega_{i\,max} = \sqrt{\frac{\pi}{8} \frac{1}{k^3 \lambda_d^3}} \omega_p \left[\frac{n_b}{n_p} \frac{T_p}{T_b} k^3 \lambda_D^3 \frac{mv_b^2}{KT_p} \times \right.$$

$$\left. \times \exp\left(-\tfrac{1}{2}\right) - \exp\left(-\frac{1}{2k^3\lambda_D^3} - \frac{3}{2}\right) \right] > 0$$

where k is the wave number, λ_D the debye length, ω_p the plasma frequency, n_p, n_b, T_p, T_b the densities and the temperatures of the plasma and the beam respectively, K the Boltzmann constant, and m is the electron mass. The first term is positive and leads to growth, while the second term represents the competition from Landau damping. This equation reveals several things. In order to increase the growth rate, the number of beam electrons must be increased, the beam temperature must be reduced, and the relative velocity v_b increased, so that in the limit we have a cold electron plasma penetrated by a monoenergetic beam. In this situation we can use the mathematically simpler multistream model, though the mechanism for instability is less obvious. since now there are no electrons moving in synchronism with the growing waves. The linear dispersion of the system can be described by the cold plasma dielectric function of beam and plasma (Briggs, 1964)

$$\varepsilon(\omega, k) = 1 - \frac{\omega_p^2}{\omega^2} - \frac{\omega_p^2 \eta}{(\omega - kv_b)^2} = 0$$

where $\eta = (n_b/n_p) \ll 1$. This describes the coupling of the slow and fast beam space charge wave $(\omega = kv_b \pm \omega_b)$ with the plasma wave $(\omega = \omega_p)$. The slow wave is a negative energy wave (see, e.g., Briggs, 1964) which is unstable in the presence of plasma. The dispersion diagram is shown in Figure 1b, where the plasma branch has been modified to include the effects of finite plasma radius.

The fastest growing wave occurs just below the plasma frequency, i.e., at

$$\omega = \omega_p \left[1 - \tfrac{1}{2} \left(\frac{\eta}{2} \right)^{1/3} \right];$$

its phase velocity v_φ is less than that of the beam by a similar factor. The growth rate, which is $\sim 0.3\, \omega_p (n_b/n_p)^{1/3}$ can therefore be very high (of the order of 0.1 ω_p), even for relatively weak beams, and bears out our introductory remarks.

Although this information is sufficient for a first approach to the analysis of the experimental results, it should be pointed out that the growth rate decreases if the beam temperature is finite. It should also be noted that, if the effects of finite plasma size and of the magnetic field ($\omega_{ce} \approx \omega_p$) are included ($\omega_{ce}$ being the electron cyclotron frequency), a further unstable coupling occurs in the region of the upper hybrid frequency. This coupling has been used by some experiments for studying beam plasma interactions.

3. Experimental

The experimental setup is shown in Figure 2. The source is a 2 cm tantalum plate, heated from behind to $\sim 1700\,°C$, from which thermionic electrons can be accelerated to energies of anywhere between 150 and 300 eV, depending on the anode-cathode potential. The electron current can be varied from a few tenths to several milliamps.

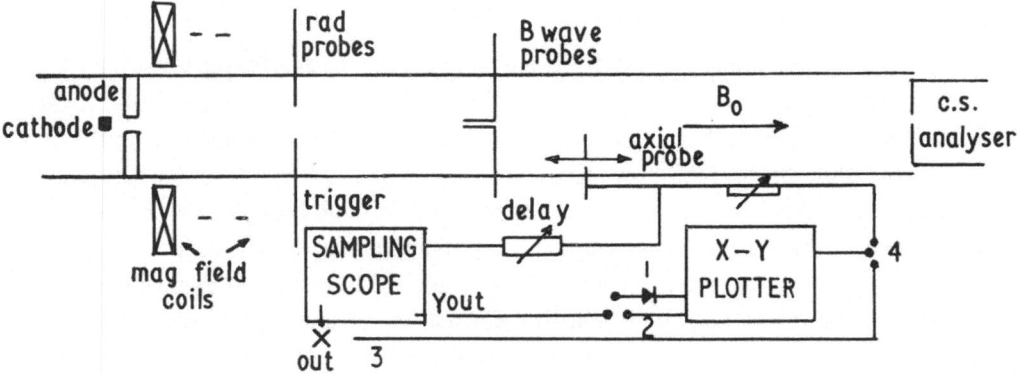

Fig. 2.   Schematic illustration of the experimental device showing electron source, various probes, and energy analyzer. The coherent detecting system is shown: connecting 1–4 gives coherent amplitude variations, 2–4 gives phase variations, 2–3 gives pulse shape.

The electrons stream axially into the experimental chamber, which contains Ar gas at low pressure (typically 10^{-4} torr). The plasma created by the rather rare electron-atom ionizing collisions is contained in a 70 cm long, 10 cm diam stainless steel tube immersed in a uniform magnetic field which can be varied in the range 0 to 700 G. A number of ports are provided for access with radial probes, and motor driven probes can be moved axially along the machine length. The system is terminated by an earthed plate, carrying an electrostatic energy analyzer, so that the emerging beam can be studied after interaction.

A typical plasma formed in this machine has a diameter of several cms, and consists of a cool thermal electron component with $T_e \sim 2$ eV (the ions are cold) penetrated by a 200 eV electron beam with small energy spread (5%) The ratio η of the beam to the plasma density is a few percent. Typically, the electron plasma frequency is $f_{pe} \sim 200$ MHz, and the electron cyclotron frequency $f_{ce} \sim 500$ MHz. Despite the low percentage ionization (0.01%), the plasma is nevertheless collisionless, at least for waves involving only electrons, since the electron-neutral mean free path is ~ 10 m.

Waves are launched by coaxial probes. They are very thin (0.5 mm in diameter) in order to minimize perturbations to the plasma. The relative spatial variations of density are measured using miniature Langmuir probes. To measure local absolute densities, we use Bernstein wave techniques. This is done by employing two identical aerials, facing each other across the magnetic field, one transmitting and the other receiving a h.f. signal. A dip in the signal received across the plasma occurs at the upper hybrid frequency $\omega_{uh}^2 = \omega_{pe}^2 + \omega_{ce}^2$, where the group velocity is zero. Since the magnitude of the magnetic field is known, the density can be deduced without affecting the beam density, which can be distorted by Langmuir probe measurements.

Our first experimental results concern the 'linear' dispersion of the beam-plasma system. By this, we mean measuring the dispersion of 'test' or small amplitude waves (launched from a thin probe) by investigating their behavior from interferometer data. The plasma is of course already unstable, so the test waves are launched into a plasma containing natural large amplitude waves which will have changed the electron beam from its original monoenergetic state. Figure 3 compares data taken be-

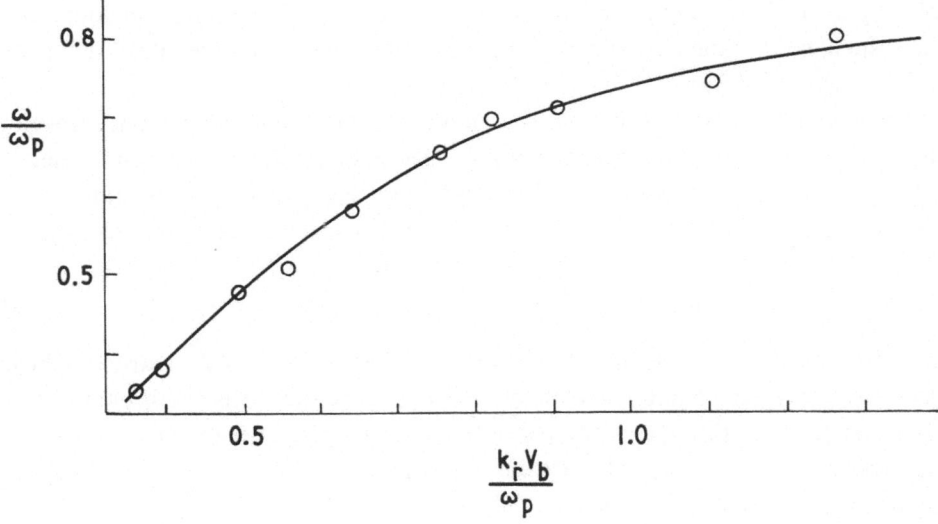

Fig. 3. Measured real part of unstable 'test wave' dispersion in the present of a large wave. The solid line is the prediction of the linear theory including non-uniform radial density $n(r)$, using experimental data for plasma and beam density. Departures from the linear theory can be seen at large k. $f_p(r = 0) = 275$ MHz, $V_b = 200$ V, $I_b = 2$ mA, $f_{ce} = 650$ MHz.

fore the test waves have grown too much, with a linear theory calculation which takes into account finite plasma effects such as non-uniform radial density and walls. For the dispersion measurements, the agreement is good in view of the fact that no parameters have been fitted and adjusted, but some genuine disagreement appears at low phase velocities. The form of the growth rate variation, while having a maximum at the predicted frequency, is also rather different. These effects are due to the presence of the large wave, and we will return to them later (the full linear theory has been verified experimentally by Malmberg and Wharton, 1969).

It is worthwhile mentioning at this stage that our experiment also shows the coupling between the beam and the slow electrostatic waves, ion sound waves and second ion cyclotron waves. This coupling directly influences the cool electron and ion components of the plasma.

4. Large Amplitude Effects

The predictions of the linear theory cannot be expected to apply to an unstable system for many wave periods. First, there is only a certain amount of 'free energy' in the plasma available to support growing waves and second, when the amplitude becomes large, non linear effects become important. The trapping of electrons in the potential wells of the unstable waves is one such effect. This phenomenon in beam plasma interactions is treated by what are called 'single wave' theories (O'Neil et al., 1971; O'Neil and Winfrey, 1972; Shapiro and Shevchenko, 1971; Matsiborko et al. 1972) which are based on the idea (and experimental observation) that in the early stage of development of the beam plasma interaction the fastest growing wave soon dominates the system, so that the next stage can be treated by considering the effects associated with a single large wave.

Conceptually, what happens is as follows: the fastest growing mode, traveling slower than the beam, has an exponentially growing amplitude, to which an electrostatic potential $\Phi(t)$ is associated. The region in velocity space which is affected by this potential has a width $\propto \Phi^{1/2}(t)$, and when this reaches the beam velocity, i.e., when

$$e\Phi(t) \simeq \tfrac{1}{4}m(v_b - v_\varphi)^2 \tag{1}$$

(where v_φ is the phase velocity of the wave) the beam is 'trapped', that is the beam electrons start to decelerate, then fall into the wave potential wells and begin to 'slosh' back and forth in the potentials, periodically exchanging energy with the wave. If the oscillation of the trapped electrons takes place at the bottom of the well, its frequency (the 'bounce frequency') is just

$$\omega_B = k\sqrt{\frac{e\Phi}{m}} \sim \omega_p \left(\frac{n_b}{n_p}\right)^{1/3}. \tag{2}$$

Details of such models can be found in the references given. Figure 4 shows computed results of such calculations, carried out for a one dimensional interaction and fixed wave number. To make the transformation to the experimentally more realistic situation of a fixed frequency, the dimensionless time τ' has been multiplied by the ratio of the phase to group velocity.

The top trace of the figure shows that the amplitude grows exponentially in accordance with the linear theory up to about $\tau'=4$, then begins to trap the beam until at $\tau'=8.6$ the beam particles are moving backwards in the wave frame, the energy transfer to the wave having reached a maximum, so that the wave amplitude saturates. The oscillation of the particles continues until at $\tau'=11.6$ the bulk of the particles is

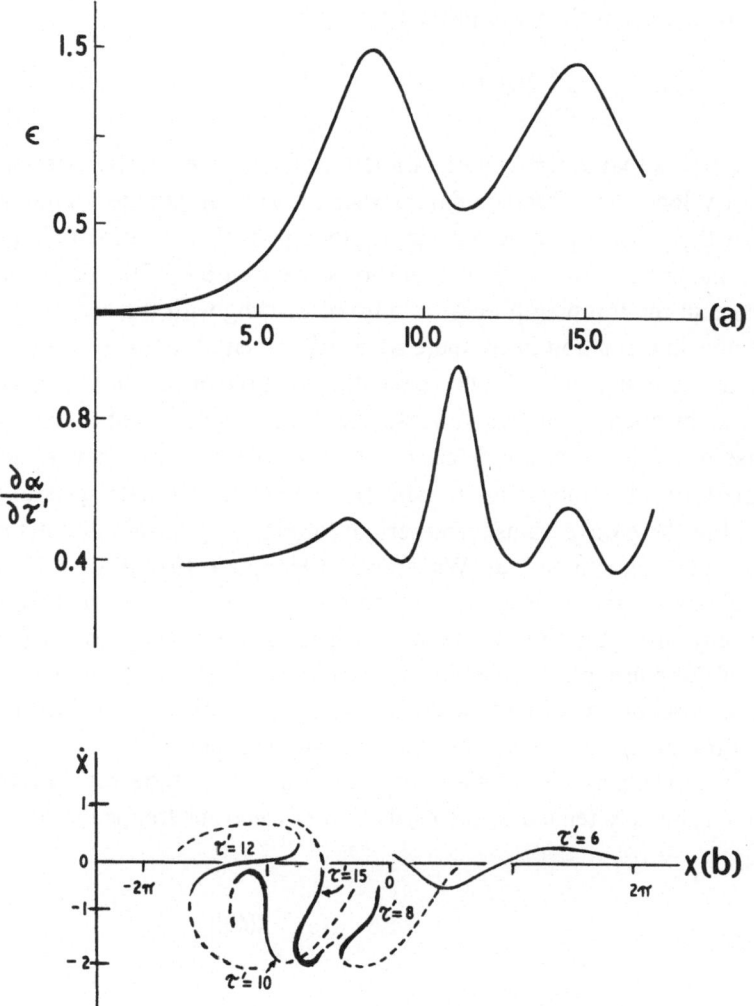

Fig. 4.   Computed prediction of single wave theory for the fastest growing mode for a fixed wave number k_0. (a) Top trace shows wave amplitude E with time τ' and lower trace the corresponding variation of $\partial\alpha/\partial\tau'$ (related to instantaneous phase velocity by Equation (3)). The time τ' is in units of $\omega_0\eta^{1/3}$ and the amplitude in units of $e\Phi/mV_b^2\eta^{2/3}$, where $\omega_0 = k_0V_b$. (b) Semischematic phase space evolution of particle distribution seen in a frame moving at the original beam velocity, i.e., position $X = k_0x$. Heavy lines are intended to convey concentration of particles and tenuous filaments have been partly omitted. Notice that at $\tau' \sim 8$ the particles are bunched in space giving rise to harmonics. The obvious rotation in phase space would continue and the distribution would eventually assume a double spiral form.

moving with maximum velocity in the potential well and the wave amplitude is a minimum; at $\tau' = 15$ the oscillation begins again.

A further consequence of the periodic energy exchange between waves and particles is shown in the lower trace of the figure. Here the time derivative of the total phase of the wave can be seen also to undergo periodic variations in time at twice the frequency of the wave amplitude. Since the wave number is fixed in the calculations, this can be related to variations in the wave phase velocity by

$$v_p = v_b \left(1 - \frac{\partial \alpha}{\partial \tau'} \eta^{1/3} \right). \tag{3}$$

The theory shows that $\partial \alpha / \partial \tau'$ is a quadratic function of the average particle velocity, so the phase velocity has a deep minimum when the wave amplitude is a minimum, and increases with increasing wave amplitude, though with a subsidiary minimum at amplitude maximum. Finally, Figure 4b shows the positions of the beam electrons at several times in the trapping process in a frame moving with the beam velocity.

We are now in a position to compare some experimental results with the prediction of the 'single wave' model. It should be mentioned at the outset that the beam-plasma discharge can be operated in several ways. The first is simply to let the system run free. In this case unstable waves are detected below the plasma frequency, in accordance with linear theory, but they come in bursts, i.e., they have a finite spectral width. In order to follow these wave trains (~ 50 periods long) as they travel through the plasma, a coherent detecting system (van Wakeren and Hopman, 1972, Figure 2) is used, in which a reference probe triggers a sampling oscilloscope. The detected signal is thus displayed only when it is coherent with the trigger, and one can measure (a) the time variation of the pulse at a fixed point in space by varying the delay in the detecting arm, (b) the amplitude as a function of distance by averaging the signal from a moving probe, and (c) the phase velocity by coherent interferometry.

Figure 5a shows periodic oscillations of the wave amplitude with distance. That this effect is genuinely related to the oscillation of electrons trapped by the wave can

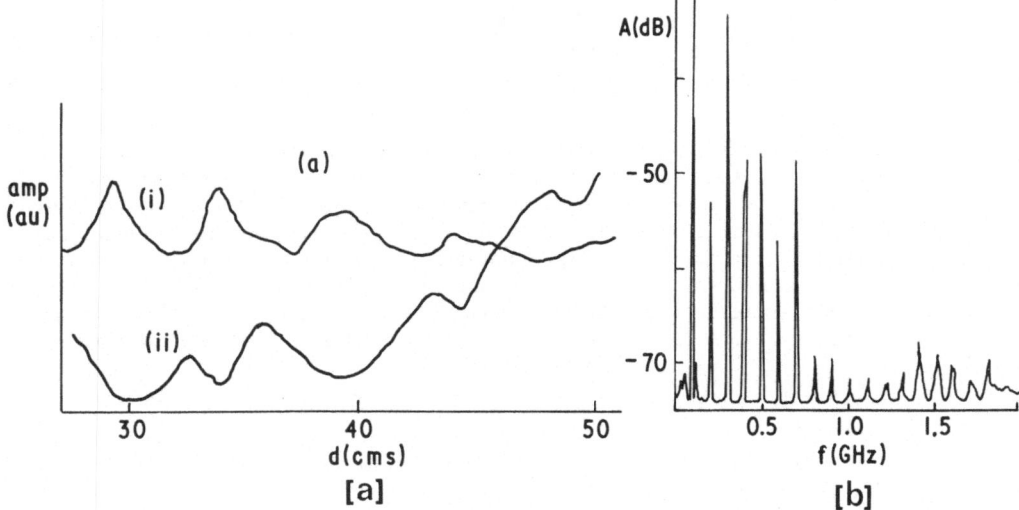

Fig. 5. (a) Comparison of wave amplitude (i) with electron flux to a miniature langmuir probe biased to just above the beam voltage, (ii) as a function of axial distance. This illustrates periodic energy exchange between waves and particles. (b) Frequency spectrum of a large amplitude wave, recorded at $\tau' \sim 7$, showing high harmonic content due to spatial bunching of trapped electrons (cf. Figure 4b).

be seen in the lower part of the figure. Here the flux of electrons traveling just faster than the beam is detected by moving a very small Langmuir probe biased to just above the beam energy. As predicted above, the oscillations of this electron flux are in antiphase with those of the wave.

The appearance of sidebands close to the main wave frequency, but separated from it by roughly ω_B, yields further evidence of trapping; we will return to this later. Since it is difficult to measure absolute wave amplitudes at the frequencies observed, we must use experimental values of the bounce distance, and of n_b, n_p, k and ω_B in Equation (1), as well as self-consistent arguments for the energy balance to deduce the amplitude (Kainer et al., 1972). These various approaches agree well with each other and give wave fields of the order of tens of V cm^{-1}, the precise value depending on the particular experimental conditions. The prediction that at the onset of trapping the wave has a high harmonic content is confirmed by the frequency spectrum shown in Figure 5b, recorded at the point where the wave field is highly distorted by spatial bunching of the trapped electrons (Figure 4b). We are confident that at least the low harmonics are not normal modes of the system, since their measured phase velocity is the same as that of the main wave (see also Apel, 1969).

Figure 6a shows a comparison of the pulse amplitude with the change of average frequency of the main wave. Another 'big wave' effect is revealed by the correlation of the variation of the frequency with that of the amplitude. A direct check with theory is more difficult here, because the single wave model predicts either a variation of the frequen :y ω for fixed wave number k, or variation of k for fixed ω. Both ω and k

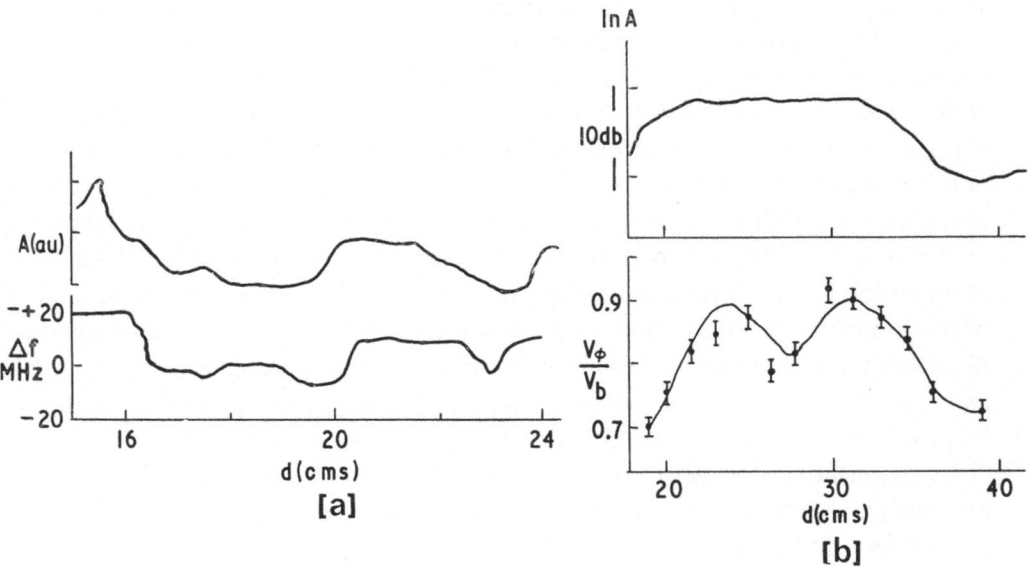

Fig. 6.   (a) Shift of average frequency of large amplitude wave pulse (lower trace) with distance, correlated with amplitude variations of wave (upper trace). (b) Observations of variation of amplitude (top curve) and phase velocity of wave (bottom curve) over one bounce length using premodulated beam $I_b = 0.85$ mA, $V_b = 200$ V, $f = 125$ MHz, $f_{ce} = 500$ MHz.

are observed to vary in these experiments, and this is a result, we believe, of the fact that wave packets are involved. Certainly, theoretical calculations of the propagation of wave packets in a non-dispersive medium, with cubic non-linearity, show variations of both ω and k (Giles, 1973).

Better measurements of the non-linear phase velocity variations can be obtained by using a premodulation technique (originally employed on beam tubes) which consists in applying a small signal from a fine coaxial probe close to the beam source. This locks the instability frequency rather in the manner of non-linear coupled oscillators, and can produce a growing wave of considerable spectral purity on which to do interferometry. The results of one such measurement are shown in Figure 6b. Notice that the phase velocity changes (changes of wavelength at a fixed frequency) are rather dramatic and, qualitatively at least, follow the theoretical predictions over a bounce period of the amplitude. In fact, the velocity variation is larger (by a factor of 2) than that predicted by the theory for the experimental parameters n_p, n_b and bounce distance. This comment also applies to some other measurements made on velocity variations at the earlier stages of the trapping process (from $\tau = 0$ to $\tau = 12$). It is worthwhile noting that the phase velocity at the end of the bounce is greater than at the beginning, so that the wave appears to have undergone, on average, an acceleration process. There are more dramatic examples of this in other measurements, and although the amplitude of the accelerated wave can be small, the possibility exists that particles that remain trapped in the wave may be carried to rather higher velocities than the original beam velocity.

Before we return to this last point, let us consider the stability of the large amplitude waves which, according to the single wave theory, reach a quasi-stationary amplitude. The possibility of such behavior exists (Bernstein et al., 1957), and has been discussed for the single wave models by Thompson (1971). However, experiments by Wharton et al. (1968), demonstrated that large amplitude waves launched in quiet (no beams) plasmas, stimulated sidebands at frequencies $\omega \pm \omega_B$, and that the frequency separation increased as $\Phi^{1/2}$ in accordance with Equation (2). More recent experiments by Franklin et al. (1972) and Ikezi et al. (1972) tend to support the theoretical line started by Kruer et al. (1969), and pursued by many authors, which predicts the existence of upper and lower sidebands at a frequency ω such that $\omega - kv_\varphi \sim \pm \omega_B$ in the limit of small numbers of trapped particles (v_φ being the phase velocity of the main wave). Physically this relation shows that the waves which were originally down in the noise, and 'see' the main wave at a doppler shifted frequency equal to the oscillation frequency of the trapped electrons, get 'pumped up' to large amplitudes.

As was mentioned earlier, trapping sidebands are also detected in beam plasma experiments (see, e.g., van Wakeren and Hopman, 1972; Katsuhiro and Shigetoshi, 1972; Gentle and Lohr, 1973) in both the free running and frequency locked mode. A characteristic of such results is that the amplitude of the upper sideband (the wave traveling *slower* than the main wave) is greater than that of the lower (and faster) sideband. The frequency separation from the main wave of the upper and lower sidebands is not equal.

These trapping sidebands are the subject of considerable discussion and experiment at the moment, and the details of the exact frequency at which they occur, as well as their growth rates, still remain somewhat mysterious. One could put the problem this way. The 'sideband to be' must interact with the main wave for at least a bounce period or so, to 'know' at what frequency it is supposed to grow.

But the trapped particle distribution changes with time (as does the wave velocity in these experiments), and over the bounce period one might expect that the growth rate of the lower sideband would be sensitive to the shape of the distribution, and to how close to the bulk of the trapped particles it was in velocity space. It is probably fair to say that the theorists have not quite managed to fuse the pumping with the time varying velocity distribution aspects of the problem (see Wong, 1972; Mima and Nishikawa, 1972; Brinca, 1972). To illustrate this point we show in Figure 7b the

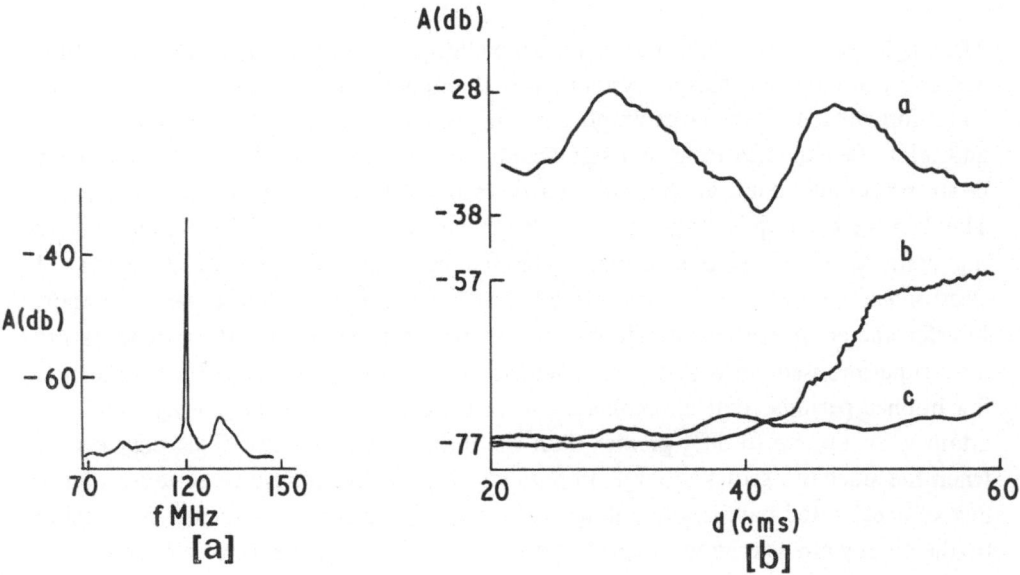

Fig. 7.   (a) Frequency spectrum of the main wave with upper and lower trapping sidebands. (b) Amplitude of main (premodulated) wave (a), $f = 96$ MHz, and upper (b) $f = 132$ MHz, and lower (c) $f = 72$ MHz sidebands, as a function of distance. $I_b = 1.1$ mA, $V_b = 250$ V, $f_{ce} = 500$ MHz.

amplitude variation of the main wave as well as the sideband amplitude as a function of distance. The upper sideband growth rate is not constant, and appears to be fastest when the trapped particles are expected to be closest to it in velocity space, i.e. when the wave amplitude is close to a minimum. We have also observed, in agreement with the theory and the 'quiet' plasma experiments, that the sideband separation is proportional to $\Phi^{1/2}$.

Before leaving this topic it is worthwhile making a few concluding remarks. Since the originally rather monoenergetic beam velocity distribution slows down and oscillates about the wave phase velocity in the trapping process, the dispersion of test wave in such a situation is considerably different from the linear predictions (Kruer *et al.*, 1969). We consider the departures from the predicted linear dispersion in Figure 3 to be a result of this effect. If the trapped particles, which look like bunches in phase space persist for a considerable time, the existence of these 'macroparticles' has profound implications for the plasma dynamics.

The formation of trapping sidebands need not necessarily stop at the stage we have discussed so far, since the possibility exists that the sidebands may 'cascade' or stimulate further sidebands each slower than its predecessor. Thode and Sudan (1973) have conjectured that this process may give rise to plasma heating.

Finally, we believe that studies of trapping sidebands along these lines may be useful in investigating VLF triggered emissions since analogous processes are involved.

5. Particle Acceleration

The single wave theory does not include the effects of growing sidebands on the time evolution of the beam plasma system. Now the upper sideband can, under favorable conditions, reach a very large amplitude comparable with that of the primary wave, and might be expected to have a significant effect on the trapped particles, and even be strong enough to untrap the slower particles and pull them to even lower velocities. This has further implications for the subsequent development of the system and, in this context, we would like to discuss briefly the results of some measurements of electron energy spectra which are correlated with the rapid growth of upper sidebands. Interferometer measurements show that the periodic phase velocity variations are now superimposed on a systematic increase of the phase velocity during the first few bounce periods; that is a coherent wave is accelerated on the average, and can attain velocities up to 50% greater than the initial beam velocity. This only occurs when the sideband grows to large amplitudes. After about 3 bounce periods, a large flux of accelerated particles traveling up to twice the initial beam energy is detected by the energy analyzer, and integral energy spectra of the type shown in Figure 8 are

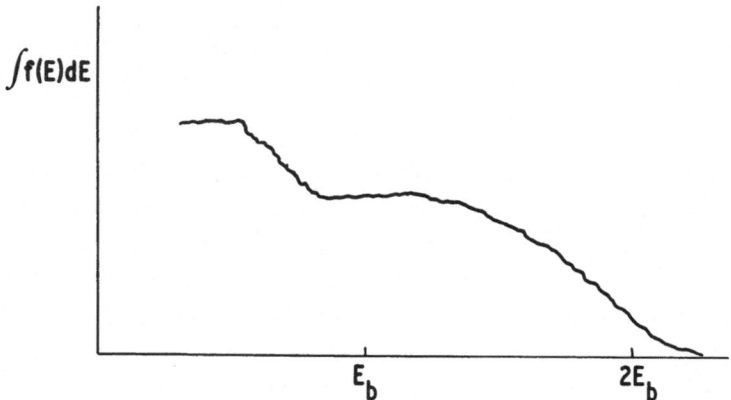

Fig. 8.   Integral electron energy spectrum recorded after about 3 bounce lengths of large wave. The initial beam energy E_b is marked. $I_b = 7.4$ mA, $V_b = 210$ V, $f_{ce} = 1$ GHz.

obtained. The spectrum shows quite a sharp cutoff at high energies (no strung-out tail). This phenomenon seems to be outside current theories of particle acceleration. The single wave model given above is of no help, since it predicts that the beam should settle down to a velocity slower than its initial velocity. Stochastic acceleration theories require much longer times to produce accelerations of this order and, moreover, they produce a long tail and require a broad band 'turbulent' wave spectrum, while in our system there are only two waves: the primary trapping wave and the slower sideband stimulated by it.

For this reason we have modified the single wave model by including a 'Maxwell's demon' to simulate the expected effect of the sideband – that is to remove some slower trapped particles from the wave by slowing them down even further. These particles are kept in the calculation of charge density and phase velocity variations, and the changes in energy and momentum which occur during the slowing down process are attributed to the sideband. The computations based on essentially experimental inputs predict, in good agreement with the experimental results, the appearance of the accelerated wave as well as the appearance and number of accelerated particles at very much the same time as the experiments show them to occur. Details will be reported elsewhere.

What the experiments certainly show is that non-stochastic processes may be important in the initial stages of the beam-plasma interactions, and that they may play a non-negligible role in particle dynamics in the ionosphere and magnetosphere.

6. Wave-Wave Interactions

In this final section we illustrate the resonant non-linear decay of the plasma waves. Many such interactions are receiving attention in the literature at the moment, and the particular case we are dealing with is described theoretically by Dysthe and Franklin (1970) and investigated in quiet plasma experiments (including Franklin et al, 1971). Just as non-linear oscillators can be parametrically coupled by a pump signal, so can waves in a plasma whose linear dispersion relations describe normal modes of the system. Thus a wave (ω_1, k_1) of amplitude sufficiently large to introduce non-linearities in the particle orbits can excite (or decay to) two other waves (ω_2, k_2) and (ω_3, k_3), providing the following conditions are reasonably closely obeyed

$$\omega_1 = \omega_2 + \omega_3, \qquad \mathbf{k}_1 = \mathbf{k}_2 + \mathbf{k}_3 .$$

In our case, wave 1 and 2 are plasma waves, and wave 3 is an ion sound wave which has a comparatively low phase velocity $C_s = (KT_e/M_i)^{1/2}$, M_i being the ion mass.

The beam plasma system already contains large amplitude plasma waves and by suitably choosing the parameters the frequency spectra in Figure 9a show the decay isolated from the other competing interactions described earlier.

The upper trace shows the primary wave at a frequency of 100 MHz with a lower sideband some 70 KHz below it, while the lower trace shows the spectrum at very low frequency with a signal at 70 KHz (and its harmonic) superimposed on a rather broad

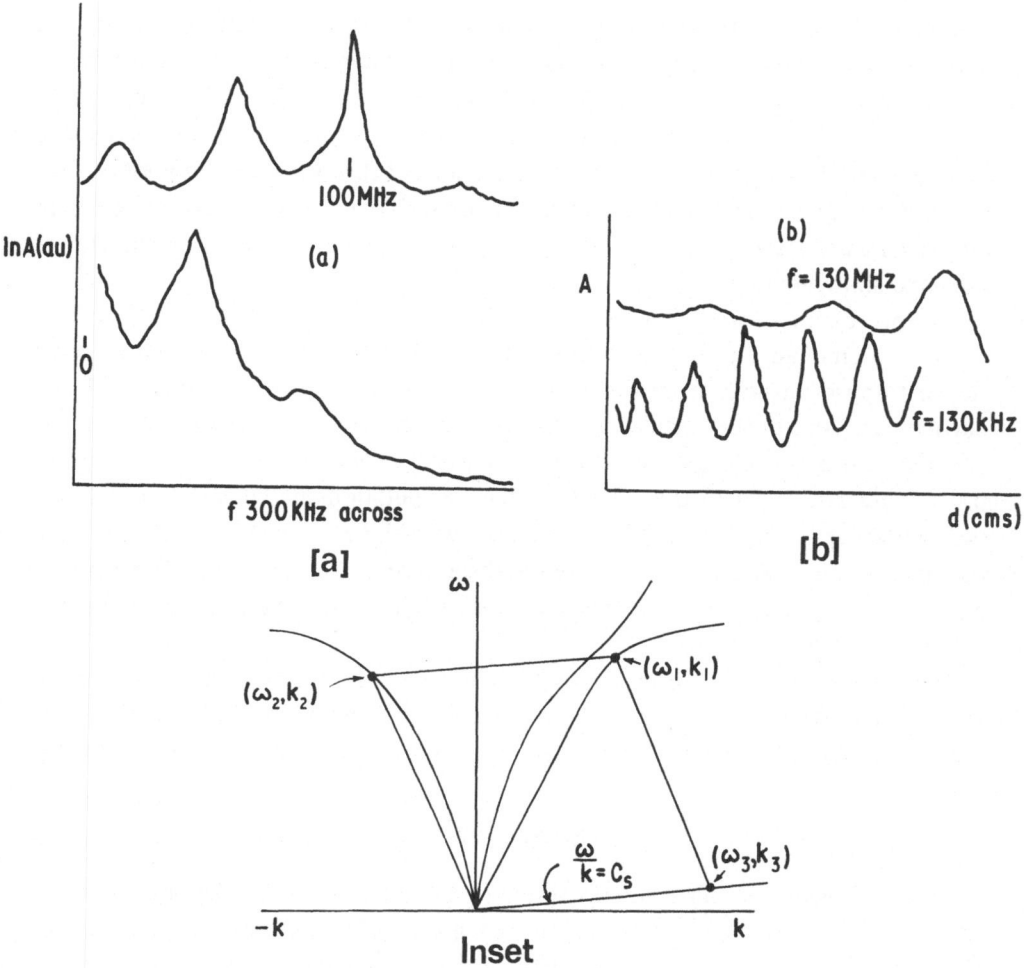

Fig. 9.   (a) Frequency spectra recorded at high (upper trace) frequency showing large wave at
100 MHz. and lower (decay) sidebands. The lower trace shows the spectrum at low frequencies.
(b) Interferometer traces of main wave, and low frequency wave recorded under slightly different
conditions to Figure 9 (a). *Inset*: Schematic diagram of decay interaction. The ion sound speed has
been exaggerated for illustrative purposes.

spectrum. Figure 9b demonstrates that the wave number conditions as well as the
frequency conditions are obeyed. The interferometer trace made under slightly different conditions from Figure 9a shows the wavelength of the primary wave to be
twice that of the ion sound wave. Since $\omega_1 \sim \omega_2$ and $\mathbf{k}_1 \sim -\mathbf{k}_2$ (as shown in the geometrical representation of the decay in the inset of Figure 9), the $k_3 = 2k_1$. The ion sound
wave in this experiment is *not* a collisionless wave, since ion-neutral collisions cannot
be neglected, and these must be taken into account when considering the threshold
amplitude at which the decay occurs.

We also observe that if the beam current is increased, the velocity of the ion sound
waves increases, and since this velocity is a measure of the electron temperature, this

fact may be evidence that the bulk of the cool electrons are also being heated. However, we stress the 'may be' in the absence of corroborative evidence at the moment. Decay interactions involving slow waves such as the ion sound are a method in which energy can be transferred to the thermal electrons in plasmas, and are certainly considered to be important in plasma heating experiments of various kinds.

7. Conclusions

Although we have only been able to describe one type of laboratory experiment in which electrostatic wave particle effects are significant, we believe such phenomena to be important. They are important not only in the laboratory but also for a greater understanding of magnetospheric and space plasmas, and would emphasize that many other laboratory experiments on turbulent heating, collisionless shocks, plasma heating, wave propagation and computer simulations have much to offer to the space physicist.

Acknowledgments

This research is supported by the Science Research Council of Great Britain, and the UKAEA, and one of us (C.A.N.) acknowledges the British Council for support. It is a pleasure to thank our colleagues R. W. Boswell, C. Christopoulos and G. J. Smith for their interest and comments, and P. N. Smith and B. J. Blackman who designed and built the machine. We wish also to acknowledge valuable conversations with F. W. Crawford and S. M. Hamberger in the early stages of this work.

References

Apel, J. R.: 1969, *Phys. Fluids* **12**, 291, 640.
Ashour-Abdalla, M. and Cowley, S. W. H.: 1974, this volume, p. 241.
Bernstein, I. B., Green, I. M., and Kruskal, M. D.: 1957, *Phys. Rev.* **108**, 546.
Briggs, R. J.: 1964, *Electron Stream Interactions in Plasmas*, MIT Press, Cambridge, Mass.
Brinca, A. L.: 1972, *J. Plasma Phys.* **7**, 855.
Dysthe, K. B. and Franklin, R. N.: 1970, *Plasma Phys.* **12**, 705.
Franklin, R. N., Hamberger, S. M., Lampis, G., and Smith, G. J.: 1971, *Phys. Rev. Letters* **27**, 1119.
Franklin, R. N., Hamberger, S. M., Ikezi, H., Lampis, G., and Smith, G. J.: 1972, *Phys. Rev. Letters* **28**, 1114.
Gentle, K. W. and Lohr, J.: 1973, *Phys. Rev. Letters.* **30**, 75.
Giles, M. J.: 1973, private communication.
Goldman, M. V. and Berk, H. L.: 1971, *Phys. Fluids* **14**, 801.
Kainer, S., Dawson, J., and Shanny, R.: 1972, *Phys. Fluids* **15**, 493.
Katsihuro, M. and Shigetoshi, T.: 1972, *Phys. Rev. Letters.* **29**, 45.
Krall, N. A. and Trivelpiece, A. W.: 1973, *Principle of Plasma Physics*, McGraw-Hill, N.Y.
Kruer, W. L., Dawson, J. M., and Sudan, R. N.: 1969, *Phys. Rev. Letters.* **23**, 839.
Landau, L. D.: 1946, *J. Phys. (U.S.S.R.)* **10**, 25.
Lashmore-Davies, C. and Martin, I. J.: 1973, *Nuclear Fusion*, **13**, 193.
Malmberg, J. H. and Wharton, C. B.: 1969, *Phys. Fluids* **12**, 2600.
Matsiborko, M. G., Onishenko, I. M., Shapiro, V. D., and Shevchenko, V. I.: 1972, *Plasma Phys.* **14**, 591.
Mima, K. and Nishikawa, K.: 1972, *J. Phys. Soc. Japan* **33**, 1699.

O'Neil, T. M. and Winfrey, J. H.: 1972, *Phys. Fluids* **15**, 1514.
O'Neil, T. M., Winfrey, J. H., and Malmberg, J. H.: 1971, *Phys. Fluids* **14**, 1.
Seidl, M.: 1970, *Phys. Fluids* **13**, 966.
Shapiro, V. D. and Shevchenko, C. I.: 1971, *Sov. Phys. JETP* **33**, 555.
Stringer, T. E.: 1964, *J. Nucl. Energy* **C6**, 269.
Thode, L. E. and Sudan, R. N.: 1973, *Phys. Rev. Letters* **30**, 732.
Thompson, J. R.: 1971, *Phys. Fluids* **14**, 1532.
Van Wakeren, J. H. and Hopman, H. J.: 1972, *Phys. Rev. Letters* **28**, 295.
Wharton, C. B., Malmberg, J. H., and O'Neil, T. M.: 1968, *Phys. Fluids* **11**, 1761.
Wong, H. V.: 1972, *Phys. Fluids* **15**, 632.

SPACECRAFT STUDIES OF VLF EMISSIONS

K. BULLOUGH, A. R. W. HUGHES and T. R. KAISER

Department of Physics, University of Sheffield, England

1. Introduction

VLF emissions, observed in satellites orbiting above the ionosphere, have their origin in the energetic particle fluxes as well as in terrestrial thunderstorms (whistlers). Man-made signals from terrestrial VLF transmitters are also detected. Their study throws light on the important role of wave-particle interactions in the dynamics of the magnetosphere and they can be useful as a diagnostic tool for the study of magnetospheric structure, especially in the vicinity of the plasmapause.

Most of the data referred to in this paper have been obtained with the U.K. satellites, Ariel 3 and Ariel 4, which were launched into near-Earth (≈ 550 km altitude) high inclination orbits and, through the use of onboard tape recorders, provided world-wide coverage. Both satellites had receivers centered on 3.2, 9.6 and 16.0 kHz and, in addition, Ariel 4 had channels at 0.75, 1.25 and 17.8 kHz. The data sampling

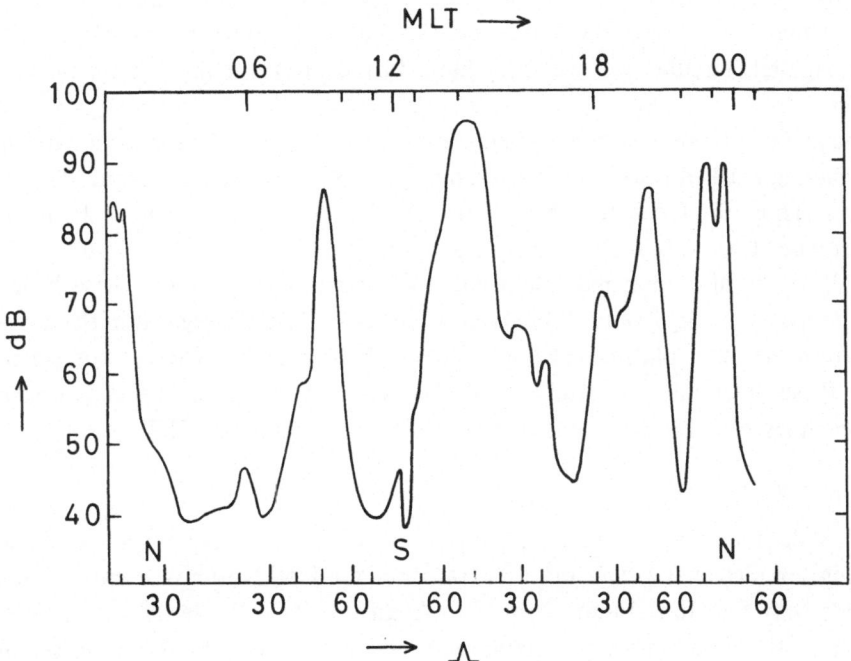

Fig. 1.   Mean signal intensity observed at 3.2 kHz with Ariel 3 on May 26, 1967, 1253–1423 UT. The abscissae are MLT and IN Lat and the ordinate is in decibels above $10^{-15}\ \gamma^2$ Hz^{-1}. The receiver noise level corresponds, approximately, to $10^{-11}\ \gamma^2$ Hz^{-1}.

B. M. McCormac (ed.), Magnetospheric Physics, 231–240. All Rights Reserved.
Copyright © 1974 by D. Reidel Publishing Company, Dordrecht-Holland.

consisted of recording the peak, mean and minimum signals in successive 28 s intervals.

Figure 1, which gives the mean signal intensity at 3.2 kHz around one orbit of Ariel 3 during a magnetically disturbed period in May 1967, illustrates the important emission regions. Strong high-latitude emissions appear between 2000 and 0000 MLT with maxima having intensity $\sim 10^{-6} \gamma^2$ Hz^{-1}* above $70° \Lambda$ (IN Lat). It is clear that the satellite passed through this zone twice, while travelling northward and southward, respectively. Medium latitude emissions are observed in each hemisphere (on both northward and southward passes) with maxima near $\pm 50° \Lambda$; enhanced emissions can also be seen at low latitudes ($< 30° \Lambda$). The local refractive index was obtained from the Birmingham University electron density experiment and used to deduce the incident power flux (assuming quasilongitudinal propagation with circular polarization).

2. The High Latitude Emissions

2.1. MORPHOLOGY

At the sensitivity level of the Ariel satellites, these emissions are observed on the majority of high latitude passes. The zone maximum lies on an oval which is polewards of the trapping boundary (as determined by the 35 keV electron fluxes) and of Feldstein's auroral oval. The location of the oval is strongly dependent on magnetic activity. In the quietest conditions the IN Lat of the emission oval varies between 73° at midnight and 79° at noon. With increasing K_p there is a general equatorwards movement which is greatest in the noon sector so that at the highest K_p the oval becomes approximately circular and even may bulge furthest towards the equator in the noon sector.

The probability of occurrence of emissions above a given intensity varies around the oval having marked peaks in the afternoon (~ 1400 MLT) and midnight (0000 MLT) sectors. Thus, for the period May 5 to July 24, 1967, the percentage frequency of occurrence at 3.2 kHz with intensity exceeding 5×10^{-16} W m^{-2} Hz^{-1} was in excess of 90% at the afternoon maximum and 80% at midnight (MLT). These have been interpreted as arising from soft electrons in the cusp (or cleft) regions in the afternoon and from the outer plasma sheet at midnight (Kaiser, 1972). There is some evidence from these morphological studies that the oval may really consist of two separate 'horse-shoes' centered in the midnight and afternoon sectors of MLT.

2.2. SPECTRA

Spectral data have been obtained from Ariel 3 by comparing the outputs of the mean-reading circuits at 3.2, 9.6 and 16 kHz for a total of 1180 hiss events during the periods May 5 to June 14 and July 17 to August 3, 1967. While there is considerable scatter in the data, some systematic features are apparent. In the noon sector the spectrum is relatively flat (within 2 to 3 dB) with maximum intensities reaching 5×10^{-14} W m^{-2} Hz^{-1}. Near midnight, while the spectrum is fairly flat between 9.6

* $1\gamma = 1$nT.

and 16 kHz, the 3.2 kHz intensity is, on the average, 8 to 10 dB above that at the higher frequencies. The maximum midnight intensity may reach 10^{-12} to 10^{-11} W m^{-2} Hz^{-1} at 9.6 kHz.

2.3. THE EMISSION MECHANISM

Various authors have attempted to explain the high latitude emissions as due to Heaviside (Cerenkov)* radiation from low energy precipitating electrons. It is generally assumed that the radiation is generated incoherently at altitudes up to several tens of thousands of km (below the level where the wave frequency equals the plasma or gyro-frequency) and is strongly ducted downwards along the field lines. Jørgensen (1968), using observed electron spectra for energies above 1000 eV with a realistic model for plasma density vs altitude, showed that intensities up to 10^{-14} W m^{-2} Hz could be obtained. More recently, English and Hughes (1974) have shown that the bulk of the radiated energy may be expected to arise from electrons of between 10 and 70 eV. Using an electron energy spectrum observed by Ackerson and Frank (1972) (extrapolated for energies less than 100 eV) they do obtain VLF intensities of 10^{-12} W m^{-2} Hz^{-1} or more but fail to reproduce the spectrum, particularly the enhancement at the lower frequencies observed in the midnight sector. Thus it seems that Heaviside radiation must be regarded as a candidate but either there is an additional process operating at frequencies somewhat below 10 kHz or, as suggested by some authors, partial coherence (involving radiation from localized bunches of electrons) must be invoked.

The association between the 'V-events' in the high altitude electron precipitation and VLF emissions which has been reported at this meeting by Gurnett (1973) is of considerable interest. If the 'V-events' are due to electric fields aligned with the magnetic field lines then we must invoke an anomalous conductivity along the field lines. This could be due to scattering of plasma particles by wave turbulence, however the waves are likely to be electrostatic. It is nevertheless possible that these could couple into the electromagnetic whistler mode and result in electromagnetic emissions observed at the foot of the field lines.

3. The Medium Latitude Emissions

These tend to maximize between 50° and 65° Λ with evidence for enhancements in the dawn and dusk sectors extending to lower latitudes. Unlike the high latitude emissions they exhibit a strong dependence on magnetic activity (Bullough et al., 1969) when the intensity at the Ariel 3 altitude is observed to reach a saturation value, typically 10^{-7} and 10^{-6} γ^2 Hz^{-1} on 9.6 and 3.2 kHz, respectively. The agreement of these values with the prediction of the weak diffusion theory of Kennel and Petschek (1966),

---

* This is radiation produced when a charged particle travels through a medium at a speed greater than the local phase velocity of electromagnetic waves. Although usually attributed to Cerenkov, the process was, in fact, first described most elegantly by Oliver Heaviside (1888).

together with evidence from the position of the high latitude boundary as a function of wave frequency (Bullough et al., 1969), lead us to conclude that the emissions are generated by cyclotron resonance with energetic electrons in the equatorial region, and are ducted along the magnetic field lines.

3.1. CO-ROTATING ZONES OF EMISSION

A striking feature of the Ariel 3 observations has been the discovery, even during magnetically quiet periods, of longitudinally localized zones of intense emissions which co-rotate with the earth and exhibit strong conjugate symmetry.

Lefeuvre and Bullough (1973) have reported a preliminary study of these emissions based on Ariel 3 observations during May 5 to October 31, 1967. The data were restricted to occasions when the signal at one or more of the three frequencies exceeded 5×10^{-14} W m^{-2} Hz^{-1}. Over this period the numbers of zone maxima (at 3.2 kHz) observed in quiet ($K_p \leqslant 2+$), moderately disturbed ($3- \leqslant K_p < 5+$) and very disturbed ($K_p \geqslant 6-$) periods were 770, 745 and 138, respectively, corresponding to a relative probability of occurrence of about 1:2:6. The intensity at the zone maxima is close to the saturation intensity obtained during magnetic storms (see above). A typical zone exhibits a maximum intensity of 5×10^{-12} W m^{-2} Hz^{-1} and is symmetrical with a width of 4° to -3 dB points, 8° to -10 dB and 17° to -20 dB. Such zones clearly co-rotate with the Earth as they may sometimes be identified 12 h or even 24 h later at the same local time and longitude (as the Earth rotates under the satellite orbit). Figure 2 represents a medium latitude zone observed on September 14, 1967. This was a rare occasion when it was possible (due to the magnetic field geometry) to observe several satellite passes over the same zone; it will be noted that this is a relatively weak zone (maximum intensity $\sim 2 \times 10^{-13}$ W m^{-2} Hz^{-1}).

The world-wide distribution of zone maxima at 3.2 kHz and $K_p \leqslant 2+$ (over the whole observing period) is shown in Figure 3. An outstanding feature is the frequent occurrence of maxima at lower latitudes ($40 \lesssim \Lambda \lesssim 50°$) in the invariant longitude interval $300 \lesssim \Phi \lesssim 360°$, to the west of South America and of the South Atlantic geomagnetic anomaly. Taking account of the number of passes and other variables, Lefeuvre and Bullough (1973) conclude that there is probably a permanent zone of emission in this region.

It is concluded that these zones are due to enhancements in the cold plasma density such that westward drifting energetic electrons encounter a region where the equatorial value of the characteristic energy $B^2/(2\mu_0 N)$ is depressed (Brice and Lucas, 1971). B is the magnetic field and N is the cold plasma electron density. The resulting particle precipitation is a likely cause of the 'ionospheric fingers', also longitudinally localized, reported by Piggott (1970). It is thus likely that diffusion upwards from these 'fingers' tends to make the zones self-sustaining. Kaiser (1972) has explained these effects in terms of a co-rotating modulation of the outer plasmasphere caused by substorm activity. In addition the South Atlantic anomaly is known to cause precipitation of energetic electrons (> 280 keV) somewhat east of but overlapping the region of the enhanced lower latitude (40 to 50°) emissions (Williams and Kohl, 1965).

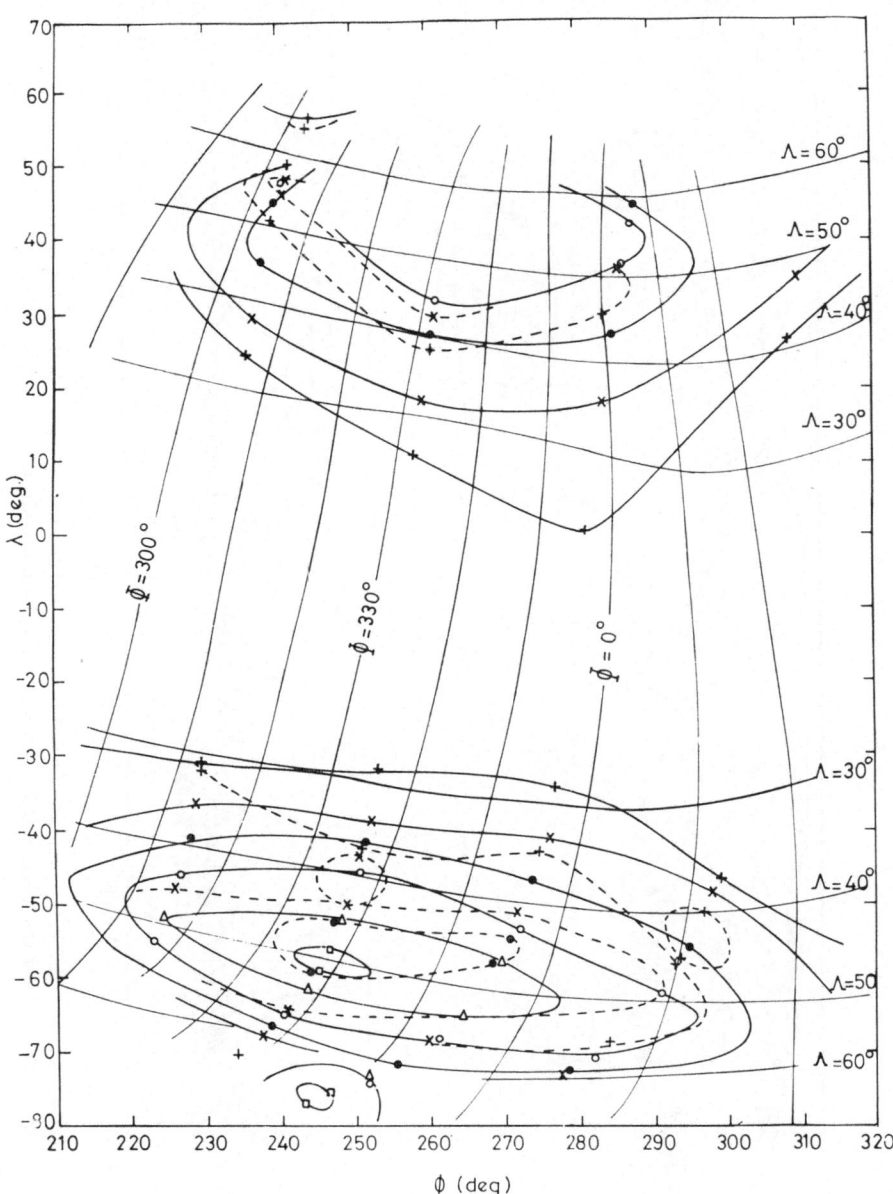

Fig. 2.   Localized emission zone, Sept. 14, 1967; ~ 0030 MLT. The contours give signal intensity in dB above 5×10^{-19} W m^{-2} Hz^{-1}: $\square \geqslant 55$, $\Delta \geqslant 50$, $\bigcirc \geqslant 45$, $\bullet \geqslant 40$, x $\geqslant 35$, $+ \geqslant 30$. ϕ, λ: geographic coordinates. Φ, Λ: invariant coordinates. ———3.2 kHz; – – – –9.6 kHz.

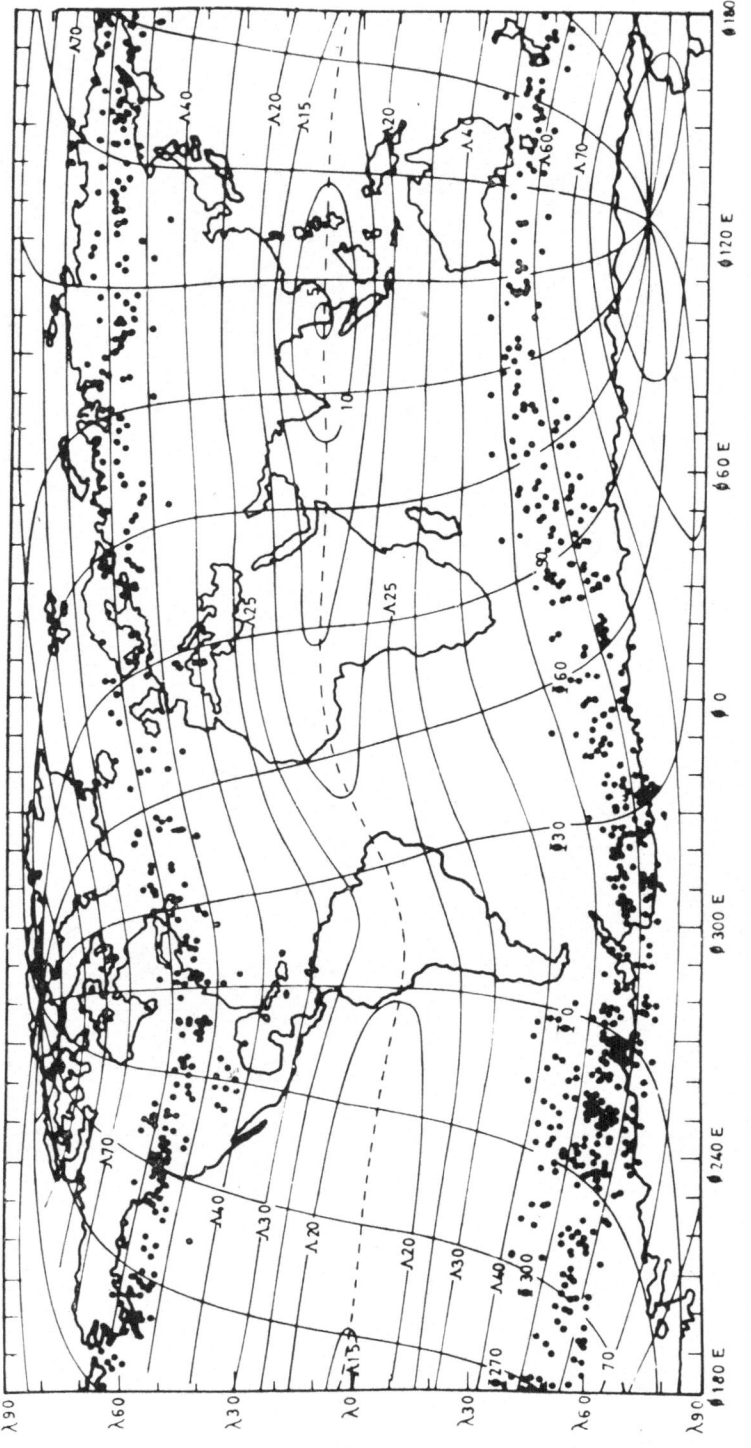

Fig. 3.   Distribution of zone maxima at 3.2 kHz (intensity $> 5 \sim 10^{-14}$ W m^{-2} Hz^{-1}) during magnetically quiet periods ($K_p \leqslant 2 +$). ϕ, λ: geographic coordinates. Φ, Λ: invariant coordinates.

3.2. SPECTRA

These, while exhibiting considerable scatter in individual events, show well defined systematic features. During quiet periods there is a marked diurnal variation. This is clear from Figure 4(a) which shows the median value of the ratio of power spectral density at 3.2 kHz to that at 9.6 kHz as a function of MLT. It varies from 35 dB at 11 h to 3 dB at 22 h. With increasing magnetic activity the diurnal variation decreases (Figure 4(b)) and in very disturbed periods, when the intensity saturates at the weak diffusion limit (see above), the spectrum tends to f^{-n} where f is the wave frequency and $n \approx 2$. Lefeuvre and Bullough (1973) explain the decrease in 9.6 kHz emissions 11 h in terms of the position of the energetic electron 'slot'. They conclude that the L value (>4.3) is then too high for the 9.6 kHz signals to propagate down from the equator (depending on whether the wave is ducted or loosely guided the wave frequency must be less than $f_H/2$ or f_H where f_H is the equatorial gyrofrequency). They suggest also that the 3.2 kHz intensity is reduced at night (~ 22 h) by reflection at the lower hybrid resonance (LHR) preventing the signals reaching low altitudes. They explain the decrease in diurnal variation with increasing K_p as due to the electron slot filling and more efficient ducting of the 3.2 kHz emissions enabling them to propagate to altitudes below that for which the wave frequency equals that of the LHR.

3.3. ELF EMISSIONS

The OGO-5 satellite has detected a steady band of ELF hiss on almost every passage through the plasmasphere. These are band-limited with a sharp lower frequency

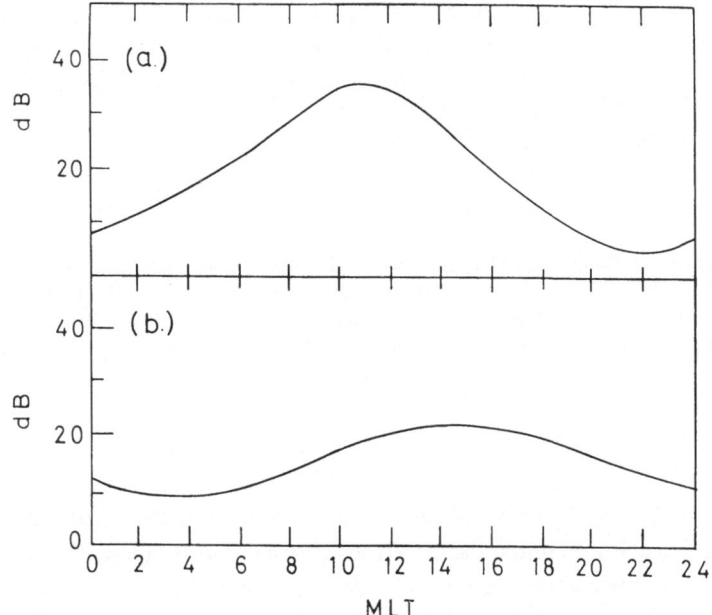

Fig. 4. Spectral variation in medium latitude emissions. The ordinate gives the median ratio (in dB) of the spectral density of 3.2 kHz emissions to that at 9.6 kHz. (a) very quiet periods ($K_p \leqslant 1$) and (b) disturbed periods ($K_p > 5 -$).

cut-off, a more diffuse upper cut-off and a maximum near a few hundred hertz with peak intensity ranging between 10^{-7} and 10^{-5} γ^2 Hz^{-1} (Thorne *et al.*, 1973). These authors find the observed properties of the hiss to be consistent with generation at all times in a restricted L range just inside the plasmapause. The most probable mechanism is the cyclotron resonance instability with electrons that continually diffuse inwards from the outer particle radiation zone. Since the properties of the hiss are essentially constant throughout the plasmasphere, they conclude that the radiation is non-ducted and propagates on complex paths to fill the plasmasphere. It is suggested that, as these waves propagate to lower L-shells, they scatter higher energy electrons thus forming the electron slot.

ELF emissions have also been observed on Ariel 4 which orbits at 550 km altitude. Again, the most striking feature is their ubiquity which is seen, not only in the occurrence of prominent zones on every record, but more especially in their enormous spatial extent. For instance, on December 23, 1971, apart from a 15° wide gap centered on the geomagnetic equator, a region of 750 Hz emission extended from 150°E, 80°N to 168°E, 80°S (geographic coordinates). Characteristic emission intensities were between 10^{-9} and 10^{-6} γ^2 Hz^{-1}. On closed field lines, the ELF power spectral

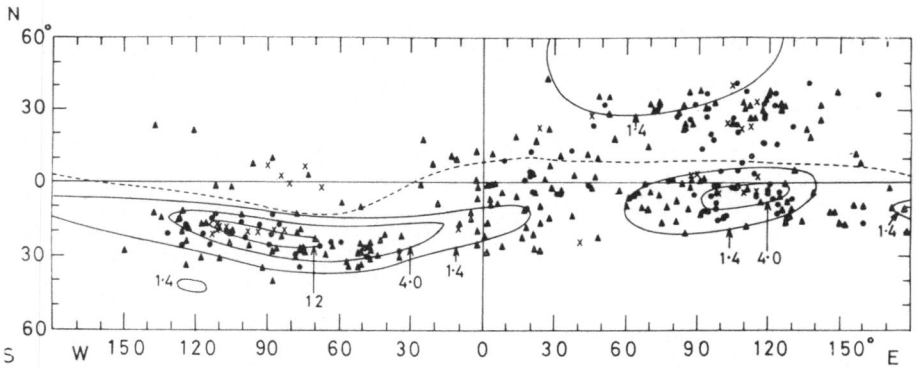

Fig. 5. Equatorial emissions (3.2 kHz) observed in the range $-30° < \Lambda < 30°$, on Ariel 3 during the period May 5, to July 24, 1967. Signal intensity (dB > 5 ~ 10^{-16} W m^{-2} Hz^{-1}): ▲, 0–5; ●, 5–10; x ≥ 10. The contours give the electron flux (in units of 10^{-6} m^{-2} s^{-1}) in the energy range 9–11 eV (Knudsen, 1968). The coordinates are geographical and the broken line locates the minimum L value at the Ariel-3 altitude.

density (like that at VLF) falls rapidly with increasing frequency; on open field lines the spectrum is generally much flatter. If, as Thorne *et al.* suggest, the plasmaspheric hiss propagates downwards on the complex ray paths predicted for non-ducted propagation we would expect it to be reflected above Ariel 4 altitudes. Nevertheless it is clearly observed, albeit with considerably lower intensities than measured on OGO-5 (especially when we convert γ^2 Hz^{-1} to W m^{-2} s^{-1}).

4. The Low Latitude Emissions

Maxima in emissions are frequently observed at latitudes $\sim 30°\Lambda$ and less (see Figure 1); they are generally less intense than those at medium and high latitudes, only rarely exceeding 5×10^{-14} W m^{-2} Hz^{-1} at 3.2 kHz. They appear to be absent during the daylight hours, 0600 to 1500 MLT ($K_p \leqslant 2 +$). The spectrum is such that the signal intensity at 3.2 kHz is characteristically between 5 and 15 dB above that at 9.6 kHz and shows less variability than at medium and high latitudes. There is little or no evidence of LHR reflection or of diurnal variation in the spectrum.

While these results are based on only limited data, one most significant feature is apparent from Figure 5. This is a close correlation with the flux of low energy electrons (10 eV) found by Knudsen (1968). It will be noted that the most intense emissions (crosses) agree well in location with the intense particle fluxes but show greater conjugate symmetry. This implies that the emissions are generated at relatively low altitudes, mainly in the southern hemisphere, again suggesting the Heaviside (Cerenkov) process as a likely mechanism.

5. Concluding Remarks

It has been possible here only to summarize a few of the significant data on wave-particle phenomena available from satellite observations. In particular, space has prevented discussion of the studies of man-made emissions, of spherics (produced by terrestrial thunderstorms) and of electrostatic waves, particularly those observed near the geomagnetic equator (Kennel *et al.*, 1970).

It must also be stated that the whole picture is not available from satellite experiments alone but requires, for its elucidation, a combination of ground based, rocket, and satellite observations. It is to be hoped that every effort will be made to achieve this, on a world-wide scale, during the forthcoming International Magnetospheric Study. We particularly draw attention, in relation to plasmaspheric studies, to the desirability of a network of Antarctic VLF stations in the vicinity of $L=4$ to support the wide range of satellite and rocket experiments planned for the IMS. One can also look forward to significant progress during this period from simultaneous observations, on the same geomagnetic field lines, from geostationary satellites, low altitude satellites, rocket and ground based studies.

References

Ackerson, K. L. and Frank, L. A.: 1972, *J. Geophys. Res.* **72**, 1128.
Brice, N. and Lucas, J.: 1971, *J. Geophys. Res.* **76**, 900.
Bullough, K., Hughes, A. R. W., and Kaiser, T. R.: 1969, *Planetary Space Sci.* **37**, 363.
English, H. W. and Hughes, A. R. W.: 1974, *Space Res.* **XIV**, 359.
Gurnett, D. A.: 1973, private communication.
Heaviside, O.: 1888, *The Electrician*, Nov. 23, p. 83.
Jørgensen, T. S.: 1968, *J. Geophys Res.* **73**, 1055.
Kaiser, T. R.: 1972, in B. M. McCormac (ed.) *Earth's Magnetospheric Processes*, D. Reidel Publishing Company, Dordrecht-Holland, p. 340.

Kennel, C. F. and Petschek, H. E.: 1966, *J. Geophys Res.* **71**, 1.

Kennel, C. F., Scarf, F. L., Fredricks, R. W., McGehee, J. H., and Coroniti, F. V.: 1970, *J. Geophys Res.* **75**, 6136.

Knudsen, W. G.: 1968, *J. Geophys. Res.* **73**, 6384.

Lefeuvre, F. and Bullough, K.: 1973, *Space Res.* **XIII**, 699.

Piggott, W. R.: 1970, *AGARD Conference Proceedings*, No. 49.

Thorne, R. M., Smith, E. J., Burton, R. K. and Holzer, R. E.: 1973, *J. Geophys. Res.* **78**, 1581.

Williams, D. J. and Kohl, J. W.: 1965, *J. Geophys. Res.* **70**, 4139.

WAVE-PARTICLE INTERACTIONS
NEAR THE GEOSTATIONARY ORBIT

M. ASHOUR-ABDALLA

Groupe de Recherches Ionosphériques, CNET, 92131, Issy-les-Moulineaux, France

and

S. W. H. COWLEY*

Cooperative Institute for Research in Environmental Sciences, University of Colorado/NOAA, Boulder, Colo. 80302, U.S.A.

1. Introduction

It has become abundantly clear during the past decade of experimental observation of the near-Earth magnetosphere that many of the gross characteristics of its particle population and their temporal behavior can at least be qualitatively understood in terms of a large-scale convection of plasma from the geomagnetic tail towards the dayside, as originally postulated by Axford and Hines (1961) and Dungey (1961). Using the very simplest of models for the convective flow i.e., a uniform dawn-to-dusk electric field across the magnetosphere associated with the tail-to-dayside motion, plus a radial electric field associated with the rotation of the Earth, leads to a basic understanding of the populations of low energy (1 to 10^4 eV) plasma within the magnetosphere, for the resulting flow pattern is divided into two parts. The flux tubes close to the Earth which corotate never become 'open' and are characterized by high densities of cold plasma from the ionosphere, while at larger distances flux tubes flow from the tail to the dayside carrying with them the hot plasma sheet particles which form the outer radiation zone and ring current. Thus the magnetosphere is divided into two regimes dominated by hot and cold plasma, respectively, and whose mutual boundary is the plasmapause (see, for example Russell and Thorne, 1970).

Time variations in the strength of the convection electric field cause changes in the position of the boundary between these two regimes and consequently a mixing of the two plasma populations. Specifically, an increase in the convection electric field causes a contraction of the region of corotating field lines. This allows some cold plasma from the plasmasphere to flow towards the dayside magnetopause, through which the hot plasma may drift, and also allows the penetration of hot plasma closer to the Earth, initially mainly on the nightside, into the region originally populated by cold plasma. The maintenance of the stronger convection field would eventually lead to the setting up of a new contracted boundary between the two populations, on a time scale of several hours. However, a subsequent electric field reduction leads to strong mixing effects. The 'forbidden zones' for the hot plasma expand, 'marooning' the near-Earth hot plasma on closed drift paths. This region then slowly fills with

* Present address: Physics Department, Imperial College, London SW7 2AZ.

B. M. McCormac (ed.), Magnetospheric Physics, 241–270. All Rights Reserved.

cold plasma from the ionosphere. Subsequent similar random fluctuations in the convective field strength acting on these 'marooned' particles over long periods of time can then cause a radial diffusion effect leading to the formation of a population of high-energy particles close to the Earth in a region to which the hot tail plasma never has access by direct flow (the inner radiation zone). In general, therefore, time variations in the convective flow lead to a 'blurring' of the boundary between hot and cold plasma regimes and a constant mixing between the two.

   Within this large-scale framework of magnetospheric convection dynamics the role of wave-particle processes may be considered as being concerned with the finer-scale details, although we do not wish *a priori* to rule out the possibility of a more fundamental role. Some of the areas in which wave-particle processes can play an important part within the scheme are the following: (a) Determination of the structure of the inner edge of the plasma sheet (outer radiation zone hot plasma) by strong pitch angle scattering (Kennel, 1969); (b) The decay of the ring current within the refilling plasmasphere following the decay of an electric field enhancement (Cornwall *et al.*, 1970, 1971); (c) Radial diffusion of the 'marooned' hot plasma within the plasmasphere produced by many random 'pumpings' of the cross tail potential, coupled with whistler-mode pitch angle scattering is thought to determine the structure of the inner electron radiation zone and 'slot' regions (Lyons *et al.*, 1972; Lyons and Thorne, 1973); (d) The localized enhancement of wave growth in the high energy plasma produced by the presence of dayside detached plasmasphere clouds (i.e., 'natural plasma seeding' (Kivelson and Russell, 1973).

   In order to discuss the important wave processes at a given location we should relate the location to the position of the plasma-pause, which is dependent on cross tail potential. OGO-6 electric field data indicate a usual range of potentials between 20 kV at quiet times and 100 kV at disturbed times (Heppner, 1972). Simplest theory then gives a plasmapause position at midnight at the equator of 7 R_E during quiet time and down to 3 R_E during disturbed periods. Therefore a proper analysis of wave processes at our chosen location, the synchronous orbit (6.6 R_E) should be in terms of the 'mixing region', in which control is transferred from cold to hot plasma and back again, time dependence being important. This description is certainly supported by the ATS-5 plasma measurements of DeForest and McIlwain (1971) and DeForest (1972). Processes listed under (a), (b), and (d) above should therefore be considered; (c) will not be further discussed. However, present theoretical status precludes us from attempting the complex time-dependent problem, so as a first step we shall consider exclusively the structure and stability of a steady-state outer zone (i.e., the region dominated by hot tail plasma) outside of the plasmasphere. This situation may correspond most closely to reality at the synchronous orbit during world-wide storms when convection may be strong for sufficiently long periods to set up a quasi-steady state with hot plasma enveloping 6.6 R_E.

   In considering a plasma stability problem it is of crucial importance to have a good idea of the expected plasma anisotropies. Only velocity-space anisotropies will be considered here. In the absence of detailed particle measurements (pitch angle and energy) much previous work has been done using an 'informed guess'. Clearly, a more satisfactory approach is to actually consider what type of anisotropies and variation of bulk parameters arises, naturally in a convecting magnetosphere, and then to analyze its stability. This problem is attempted here, and is used as a framework within which to review and hopefully constructively criticize the few previous analyses.

In the second section we therefore consider the anisotropies expected in the convecting outer zone hot plasma, while the remainder of the paper consists of various stability analyses (Section 3 the electron whistler mode; Section 4 the electron-electrostatic modes and Section 5 the ion-cyclotron mode.)

2. A Model for the Outer Zone Distribution Functions

A model for the distribution function anisotropies and variation of plasma bulk parameters with L value in the outer zone may be reasonably simply constructed if the particles conserve the first two adiabatic invariants:

$$\mu = \frac{W_\perp}{B} \quad \text{and} \quad J = m \int_{-s_m}^{s_m} v_\parallel \, ds \tag{1}$$

where $W_\perp$ is the perpendicular thermal energy of the particle, B the total field strength, s is distance along a field line from the equator and s_m is the distance of the mirror-point from the equator. In order to produce a manageable first problem we shall take the magnetic field structure to be given by a dipole, restricting ourselves to a L range of interest $1 \leqslant L \leqslant 10$. After development of some rough criteria for the validity of the conservation of μ and J in a dipole field we find that we are restricted to the following energy ranges; 10^{-5} to 10^8 eV for electrons and 10^{-2} to 10^6 eV for protons. The upper limit is provided by μ-conservation, the lower limit by J-conservation. This range of validity spans with comfort the entire spectrum of magnetospheric plasma energies, from the plasmasphere (few eV) to plasma sheet (few 100 eV to few keV) and storm time ring current (few 10's keV).

We therefore consider the distribution functions at the equator which are produced by plasma transport from a given L_0 (taken to be $L_0 = 10$ throughout the work) for particles which conserve μ and J. We consider equatorial single-particle properties first, subscript '0' denotes equatorial values at L_0. Conservation of μ gives

$$v_\perp = v_{\perp_0} \left(\frac{L_0}{L} \right)^{3/2}. \tag{2}$$

Dropping non-relativistic constants in J we define the invariant I

$$I = \int_0^{s_m} v_\parallel \, ds = vL R_E S(\alpha_{eq}) \tag{3}$$

where

$$S(\alpha_{eq}) = \int\limits_0^{\lambda_m(\alpha_{eq})} d\lambda \cos\lambda (1 + 3\sin^2\lambda)^{1/2} \times$$

$$\times \left\{ 1 - \sin^2\alpha_{eq} \frac{(1 + 3\sin^2\lambda)^{1/2}}{\cos^6\lambda} \right\}^{1/2}. \quad (4)$$

In Equation (4), α_{eq} is equatorial pitch angle, λ is latitude measured from the equator, λ_m is the mirror latitude and we have used μ-conservation along the field line. Function $S(\alpha_{eq})$ decreases monotonically from a value of 1.3082 at $\alpha_{eq} = 0°$ to zero at $\alpha_{eq} = 90°$. Conservation of I can then be written as

$$v = v_0 \left(\frac{L_0}{L}\right) \frac{S(\alpha_{eq0})}{S(\alpha_{eq})}$$

or by use of Equation (2) we have for the change in pitch angle with L, following Nakada et $al.$ (1965)

$$G(\alpha_{eq}) = \left(\frac{L}{L_0}\right)^{1/2} G(\alpha_{eq0}) \quad (5)$$

where

$$G(\alpha_{eq}) = \frac{S(\alpha_{eq})}{\sin\alpha_{eq}}.$$

Function G decreases monotonically from infinity at $\alpha_{eq} = 0°$ to zero at $\alpha_{eq} = 90°$, so that from Equation (5) we see that on convecting to smaller L values, the particle equatorial pitch angles will increase.

The change in α_{eq} with L is independent of energy in this nonrelativistic regime, a great simplifying feature. In the two special cases $\alpha_{eq} = 0°$ and $\alpha_{eq} = 90°$ there is no change in α_{eq} with L; for $\alpha_{eq} = 0°$ we have

$$\frac{v}{v_0} = \frac{v_\parallel}{v_{\parallel 0}} = \left(\frac{L_0}{L}\right) \quad (6)$$

i.e., the particle energy varies as $1/L^2$, while for $\alpha_{eq} = 90°$ we have

$$\frac{v}{v_0} = \frac{v_\perp}{v_{\perp 0}} = \left(\frac{L_0}{L}\right)^{3/2} \quad (7)$$

i.e., the energy varies as $1/L^3$. For large pitch angles we find

$$G(\alpha_{eq}) \simeq \frac{\pi}{6\sqrt{2}} \cot^2\alpha_{eq} \quad (8)$$

so that the change in equatorial pitch angle with L is given by, from Equation (5)

$$\tan\alpha_{eq} = \tan\alpha_{eq0} \left(\frac{L_0}{L}\right)^{1/4} \quad (9)$$

and the change in parallel velocity is given by

$$\frac{v_{\|}}{v_{\|0}} = \left(\frac{L_0}{L}\right)^{5/4} \quad \text{or} \quad \frac{E_{\|}}{E_{\|0}} = \left(\frac{L_0}{L}\right)^{5/2}. \tag{10}$$

This is a more rapid rise in parallel energy $(E_{\|})$ than occurs at $\alpha_{eq}=0°$, but since $E_{\perp}$ varies as L^{-3} the pitch angles increases with decreasing L. In Figure 1 we show the variation of $E_{\perp}$ and $E_{\|}$ with L for individual particles with differing α_{eq_0} obtained by numerical evaluation of Equation (5) and the use of Equation (2).

Let us now suppose that the velocity-space contours of constant distribution function value f are spheres at $L=L_0$ (e.g., an isotropic Maxwellian), that is, we assume that the plasma entering the dipolar region from the tail is isotropic in pitch angle. From the above results we may find the shape of the contours at any other L value if μ and I are conserved. In Figure 2 we show the deformation of a single sphere when convected to various L values. Apart from the increase in 'size' the major effect is a

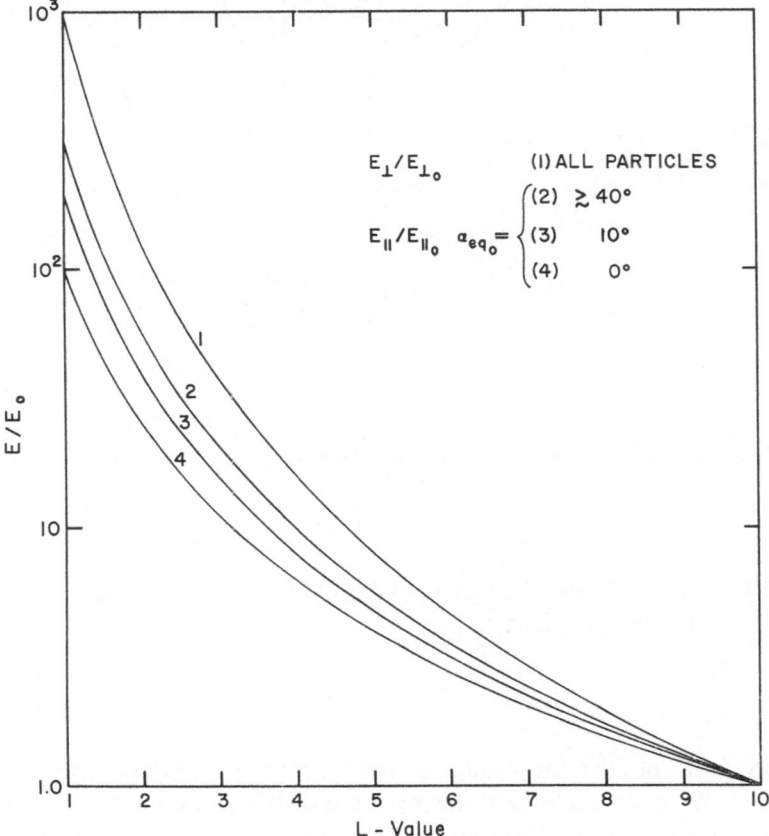

Fig. 1. The variation of $E_{\perp}/E_{\perp 0}$ and $E_{\|}/E_{\|0}$ with L for individual particles conserving μ and J, starting with differing a_{eq0} at L_0.

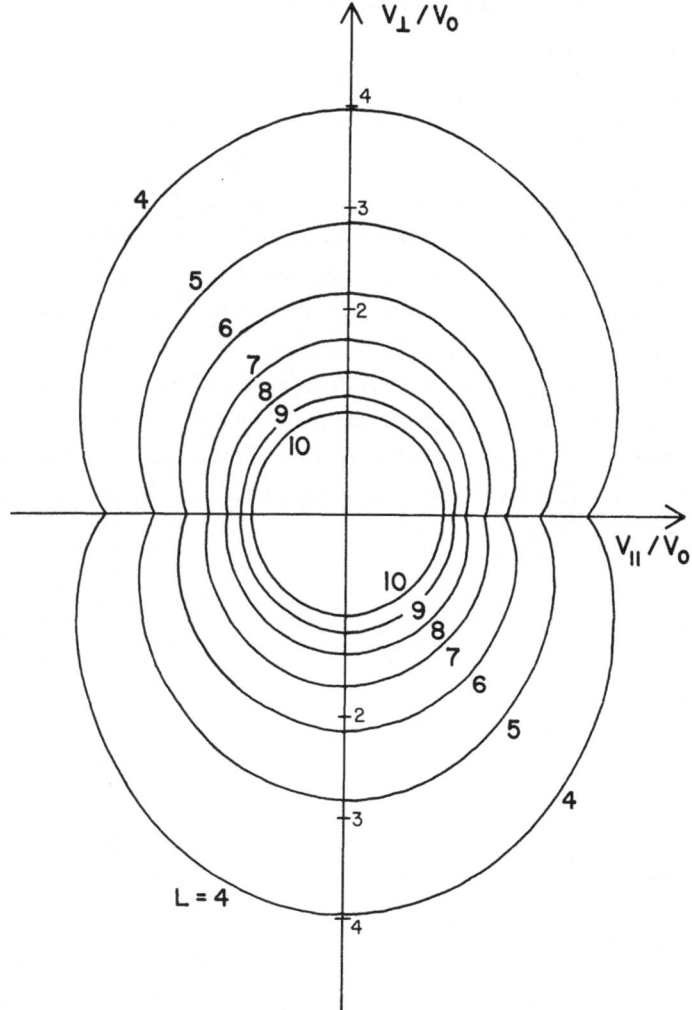

Fig. 2. Deformation of a spherical shell in velocity space with change in L value during convection, $L = 10$ to 4.

temperature anisotropy with $T_\perp/T_\parallel > 1$. In addition a 'cusp' is formed at small pitch angles, producing regions where

$$\frac{\partial f}{\partial v_\perp} > 0.$$

We may now consider the change in macroscopic parameters with L for these distributions. We have seen how a velocity-space shell deforms on inward convection, and all shells deform in the same way due to the previously noted fact that the change in α_{eq} with L is independent of energy for this model. Let the contribution of a given shell to the macroscopic parameters be ΔM at L and ΔM_0 at L_0. The ratio $\Delta M/\Delta M_0$

is independent of the shell considered since all deform in the same way. Since the final macroscopic parameters are obtained by a sum over the shells the ratio $\Delta M/\Delta M_0$ holds for the entire distribution. This presupposes that all shells from L_0 are present at L, a point we shall reconsider later. Assuming that this is true, we then find that

$$\frac{n}{n_0} = \left(\frac{L_0}{L}\right)^{9/2} \int_0^{\pi/2} d\alpha_{eq} \sin\alpha_{eq} \left(\frac{\sin\alpha_{eqo}}{\sin\alpha_{eq}}\right)^3$$

$$\frac{P_\parallel}{P_{\parallel 0}} = 3\left(\frac{L_0}{L}\right)^{15/2} \int_0^{\pi/2} d\alpha_{eq} \sin\alpha_{eq} \cos^2\alpha_{eq} \left(\frac{\sin\alpha_{eqo}}{\sin\alpha_{eq}}\right)^5$$

$$\frac{P_\perp}{P_{\perp 0}} = \frac{3}{2}\left(\frac{L_0}{L}\right)^{15/2} \int_0^{\pi/2} d\alpha_{eq} \sin^3\alpha_{eq} \left(\frac{\sin\alpha_{eqo}}{\sin\alpha_{eq}}\right)^5$$

$$\frac{P}{P_0} = \left(\frac{L_0}{L}\right)^{15/2} \int_0^{\pi/2} d\alpha_{eq} \sin\alpha_{eq} \left(\frac{\sin\alpha_{eqo}}{\sin\alpha_{eq}}\right)^5$$

(12)

where n is density, $P_\perp$ and $P_\parallel$ perpendicular and parallel pressures respectively, and P is the total pressure (total energy density). The L dependence in the integrals enters through the function

$$\left(\frac{\sin\alpha_{eqo}}{\sin\alpha_{eq}}\right)$$

which is obtained from Equation (5). From Equation (12) we have the further ratios

$$\frac{\beta}{\beta_0} = \left(\frac{L}{L_0}\right)^6 \frac{P_\perp}{P_{\perp 0}}\left(\frac{n/B}{n_0/B_0}\right) = \left(\frac{L}{L_0}\right)^3 \frac{n}{n_0}\frac{\bar{E}}{E_0} = \left(\frac{p/n}{p_0/n_0}\right)$$

where β is the ratio of perpendicular energy density to magnetic field density, $\bar{E}$ is the average particle energy. The numerical results are shown in Figures 3 and 4 and are thought to be accurate to within about 2%, (we note that the CGL invariants are obeyed to this accuracy). It can be seen that both β and n/B increase with decreasing L.

We now turn to consider in more detail the actual quantities required for the stability analysis. These are first, the velocity space anisotropy and second the absolute values of the plasma bulk parameters. We assume that the distribution function at L_0 is Maxwellian in energy as well as isotropic in pitch angle (Hones *et al.* (1971) indicate that the thermal part of the plasma sheet electron distribution is often of Maxwellian form.) We note that this is the most stabilizing choice which can be made. As previously noted, the velocity-space anisotropy produced by inward convection is basically a temperature anisotropy with $T_\perp/T_\parallel > 1$. In order to simplify the stability analysis of these distributions we shall attempt to approximate them by bi-Maxwellians which reflect this basic property. Let the temperature of the Maxwellian at L_0 be T_0, then

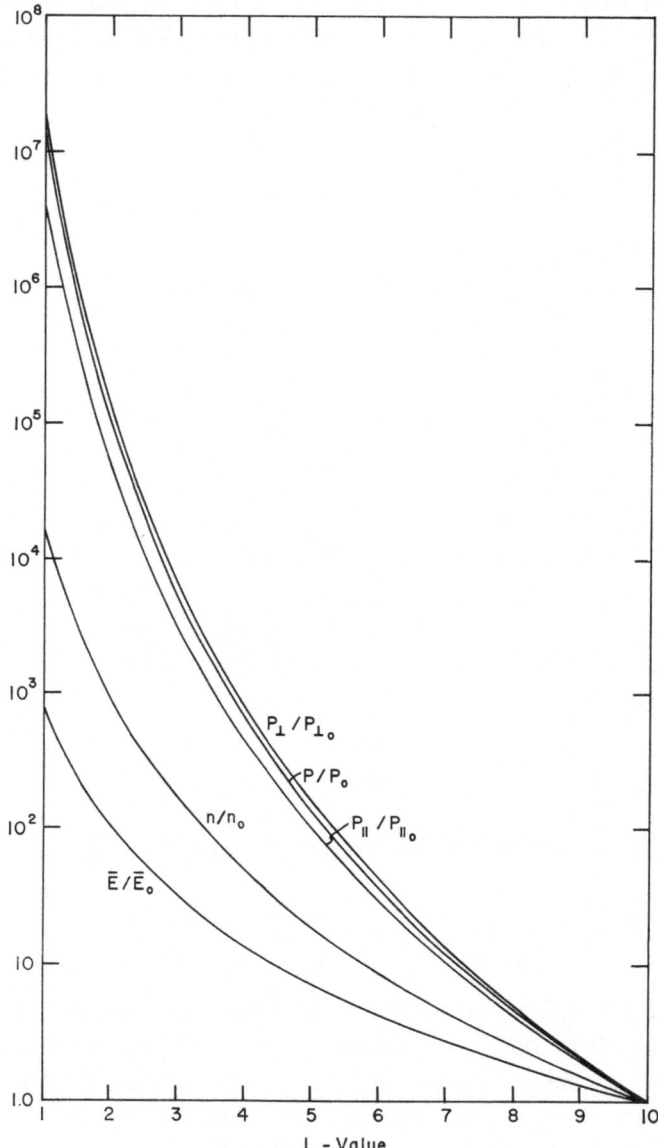

Fig. 3. Variation with L of the macroscopic plasma parameters, pressure ($P_\perp$ and $P_\parallel$), density and average energy.

Equation (2) implies that the $v_\perp$ distribution at any other L will be Maxwellian with temperature

$$T_\perp = \left(\frac{L_0}{L}\right)^3 T_0 . \tag{13}$$

However, the parallel temperature for small pitch angles from Equation (6) should

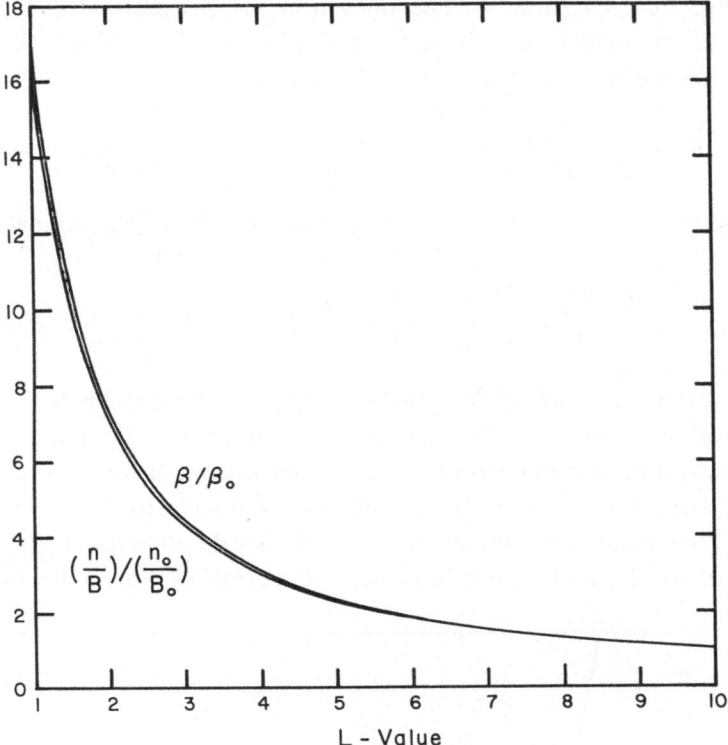

Fig. 4. Variation with L of β and n/B.

be given by

$$T_{\parallel} = \left(\frac{L_0}{L}\right)^2 T_0 \tag{14}$$

while for large pitch angles Equation (10) implies

$$T_{\parallel} = \left(\frac{L_0}{L}\right)^{5/2} T_0. \tag{15}$$

If either Equations (6) or (10) held at all pitch angles then the bi-Maxwellian approximation would become exact, in the case of Equation (6) and (14) corresponding to a temperature anisotropy

$$A = \frac{T_{\perp}}{T_{\parallel}} - 1 = \left(\frac{L_0}{L}\right) - 1 \tag{16}$$

or in the case of Equations (10) and (15)

$$A = \left(\frac{L_0}{L}\right)^{1/2} - 1. \tag{17}$$

These then represent the upper lower limits of the temperature anisotropy of the real functions. However, since we integrated over the distributions to find $P_\perp/P_{\perp_0}$ and $P_\parallel/P_{\parallel_0}$ we may simply consider the bi-Maxwellian which would give these ratios. We therefore define 'average' temperature anisotropy by

$$\langle A \rangle = \left(\frac{P_\perp}{P_{\perp_0}} \cdot \frac{P_{\parallel_0}}{P_\parallel} \right) - 1 \tag{18}$$

i.e., the perpendicular temperature is given by Equation (13) and the parallel temperature by

$$\frac{T_\parallel}{T_0} = \left(\frac{P_\parallel}{P_{\parallel_0}} \cdot \frac{P_{\perp_0}}{P_\perp} \right) \frac{T_\perp}{T_0} = \left(\frac{P_\parallel}{P_{\parallel_0}} \cdot \frac{P_{\perp_0}}{P_\perp} \right) \left(\frac{L_0}{L} \right)^3. \tag{19}$$

The averaged anisotropy defined by Equation (18) with the upper and lower limits of Equations (16) and (17) is shown in Figure 5. As can be seen, the anisotropies are generally below unity for outer-zone L values, but become large for small L. The lower limit (Equation (17)) provides a better approximation to $\langle A \rangle$ than does the upper limit. This is expected because Equation (6) is valid only for small α_{eq}, while Equation (10) is valid to 10% over a considerable range of pitch angle ($\alpha_{eq} \gtrsim 40°$).

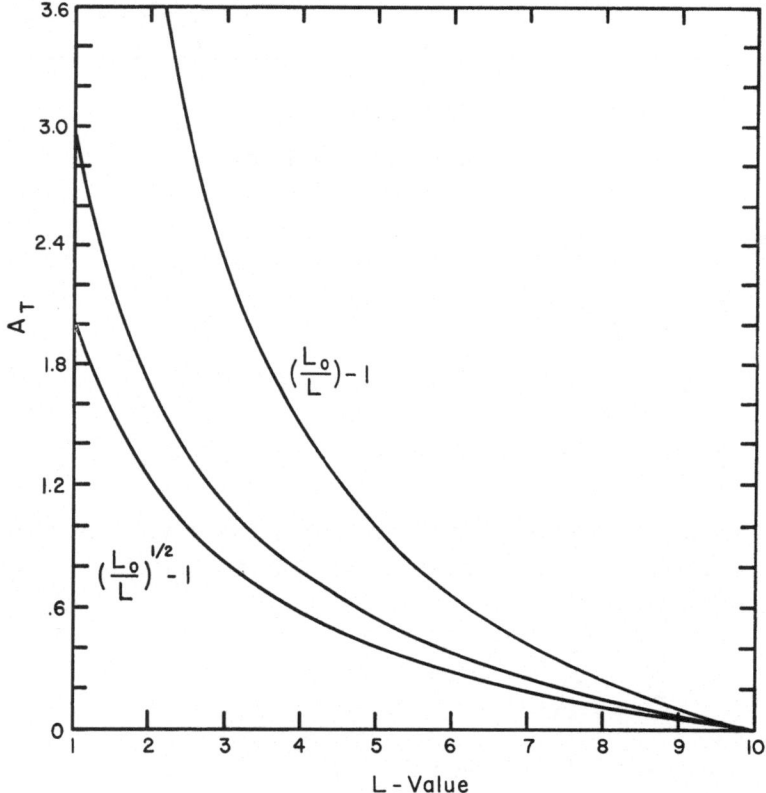

Fig. 5.   Variation with L of the plasma thermal anisotropy, compared with its expected upper and lower limits.

Turning now to the absolute values of the bulk parameters, since the ratios n/n_0 etc., have been calculated, we just need to set their values at $L_0 = 10$. We have used the following pieces of observational material:

(a) The plasma β-value near the equator at $L \sim 10$ is usually near or a little less than unity, even for quiet times (Sugiura, 1972).

(b) Proton energy densities are larger than electron energy densities by factors of 2 to 5 in both the plasma sheet and outer zone (Bame, 1968; Frank, 1971).

(c) Number densities at $L \sim 10$ are of the order of $n \sim 1$ cm^{-3}, with average particle energies a few keV (Schield and Frank, 1970). There will be an incompatibility between these requirements and our model since the real magnetic field at $L = 10$ will differ from a dipole. In this situation we choose to represent the plasma β-value in a reasonable way, and then adjust particle densities and average energies to fit these chosen values. In the calculations presented here the following values were taken:

$$\beta_{0_e} = 0.2 \qquad \beta_{0_p} = 0.6 \quad n_0 = 0.5 \text{ cm}^{-3}$$

leading to average energies

$$\bar{E}_e = 0.958 \text{ keV} \quad \bar{E}_p = 2.88 \text{ keV}$$

which would appear to be reasonable. The above graphs can then be used to predict the equatorial bulk parameters at any other L value. For example, at the synchronous orbit (6.6 R$_E$) we find that

$$\beta \simeq 1.33 \quad n = 2.9 \text{ cm}^{-3} \quad \bar{E}_e \simeq 3.1 \text{ keV} \quad \bar{E}_p \simeq 9.4 \text{ keV}$$

which appear to be in reasonable agreement with the observations of DeForest and McIlwain (1971). They report β-values in the nighttime sector between 0.1 and 3, densities between 1 and 4 cm^{-3} (typically about 2 cm^{-3}) and average energies $\bar{E}_e \sim 5$ keV, $\bar{E}_p \sim 15$ keV.

The contours of constant distribution function should be valid for any transport process which conserves μ and J, provided the magnetic field is assumed a dipole. However, as previously noted the bulk parameter ratios are only valid provided that all of the 'shells' at L_0 are in fact transported to any other L-value under consideration. For transport by convection electric fields this will not be true. As the plasma convects towards the Earth the finite-temperature magnetic and corotational electric field azimuthal drifts become important, and lead to the formation of 'forbidden zones' around the Earth for particles of fixed μ and J. The calculated distribution function will therefore be 'cut-off' above (and for protons inside the plasmasphere, below) a certain energy at all L values. These 'cut-off' energies depend on pitch angle and L-value for a given convection electric field. As we move from L_0 towards the Earth the cut-off energies will steadily decrease, going to zero at all pitch angles at or near the plasmapause. The ratios based on our distribution which neglected the cut-offs will therefore have a larger value than the 'real' value, but this is negligible at L values sufficiently far away from the plasmapause such that 'most' of the original distribution is contained within the upper cut-off energy. That is, if E_c and E_{th} are the cut-off and thermal energies at a given L, then we require, say $E_c \geqslant 2E_{th}$ for the ratio values to be of reasonable accuracy.

In order to obtain a quantitative estimate of the upper cut-off energy relative to the model thermal energy, we consider a simple convection model consisting, at the equator, of a uniform dawn-to-dusk convection electric field plus a radial corotation electric field in a dipole magnetic field. As far as is known to us, only the motion of $90°$ pitch angle particles has been investigated in such a configuration (Chen, 1970). We follow the latter work here. The effective potential for flow in the equatorial plane is

$$\Phi = -\left\{ \frac{\Phi_c L \sin \varphi}{D_T} + \frac{2\pi R_E^2 B_E}{c T_D L} \pm \frac{\mu B_E}{e L^3} \right\} \tag{20}$$

where Φ_c is the total cross tail convection potential, D_T the tail diameter in Earth radii ($D_T = 40$); B_E the idealized magnetic field at the equator on the Earth's surface ($B_E = 0.3106$ G); T_D the length of a day in seconds; φ the solar magnetospheric azimuthal angle (measured positive anticlockwise from noon when looking on the north pole) and μ the particle magnetic moment. The drift is directed perpendicular to $\mathbf{B}$ and $\nabla \varphi$ and is proportional to $|\nabla \Phi|$. The first two terms on the right hand side of Equation (20) are from the $\mathbf{E} \wedge \mathbf{B}$ drift in the convection and corotation electric fields, respectively, while the third term represents the ∇B-drift. The upper sign is valid for electrons, the lower for protons. For electrons the ∇B and corotation azimuthal drifts are in the same direction and the resulting hot-electron forbidden zones are simple extensions of the well-known plasmasphere structure. A stagnation point forms on the dusk meridian ($\varphi = \pi/2$) at an L value given by requiring $(\partial \Phi / \partial L) = 0$ from Equation (20), giving

$$L_s = \frac{L^*}{\sqrt{2}} \left\{ 1 + \left(1 + \frac{12\mu B_E D_T}{e\Phi_c L^{*4}} \right)^{1/2} \right\}^{1/2} \tag{21}$$

where

$$L^* = \left(\frac{2\pi R_E^2 B_E D_T}{c\Phi_c T_D} \right)^{1/2}$$

is the plasmapause ($\mu = 0$) stagnation point L value. Note that L_s increases with increasing μ. The forbidden zone boundary for electrons of given μ as a function of azimuth φ is then given by the cubic equation

$$\Phi(L, \varphi) = \Phi(L_s, \varphi = \pi/2) \tag{22}$$

in an obvious manner. The boundary for particles whose perpendicular energy is K times the perpendicular thermal energy at that L value is simply obtained. If we consider a particle at L_0 with energy $E_{\perp_0} = KE_{th_0}$ then its perpendicular energy at L is

$$E_\perp(L) = \left(\frac{L_0}{L} \right)^3 E_{\perp_0}. \tag{22}$$

However, from our model calculation we found that

$$E_{\perp_{\mathrm{th}}}(L) \simeq \left(\frac{L_0}{L}\right)^3 E_{\mathrm{th}0}$$

so that $E_{\perp}(L) \simeq K E_{\perp_{\mathrm{th}}}(L)$ i.e., the proportionality is maintained in the convective flow. Therefore the position of the boundary where the cut-off energy is K times the thermal energy at the boundary position is given by the forbidden-zone boundary of the particle whose magnetic moment corresponds to an energy K times the thermal energy at L_0, i.e., whose μ-value is

$$\mu = \frac{E_\perp}{B_0} = \frac{K E_{\mathrm{th}0} L_0^3}{B_{\mathrm{E}}} = \frac{K \beta_0 B_{\mathrm{E}}}{8 \pi n_0 L_0^3}. \tag{22}$$

We consider here the cut-off energies only along the midnight meridian. For $\mu = 0$ the boundary (plasmapause) is given by $L^*/2$ and is shown in Figure 6 as a function of Φ_{c}. As previously noted, the results of Heppner (1972) indicate that $\Phi_{\mathrm{c}} \sim 20$ kV may correspond to quiet times, with $\Phi_{\mathrm{c}} \sim 50$ kV a 'typical' value, while $\Phi_{\mathrm{c}} \sim 100$ kV may correspond to periods of strong disturbance. The electron upper cut-off energy $\mathscr{E}_{\perp_{\mathrm{c}}}$ compared with $\mathscr{E}_{\perp_{\mathrm{th}}}$ (i.e., $K = \mathscr{E}_{\perp_{\mathrm{c}}}/\mathscr{E}_{\perp_{\mathrm{th}}}$) is shown as a function of L for these three

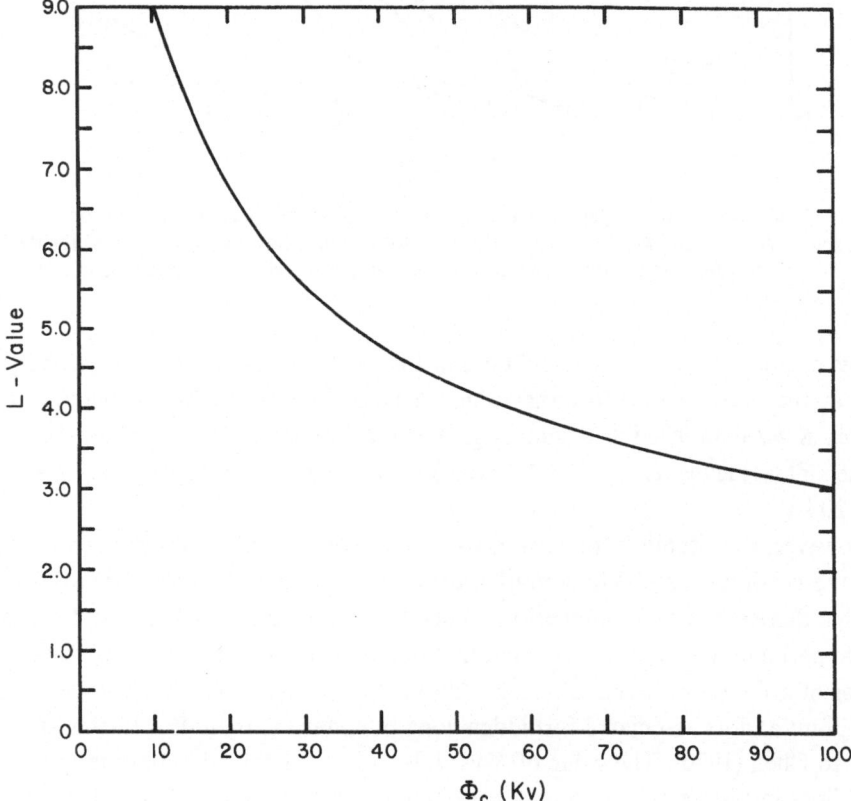

Fig. 6. The midnight position of the plasmapause vs. the cross-tail potential.

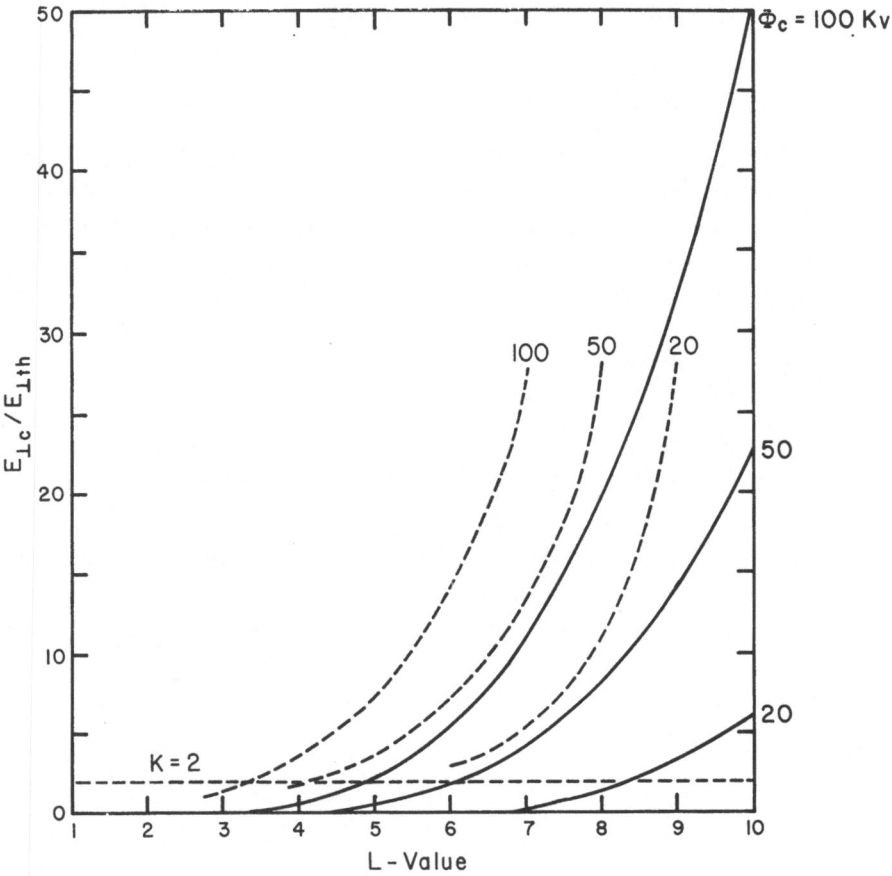

Fig. 7.   Solid lines show the perpendicular cut-off energy in the electron distribution functions at midnight vs. L value, for $\Phi_c = 20$, 50 and 100 kV. Dotted lines show the upper cut-off for the further criterion for population of the drift-paths with plasma sheet particles (Section 2).

values of potential Φ_c by the solid lines in Figure 7. The value of K rises rapidly from zero at the plasmapause to large values at large L. The criterion of validity of our model, $K \gtrsim 2$ is satisfied at distances greater than some 1.5 to 2 R_E from the plasmapause. This means $L \gtrsim 5$ for $\Phi_c \sim 100 \, \text{kV}$; $L \gtrsim 6$ for $\Phi_c \sim 50 \, \text{kV}$ and $L \gtrsim 8.5$ for $\Phi_c \sim 20 \, \text{kV}$.

However, the situation for protons is rather more complex, resulting from the fact that the corotation and ∇B azimuthal drifts are oppositely directed. Below a certain μ-value there exists three stagnation points, two on the dusk meridian and one on the dawn meridian. One of the dusk meridian points is an 'O' type stagnation point in the center of a region of 'banana' orbits. Above this μ-value only the dawn-side stagnation point and its associated forbidden-zone boundary remain. For further details we refer to Chen (1970). The value of K vs. L is shown for the midnight meridian by the solid lines in Figure 8. Protons can penetrate much closer to the Earth than electrons, for there exists within the plasmasphere a region with both an upper and lower cut-

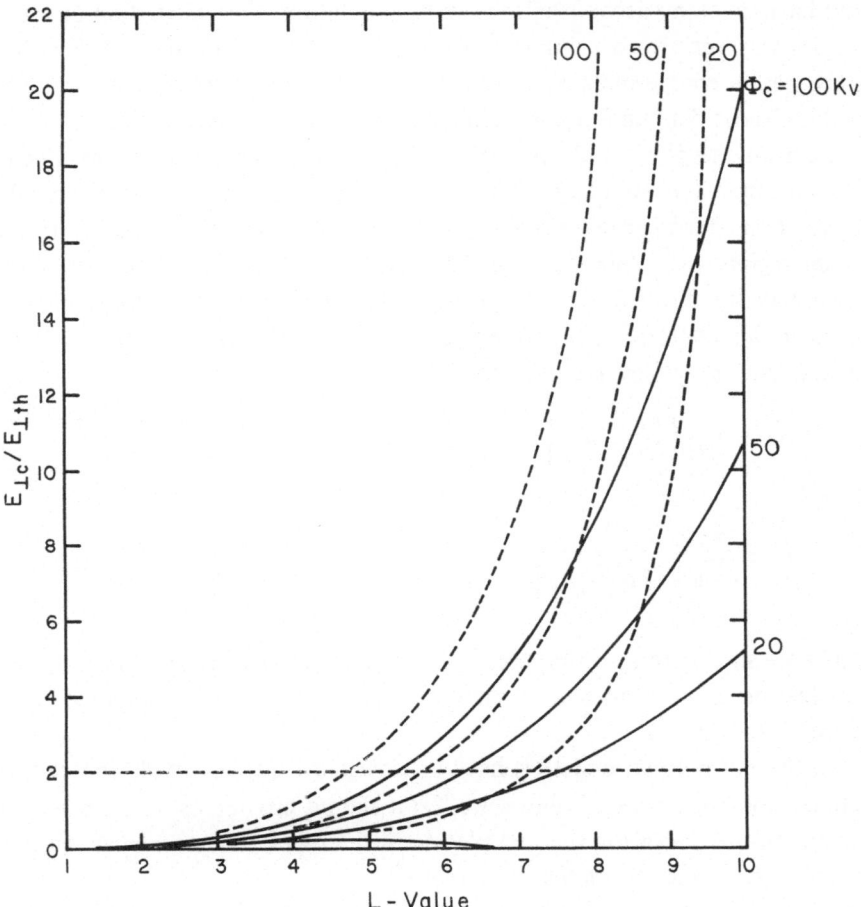

Fig. 8. Same as Figure 7, exception for protons.

off in $\mathscr{E}_\perp$, resulting in a ring-like distribution function. It results from the splitting up of the region into two distinct forbidden-zones associated with the dawn and dusk-side stagnation points with an open-orbit 'chanel' running between, above a critical value of μ. Comparing Figure 8 with Figure 7 we note that the upper cut-off K-value for protons is generally rather smaller at a given L, Φ_c value than that for electrons. This is due to the higher thermal energy at L_0 chosen for protons ($E_{th_p} = 3E_{th_e}$). If we had taken equal thermal energies at L_0, then the proton upper cut-off K value would have been somewhat higher than that of the electrons.

Before proceeding to a discussion of the cold ionspheric plasma in the outer zone we would like to make some comments about the validity of the model for the hot plasma which we have constructed above. The model has some shortcomings and is clearly not self-consistent for a number of reasons. First, a dipole field has been assumed and no account has been taken of the day-night asymmetry due to the mag-

netopause and tail current systems. Then we introduced a high-β plasma into this
field and have taken no account of the resulting inflation. The calculation of the cut-
off energies was performed using a uniform convection electric field. This will immediately yield a charge imbalance between electrons and protons, and a modification of this electric field until $n_p = n_e$. This latter modification may not be too serious
until the cut-off energies are of the order of the thermal energies, however. Furthermore, it was implicitly assumed that all 'open' orbits were populated by plasma from
the plasma sheet. It is clear that some regions of open orbits will in fact originate from
the dipolar region magnetopause, especially at high μ-values. As an estimate of these
effects we may simply write that the energy gained by a particle in convecting to a
given L at midnight from L_0 cannot exceed one half the cross-tail potential. This is
a necessary, but not sufficient condition. Very roughly, we write

$$\Delta E = E_0 \left[\left(\frac{L_0}{L} \right)^3 - 1 \right] \leq \frac{e\Phi_c}{2}$$

or if $E_0 = KE_{th}$

$$K \lesssim \frac{e\Phi_c}{2E_{th}} \cdot \frac{1}{\left[\left(\dfrac{L_0}{L} \right)^3 - 1 \right]}.$$

These lines are shown dotted in Figures 7, 8 and 9, and as can be seen, this very rough
criterion does not impose any new restrictions on the upper cut-offs except for protons
at very low values of Φ_c.

Though the hot plasma model we have constructed is rather rough and certainly
has its limitations, as discussed above, we feel it should provide us with a reasonable
guide to the properties of the thermal plasma in the outer zone and hence an initial
framework within which to discuss its stability.

We shall now briefly consider the properties of the cold plasma (few eV) of ionospheric origin which is present in the outer zone. Within the plasmasphere (i.e., the
corotation region) the cold plasma is in equilibrium with the ionosphere and has high
density, 10^2 to 10^3 cm^{-3} (Harris et al., 1970). In the region where field lines periodically become 'open' (i.e., the outer zone) the ionosphere is continuously in a state of
trying to fill the flux tubes. The density corresponding to this flow situation is very
low. For example, if we take a topside light-ion flux of $\sim 10^8$ ions cm^{-2} s^{-1} (Banks
and Holzer, 1968) and a flow speed of ~ 40 km s^{-1} ($\mathscr{E}_p$ (bulk) ~ 10 eV) then the particle
density on an open tail field line where the field strength is ~ 30 γ is estimated at
~ 0.01 cm^{-3}. As field lines reconnect in the tail this streaming plasma becomes trapped and probably thermalized on closed field lines and its subsequent dynamics may
be complex (Banks et al., 1971). Transport of the reconnected flux tubes towards the
Earth will compress and heat the trapped cold plasma in the same way as for the hot
plasma and in addition there will be continuous infilling from the ionosphere during
the convection, raising the value of n_c. In this complicated situation we have compared
the value of n_H from our model with the direct measurements of n_c in the outer zone,
taken from Sharp et al. (1972), and found that a reasonable model was to take

$n_c/n_H \simeq 0.1$ throughout the outer zone (e.g., $n_c \simeq 0.05$ cm^{-3} at $L = 10$). The fact that a constant ratio turns out to be reasonable implies that the cold plasma already present on the field line at $L_0 = 10$ is a large contributor to n_c throughout the outer zone, and that infilling from the ionosphere is not particularly significant during convection from $L_0 = 10$ up to the plasmapause. (That is, the infilling may provide, say, about an equal amount of plasma to that already present at $L_0 = 10$ during the convection, but not say, ten or twenty times as much.) This tentative conclusion was confirmed by an order of magnitude estimate of the infilling rate, based on a topside flux of 10^8 ions cm^{-2} s^{-1}. The temperature ratio between hot and cold species is about 10^{-2} (i.e., about 10 eV for the thermalized polar wind, at $L_0 \simeq 10$ as opposed to $\simeq 1$ keV for the hot particles). Thus, we will use in our analyses $n_c/n_H \simeq 0.1$ and $\bar{E}_c/\bar{E}_H \simeq 0.01$, taken to be rough constants throughout the outer zone.

3. The Whistler Instability in the Outer Zone

Resonant interactions between particles and waves violate the conservation of the magnetic moment and cause diffusion in energy and pitch angle. If the waves are sufficiently strong and within the right frequency band to resonate with the bulk of the particles, then the resulting pitch angle scattering and precipitation can be extremely important in modifying the structure of the outer zone plasma (Kennel and Petschek, 1966; Kennel, 1969; Coroniti and Kennel, 1971) in addition to the possibility of producing high energy particles within their 'forbidden zones'.

The obvious initial candidates for consideration are the electromagnetic whistler-mode wave for gyroresonance with electrons and the ion-cyclotron wave for protons. In this section the whistler-mode wave is considered, while in Section 5 similar analysis is performed for the ion-cyclotron wave. For simplicity, we consider waves propagating parallel to the magnetic field near the equator. This is expected to be the most unstable region in space and $\mathbf{k}$ spectrum, waves propagating at large angles suffer from Landau damping. The real and imaginary parts of the dispersion relation for small growth rate are (Kennel and Petschek, 1966)

$$\frac{k_\parallel^2 c^2}{\omega^2} = 1 - \frac{\pi \omega_{pe}^2}{k_\parallel \omega} \int_0^\infty dv_\perp v_\perp^2 P \int_{-\infty}^\infty dv_\parallel \frac{\left[\dfrac{\partial f}{\partial v_\perp} - \dfrac{k_\parallel}{\omega}\left(v_\parallel \dfrac{\partial f}{\partial v_\perp} - v_\perp \dfrac{\partial f}{\partial v_\parallel} \right) \right]}{\left(v_\parallel + \dfrac{(\Omega_e - \omega)}{k_\parallel} \right)} \quad (23)$$

$$\gamma \left[2\omega - \frac{\pi \omega_{pe}^2}{k_\parallel} \int_0^\infty dv_\perp v_\perp^2 P \int_0^\infty dv_\parallel \frac{\dfrac{\partial f}{\partial v_\parallel}}{\left(v_\parallel + \dfrac{(\Omega_e - \omega)}{k_\parallel} \right)} \right]$$

$$= \pi \omega_{pe}^2 \eta (v_{res}) \left[A_e(v_{res}) - \frac{1}{\left(\dfrac{\Omega_e}{\omega} - 1 \right)} \right] \quad (24)$$

where ω and γ are, respectively, the real and imaginary parts of the frequency($\gamma > 0$ means wave growth); ω_{pe} is the electron plasma frequency; $k_{\parallel}$ is the wave vector; Ω_e the electron gyrofrequency, and P indicates that the principal value of the singular integrals are to be taken. Throughout this work the distribution function f for a given species is normalized such that

$$\int\int_{-\infty}^{\infty}\int d^3vf = 1 .$$

The proton terms have been omitted from Equation (23) and (24) since we are considering waves with $\omega \sim \Omega \gg \Omega_p$. The resonant particle parallel velocity is

$$v_{res} = \frac{\omega - \Omega_e}{k_{\parallel}} \tag{25}$$

and the anisotropy A_e at that velocity is defined by

$$A_e(v_{res}) = \frac{\dfrac{1}{v_{res}}\displaystyle\int_0^{\infty} dv_{\perp}v_{\perp}^2\left(v_{\parallel}\dfrac{\partial f}{\partial v_{\perp}} - v_{\perp}\dfrac{\partial f}{\partial v_{\parallel}}\right)\bigg|_{v_{\parallel}=v_{res}}}{2\displaystyle\int_0^{\infty} dv_{\perp}v_{\perp}f\bigg|_{v_{\parallel}=v_{res}}} . \tag{26}$$

The function $\eta(v_{res})$ is a measure of the number of particles near resonance, given by

$$\eta(v_{res}) = 2\pi|v_{res}|\int_0^{\infty} dv_{\perp}v_{\perp}f\bigg|_{v_{\parallel}=v_{res}} \tag{27}$$

The real part of the dispersion function (23) determines the dependence of ω on $k_{\parallel}$, while the imaginary part of Equation (24) determines the growth rate γ. If we use our simplified outer zone model distributions, i.e., bi-Maxwellians, then the anisotropy is

$$A_T = \frac{T_{\perp}}{T_{\parallel}} - 1 \tag{28}$$

independent of v_{res} and the real part of the dispersion function is, for $c \gg \omega/k_{\parallel}$

$$\frac{k_{\parallel}^2 c^2}{\omega^2} = \frac{\omega_{pc}^2}{\omega(\Omega_e - \omega)} + \frac{\omega_{pH}^2}{\omega^2}\left[\frac{2}{v_{\parallel th}}\left(-A_T\left(\frac{\Omega_e - \omega}{k_{\parallel}}\right) + \frac{\omega}{k_{\parallel}}\right)S\left(\frac{\Omega_e - \omega}{v_{\parallel th}k_{\parallel}}\right) + A_T\right].$$

In Equation (28) ω_{pc} and ω_{pH} are the electron plasma frequencies corresponding to cold and hot species respectively, the function $S(x)$ is defined by

$$S(x) = e^{-x^2}\int_0^x e^{t^2} dt$$

(Stix, 1962) and is related to the plasma dispersion function $Z(x)$ of Fried and Conte (1961) by the relation Re $Z(x) = -2S(x)$. For large $x (x \gtrsim 2)$ we have $S(x) \simeq (1/2x)$, so that for $v_{res} \gtrsim 2v_{\parallel th}$ Equation (28) reduces to

$$\frac{k_{\parallel}^2 c^2}{\omega^2} \simeq \frac{\omega_{pc}^2}{\omega(\Omega_e - \omega)} + \frac{\omega_{pH}^2}{\omega(\Omega_e - \omega)} = \frac{\omega_{pe}^2}{\omega(\Omega_e - \omega)} \tag{30}$$

which is the cold plasma dispersion equation for the whistler mode. Assuming that approximation (30) is valid the parallel particle energy for resonance with a wave of frequency ω is, from Equation (25)

$$E_{res} = \frac{m_e c^2}{2} \frac{(\Omega_e - \omega)^3}{\omega \omega_{pe}^2}. \tag{31}$$

The frequency function on the right side of Equation (31) is infinite at $\omega = 0$ and goes monotonically to zero at $\omega = \Omega_e$. Thus particles at all energies can resonate with whistler waves, the resonant energy decreasing with increasing frequency. However, only those waves which are unstable will be present in the spectrum, hence we must examine Equation (24). Since the bracketed quantity on the left side of Equation (24) will be positive, the instability criterion is (following Kennel and Petschek, 1966)

$$A_e(v_{res}) \geq \frac{1}{\left(\dfrac{\Omega_e}{\omega} - 1\right)} \quad \text{or} \quad \frac{\omega}{\Omega_e} \leq \frac{A_e}{(1 + A_e)}. \tag{32}$$

Thus for a given A_e value, only waves with frequencies below a critical value given by Equation (32) will be unstable. From the above discussion of E_{res} this in turn implies that only particles above a certain critical energy will be involved in unstable resonant interactions. Using the cold plasma approximation given by Equation (31) we find that the minimum unstable resonant energy $E_{res_{min}}$ is

$$\frac{E_{res_{min}}}{E_{\perp th}} = \frac{1}{\beta_{\perp e}\left(1 + \dfrac{n_c}{n_H}\right) A_e (1 + A_e)^2} \tag{33}$$

where $\beta_{\perp e} = n_H E_{\perp th}/(B^2/8\pi)$. Note that use of the cold plasma approximation Equation (31) used to derive Equation (33) will only be valid if Equation (33) yields $E_{res_{min}}/E_{\perp th}$ somewhat greater than unity. The full anisotropy for our bi-Maxwellians with an empty loss-cone of half-angle α_0 is

$$A_e(v_{res}) = A_T + A_{LC}(v_{res}) \tag{34}$$

where A_T is the thermal anisotropy (Equation (28)) and the loss-cone anisotropy A_{LC} is given by

$$A_{LC}(v_{res}) = \frac{v_{res}^2}{v_{\perp th}^2} \tan^2 \alpha_0 (\sec^2 \alpha_0 + A_T) \tag{35}$$

or, for

$$\alpha_0 \simeq \frac{1}{\sqrt{2}\, L^{3/2}} \quad \text{(valid for } L \geq 3\text{)}$$

$$A_{\text{LC}}(E_{\text{res}}) \simeq \frac{E_{\text{res}}}{E_{\perp_{\text{th}}}} \frac{\left(\dfrac{T_\perp}{T_\parallel}\right)}{2L^3} \tag{36}$$

Substituting Equation (34) into (33), and taking the value of A_{T} from our model gives

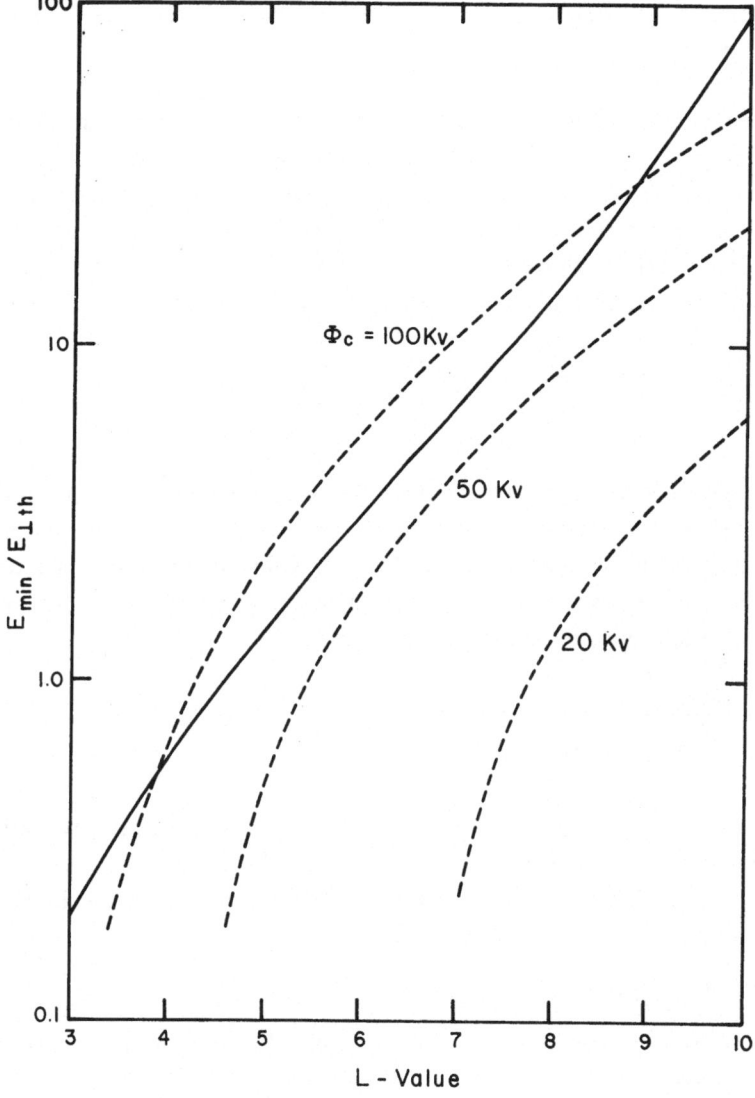

Fig. 9. The solid line shows the minimum electron energy for involvement in unstable resonant iterations with whistlers for our plasma model. This is compared with (dotted lines) the forbidden zor.e upper cut-offs for various cross tail potentials from Figure 7.

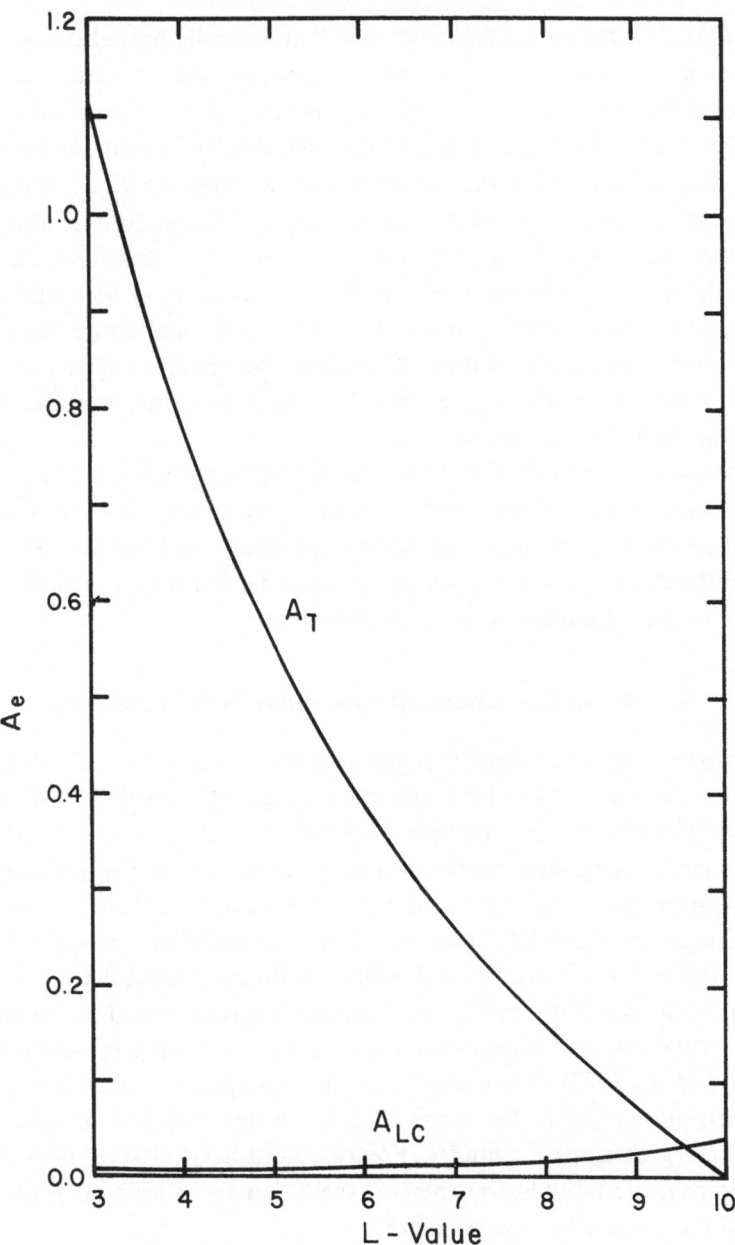

Fig. 10.   The thermal, A_T and loss-cone A_{LC} anisotropies compared at the maximum unstable whistler mode frequency, vs. L.

the results for $E_{\mathrm{res_{min}}}/E_{\perp \mathrm{th}}$ versus L shown by the solid line in Figure 9. In Figure 10 we show the contribution of the loss cone anisotropy at this critical value compared with thermal anisotropy, from which it can be seen that the effect of the loss cone is small except near L_0 where $A_T \simeq 0$. Returning now to Figure 9 we see that $E_{\mathrm{res_{min}}} > > E_{\perp \mathrm{th}}$ for $L \geqslant 4.5$ and rises steeply with increasing L value (hence Equation (33) is in general valid). This means for the outer zone that generally only electrons well above the thermal energy are able to take part in unstable resonant interactions (except possibly near the plasmapause at very large potentials) and hence with this model whistlers cannot be expected to play an important role in determining the outer zone structure. We recognize here that we should have compared $E_{res_{min}}$ with $E_{\parallel th}$, but the difference between $E_{\perp th}$ and $E_{\parallel th}$ is not large. This was done so that we could compare the results with the upper cut-off energies in $E_\perp$ taken from Figure 7, and plotted as dotted lines in Figure 9. From these curves it can be seen that not only is $E_{res_{min}}$ generally well above $E_{\perp th}$ but it will be generally well above the cut-off too, except for large cross tail potentials. Therefore, the above analysis was in general invalid, since we are not able to prescribe the anisotropy of any particles which may be present above the cut-off energy.

Our conclusion therefore is that if the distribution function is an isotropic Maxwellian at the outer limits of the dipole field then the anisotropies developed during inward convection are not sufficiently large to produce a whistler instability affecting the bulk of the electron plasma. Therefore we consider in the next section the possible importance of gyroresonance with electrostatic waves.

4. The Electron-Electrostatic Instability in the Outer Zone

It is well known from recent satellite results that there frequently exists noise bands in the electric field wave data above the electron gyrofrequency in the outer zone ($4 \leqslant L \leqslant 10$). These waves are thought to be electrostatic in nature, no magnetic component ever having been observed, and occur at or near the half-harmonics of the electron gyrofrequency. The second half-harmonic $\omega \simeq 3\Omega_e/2$ is the most frequently detected (Kennel *et al.*, 1970). These half-harmonic emissions are typically narrow-band (few 100 Hz at a few kHz) and exhibit uniform spectral density across their bandwidths (Coroniti *et al.*, 1971). Amplitudes are typically in the 1 to 10 mV m^{-1} range, but ~ 100 mV m^{-1} bursts have been observed during a substorm expansion phase (Scarf *et al.*, 1973). These emissions have no apparent local time preference, and are seen most often in the range $5 \leqslant L \leqslant 7$, where they are almost continually present at low geomagnetic latitudes (Fredricks and Scarf, 1973). They also occur at high geomagnetic latitudes on these L shells (up to at least 50°), but here are weaker and more sporadic in appearance.

To date the literature concerning the theory of magnetospheric electrostatic waves is very sparce. Fredricks (1971) considered a 'ring' anisotropy in velocity space as being the driving anomaly of such waves. This work was followed up by Young *et al.* (1973) who performed some linear stability analyses on more realistic distributions, including 'ring' and temperature anisotropies, and by Chanteur *et al.* (1973). In keeping with our model, we consider only temperature anisotropies here, and then comment briefly about the production of 'ring' anisotropies in the outer zone.

The real and imaginary parts of dispersion relation for electron-electrostatic waves for small growth rate are

$$\mathscr{R}_e D(\omega, k) = 1 - \frac{2\pi\omega_{pe}^2}{k^2} \sum_{n=-\infty}^{\infty} \int_0^{\infty} dv_\perp v_\perp J_n^2\left(\frac{k_\perp v_\perp}{\Omega_e}\right) \times$$

$$\times P \int_{-\infty}^{\infty} dv_\| \frac{\left[\frac{\partial f}{\partial v_\|} + \frac{n\Omega_e}{k_\| v_\|}\frac{\partial f}{\partial v_\perp}\right]}{\left(v_\| + \frac{(n\Omega_e - \omega)}{k_\|}\right)} \tag{37}$$

$$\mathrm{Im}\, D(\omega, k) = - \frac{2\pi^2\omega_{pe}^2}{k^2}\frac{k_\|}{|k_\||} \sum_{n=-\infty}^{\infty} \int_0^{\infty} dv_\perp v_\perp J_n^2\left(\frac{k_\perp v_\perp}{\Omega_e}\right) \times$$

$$\times \left[\frac{\partial f}{\partial v_\|} + \frac{n\Omega_e}{k_\| v_\perp}\frac{\partial f}{\partial v_\perp}\right]\Bigg|_{v_\| = v_{res \atop n}} \tag{38}$$

where

$$v_{res \atop n} = \left(\frac{\omega - n\Omega_e}{k_\|}\right). \tag{39}$$

Here $k_\perp$ and $k_\|$ are the components of the wave vector **k** perpendicular and parallel to the static magnetic field, J_n is the Bessel function of the first kind, and the proton term has been neglected for waves with $\omega \gg \Omega_i$. The condition for marginal stability ($\gamma = 0$) is $\mathscr{R}e\, D = \mathrm{Im}\, D = 0$. If we have a plasma consisting of a cold and a warm component, where the cold component is assumed an isotropic Maxwellian, while the warm component is bi-Maxwellian then we find

$$\mathscr{R}e\, D(\omega, k) \simeq 1 - \left(\frac{k_\perp}{k}\right)^2\left(\frac{\omega_{pc}}{\Omega_e}\right)^2\left[\left(\frac{k_\|}{k_\perp}\right)^2\frac{1}{\left(\frac{\omega}{\Omega_e}\right)} + \frac{1}{\left(\frac{\omega}{\Omega_e}\right)^2 - 1}\right] +$$

$$+ \frac{e^{-\lambda_w}}{\lambda_w}\left(\frac{k_\perp}{k}\right)^2\left(\frac{\omega_{pw}}{\Omega_e}\right)^2 \sum_{n=-8}^{\infty}\left[\left(\frac{v_{\perp thw}}{v_{\| thw}}\right)^2 S'(Z_n) + \frac{2n\Omega_e}{k_\| v_{\| thw}} S(Z_n)\right] I_n(\lambda_w) \tag{40}$$

$$\mathrm{Im}\, D(\omega, k) = \pi^{1/2}\frac{e^{-\lambda_w}}{\lambda_w}\left(\frac{k_\perp}{k}\right)^2\left(\frac{\omega_{pw}}{\Omega_e}\right)^2\left(\frac{\Omega_e}{|k_\|| v_{\| thw}}\right) \times$$

$$\times \left\{e^{(\lambda_w - \lambda_c)}\left(\frac{n_c}{n_w}\right)\left(\frac{v_{\| thw}}{v_{\| thc}}\right)\left(\frac{v_{\perp thw}}{v_{\perp thc}}\right)^2\frac{\omega}{\Omega_e}\sum_{n=-\infty}^{\infty}\exp\left[-Z_{n_c}^2\right] I_n(\lambda_c) + \right.$$

$$\left. + \sum_{n=-\infty}^{\infty}\left[\frac{T_\perp}{T_\|}\frac{\omega}{\mho_e} - \left(\frac{T_\perp}{T_\|} - 1\right)n\right]\exp\left[-Z_{n_w}^2\right] I_n(\lambda_w)\right\} \tag{41}$$

where subscripts c and w denote cold and warm plasma, respectively, S is the function defined in the previous section and S' is its first derivative with respect to its argument

$$S'(x) = 1 - 2xS(x)$$

I_n is the Bessel function of imaginary argument, parameter λ is defined by

$$\lambda = \frac{k_\perp^2 v_{\perp_{th}}^2}{2\Omega_e^2}$$

and

$$Z_n = \left(\frac{n\Omega_e - \omega}{k_\| v_{\|_{th}}}\right) = -\frac{v_{res_n}}{v_{\|_{th}}}.$$

In deriving Equations (40) and (41) from Equations (37) and (38) use has been made of Weber's second exponential integral (e.g., Watson, 1944). We now consider the imaginary part of the dispersion function in detail. For isotropic Maxwellians (cold as well as hot) all terms in the series of Equation (41) are positive since $I_n(\lambda) \geqslant 0$ for all λ and n, so that $\mathrm{Im}\, D(\omega, k) > 0$ for all wave parameter values ($\omega > 0$). Thus, the marginally-stable state $\mathrm{Im}\, D = 0$ cannot be reached; this situation corresponds to wave damping. We must therefore consider the conditions under which $\mathrm{Im}\, D$ can be made zero and negative, which is a potentially unstable regime. The form of the above harmonic contributions in the series for both hot and cold components is reasonably clear. The nth contributions are modulated by the term

$$e^{-Z_n^2} = \exp\left\{-\frac{(\omega - n\Omega_e)^2}{(k_\| v_{\|_{th}})^2}\right\}$$

so that in general the nth term peaks near $\omega = n\Omega_e$ and falls away rapidly from this value of ω over a half-width of $k_\| v_{\|_{th}}$. The smaller is $k_\|$ the more localized are the functions about their respective gyrofrequency harmonics. If we consider $k_\| v_{\|_{th_w}} \lesssim \Omega_e$ so that the warm plasma terms are reasonably localized about their harmonics, then the cold plasma terms will become negligible everywhere except very close to the harmonics, and can be ignored. If this criterion is not satisfied, such that for example $k_\| v_{\|_{th_c}} \gtrsim \Omega_e$ (and $v_{th_w} \sim 10 v_{th_c}$ in our model) then the cold plasma provides large damping (positive) contributions which are overwhelmingly dominant for the low n terms (as Young et al., 1973). In looking for instability we must therefore have $k_\|$ small enough so that the cold terms are negligible except very close to $\omega = n\Omega_e$ and in which case we also expect the warm plasma terms to be reasonably well localized. If this is satisfied then we need only consider those terms in the warm plasma series with n near the (ω/Ω_e) range of consideration, i.e., in looking for possible marginal states with $k_\| v_{\|_{th_w}} \lesssim \Omega_e$ between frequencies $n\Omega_e$ and $(n+1)\Omega_e$ we need only consider the zeros and negative value ranges of the function.

$$F_n = \frac{T_\perp}{T_\parallel} \exp\left\{ -\left(\frac{\omega - n\Omega_e}{k_\parallel v_{\parallel\text{th}}}\right)^2 \right\} I_n(\lambda_w) \left[\left(\frac{\omega}{\Omega_e} - \left(1 - \frac{T_\parallel}{T_\perp}\right)n\right) + \right.$$
$$\left(\frac{\omega}{\Omega_e} - \left(1 - \frac{T_\parallel}{T_\perp}\right)(n+1)\right) \frac{I_{n+1}(\lambda_w)}{I_n(\lambda_w)} \times$$
$$\left. \times \exp\left(\frac{2\Omega_e^2}{(k_\parallel v_{\parallel\text{th}})^2}\left(\frac{\omega}{\Omega_e} - \right)n + \frac{1}{2}\right)\right)\right] \tag{42}$$

where the cold plasma terms and constant factors have been dropped from Equation (41).

The first term in the square bracket is always positive in the frequency range $n\Omega_e$ to $(n+1)\Omega_e$, and the second can only be negative if

$$A_T = \left(\frac{T_\perp}{T_\parallel} - 1\right) \geq n \tag{43}$$

i.e., instability between n and $(n+1)$ is only possible if the thermal anisotropy is greater than n, but it is not guaranteed above this anisotropy. Specializing to the most frequently observed mode i.e. $n=1$ (the $3/2\Omega_e$ emission), we require, at the least $A_T \geq 1$. According to our model this means that the outer zone plasma must be stable to the $3/2\Omega_e$ emission for $L \gtrsim 3$ and always stable to the higher half-harmonic modes. Instability is possible for $0 \leqslant (\omega/\Omega_e) \leqslant 1$, and such modes have been infrequently observed (Coroniti et al., 1971), but probably are usually suppressed by the huge cold-plasma Landau damping of the $n=0$ term.

Thus, our outer zone plasma model is stable to the observationally dominant electrostatic mode of the outer zone, the $3/2\Omega_e$ emission. However, we know these waves exist, and following Young et al. (1973) and Chanteur et al. (1973) it would appear that their most likely cause is the existence of non-monotonic ('ring'-like) velocity distributions. These can give instability above, below, and across $3/2\Omega_e$. Such non-monotonic distributions could perhaps arise through several processes, and these require further observational and theoretical study. Some possibilities which occur to us are listed below.

(a) The electron distribution at L_0 from the plasma sheet could be non-monotonic. Why this should be so, if the spectra at the Vela orbit are usually of Maxwellian form, is not known; however, studies of the spectrum at large L value within the dipolar region should be undertaken.

(b) They could be produced by selective precipitation of the low energy $(E \leqslant E_{\text{th}})$ part of the plasma during inward convection. No mechanism for such a process is known, and we note that the minimum precipitation lifetime is largest for the low energy particles.

(c) Perhaps most reasonably they could be produced by time dependent processes outside the scope of the present work, for example, by the drifting plasma clouds of DeForest and McIlwain (1971). Following a short-lived injection event due to enhanced cross tail electric field differential azimuthal drifts at differing particle energies could readily produce 'ring' distributions, like those shown in the above paper. This suggestion of the basic source of 'ring' distributions being due to time dependent processes seems to have been implied by Young *et al.* (1973) who took data from DeForest and McIlwain (1971) as justification for the consideration of such functions. The suggestion is perhaps further strengthened by noting that the waves are most often observed in the $5 \leqslant L \leqslant 7$ regime, which is just the usual mixing region.

In summary we conclude with Young *et al.* (1973) that the observed electrostatic wave spectra of the outer zone are most reasonably interpreted in terms of a non-monotonic distribution function. A pure thermal anisotropy generated during inward convection in the dipolar region is stable to the most frequently observed modes.

5. The Ion-Cyclotron Instability in the Outer Zone

We shall now discuss essentially the same story for protons vis-à-vis the ion-cyclotron mode as we did for electrons vis-à-vis whistlers. The instability criterion for the ion-cyclotron mode propagating parallel to the magnetic field is (Kennel and Petschek, 1966)

$$A_p \left(v_{\substack{res \\ p}} \right) \geq \frac{1}{\left(\dfrac{\Omega_p}{\omega} - 1 \right)} \tag{44}$$

where Ω_p is the ion gyrofrequency, the distribution anisotropy A_p is defined as in Equation (26), and the proton resonant velocity is

$$v_{\substack{res \\ p}} = \frac{\omega - \Omega_p}{k_\parallel}. \tag{45}$$

As in the case of the whistler instability Equation (44) implies an upper cut-off to the unstable frequency spectrum i.e.,

$$\left(\frac{\omega}{\Omega_p} \right)_{max} = \frac{A_p}{(1 + A_p)} \tag{46}$$

and a minimum proton energy for resonance with unstable waves

$$\frac{E_{\substack{res \\ min}}}{E_{\perp th}} = \frac{1}{\beta_p \left(1 + \dfrac{n_c}{n_H} \right) A_p^2 (1 + A_p)}. \tag{47}$$

As in the similar electron formula (Equation (33)) we have taken the real part of the dispersion function as being approximated by the cold dispersion relation (with $\omega \ll \Omega_e$, $c \gg \omega/k_\parallel$)

$$\frac{c^2 k_\parallel^2}{\omega^2} = \frac{\omega_{pp}^2}{\Omega_p (\Omega_p - \omega)} \tag{48}$$

where ω_{pp} is the proton plasma frequency. This is only valid if $E_{\mathrm{res}}/E_{\mathrm{th}}$ is somewhat larger than unity. Our formulation is therefore valid only if Equation (47) results in values of $E_{\mathrm{res}\atop\mathrm{min}}/E_{\perp\mathrm{th}}$ larger than unity in our range of interest. Taking the same proton anisotropy as we did for the electrons (Equations (34), (35) and (36)), $n_{\mathrm{c}}/n_{\mathrm{H}} = 0.1$ $\beta_{\mathrm{op}} = 0.6$ as per our model we compute $E_{\mathrm{res}\atop\mathrm{min}}/E_{\perp\mathrm{th}}$ as a function of L from Equation (47). The results are shown by the solid line in Figure 11. As for the whistler instability we find that $E_{\mathrm{res}\atop\mathrm{min}} > E_{\perp\mathrm{th}}$ for $L \gtrsim 5$ and rises rapidly with increasing L value. $E_{\mathrm{res}\atop\mathrm{min}}$ is also generally larger than the cut-off energy produced by the forbidden zones (dotted lines). Therefore we conclude that the outer-zone protons are stable to ion-cyclotron mode for our model, except perhaps in the low L region near the plasmapause for large potentials $\Phi_{\mathrm{c}} > 100$ Kv.

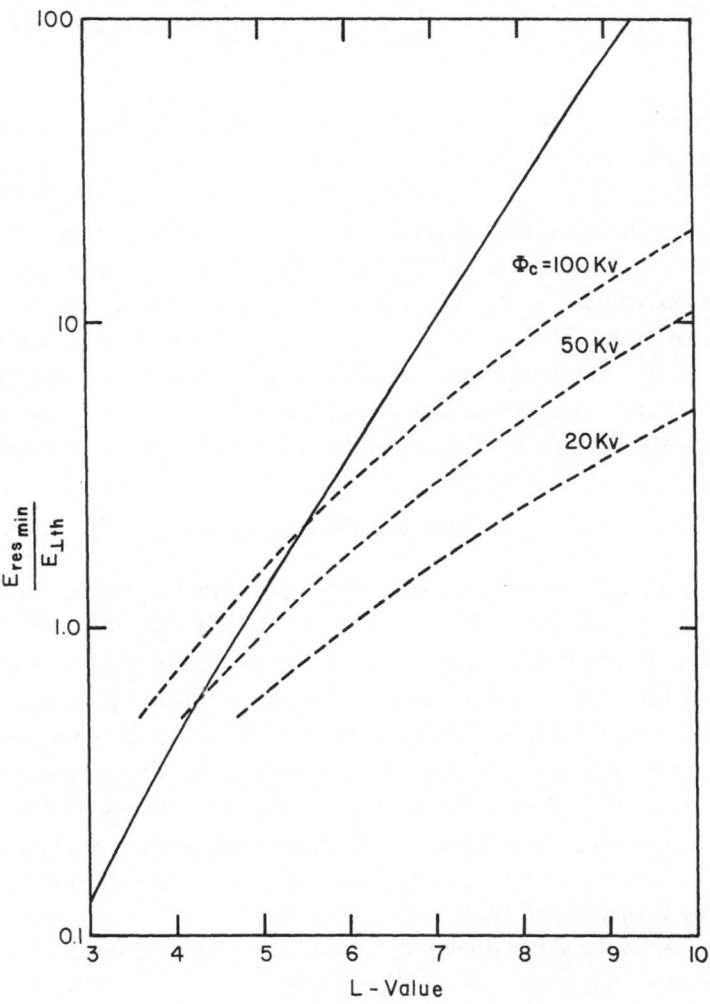

Fig. 11. Same as Figure 9, except for protons vis-a-vis the ion-cyclotron mode.

We note, however, from Equation (47) that the minimum resonant energy would be much reduced if $n_c \gg n_H$. This observation, which applies also to the hot electrons, has formed the basis of theories of the decay of the ring-current following geomagnetic storms, when cold-plasma infilling from the ionosphere can raise the value of n_c in the region containing 'marooned' hot plasma (Cornwall *et al.*, 1971; Russell and Thorne, 1970; Thorne, 1972). Recent satellite data, however, have appeared to cast doubt on whether this expected strong ion-cyclotron turbulence and associated proton precipitation does, in fact, occur inside the plasmasphere (Coroniti, 1973). Rather, emphasis has been shifted to the consideration of the high-β regime outside the plasmapause, which we have been considering here. This was investigated by Coroniti *et al.* (1972) who took a 'loss cone' distribution of the form

$$f_p(v_\parallel, v_\perp) = F_0 \left(\frac{v_\perp}{v_{\perp th}} \right)^{2m} \exp\left(-\frac{v_\perp^2 + v_{\parallel\perp}^2}{v_{th}^2} \right)$$

where the F_0 represents normalizing factors. The anisotropy is given by

$$A = m$$

and $m = 2$ was considered as representative of conditions during substorm-associated proton injections at $L \approx 4$. As may be expected for such an anisotropy (we found $A_T \simeq 0.8$ at this L value), the authors found a low value of $E_{min}/E_{th} \simeq 0.5$ and hence concluded that the ion-cyclotron mode can be important in determining ring-current dynamics even outside of the plasmasphere. This may be perfectly true, but as we have tried to emphasize in this paper, the important question is why is the anisotropy so large? The 'loss cone' distribution anisotropy is basically due to its 'ring-like nature about the $v_\parallel$ axis in velocity space, so we are again requiring to ask how such distributions come about. Some suggestions have already been presented in Secton 4 vis-à-vis the electron-electrostatic modes, and will not be repeated here.

6. Summary and Conclusions

In writing this paper we are emphasizing the point that wave calculations for the magnetosphere should be performed using distribution functions which we expect to be present, or which have been observed. If observed spectra are used then consideration of how these came about should rank in equal importance with the instability calculations. Only by consideration of both aspects of the problem can one hope to form a self-consistent view of the role of wave-particle interactions in the magnetosphere. The 'backreaction' of the expected instabilities on the distribution functions then forms a second phase of the investigations, maybe along the lines of diffusive equilibria initially considered by Kennel and Petschek (1966) and subsequently developed by Etcheto *et al.* (1973).

As a first step in the above direction we have considered the steady-state properties of the convecting tail plasma in the dipolar region i.e., in the outer-zone which includes the synchronous orbit for moderate cross tail electric fields. By choosing an isotropic Maxwellian at the 'source' L value from the tail (i.e., at the inner boundary of the tail, taken to be $L_0 = 10$) we then calculated the anisotropy and variation of plasma parameters with L. A linear instability analysis of these distributions showed them to be stable in the usual region of interest to the ion-cyclotron, whistler and electron-electrostatic $3/2\ \Omega_e$ modes. We cannot therefore, on this basis, understand the structure of the inner edge of the plasma sheet, the production of high-energy particles in the outer zone (see, for example, Kennel 1969 and references therein) or the actual observations of the waves.

However, the choice of isotropic Maxwellian at L_0 was the most stabilizing we could have made. As pointed out above, it may be profitable to look at the observed spectra in the near-Earth region of the tail plasma sheet to determine whether the above choice was appropriate or whether the plasma does consistently carry with it any form of anisotropy from the tail. If this is so, then the electromagnetic as well as electrostatic instabilities would require reconsideration. If not, then the next step would seem to be the time-dependent problem, which would appear from observation to be able to produce highly anisotropic distributions, like the 'ring' or 'loss cone' forms discussed above. A self-consistent theoretical treatment of these phenomena would not appear to be an easy problem, but nevertheless probably represents the next large step forward to be taken.

Acknowledgements

One of us (MAA) would like to thank Mr G. Chanteur and Dr R. Pellat for discussions on electrostatic waves; and would also like to acknowledge the financial support of CNES (French National Centre for Space Research). The other author (SWHC) would like to thank Dr L. R. Lyons for several useful discussions; the work was performed while he was a Visiting Fellow at CIRES, University of Colorado/ NOAA. We would both like to thank Dr R. Gendrin for his encouragement, and careful and penetrating criticism of our manuscript; also Dr R. W. Fredricks for his help in its reorganization.

References

Axford, W. I. and Hines, C. O.: 1961, *Can. J. Phys.* **39**, 1433.
Bame, S. J.: 1968, in B. M. McCormac (ed.), *Earth's Particles and Fields*, Reinhold Book Corporation, New York, p. 373.
Banks, P. M. and Holzer, T. E.: 1968, *J. Geophys. Res.* **73**, 6846.
Banks, P. M., Nagy, A. F., and Axford, W. I.: 1971, *Planetary Space Sci.* **19**, 1053.
Chanteur, G., Ashour-Abdalla, M., and Pellat, R.: 1973, Preprint, GRI, Paris, France.
Chen, A. J.: 1970, *J. Geophys. Res.* **75**, 2458.
Cornwall, J. M., Coroniti, F. V., and Thorne, R. M.: 1970, *J. Geophys. Res.* **75**, 4699.
Cornwall, J. M., Coroniti, F. V., and Thorne, R. M.: 1971, *J. Geophys. Res.* **76**, 4428.
Coroniti, F. V.: 1973, *Radio Sci.*, to be published.
Coroniti, F. V. and Kennel, C. F.: 1972, *J. Geophys. Res.* **77**, 2835.
Coroniti, F. V., Fredricks, R. W., Kennel, C. F., and Scarf, F. L.: 1971, *J. Geophys. Res.* **76**, 2366.
Coroniti, F. V., Fredricks, R. W., and White, R.: 1972, *J. Geophys. Res.* **77**, 6243.
De Forest, S. E.: 1972 *J. Geophys. Res.* **77**, 65.

De Forest, S. E. and McIlwain, C. E.: 1971, *J. Geophys. Res.* **76**, 3587.
Dungey J. W.: 1961, *Phys. Rev. Letters* **6**, 47.
Etcheto, J., Gendrin, R., Solomon, J., and Roux, A.: 1973, *J. Geophys. Res.*, to be published.
Frank, L. A.: 1971, *J. Geophys. Res.* **76**, 2265.
Fredricks, R. W.: 1971, *J. Geophys. Res.* **76**, 5344.
Fredricks, R. W. and Scarf, F. L.: 1973, *J. Geophys. Res.* **78**, 310.
Fried, B. D. and Conte, S. D.: 1961, *The Plasma Dispersion Function*, Academic Press, New York.
Harris, E. G.: 1959, *Phys. Rev. Letters* **2**, 34.
Harris, K. K., Sharp, G. W., and Chappell, C. R.: 1970, *J. Geophys. Res.* **75**, 219.
Heppner, J. P.: 1972, *Planetary Space Sci.* **20**, 1475.
Holnes, E. W. Jr., Asbridge, J. R., Bames, S. J., and Singer, S.: 1971, *J. Geophys. Res.* **76**, 63.
Kennel, C. F.: 1969, *Rev. Geophys. Space Phys.* **7**, 379.
Kennel, C. F. and Petschek, H. E.: 1966, *J. Geophys. Res.* **71**, 1.
Kennel, C. F., Scarf, F. L., Fredricks, R. W., McGehee, J. H., and Coroniti, F. V.: 1970, *J. Geophys. Res.* **75**, 6136.
Kivelson, M. G. and Russell, C. T.: 1973, *Radio Sci.*, to be published.
Lyons, L. R., Thorne, R. M., and Kennel, C. F.: 1972, *J. Geophys. Res.* **77**, 3455.
Lyons, L. R. and Thorne, R. M.: 1973, *J. Geophys. Res.* **78**, 2142.
Nakada, M. P., Dungey, J. W., and Hess, W. N.: 1965, *J. Geophys. Res.* **70**, 3529.
Russell, C. T. and Thorne, R. M.: 1970, *Cosmic Electrodyn.* **1**, 67.
Scarf, F. L., Fredricks, R. W., Kennel, C. F., and Coroniti, F. V.: 1973, *J. Geophys. Res.* **78**, 3119.
Schield, M. A. and Frank, L. A.: 1970, *J. Geophys. Res.* **75**, 5401.
Sharp, G. W., Chappell, C. R, and Harris, K. K.: 1972, in K. Folkestad (ed.), *Magnetosphere-Ionosphere Interactions*, Universitetsforlaget, Oslo, Norway, p. 169.
Stix, T. H.: 1962, The Theory of Plasma Waves, MacGraw-Hill Book Company, New York, pp. 179–180.
Sugiura, M.: 1972, *J. Geophys. Res.* **77**, 6093.
Thorne, R. M.: 1972, in K. Folkestad (ed.), *Magnetosphere-Ionosphere Interactions*, Universitetsforlaget, Oslo, Norway, p. 185.
Watson, G. N.: 1944, *A Treatise on the Theory of Bessel Functions*, (2nd Edition), Cambridge University Press.
Young, T. S. T., Callen, J. D., and McCune, J. E.: 1973, *J. Geophys. Res.* **78**, 1082.

CRUDE APPROXIMATIONS TO SOME ASPECTS
OF THREE-DIMENSIONAL MAGNETOSPHERIC DYNAMICS

JOHN M. CORNWALL

Space Physics Laboratory, The Aerospace Corporation, Los Angeles, Calif., U.S.A.

and

Department of Physics, University of California, Los Angeles, Calif., U.S.A.

1. Introduction

It is quite well understood in principle how to formulate and solve dynamical problems in a three-dimensional magnetosphere (i.e., one in which all three adiabatic invariants M, J, Φ come into play), given the relevant diffusion coefficients, loss rates, and so forth (Haerendel, 1968; Lanzerotti and Schulz, 1973). But there are very severe practical difficulties in carrying out a truly three-dimensional calculation, and practically none exist in the literature. Two-dimensional calculations abound: radial diffusion of equatorially-mirroring ($J = 0$) particles, pitch angle scattering at fixed L (or fixed Φ). In many cases this is not good enough; for example, most data on energetic trapped α-particles (e.g., Fennell *et al.*, 1973) are off-equatorial, but theory (e.g., Cornwall, 1972) has concentrated on $J = 0$ particles.

The purpose of the present work is to go one small step beyond purely qualitative discussions of three-dimensional problems by providing a crude, semi-quantitative overview of the essential features of such problems. This work is not intended in any way to replace real three-dimensional calculations, but it should provide space physicists with a road map for following the results of such calculations, should they be carried out. Our main emphasis is on the change of pitch angle anisotropy due to radial diffusion, with or without pitch angle diffusion.

In Section 2, an approximate formula for the variation of equatorial pitch angle with L during M and J conserving transport processes is given. The virtue of this approximation is that it yields a simple, immediately interpretable scaling law for distribution functions which are power laws in energy and pitch angle, when the transport processes are independent of M and J. Section 3 discusses a simple problem where M and J dependent transport processes are involved, and Section 4 discusses an approximation to the problem of coupled radial diffusion pitch angle instability. Here radial diffusion tries to increase the pitch angle anisotropy, and pitch angle diffusion tries to decrease it.

2. Approximate Kinematics

The theoretician formulates theories in terms of the adiabatic invariants M, J, and Φ while experimentalists measure energy E, equatorial pitch angle α, and L. In a dipole field, we may (and do) choose $\Phi = L^{-1}$, but the relationship between L, M and J, is on the one hand, and E and $y \equiv \sin\alpha$ on the other, is usually given numerically

B. M. McCormac (ed.), Magnetospheric Physics, 271–280. All Rights Reserved.

(e.g., Nakada *et al.*, 1965). For non-relativistic motion, y is a function of L only at fixed M and J. In the notation of Lanzerotti and Schulz (1973) the first invariant is

$$M = \frac{p_\perp^2}{2mB} = \frac{Ey^2L^3}{2B_0} \tag{1}$$

and the second invariant is

$$J = \oint p_\parallel \, ds = 2pLaY(y) \tag{2}$$

where $Y(y)$ is a complicated function of pitch angle. Here $p_\perp$, $p_\parallel$, p are momentum variables, m is mass, $E = p^2/2m$, $B_0 = 0.31$ G, $a = 1$ R_E. It follows upon elimination of p that

$$\frac{Y(y)}{L^{1/2}y} = \text{const at fixed } M, J. \tag{3}$$

It is not hard to see that Equation (3) can be written in the equivalent form

$$y^{h(L, y_0)} = \text{const} = y_0^{h(L_0, y_0)} \tag{4}$$

where the exponent h, depending on both L and y_0, can be expressed in terms of the function Y/y and its inverse. (The reference value y_0 depends on L_0, and on J^2/M, from Equations (1) and (2).) Equation (4) would be much more useful if the dependence of h on y_0 could be suppressed, at least approximately, for then an interesting scaling law holds for the adiabatic transformation of a distribution function which is the product of a power law in energy and in y. A numerical study shows that for a wide range of values of y_0, the exponent h is indeed roughly independent of y_0. The expression

$$y^{h(L)} = y_0^{h(L_0)} \tag{5}$$

is valid to within 3% or better in the range $0.2 \leqslant y_0 \leqslant 0.8$ (roughly $10°$ to $60°$ in equatorial pitch angle) with $L_0 = 7$, and $2 \leqslant L \leqslant 7$. The function $h(L)/h(L_0)$ is shown in Figure 1. For larger or smaller values of y_0, the approximation that h is independent of y_0 becomes progressively worse, but even for $0.8 \leqslant y_0 \leqslant 1$, Equation (5) is usefully accurate. The reason is that if y_0 is sufficiently close to 1, y_0^h depends insensitively on h; thus Equation (5) correctly predicts that $y = 1$ if $y_0 = 1$, no matter what h is.

With the approximation that h is independent of y_0, we can extend the fundamental energy scaling laws of Nakada *et al.* (1965) to pitch angle scaling laws. Let a particle, initially at L_0, have energy E_0 and pitch angle variable y_0 there. The conservation of M as expressed in the second form of Equation (1), plus the approximate formula (5), yields

$$E = E_0 (L_0/L)^3 \, y^{2(h/h_0)-2} \tag{6}$$

where E and y are the transformed values after the particle has undergone adiabatic transport (conserving M and J), and $h_0 \equiv h(L_0)$. Equation (6) expresses the well known

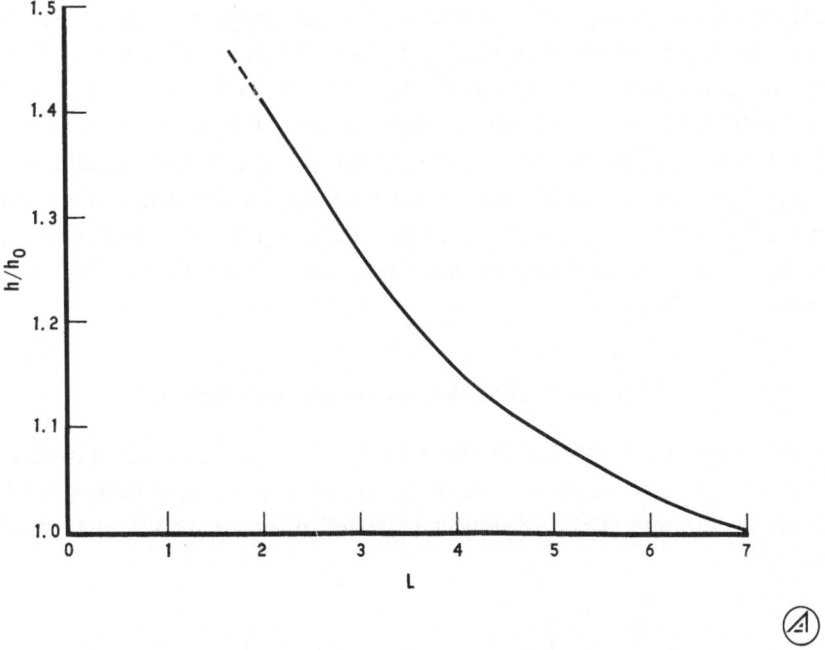

Fig. 1.   Variation of the pitch angle exponent with L.

result that particles with small pitch angles gain less energy than those with large pitch angles.

Let a group of particles have an initial momentum space distribution function at $L=L_0$ of the type

$$f_0 = E_0^{-a} y_0^b \tag{7}$$

and let these particles be acted upon by processes which (a) preserve M and J, and (b) have transport coefficients independent of M and J. Then, according to the single particle laws Equations (5) and (6), the distribution function maps into a similar form

$$f_0 \rightarrow f \sim E^{-a} y^{+\beta} \tag{8}$$

(leaving out a coefficient depending on L only) where

$$\beta(L) = -2a + (b + 2a)(h/h_0). \tag{9}$$

Because h/h_0 increases with decreasing L, the anisotropy β increases, as is well known, but what may not be so well known is that most of the increase in anisotropy comes from the decreasing power law in energy. For example, the values $a=3$, $b=1$ roughly characterize energetic (>50 keV) ring current protons at $L=7$. At $L=4$, Equation (9) yields $\beta=2$ and at $L=2$, $\beta=4$. However, for the less energetic protons at the peak (10 to 20 keV) of the ring current distribution at $L=7$, $a\approx0$, and $\beta\approx1.1$ at $L=4$, 1.4 at $L=2$. Thus processes of the sort considered in this paragraph do not lead to much increase of anisotropy of the particles at the flux peak, which may very well be significant for the dynamical role of instabilities driven by pitch angle anisotropy.

In fact, no known transport processes are independent of M and J, so the single particle laws (Equations (5) and (6) cannot be promoted to a distribution function law such as Equation (8). Moreover, no distribution function is really of the factorizable power law type in Equation (7). Nonetheless, Equations (8) and (9) be should a useful and rapid way of characterizing the zeroth-order change in the distribution function. In Sections 3 and 4, we go beyond the simple rule in Equation (9) to discuss more or less realistic dynamical processes, and interpret a and β in terms of suitable moments of the distribution function.

3. Quasi-Realistic Radial Diffusion Dynamics

Let f be the phase-space distribution function averaged over cyclotron phase, bounce phase, and longitude or what amounts to the same thing, the distribution function in M, J, and Φ. Suppose that f is subject to radial diffusion and a loss process, as described by:

$$\frac{\partial f}{\partial t} = \frac{\partial}{\partial \Phi}\left(D_{\Phi\Phi}\frac{\partial f}{\partial \Phi}\right) - \lambda f. \tag{10}$$

(With $\Phi = L^{-1}$, $D_{\phi\phi} = L^{-4}D_{LL}$.) Here the diffusion itself is one dimensional, but if λ and D_{LL} depend on M and J, one has a non-trivial complication of the sort discussed in Section 2, in converting from M and J to E and y.

Let us parametrize D_{LL} and λ by power laws in L, E_0, and y_0 where E_0 and y_0 (equivalent to M and J) are the energy and pitch angle variable at a reference L value L_0 (we choose $L_0 = 7$ in what follows):

$$D_{LL} = D_0 L^z E_0^s y_0^t, \qquad \lambda = \lambda_0 L^{-\beta} E_0^p y_0^q. \tag{11}$$

As Haerendel (1968) has indicated, the time stationary solution of Equation (10) is a linear combination of the functions

$$L^{-m/2}K_\nu(z), \qquad L^{-m/2}I_\nu(z) \tag{12}$$

with

$$m = \alpha - 3, \qquad n = \alpha + \beta - 2, \qquad \nu = -m/n,$$

$$z^2 = \left(\frac{4\lambda_0}{n^2 D_0}\right)L^{-n}E_0^{p-s}y_0^{q-t}. \tag{13}$$

Here K_ν, I_ν are the usual modified Bessel functions. To express the solution in terms of E and y, one uses the relations (5) and (6).

There are two main processes for radial diffusion: magnetic impulses and electrostatic fluctuations. For both, we take the usual value $\alpha = 10$. With a fluctuation power spectrum falling like ω^{-2} (for both electric and magnetic variations), magnetic diffusion has $s \approx 0$, $t \approx 2.7$, while for electrostatic diffusion, $s \approx -2$, $t \approx 0$. Less is known about the dominant loss processes; for simplicity, we take $p = q = \beta = 0$, hence $n = 8$,

$v = -7/8$. Other cases are easily worked out. The positive power t for magnetic diffusion reflects the well-known fact that magnetic diffusion is weak at high latitudes, i.e., small pitch angles.

First, we discuss electrostatic diffusion. For not-too-relativistic electrons and for protons, Cornwall (1972) has estimated that for electrostatic fluctuations $D_0 \sim$ $\sim 10^{-4}$ (keV)^2 day^{-1}. Ring current protons have an effective lifetime of a day or so, thus $\lambda_0^{-1} \sim 1$ day. In this case, Equation (13) gives $z \sim 25 L^{-4} E_0$ with E_0 in keV. The decrease of z with increasing L indicates that the dynamics are diffusion dominated at large L ($z \ll 1$) and loss dominated at small L ($z \gg 1$). The boundary condition $f \simeq 0$ at $L \simeq 1$, where $z \gg 1$, requires us to use only the K_v solution, with the asymptotic behavior

$$K_v(z) \sim (\pi/2z)^{1/2} e^{-z}, \quad z \gg 1. \tag{14}$$

The full solution, satisfying the appropriate boundary condition at $L = L_0$, is:

$$f(L, E_0, y_0) = \left(\frac{L_0}{L}\right)^{m/2} f(L_0, E_0, y_0) K_v(z)/K_v(z_0) \tag{15}$$

where $z_0 = z(L = L_0)$.

In the diffusion-dominated regime $K_v(z)/K_v(z_0) \approx 1$, and the remarks of Section 2 hold: power law distribution functions map into power laws, according to Equation (9). However, in the loss dominated regime there are new effects: power law scaling breaks down, but the local pitch angle anisotropy at fixed energy (defined in Equation (19)) increases over the value given in Equation (9). The reason for the increase is that particles with high energy diffuse electrostatically more slowly than low energy particles, so that the energy spectrum becomes steeper; that is, in effect the parameter a of Equation (9) is not constant, but increasing with decreasing L.

With the aid of Equation (14), the loss dominated solution is (aside from an overall multiplicative function of L)

$$f \sim E_0^{-1/2} e^{z_0 - z} f(L_0, E_0, y_0). \tag{16}$$

Take the initial distribution to be of the power law type Equation (7), and apply the transformation laws (5) and (6) to find

$$f \sim E^{a - 1/2} y^{\beta + r} \exp\left(-0.07 \left(\frac{E}{L}\right) y^{-2r}\right) \tag{17}$$

where β is given in Equation (9), and

$$r = (h/h_0) - 1. \tag{18}$$

Define the local pitch angle anisotropy as

$$\gamma \equiv (y/f) \, \partial f/\partial y. \tag{19}$$

This definition has physical significance; the growth rate of the electromagnetic cyclotron instability is essentially an integral over energy of $f\gamma$ as given in Equation (19).

For a power law distribution, γ is just the power of y. For the distribution function Equation (17), it is easy to find

$$\gamma(E, y) = \beta + r + 0.14 \, r \, (E/L) \, y^{-2r}. \tag{20}$$

The anisotropy has significantly increased over the value β derived from the scaling law Equation (9). At $L = 3, 5$, where from Figure 1 $r = 0.2$, $\gamma \simeq \beta + 0.2 + 9 \times 10^{-3} \, E$ and a 100 keV particle has $\gamma \simeq \beta + 1.1$; at $L = 2$, $\gamma \approx \beta + 2.2$. This rapid increase in anisotropy would represent a significant increase in the free energy available to drive instabilities, were it not for the fact that it is the loss mechanism which is responsible for the anisotropy increase; the available free energy may be increased or decreased as a result of the process discussed here.

The increase of anisotropy can be directly traced to the fact that z decreases with increasing y, at fixed L and E. Thus the same phenomenon occurs for magnetic diffusion. The general condition for the anisotropy to increase above β with decreasing L is

$$2r \, (p - s) + (t - q) \, (r + 1) > 0 \tag{21}$$

which is satisfied for magnetic diffusion and constant λ for $t > 0$. Using a small magnetic diffusion coefficient, such as $D \sim 10^{-10} L^{10}$, leads to tremendous increases in anisotropy, but far more tremendous decreases in total flux. An interesting balance between anisotropy increase and flux decrease can only be achieved when z is not much larger than, or much smaller than, one, that is, on the boundary of loss dominated and diffusion dominated transport.

It is worth noticing that this sort of process may actually *decrease* the anisotropy (compared to that given in Equation (9)) for low energy protons. Here charge exchange losses are important, and for them p is negative which works the wrong way for condition (21) to be satisfied. Moreover, for very low energy protons ($\lesssim 10$ keV at $L = 4$) Cornwall (1972) has estimated that $s \approx 0$, $\alpha = 6$. Again, this behavior of the diffusion coefficient works the wrong way in Equation (21).

4. Coupled Radial Diffusion and Pitch Angle Diffusion

In the presence of transport processes which violate M and J as well as Φ, Equation (10) is extended to (the loss term λ is dropped, for the sake of brevity):

$$\frac{\partial f}{\partial t} = \frac{\partial}{\partial \Phi} \left(D_{\Phi\Phi} \frac{\partial f}{\partial \Phi} \right) + Q^{-1} \sum_{ij} \frac{\partial}{\partial x^i} \left(Q D^{ij} \frac{\partial f}{\partial x^j} \right) \tag{22}$$

where the variables x^1, x^2 are equivalent to M and J, and

$$Q = \frac{\partial (M, J)}{\partial (x^1, x^2)}. \tag{23}$$

The presence of the Jacobian Q is demanded by the canonical nature of the variables

M, J, Φ. It is convenient to choose the variables x^i to be the velocity components at the equator:

$$(x^1, x^2) = (v_\perp, v_\parallel); \qquad v_\perp = \left(\frac{2E}{m}\right)^{1/2} y, \qquad v_\parallel = \left(\frac{2E}{m}\right)^{1/2} (1 - y^2)^{1/2}.$$

(24)

Then one readily finds that

$$Q \sim v_\perp (1 - y^2)^{1/2} T(y) \qquad (25)$$

(a factor independent of M and J is omitted), where $T(y)$ is the normalized bounce time:

$$T_B = (4mLa/p) T(y). \qquad (26)$$

The diffusion coefficients D^{ij} are averaged over bounce phase, cyclotron phase, and longitude; thus they differ from the usual locally defined diffusion coefficients of quasi-linear theory. For electromagnetic cyclotron waves propagating parallel to the static field, the *local* equivalent of the second term on the right of Equation (22) is in the resonant approximation (Kennel and Engelmann, 1966)

$$\sum_k v_\perp^{-1} G_k (v_\perp D_k G_k f) \qquad (27)$$

$$G_k = \left(\frac{\omega}{k} - v_\parallel\right) \frac{\partial}{\partial v_\perp} + v_\perp \frac{\partial}{\partial v_\parallel} \qquad (28)$$

$$D_k = \frac{\pi}{2} \left|\frac{eb_k}{mc}\right|^2 \delta(kv_\parallel - \omega_k + \Omega). \qquad (29)$$

In these equations, $v_\perp$ and $v_\parallel$ refer to local components, not components of the equatorial velocity. By comparison with Equation (22), it is reasonable to assume that the bounce-phase averaging process yields equations like (27) and (28), except that the explicit $v_\perp$ in Equation (26) is replaced by Q (see Equation (25)), and $v_\perp$, $v_\parallel$ are taken to mean equatorial components, as in Equation (24). Furthermore, D_k must be defined as an average over the actual wave fields, so the form (29) is not really appropriate. As the reader will soon see, the use of Equation (29) does not invalidate the formulas below.

The quasi-linear description is completed with the equation for wave energy:

$$\frac{\partial W_k}{\partial t} + \mathbf{\nabla} \cdot (\mathbf{V}_G W_k) = 2\gamma_k W_k \qquad (30)$$

where $\mathbf{V}_G$ is the group velocity, γ_k the growth rate, and the wave energy W_k is given by

$$W_k = \frac{|b_k|^2}{8\pi} \left[1 + n^{-2} \frac{\partial}{\partial \omega} (\omega n^2)\right] = \frac{|b_k|^2}{8\pi} \frac{2v_{ph}}{V_G}. \qquad (31)$$

Here n is the index of refraction, and v_{ph} the phase velocity. Again, these equations must be interpreted as suitable averages over the bounce phase and longitude, which amounts to replacing $\mathbf{V} \cdot (\mathbf{V}_G W_k)$ by $-(V_G/l) \, W_k$, where $l \sim La$ is a length characteristic of the field line.

The expression for the growth rate γ_k must be such that it correctly accounts for the transfer of energy between waves and particles. As with all the other formulas here, a spatial average is carried out, so γ_k is not given by the usual local expression. The only difference is the appearance of the Jacobian Q:

$$\gamma_k = \pi^2 V_G \frac{e^2}{mc^2} \int d^3v Q \delta (k v_\parallel - \omega_k + \Omega) \, G_k f \tag{32}$$

Here, of course, the multiplicative constant omitted in the definition (25) of Q matters. Since Equation (32) gives the correct local growth rate if Q is set equal to $v_\perp$ with the usual velocity space distribution function, and since $(1-y^2)^{1/2} \, T(y)$ is of $0(1)$ if y is not too close to one, this multiplicative constant is itself nearly one. The final value of this constant should be chosen on phenomenological grounds (which go beyond the scope of this paper) having to do with the bounce phase averaging, loss of resonance in the inhomogeneous magnetic field of the Earth, etc. For illustrative purposes, we take this constant to be exactly one.

Equations (22) through (32) are a truly formidable set of non-linear partial differential equations in four variables, with two unknowns. It is to be hoped that someone will tackle these equations in their full complexity some day with the help of computers, but even if this is done, the results will be as difficult to interpret simply as if they were experimental data. It appears useful to extract from these equations a simpler, approximate set of new equations, which is both easier to interpret and compute. In view of the fairly simple results of Sections 2 and 3, such a simplified set of equations might make reference to an effective power law in energy and pitch angle. Equivalently, one may take velocity space moments of the quasi-linear particle Equation (22) (e.g., Roux and Solomon, 1971; Hamasaki and Krall, 1973). Three such equations suffice: for the number density, the density of total energy, and the density of perpendicular energy.

Because the derivatives with respect to Φ (or L) in Equation (22) are to be taken at fixed M and J, it is necessary at any given L to express the distribution function f in terms of M, J, and L, and then to integrate over M and J. It is not hard to show that

$$dM \, dJ = 2\pi CL^4 \, Q \, dv_\perp \, dv_\parallel \tag{33}$$

where $Q = v_\perp (1-y^2)^{1/2} \, T(y)$, and C is a universal constant, independent of M, J, and L. By appropriate choice of units for M and J, we may choose $C=1$. Define

$$N(L) = 2\pi \int Q \, dv_\perp \, dv_\parallel f = L^{-4} \int dM \, dJ f . \tag{34}$$

Since $\int dM \, dJ f$ is (proportional to) the total number of particles per unit Φ, $L^2 N(L) \, dL$ is the total number of particles between two L shells separated by dL, modulo a universal constant. Similarly we define moments of any function $G(M, J)$:

$$\langle NG \rangle = 2\pi \int Q \, dv_\perp \, dv_\parallel Gf. \tag{35}$$

If D_{LL} in Equation (22) depends on M and J, the procedure of taking moments of Equation (22) yields more unknowns than there are equations. The only way around this drawback is to assume a functional form for the distribution function f; in order to make contact with Sections 2 and 3, we assume a double power law as in Equation (8), where *both* powers may depend on L:

$$f(L, E, y) = F(L) E^{-a} y^b. \tag{36}$$

Now given a particular form of D_{LL} (e.g., Equation (11)), all moments are expressed in terms of the three quantities F, a, b, as well as the lower cut off energy at which the energy integration is terminated. To express f in terms of L, M, J instead of L, E, y, the adiabatic relations (5) and (6) are used. It is a simple matter to multiply Equation (22) by 1, E, and $E_\perp$ and integrate to come to the three moment equations

$$\frac{\partial N}{\partial t} = \frac{\partial}{\partial \Phi} \left\langle D_{\Phi\Phi} \frac{\partial N}{\partial \Phi} \right\rangle \tag{37}$$

$$\frac{\partial \langle NE \rangle}{\partial t} = \frac{\partial}{\partial \Phi} \left\langle D_{\Phi\Phi} E \frac{\partial N}{\partial \Phi} \right\rangle - \sum_k 2\gamma_k W_k \tag{38}$$

$$\frac{\partial}{\partial t} \langle NE_\perp \rangle = \frac{\partial}{\partial \Phi} \left\langle D_{\Phi\Phi} E_\perp \frac{\partial N}{\partial \Phi} \right\rangle - \sum_k 2\gamma_k \left(\frac{\Omega}{\omega_k} \right) W_k. \tag{39}$$

In deriving Equations (38) and (39), Equations (27) to (29), (30), and (31) were used. In the absence of energy transport by radial diffusion or convective loss of waves, Equation (38) plus (30) would express conservation of total energy $\langle NE \rangle + \sum W_k$; it is for this reason that the expression (32) for γ_k is used.

The set of equations must be completed with some reference to the wave equation (Equation (30)). The shortest time scale in the problem corresponds to the inverse of the maximum growth rate (the maximum being taken as frequency ω_k or wave number k is varied), which we denote $\bar{\gamma}$. Of course, $\bar{\gamma}$ no longer depends on k, but only on such quantities as a, b, N, etc. Because the wave growth time is short compared to the radial diffusion time, Equation (30) can be replaced by the simpler equation

$$\bar{\gamma} = \bar{V}_g / l \tag{40}$$

where $\bar{V}_g$ is the group velocity for the wave number of maximum growth, and l a length characteristic of the field line. Equation (40) is an algebraic constraint on a, b, and F; added to the three Equations (37) to (39), one has a set of four equations for four unknowns a, b, F, $\bar{W}$ (where $\bar{\gamma} \bar{W} = \sum \gamma_k W_k$). In the time stationary case ($\partial/\partial t = 0$), these equations are non-linear ordinary differential equations in the single variable L, thus considerably simpler than the original non-linear partial differential equations Equations (22), (30), and (32). It is even possible to make some analytic progress with these simpler equations, but it would be premature to report on this work now.

References

Cornwall, J. M.: 1972, *J. Geophys. Res.* **77**, 1756.
Fennell, J. F., Blake, J. B., and Paulikas, G. A.: 1973, *J. Geophys. Res.* to be published.
Haerendel, G.: 1968, in B. M. McCormac (ed), *Earth's Particles and Fields*, Reinhold Publishing Company, New York, p. 171.
Hamasaki, S. and Krall, N.: 1973, *Phys. Fluids* **16**, 145.
Kennel, C. F. and Engelmann, F.: 1966, *Phys. Fluids* **9**, 2377.
Lanzerotti, L. J. and Schulz, M.: 1973, *Particle Diffusion in the Radiation Belts*, Springer-Verlag, Berlin.
Nakada, W., Dungey, J. W., and Hess, W.: 1965, *J. Geophys. Res.* **70**, 3529.
Roux, A. and Solomon, J.: 1971, *J. Atmospheric Terrest. Phys.* **33**, 1457.

PLASMA INSTABILITY MODES RELATED TO THE EARTH'S BOW SHOCK

E. W. GREENSTADT and R. W. FREDRICKS

Space Sciences Department, TRW Systems Group, Redondo Beach, Calif., U.S.A.

1. Introduction

The present paper is an attempt to give a very brief outline of the status of physical interpretations of some of the microscopic plasma physical phenomena occurring in bow shock structures. We shall not give any extensive discussion of collisionless shock models and theories in the short space available here, since there exist excellent detailed sources for such information. For general collisionless shock theories, we refer the reader to the recent monograph by Tidman and Krall (1971), who in addition to their own contributions, give bibliographies containing all of the important source material up to perhaps 1970. The main advances in collisionless shock theory since that time have come primarily from computer simulation work.

Because of the severe limitations imposed by single-point spacecraft measurements, namely, the often unknown state of motion of the bow shock structure relative to the spacecraft inertial frame, and the frequent lack of knowledge of upstream solar wind velocity, density, temperature, and interplanetary field vector, the interpretations of observations are often ambiguous. Nevertheless, a great deal of useful interpretation may be effected using such data. With this in mind, we have chosen to enumerate the microscopic phenomena, and examine how various modes may or may not be invoked in explaining spacecraft measurements on bow shock structure. We first define the macroscopic and observational context of the bow shock as presently understood, then outline some of the microscopic plasma physical phenomena one may expect to find associated with certain macroscopic structures. We then note the current status of spacecraft observations, and discuss the issues to be settled in using the bow shock to test plasma shock theory.

2. Classification of Macroscopic Structures

It is shown in treatments of collisionless plasma shocks that a number of plasma parameters influence the type of macroscopic shock structure one expects to observe. This leads to a loose but useful classification of shocks into two broad categories: laminar; turbulent (Tidman and Krall, 1971). Within these categories, one finds a sub-categorization into: perpendicular; oblique (Tidman and Krall, 1971).

In dealing specifically with the bow shock, two systems of categorization have been devised in recent years. The first was defined by Greenstadt *et al.* (1970a, b) on the basis of purely empirical considerations and fairly broad time scales. Observations with magnetometers alone disclosed that for averaging intervals of, say, 10 s or longer,

B. M. McCormac (ed.), Magnetospheric Physics, 281–290. All Rights Reserved.
Copyright © 1974 by D. Reidel Publishing Company, Dordrecht-Holland.

the shock was represented at different times by two grossly distinct signatures. In one, a sensibly quiescent upstream condition persisted outside the shock, which was recognized as a more or less abrupt, monotonic gradient in total magnetic field at a definite time. In the other signature, large amplitude ($\delta B \approx \frac{1}{2} B$), often quasi-periodic ($T \approx 30$ s), magnetic field oscillations appeared, sometimes for hours, outside of and adjacent to still larger oscillations ($\delta B \approx B$) leading eventually to elevated 'postshock' mean fields, but often without a definite time of transition from upstream to downstream field levels. The latter case was dubbed a 'pulsation' or 'multigradient' shock. The term pulsation suggested itself because of the often quasi-periodic character of the large amplitude 'shock' oscillations. Figure 1 shows the field magnitude record from a pass by OGO-5 in which both signatures were observed within a span of a few hours;

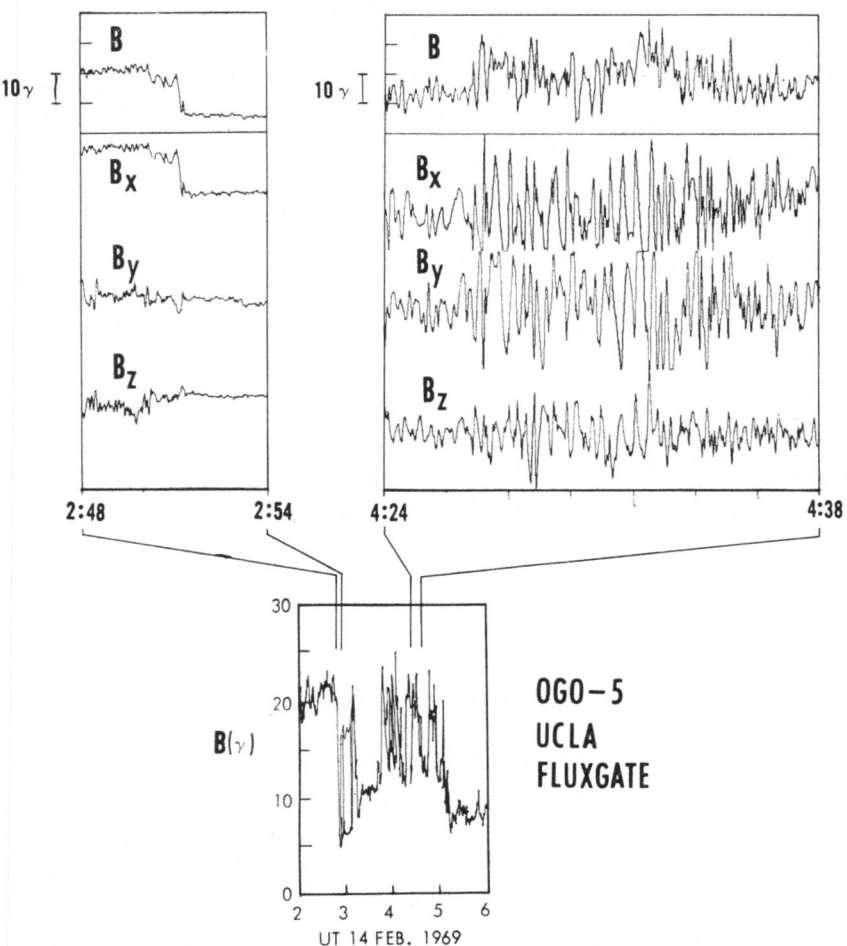

Fig. 1.   Two contrasting bow shock structures recorded on a single pass by OGO-5. Below, 1 min averages of field magnitude. Above, two details of the magnitude and three components sampled every 1.15 s; the Z component is approximately along the shock normal:
Field values uncorrected for spacecraft bias.

details of the distinction are preserved in the two inserts displaying short sections of the data at one-second time resolution. A full description of the pulsation phenomenon can be found in the sources cited above or in their references. We note explicitly one important finding: pulsation shocks and their associated upstream waves appear when the local nominal shock normal and the average upstream field make an angle less than $50° \pm 10°$ with each other, a state rare in the laboratory.

The second system of categorization has been introduced by Dobrowolny and Formisano (1973) (see also Formisano and Hedgecock, 1973). It is a refinement of the Laminar-Turbulent system of Tidman and Krall (1971). In this refinement, bow shock structures are classified as laminar, quasi-laminar, quasi-turbulent, turbulent, mixed, and unclassified. These categories are based on the values of upstream parameters β, M_A, and $\mathbf{B}_0 \cdot \mathbf{n}$. The unclassified group contains all shock observations for which these parameters are unknown. According to this scheme, the shock goes from laminar to turbulent extremes as β and M_A increase independently from low to high levels. Low and high ranges of these parameters are separated, roughly at the values $\beta = 0.1$, $M_A = 3$. Perpendicular means strictly that $\pi/2 - \arccos (\mathbf{B} \cdot \mathbf{n}/\mathbf{B}) \lesssim (M_e/M_i)^{1/2}$ radians; oblique means any other angle between $\mathbf{B}$ and $\mathbf{n}$.

An effort is underway, as this is written, to document the refined categorization with observational examples. Some of the distinctions are comparatively subtle from an experimental viewpoint. The second system seems to encompass that of Greenstadt et al., but the relationship between the latter's empirical classification and the more theoretically-oriented one of Dobrowolny and Formisano is not yet established. Most pulsation cases examined so far probably, although not necessarily, correspond to the mixed structure category ($\beta > 0.1$, $M_A > 3.0$, $\mathbf{B} \cdot \mathbf{n}/\mathbf{B} > \cos 45°$), but the pulsation behavior is undoubtedly present for other values of β and M_A as well.

In a sense, the problem of macroscopic structure of the bow shock consists in determining how the observed phenomena are most appropriately organized into a hierarchy of categories and subcategories. Certainly the most striking categorization on all but the most refined time scales is in the division represented by Figure 1. A sensible effort to match theory with observation begins by examining the relatively simple jump shock at the upper left of Figure 1, in which a single major field gradient predominates and the dependence of microstructure on plasma parameters can be discerned. Measurement of these parameters becomes unreliable in the midst of the pulsation phenomenon. In the following section, the term ΔB is assumed to apply to any clearly-definable field jump accompanied by irreversible heating. We do not exclude the possibility that it may apply to a selected gradient surrounded by others.

3. Instabilities in Magnetic Gradients

3.1. DRIFT SPEEDS AND SCALE LENGTHS

By examining a single magnetic field gradient within a macroscopic shock structure, the net drift speed between ions and electrons can be estimated from

$$\Delta B/\Delta x \simeq 4\pi J/c = (4\pi/c) Ne(\bar{v}_i - \bar{v}_e) \tag{1}$$

and of course drift velocity $\bar{v}_\mathrm{D} \equiv \bar{v}_\mathrm{i} - \bar{v}_\mathrm{e}$. For a jump $\Delta B = B_1 - B_0$ over a scale length $L_\mathrm{B} \sim \Delta x$, one finds

$$v_\mathrm{D} = c\Delta B/4\pi N e L_\mathrm{B}. \tag{2}$$

In Table I we give some expressions for drift speed normalized to either the electron thermal speed a_e, or that of ions, a_i, where $a_\mathrm{j}^2 = 2KT_\mathrm{j}/M_\mathrm{j}$ defines these speeds. We have chosen two limits, the 'thinnest' gradient length $L_\mathrm{B} = c/\omega_\mathrm{pe}$, the thicker one,

TABLE I

Normalized drift speeds in magnetic field gradients having scale lengths L_B indicated

Drift Speed	$L_\mathrm{B} = c/\omega_\mathrm{pe}$	$L_\mathrm{B} = c/\omega_\mathrm{pi}$	$L_\mathrm{B} = a_\mathrm{i}/\Omega_\mathrm{i} = R_\mathrm{i}$
$v_\mathrm{D}/a_\mathrm{e}$	$(T_\mathrm{i}/\beta_\mathrm{i}T_\mathrm{e})^{1/2}\,\Delta B/B_0$	$(m_\mathrm{e}T_\mathrm{i}/m_\mathrm{i}T_\mathrm{e}\beta_\mathrm{i})^{1/2}\,\Delta B/B_0$	$(m_\mathrm{e}T_\mathrm{i}/m_\mathrm{i}T_\mathrm{e}\beta_\mathrm{i}^2)^{1/2}\,\Delta B/B_0$
$v_\mathrm{D}/a_\mathrm{i}$	$(m_\mathrm{i}/m_\mathrm{e}\beta_\mathrm{i})^{1/2}\,\Delta B/B_0$	$\beta_\mathrm{i}^{-1/2}\,\Delta B/B_0$	$\beta_\mathrm{i}^{-1}\,\Delta B/B_0$

c/ω_pi, as well as the often-suggested scale length $R_\mathrm{i} =$ ion gyroradius. Each entry in the body of Table I is a drift speed formula corresponding to the normalized drift speed at the far left column and the gradient length at the top. Note that these normalized drift speeds are all directly proportional to the jump $\Delta B/B_0$ in magnetic field, and to either $\beta_\mathrm{i}^{-1/2}$ or β_i^{-1}. Thus, for fixed $\Delta B/B_0$, decreases in β_i enhance the drifts and thus would appear to enhance the potential for a current-driven instability.

We now can discuss some of the linear instabilities which may arise when the drift speeds in Table I exceed certain threshold values.

3.2. Microinstabilities in Current Layers

There are a host of possible microinstabilities that may arise in current-carrying plasmas. The modes which have been postulated include the following: *Buneman modes*, suggested by Kellogg (1964); *ion-acoustic modes* (Fried and Gould, 1961) suggested by Fredricks *et al.* (1970); *drift modes*, suggested by Krall and Book (1969); the *cyclotron-drift modes* (Forslund *et al.*, 1970; Wong, 1970; Gary and Sanderson, 1970; Gary, 1970a, b; Lashmore-Davies, 1970; Lampe *et al.*, 1971a, b; Gary and Biskamp, 1971) which were discussed extensively in the framework of the bow shock by Wu and Fredricks (1972); *ion-ion counterstreaming modes*, suggested by Tidman (1967), and later elaborated by Davidson *et al.* (1970), Auer *et al.* (1971), and Papadopoulos (1971).

Of these possible modes, it can be shown that for a plasma with significant β, the magnetization of the electrons almost certainly rules out the applicability of the usual Buneman and ion-acoustic dispersion relations. However, the drift, cyclotron-drift, and ion-ion counterstreaming modes remain viable candidates. For arguments against the Buneman and the ion-acoustic cross field instabilities see Wu and Fredricks (1972).

In Table II we have listed some of the modes yielding microinstabilities and their properties. Each entry gives the criterion the corresponding quantity at the top of its

TABLE II

Criteria for the establishment of various unstable modes

Mode	ω_r	γ_{max}	$v_D(min)$
Buneman	$\sim (m_e/m_i)^{1/3}\,\omega_{pe}$	$\sim (m_e/m_i)^{1/3}\,\omega_{pe}$	$\gtrsim a_e$
Ion-Acoustic	$\gtrsim \omega_{pi}$	$\gtrsim 0.1\,\omega_{pi}$	$\gtrsim 20\,a_i$
Drift	$\sim (\Omega_e\Omega_i)^{1/2}$	$\ll (\Omega_e\Omega_i)^{1/2}$	$\gg a_i$
Cyclotron-Drift $(kR_E/n=3)$	$\sim n\Omega_e$	$0.116\,\Omega_e$	$0.35\,a_e$
Cyclotron-Drift $(kR_E/n=15)$	$\sim n\Omega_e$	$0.023\,\Omega_e$	$0.083\,a_e$
Ion-Ion Streaming	$\dfrac{\omega_{pi}}{2(1+\omega_{pe}^2/\Omega_e^2)}$	$\dfrac{\omega_r}{2}$	ω_r/k
Non-Linear Modes	?	?	
Fluid Turbulence	?	?	

Mode	$\dfrac{\Delta B}{B_0},\ c/\omega_{pe}$	$\dfrac{\Delta B}{B_0},\ c/\omega_{pi}$	$\dfrac{\Delta B}{B_0},\ R_i$
Buneman	$\gtrsim \left(\dfrac{T_e}{T_i}\beta_i\right)^{1/2}$	$\gtrsim \left(\dfrac{m_i T_e}{m_e T_i}\beta_i\right)^{1/2}$	$\gtrsim \left(\dfrac{m_i T_e}{m_e T_i}\right)^{1/2}\beta_i$
Ion-Acoustic	$\gtrsim 20\left(\dfrac{m_e}{m_i}\beta_i\right)^{1/2}$	$\gtrsim 20\beta_i^{1/2}$	$\gtrsim 20\beta_i$
Drift	Stable	$\gtrsim \beta_i^{1/2}$	$\gtrsim \beta_i$
Cyclotron-Drift $(kR_E/n=3)$	$0.35\left(\dfrac{T_e}{T_i}\beta_i\right)^{1/2}$	$0.35\left(\dfrac{m_i T_e}{m_e T_i}\beta_i\right)^{1/2}$	$0.35\left(\dfrac{m_i T_e}{m_e T_i}\right)^{1/2}\beta_i$
Cyclotron-Drift $(kR_E/n=15)$	$0.083\left(\dfrac{T_e}{T_i}\beta_i\right)^{1/2}$	$0.083\left(\dfrac{T_e m_i}{T_i m_e}\beta_i\right)^{1/2}$	$0.083\left(\dfrac{m_i T_e}{m_e T_i}\right)^{1/2}\beta_i$
Ion-Ion Streaming			
Non-Linear Modes			
Fluid-Turbulence			

Mode	$A,\ c/\omega_{pe}$	$A,\ c/\omega_{pi}$	$A,\ R_i$	Heats
Buneman	$\left(\dfrac{m_e}{m_i}\right)^{1/3} c/V_F$	$\left(\dfrac{m_i}{m_e}\right)^{1/6} c/V_F$	$\left(\dfrac{m_i}{m_e}\right)^{1/6}\beta_i^{1/2} c/V_F$	Electrons
Ion-Acoustic	$0.1\left(\dfrac{m_e}{m_i}\right)^{1/2} c/V_F$	$0.1\,c/V_F$	$0.1\,\beta_i^{1/2}\,c/V_F$	Electrons
Drift	Stable	$\gg 1$	$\gg 1$	Electrons
Cyclotron-Drift $(kR_E/n=3)$	$0.116\left(\dfrac{m_i}{m_e}\right)^{1/2} c_A/V_F$	$0.116\left(\dfrac{m_i}{m_e}\right) c_A/V_F$	$0.116\,\dfrac{m_i}{m_e}\dfrac{a_i}{V_F}$	Electrons
Cyclotron-Drift $(kR_E/n=15)$	$0.023\left(\dfrac{m_i}{m_e}\right)^{1/2} c_A/V_F$	$0.023\left(\dfrac{m_i}{m_e}\right)^{1/2} c_A/V_F$	$0.023\,\dfrac{m_i}{m_e}\dfrac{a_i}{V_F}$	Electrons
Ion-Ion Streaming				Ions
Non-Linear Modes				Both
Fluid-Turbulence				Both

column must meet (for the associated scale length following the comma) to support the corresponding instability named at the far left. Entry (2, 5), for example, says that $\Delta B/B_0$ must exceed $20\,\beta_i^{1/2}$ in distance c/ω_{pi} in order that the ion-acoustic mode be generated. The rest-frame frequencies and growth rates listed are those given by linear theory. If these modes grow several e-folds in one shock scale length L_B, they will become either quasi-linearly or non-linearly 'saturated', and their rest-frame frequencies and growth rates will be different from those in the table. The amplification factor A over a scale length L_B is (Kellogg, 1964),

$$A = \int_0^{L_B} \gamma\,(v_D)\,\frac{dx}{V_F} \tag{3}$$

where $\gamma(V)$ is the growth rate, $v_D(x)$ is the drift velocity profile, and $V_F(x)$ is the flow speed profile though the gradient $\Delta B/\Delta x$. The functions $v_D(x)$ and $V_F(x)$ are found from shock theory or from measurements (the latter are not really available on the time scales required for use in the above formula). An expression for $\gamma_{max}(v_D)$ must be found from solutions to the dispersion relations governing the modes assumed. An order of magnitude approximation to the formula (3) is

$$A \sim \gamma_{max}\,(\bar{v}_D)\,\frac{L_B}{\bar{V}_F}, \tag{4}$$

where $\bar{v}_D$ and $\bar{V}_F$ are averaged through the gradient, or $\bar{v}_D$ is selected from Table I. The values given in Table II for A were estimated using Equation (4). The ratio $c/\bar{V}_F$ is typically $\sim 10^3$, while $c_A/V_F \sim M_A \sim 1$ to 20. In order to generate the instabilities listed, the drift velocity in the gradient length L_B must exceed a threshold value (Table II). This in turn requires the jump $\Delta B/B_0$ to exceed the critical values shown for the scale length quoted. The reader can choose values of the parameters and investigate the feasibility of mode excitation for given situations. We note in passing only that some modes are favored at low β_i, and all of the first five modes preferentially heat electrons, thus leading to an increase in T_e/T_i if a many-gradient precursor region precedes the main shock gradient.

The main shock gradient can be expected to contain large, probably time-dependent electric potentials. Thus, one can expect for $\beta \gg 0$, that this region can produce a very non-Maxwellian, multi-peaked proton distribution. Under these conditions, a host of ion-ion beam interactions very likely go on, and results of several calculations (Davidson et al., 1970; Papadopoulos, 1971) including the magnetization of electrons show that one could expect ion thermalization in a length $L \sim (5–10)\bar{V}_F/\gamma_{max} \sim 40$ to 80 km for typical solar wind conditions. This length is comparable to the scales c/ω_{pi} or R_i for $\beta \sim 1$. Papadopoulos (1971) has argued that the calculated saturation spectral density $E_w^2/8\pi \lesssim 10^{-2}\,NKT$, which would be in agreement with observations of the bow shock by Fredricks et al. (1970).

The 'non-linear instabilities' and the 'fluid turbulence' listed in Table II represent little-understood phenomena, which need exploration by (perhaps) properly-designed computer simulations of shocks.

4. Status of the Observations

Experimental investigation of the relation of the modes tabulated above to individual bow shock structures has been elementary and superficial. Comparison of shock dissipation theory with measurement depends on the detection of certain wavemodes which must be verified by concurrent measurement of critical plasma parameters and on the observation of certain characteristic scale lengths in the shock structure. Clearly, data must be acquired by properly instrumented satellites, and some means of ascertaining shock velocities, hence thicknesses, must be available. In general, high resolution particle (proton and electron) spectra, plasma wave noise, EM noise, and magnetic field data are all required simultaneously, both in the shock structure and in the unshocked plasma.

So far, almost all space observations have been useful chiefly for what might be termed 'existence proofs', where the presence of some wavemodes has been established, but the precise connection between these modes and particle heating is incomplete. Since the bow shock's gradient regions have associated currents, it is not surprising that electrostatic waves have been observed in these gradient regions (Fredricks and Coleman, 1969; Fredricks et al., 1968, 1970; Rodriguez and Gurnett, 1973). Unfortunately, to date no accurate correlations of these electrostatic wave observations with upstream Mach number, M_A, temperature ratio, T_e/T_i, ion beta, or angle between B_0 and the shock normal n have been published. Thus, it is not certain exactly what macroscopic structures were producing the observed microinstabilities. However, it has been suggested that most of the shocks in which strong electrostatic wave spectra were observed were turbulent, with high M_A and appreciable obliquity (Formisano, 1973). There is reason to believe that both unigradient and incipient pulsation structures have been resolved, but the correct structures have not been established.

Beyond the question of their existence and compatibility with ambient plasma conditions, instability modes can be tested for their consistency with shock scale lengths if the latter can be estimated. Such estimation requires measurement of shock velocity V_{SH} and gradient passage time Δt to obtain scale $L = V_{SH}\Delta t$. Two methods of estimating shock velocities have been employed. One depends on applying the Rankine-Hugoniot conditions using averaged upstream and downstream plasma parameters, and solving for the shock velocity; this can be done with one spacecraft (Lepping and Argentero, 1971). The other method uses two or more spacecraft to obtain a kinematic estimate of shock velocity derived from elapsed time measurements (Greenstadt et al., 1972). A combined, hybrid method has also been devised. The difficulty of assessing 'shock velocity' by either method in the presence of pulsation structure is self-evident from Figure 1: There is no unambiguously-defined 'front' whose motion can be followed, and average upstream and downstream parameters, especially the latter, are problematical quantities. The only established dimension or scale associated with the pulsation structure is its overall thickness which is known to exceed 2 or 3 R_E on occasion.

Velocities of the more easily identified unigradient shocks have been measured by both methods, yielding a range of speeds from 0 to 150 km s^{-1} (Formisano et al., 1973). The most reliable velocities and scale lengths obtained at the bow shock so far have been for cases of laminar, oblique (but almost perpendicular) structure, where magnetic ramp thicknesses on the order of 2 to 3 c/ω_{pi} have been estimated (Greenstadt et al., 1973). Examples of the laminar case with broadband electric field noise appearing in and close to the shock's magnetic ramp have been shown by Greenstadt et al. (1973). The relationship of particle heating to the ramp of their examples is presently being investigated. Unfortunately, even there, the analysis is condemned to remain incomplete because the electric field sampling interval was only on the order of the ramp duration, and only broadband coverage was obtained.

5. Discussion

The preceding section indicates that the process of relating theoretically-significant quantities to measured properties of the shock is in its infancy. The process is complicated by the existence of the whole class of pulsation shocks for which no macroscopic theory has been fully developed.

The phenomena described in Section 3 are dissipative mechanisms capable of producing irreversible particle heating across the bow shock structure. The electron-heating instabilities can be expected in the precursor region of the bow shock, and indeed have been observed there on many occasions by the OGO-5 Plasma Wave Detector Experiment. Examples have been shown, some of the clearest being in the paper by Fredricks et al. (1970). These precursor gradients need only exceed values given in Table II. We note that the most likely candidate is the cyclotron-drift mode with ratio of $kR_e/n = 15$, i.e., a large electron Larmor radius, low harmonic number ($n = 1, 2, 3 \ldots$ integer) mode. This mode can have significant growth even in small gradients and large scale lengths ($L_B = c/\omega_{pi}$ or R_i), as one can verify by numerically evaluating expressions contained in Table II.

However, the main proton heating mechanism cannot be found in this mode. The evidence is that ion heating occurs rapidly in the main gradient of a well-defined shock structure, and probably gradually in an ill-defined 'pulsation' shock. Thus, ion-ion streaming, non-linear instabilities, or 'fluid turbulence' (whatever it is) are still the mysterious and poorly-defined, probable dissipation mechanism responsible for proton heating in the bow shock. In the 'pulsation' shock, one may be dealing with a very gradual particle mechanism extending over broad regions of space, and probably with multiple, partial heating episodes.

Another poorly understood phenomenon is that of reflected particles streaming back along field lines intersecting the shock front. These particles must be accounted for as a part of the dissipation mechanism. They also produce interesting side effects, and are doubtless the generating agent for upstream hydromagnetic waves (Greenstadt *et al.*, 1967; Fairfield, 1969; Russell *et al.*, 1971; Scarf *et al.*, 1970; Barnes, 1970). The presence of these reflected particles indicates existence of some electric potential barrier at the shock front capable not only of classical retarding-potential reflection, but also of some energization process as well (Sonnerup, 1969).

At the present time, we can formulate some opinions about the state of interpretation of phenomena in the Earth's bow shock.

It is probable that we have at least a working understanding of the electron heating mechanism in certain bow shock structures. Once reliable input (upstream) parameters, such as β_i, M_A, T_e/T_i, and $\mathbf{B}_0 \cdot \mathbf{n}/B_0$ are known, and the shock speed V_S relative to the satellite measurement frame can be accurately estimated, we shall be able to predict instabilities from Table I and II. This will provide us with an understanding of why, in some structures, electrostatic noise is observed, while in others of similar macroscopic appearance, it is not. This is almost surely M_A- and β-dependent.

Because of inherent limitations on spacecraft-borne particle probes (time-resolution limitations) however, it is not so clear that we shall, in the near future, have a much better experimental basis for understanding the ion heating mechanism, which, as Table II shows, is still a theoretical mystery. Improvement in plasma wave detection equipment may be necessary to solve this problem.

Acknowledgments

This work was partially supported by the National Aeronautics and Space Administration under Contracts NASW-2398 (EWG) and NAS 5-9278 (RWF).

Partial Glossary of Symbols

Plasma Frequency: $\omega_{pe,i}^2 = 4\pi N_e^2/m_{e,i}$ Ion Beta: $\beta_i = 8\pi NKT_i/B^2$

Gyrofrequency: $\Omega_{e,i} = eB/m_{e,i}c$ Thermal Speed: $a_{e,i}^2 = 2KT_{e,i}/m_{e,i}$

Alfvén Speed: $c_A^2 = B^2/4\pi m_i$ Gyroradius: $R_{E,i} = a_{e,i}/\Omega_{e,i}$

Flow Speed: V_F Real Frequency: ω_r

Mach Number: $M_A = V_F/c_A$ Growth Rate: γ

References

Auer, P. L., Kilb, R. W., and Crevier, W. F.: 1971, *J. Geophys. Res.* **76**, 2927.

Barnes, A.: 1970, *Cosmic Electrodyn.* **1**, 90.

Davidson, R. C., Krall, N. A., Papadopoulos, K., and Shanny, R.: 1970, *Phys. Rev. Letters* **24**, 579.

Dobrowolny, M. and Formisano, V.: 1973, CNR Preprint LPS-73-11, Rome.

Fairfield, D. H.: 1969, *J. Geophys. Res.* **74**, 3541.

Formisano, V. and Hedgecock, P. C.: 1973, *J. Geophys. Res.* **78**, 3745.

Formisano, V., Hedgecock, P. C., Moreno, G., Palmiotto, F., and Chao, J. K.: 1973, *J. Geophys. Res.* **78**, 3731.

Forslund, D. W., Morse, R. L., and Nielsen, C. W.: 1970, *Phys. Rev. Letters* **25**, 1266.

Fredricks, R. W. and Coleman, P. J. Jr.: 1969, in D. G. Wentzel and D. A. Tidman (eds.), *Plasma Instabilities in Astrophysics*, Gordon and Breach, New York.

Fredricks, R. W., Kennel, C. F., Scarf, F. L., Crook, G. M., and Green, I. M.: 1968, *Phys. Rev. Letters* **21**, 1761.

Fredricks, R. W., Crook, G. M., Kennel, C. F., Green, I. M., Scarf, F. L., Coleman, P. J., Jr., and Russell, C. T.: 1970, *J. Geophys. Res.* **75** 3751.

Fried, B. D. and Gould, R. W.: 1961, *Phys. Fluids* **4**, 139.

Gary, S. P.: 1970a, *J. Plasma Phys.* **4**, 753.

Gary, S. P.: 1970b, *J. Plasma Phys.* **5**, 561.

Gary, S. P. and Biskamp, D.: 1971, *J. Phys.* **A4**, L27.

Gary, S. P. and Sanderson, J. J.: 1970, *J. Plasma Phys.* **4**, 739.

Greenstadt, E. W., Inouye, G. T., Green, I. M., and Judge, D. L.: 1967, *J. Geophys. Res.* **72**, 3855.

Greenstadt, E. W., Green, I. M., Inouye, G. T., Colburn, D. S., Binsack, J. H., and Lyon, E. F.: 1970a, *Cosmic Electrodyn.* **1**, 160.

Greenstadt, E. W., Green, I. M., Inouye, G. T., Colburn, D. S., Binsack, J. H., and Lyon, E. F.: 1970b, *Cosmic Electrodyn.* **1**, 316.

Greenstadt, E. W., Hedgecock, P. C., and Russell, C. T.: 1972, *J. Geophys. Res.* **77**, 1116.

Greenstadt, E. W., Russell, C. T., Formisano, V., and Hedgecock, P. C.: 1973, *J. Geophys. Res.*, to be submitted.

Kellogg, P. J.: 1964, *Phys. Fluids* **7**, 1555.

Krall, N. A. and Book, D. L.: 1969, *Phys. Fluids* **12**, 347.

Lampe, M., McBride, J. B., Orens, J. H., and Sudan, R. N.: 1971a, *Phys. Rev. Letters* **35**, 131.

Lampe, M., Manheimer, W. M., McBride, J. B., Orens, J. H., Papadopoulos, K., Shanny, R., and Sudan, R. N.: 1971b, *Memo Rept.* **2358**, Naval Research Lab., Washington, D.C.

Lashmore-Davies, C. N.: 1970, *J. Phys.* **A3**, L40.

Lepping, R. P. and Argentiero, P. D.: 1971, *J. Geophys. Res.* **76**, 4349.

Papadopoulos, K.: 1971, *J. Geophys. Res.* **76**, 3806.

Rodriguez, P., and Gurnett, D. A.: 1973, *Trans Amer. Geophys. Union* (*EOS*) *Abstract* **54**, 433.

Russell, C. T., Childers, D. D., and Coleman, P. J.: 1971, *J. Geophys. Res.* **76**, 845.

Scarf, F. L., Fredricks, R. W., Frank, L. A., Russell, C. T., Coleman, P. J., Jr., and Neugebauer, M.: 1970, *J. Geophys. Res.* **75**, 7316.

Sonnerup, B. U. Ö.: 1969, *J. Geophys. Res.* **74**, 1301.

Tidman, D. A.: 1967, *J. Geophys. Res.* **72**, 1799.

Tidman, D. A. and Krall, N. A.: 1971, *Shock Waves in Collisionless Plasmas*, Wiley-Interscience, New York.

Wong, H. V.: 1970, *Phys. Fluids* **13**, 757.

Wu, C.-S. and Fredricks, R. W.: 1972, *J. Geophys. Res.* **77**, 5585.

PERTURBATION OF A WAVE FIELD BY AN IRREGULARITY
IN A MAGNETO-PLASMA

T. R. KAISER

Department of Physics, University of Sheffield, Sheffield, England

P. C. KENDALL* and G. T. VICKERS

Department of Applied Mathematics, University of Sheffield, Sheffield, England

1. Introduction

When a satellite is in an environment such that the order of the Debye length exceeds the lengths of the aerials or booms, the wave field which is recorded may be perturbed by the irregularity of plasma around the satellite. This may happen, for example, in a satellite such as GEOS. The region of non-uniformity may be electron-depleted at night, or electron-enhanced during the day. In order to study the qualitative effects of an irregularity on an incident wave field we have formulated an idealized model. In the simplified problem, electromagnetic radiation propagating through a cold magneto-plasma falls upon an irregularity in the form of an ellipsoid of different density whose axis is parallel to the unperturbed magnetic field. This idealization is treated in the quasi-static approximation. We use spheroidal coordinates, which have previously been successful in dealing with the problem of an evacuated cavity in a magneto-plasma. The results show that the wave field is perturbed in both amplitude and phase within the irregularity and its neighborhood. Some of the resonances found in the case of a cavity now disappear. The longitudinal and split transverse Tonks-Langmuir resonances still remain, but become separated in frequency when the ellipsoid varies from the spherical shape.

2. Equations of the Problem

We assume that the plasma is collisionless and cold. If ω is the angular frequency of the wave incident upon the irregularity, $\mathbf{B}$ is the magnetic induction field and, using the *quasistatic approximation*, Ω is the electrical potential, then Maxwell's equations give

$$\nabla \times \mathbf{B} = - i\omega \mathcal{K} \nabla \Omega, \tag{1}$$

where $\mathcal{K}$ is a complex dielectric tensor,

$$\mathcal{K} = \begin{pmatrix} \kappa_1 & -i\kappa_2 & 0 \\ i\kappa_2 & \kappa_1 & 0 \\ 0 & 0 & \kappa_0 \end{pmatrix}, \tag{2}$$

* Now at the University of Keele

B. M. McCormac (ed.), Magnetospheric Physics, 291–296. All Rights Reserved.
Copyright © 1974 by D. Reidel Publishing Company, Dordrecht-Holland.

as defined by Ratcliffe (1959). Equation (1) shows that the basic equation satisfied by Ω is

$$\nabla \cdot (\mathscr{K} \nabla \Omega) = 0. \tag{3}$$

Let the suffix 1 distinguish quantities outside the ellipsoidal irregularity and 2 distinguish quantities inside. The suffix $q = 1, 2$ is reserved for this purpose, but it will be omitted if there is no possibility of any ambiguity. Then the wave frequency is $f = = \omega/2\pi$, and the plasma, ion-gyro, and electron-gyro frequencies are f_{pq}, f_i and f_e respectively. We find also that the following dimensionless constants α, β appear repeatedly in the analysis:

$$\alpha = \frac{\kappa_1}{\kappa_0} = \frac{f^2}{f^2 - f_p^2} \left\{ 1 - \frac{f_p^2 (f^2 - f_i f_e)}{(f^2 - f_e^2)(f^2 - f_i^2)} \right\} \tag{4}$$

and

$$\beta = \frac{\kappa_2}{\kappa_0} = \frac{f^3 f_p^2 (f_e - f_i)}{(f^2 - f_p^2)(f^2 - f_e^2)(f^2 - f_i^2)}. \tag{5}$$

Choosing cylindrical polar coordinates (R, ϕ, z) whose axis and origin coincide respectively with the axis and center of the ellipsoidal irregularity, whose axis is also parallel to the background magnetic field, Equation (3) reduces to

$$\frac{\partial^2 \Omega}{\partial z^2} + \alpha_q \left\{ \frac{1}{R} \frac{\partial}{\partial R} \left(R \frac{\partial \Omega}{\partial R} \right) + \frac{1}{R^2} \frac{\partial^2 \Omega}{\partial \phi^2} \right\} = 0, \quad q = 1, 2. \tag{6}$$

As α_q may be either positive or negative this equation may be either elliptic or hyperbolic in character, either inside or outside independently of each other. If $\alpha_q < 0$ waves can only propagate along rays making angles of less than $\tan^{-1} |\alpha_q|^{1/2}$ with the background magnetic field.

The problem is to solve Equations (6) both outside and inside the irregularity bounded by the ellipsoidal surface S represented by the equation

$$z^2/a^2 + R^2/b^2 = 1. \tag{7}$$

The boundary conditions are:
(a) as $r \to \infty$, $-\nabla \Omega \to \mathbf{E}_\infty$, a constant electric field,
(b) as $r \to 0$, $-\nabla \Omega$ remains finite,
    (Here, $r = (R^2 + z^2)^{1/2}$ and the time factor $e^{i\omega t}$ has been omitted),
(c) at S, Ω is continuous, and
(d) at S, the normal component of displacement, $-\hat{\mathbf{n}} \cdot \mathscr{K} \nabla \Omega$ is continuous.

3. Solutions

Using scaled coordinates (Kaiser, 1962) defined by

$$z' = z, \qquad R' = R/\alpha_q^{1/2}, \quad q = 1, 2, \tag{8}$$

different in the two regions, we may reduce Equation (6) to

$$(\nabla^2)'\Omega = 0. \tag{9}$$

The ellipsoid then becomes two ellipsoids:

$$z'^2/a^2 + \alpha_q R'^2/b^2 = 1, \quad q = 1, 2. \tag{10}$$

Using *two* sets of spheroidal coordinates instead of the one set used by Kaiser and Kendall (1973) we introduce μ_q and ζ_q through the definitions

$$R' = \sigma_q(1 - \mu_q^2)^{1/2} (\zeta_q^2 - 1)^{1/2}, \qquad z' = \sigma_q\mu_q\zeta_q, \tag{11}$$

where σ_q, $q = 1, 2$ are constants. These coordinates were used by Lamb (1932) in the solution of fluid flows past an ellipsoid of revolution. The ellipsoids (Equation (10)) then become surfaces $\zeta = \zeta_{q0}$, a constant in each case, provided that

$$\zeta_{q0}^2 = \frac{\alpha_q}{\alpha_q - (1 - e^2)} \quad \text{and} \quad \sigma_q = \frac{a}{\zeta_{q0}}, \tag{12}$$

where e is the eccentricity defined by $b^2 = a^2(1 - e^2)$: for a sphere $e = 0$, for an ellipsoid elongated along the magnetic field $e^2 > 0$, and for an oblate ellipsoid $e^2 < 0$.

If x, y, z are rectangular Cartesian coordinates such that

$$x = R\cos\phi, \qquad y = R\sin\phi, \qquad z = z \tag{13}$$

we find that on the ellipsoid S

$$\begin{aligned} x &= b(1 - \mu_q^2)^{1/2}\cos\phi \\ y &= b(1 - \mu_q^2)^{1/2}\sin\phi \\ z &= a\mu_q. \end{aligned} \tag{14}$$

This shows that on the ellipsoid S, $\mu_1 = \mu_2 = \mu$ (say) and gives the link between the two regions.

The boundary condition on the normal displacement at S becomes a requirement that the expression

$$\left(1 - \frac{\omega_p^2}{\omega^2}\right)\left\{z\frac{\partial\Omega}{\partial z} + \frac{\alpha^2}{1 - e^2}\left(x\frac{\partial\Omega}{\partial x} + y\frac{\partial\Omega}{\partial y}\right) - \frac{i\beta}{1 - e^2}\left(x\frac{\partial\Omega}{\partial y} - y\frac{\partial\Omega}{\partial x}\right)\right\} \tag{15}$$

is continuous at S.

In the rectangular Cartesian coordinates (x, y, z) we write $\mathbf{E}_\infty = (E_1, E_2, E_3)$. Then, for convenience put

$$\Omega_1 = \Omega_1' - (xE_1 + yE_2 + zE_3), \tag{16}$$

so that $\Omega_1' = O(1/r^2)$ as $r \to \infty$.

In previous work (Kaiser and Kendall, 1973) involving only one set of spheroidal coordinates we have found the solution for Ω_1' to be of the form

$$\Omega_1' = A_1\{Q_1(\zeta_1)/Q_1(\zeta_{10})\}P_1(\mu_1) + C_1\{Q_1^1(\zeta_1)/Q_1^1(\zeta_{10})\}P_1^1(\mu_1)\exp(j\phi), \tag{17}$$

where $P_n^m(\mu)$ denotes the Legendre function of degree n and order m and $Q_n^m(\zeta)$ denotes the associated Legendre function defined such that $Q_n^m(\zeta) \to 0$ as $\zeta \to \infty$. Here A_1 and C_1 are constants determined by the boundary conditions. This suggests that we might try as a solution for Ω_2 the form

$$\Omega_2 = A_2 \{P_1(\zeta_2)/P_1(\zeta_{20})\} P_1(\mu_2)$$
$$+ C_2 \{P_1^1(\zeta_2)/P_1^1(\zeta_{20})\} P_1^1(\mu_2) \exp(j\phi), \tag{18}$$

where A_2 and C_2 are constants. The field inside the irregularity will then be constant. The complex number $j = \sqrt{(-1)}$ is taken to be independent of i in order to allow complete generality. Substituting for Ω_1 and Ω_2, using the continuity of Ω and expression (15) at the surface S, gives lengthy expressions for A_2 and C_2 which will not be reproduced here. If the electric field within the ellipsoid is (E_x, E_y, E_z) we find

$$E_z = \frac{(1 - f_{p1}^2/f^2)(\lambda_0 - 1) E_3}{\lambda_0(1 - f_{p1}^2/f^2) - (1 - f_{p2}^2/f^2)} \tag{19}$$

and

$$E_x - jE_y = \frac{(1 - f_{p1}^2/f^2)(\lambda_1 - \alpha_1)(E_1 - jE_2)}{\lambda_1(1 - f_{p1}^2/f^2) - \alpha_2(1 - f_{p2}^2/f^2) - ij\{\beta_1(1 - f_{p1}^2/f^2) - \beta_2(1 - f_{p2}^2/f^2)\}} \tag{20}$$

where

$$\lambda_0 = \zeta_{10} Q_1'(\zeta_{10})/Q_1(\zeta_{10}) \tag{21}$$

and

$$\lambda_1 = (1 - e^2)\zeta_{10} Q_1^{1\prime}(\zeta_{10})/Q_1^1(\zeta_{10}). \tag{22}$$

Here, the primed functions such as Q_1' (say) denote derivatives, also

$$\zeta_{10}^2 = \alpha_1/(\alpha_1 - 1 + e^2), \tag{23}$$

where 1 denotes the inside and 2 the outside of the ellipsoid.

4. Results and Discussion

Without loss of generality we may put $E_2 = 0$ in Equation (20). It is now convenient to let

$$R = \frac{E_z}{E_3}, \qquad R_1 = \frac{E_x}{E_1}, \qquad R_2 = \frac{E_y}{E_1}. \tag{24}$$

The positive values of f which make $\alpha = 0$ (see Equation (4)) are defined to be the lower and upper-hybrid frequencies denoted by f_{LH} and f_{UH} respectively. Thus, neglecting f_i in comparison with f_e,

$$f_{UH}^2 \approx f_p^2 + f_e^2 \tag{25}$$

and

$$f_{LH}^2 \approx f_p^2 f_i f_e/(f_e^2 + f_p^2). \tag{26}$$

Then

$$\alpha \approx \frac{f^2(f^2 - f_{\text{LH}}^2)(f^2 - f_{\text{UH}}^2)}{(f^2 - f_i^2)(f^2 - f_e^2)(f^2 - f_p^2)}. \tag{27}$$

The results will be compared with those obtained for a spherical void, and so we first give a summary of that case. There are six regions for which the behavior of R, R_1 and R_2 (as functions of f) are particularly interesting.

(a) As $f \to 0$; $R \to \infty$, $R_1 \to 1$ and $R_2 \to 0$.

(b) At $f = f_i$ there are resonances in R_1 and R_2; $R = 1.0$.

(c) Near the lower-hybrid frequency there are resonances in R_1 and R_2 at

$$f \approx \frac{f_{\text{LH}}}{\sqrt{(1 + \pi^2/16)}}$$

When $f = f_{\text{LH}}$, then $R_1 = R_2 = 0$ and R has a resonance.

(d) At $f = f_e$ there are resonances in R_1 and R_2; $R = 1.0$;

(e) Near the plasma frequency there are split resonances in R_1 and R_2 at

$$f \approx f_p \sqrt{\tfrac{2}{3}} \pm \frac{f_e}{4},$$

also R is resonant at $f \approx f_p \sqrt{(\tfrac{2}{3})}$; and

(f) When $f > f_{\text{UH}}$ the results are like the isotropic case (i.e. with $\mathbf{B} = 0$). For $f \geqslant 1.5 f_{\text{UH}}$, R, R_1 and R_2 differ from the isotropic values by less than 0.01.

When the irregularity is an ellipsoid of different density, the behavior of R, R_1 and R_2 is modified as follows:

(a) As $f \to 0$; $R \to (f_{p2}/f_{p1})^2$, $R_1 \to 1$ and $R_2 \to 0$;

(b) There are no resonances near $f = f_i$, however R_1 and R_2 have local maxima and R has a local minimum close to $f = f_i$.

At $f = f_i$ we have

$$R_1 = \frac{1}{2}\left\{\left(\frac{f_{p1}}{f_{p2}}\right)^2 + 1\right\}, \qquad R_2 = \frac{1}{2}\left\{\left(\frac{f_{p1}}{f_{p2}}\right)^2 - 1\right\}, \quad R = 1.0;$$

(c) Near $f = f_{\text{LH1}}$ there are resonances in R_1 and R_2 at

$$f \approx \frac{f_{\text{LH1}}}{\left[1 + \dfrac{\pi^2}{16(1 - e^2)}\left\{1 - \left(\dfrac{f_{p2}}{f_{p1}}\right)^2\right\}^2\right]^{1/2}}.$$

When $f = f_{\text{LH1}}$, then $R_1 = R_2 = 0$ and $R = (f_{p1}/f_{p2})^2$.

(d) when $f \approx f_e$ the behavior of R_1, R_2 and R is similar to that at $f \approx f_i$ (see b);

(e) the resonance in R occurs at nearly the same frequency as in the isotropic case, of no magnetic field, namely where

$$f^2 \approx (v_0 f_{p1}^2 + f_{p2}^2)/(1 + v_0).$$

Here

$$v_0 = -1 + e^2/\{(1 - e^2)(\ell - 1)\},$$

where

$$\ell = \frac{1}{2e} \log\left(\frac{1+e}{1-e}\right), \quad e^2 > 0$$
$$= |e|^{-1} \tan^{-1}|e|, \quad e^2 < 0.$$

The split resonance in R_1 and R_2 occurs with the mid-point of the split where there is a resonance in R_1 in the isotropic case, namely where

$$f^2 \approx (v_1 f_{p1}^2 + f_{p2}^2)/(1 + v_1).$$

Here

$$v_1 = 1 + 2/v_0,$$

where v_0 is defined above.

The width of the split is approximately

$$\frac{f_e |f_{p1}^2 - f_{p2}^2|}{v_1 f_{p1}^2 + f_{p2}^2}.$$

(f) when $f > \max(f_{UH1}, f_{UH2})$, then the results are again like the isotropic case. Note that the value of e^2 will affect the isotropic state.

References

Kaiser, T. R.: 1962, *Planetary Space Sci.* **9**, 639.
Kaiser, T. R. and Kendall, P. C.: 1973, in R. J. L. Grard (ed.), *Photon and Particle Interactions with Surfaces in Space*, D. Reidel, Publishing Company, Dordrecht-Holland, p. 91.
Lamb, H.: 1932, *Hydrodynamics*, Cambridge University Press.
Ratcliffe, J. A.: 1959, *Magneto-Ionic Theory and its Applications to the Ionosphere*, Cambridge University Press.

MONOCHROMATIC WAVES AND INHOMOGENEITY EFFECTS

ALAIN ROUX

Groupe de Recherches Ionosphériques, 3 Avenue de la République,
92131 Issy-les-Moulineaux, France

1. Introduction

Wave-particle interactions (WPI's) are known to play an important role in the loss or acceleration of magnetospheric plasma. A wide variety of waves, observed either on satellites, or from ground based observatories, is due to WPI's. We will be mainly concerned here with circularly polarized waves propagating along field lines.

The linear theory of the cyclotron instability (Sagdeev and Shafranov, 1961) is well known, many geophysical consequences have been stressed (see for instance Gendrin (1972) and references therein). As far as particle losses are concerned, it is essential to study the non-linear saturation level of these waves. The equilibrium between energetic electrons and VLF or ELF wide band hiss has been extensively studied (see Kennel and Petschek (1966) or a more recent work by Etcheto *et al.* (1973)). These studies were made in the framework of the quasi-linear approximation (QLA).

Meanwhile, many observed waves have a relatively narrow spectrum, and very large amplitudes, and for these waves QLA breaks. ELF chorus, Artificially Stimulated Emissions (ASE), and 'pearls' (in the Pc1 range) do belong to this class of relatively narrow and intense phenomenon. We will study here the non-linear development of these monochromatic waves.

For the monochromatic waves we are considering, cyclotron instability can be limited by three different processes:

(a) For small amplitude waves, trapping can be neglected. We will show in the third part that cumulative 'collisions' between particles and a monochromatic wave provide a diffusion in velocity space which strongly depends on the inhomogeneity along field lines. This mechanism was shown by Ashour-Abdalla (1972) and Welti *et al.* (1973) to give rise to sideband waves of the initial one. If the wave continues to grow, it will reach a trapping dominated regime.

(b) Then, the magnetic field of the wave can no longer be considered as small. As an effect of its finite size, particles can be trapped in the potential wells of the wave. Such an effect strongly reduces the amplification process, even in an inhomogeneous medium. This saturation effect by trapping oscillations will be studied in the second part.

(c) When trapping is efficient over many trapping lengths along the field line, trapped particle motions can couple the fastest growing mode to waves at slightly different frequencies. This is the so called trapped particle instability. It can also modify wave growth and give rise to sidebands of the initial wave.

In the last part, we will study in detail proper geophysical consequences of these non-linear effects.

B. M. McCormac (ed.), Magnetospheric Physics, 297–312. All Rights Reserved.
Copyright © 1974 by D. Reidel Publishing Company, Dordrecht-Holland.

2. Trapping Effects in an Inhomogeneous Medium

We will first examine trapping effects on particles which resonate with a given mono-chromatic wave propagating in an inhomogeneous medium. Secondly, the conse-quences of these trapping oscillations on wave instability (or damping) will be stressed.

2.1. Motion of Trapped Particles

Assuming a varying number, Gary et al. (1968) have computed trapped particle trajectories and shown that an electrostatic wave can accelerate trapped particles when its phase velocity increases.

Laval and Pellat (1970) analytically studied the same problem. They calculated an adiabatic invariant for trapped particles and showed that acceleration takes place as long as it is conserved.

For particles resonating with a circularly polarized electromagnetic wave, trapped particle motion can also be reduced to a one dimensional oscillation. Following for instance Dysthe (1971), one obtains (*in the resonant frame*):

$$\frac{d^2\psi}{dt^2} + \omega_T^2\psi = -k\Gamma \tag{1}$$

where

$$
\begin{cases}
\omega_T = \left\{ k_\parallel V_\perp \frac{eB_1}{m} \right\}^{1/2} \text{ is the trapping frequency} \\[2mm]
V_\perp = [V_x^2 + V_y^2]^{1/2} \text{ and } V_\parallel \text{ are the perpendicular and parallel compo-} \\
\hspace{5cm} \text{nents of the velocity} \\[2mm]
\psi = \varphi - \omega t + \int k(z)\,dz \\[2mm]
\varphi = tn^{-1} V_y/V_x \text{ is the phase angle} \\[2mm]
B_1 \text{ is the wave amplitude and } k_\parallel \text{ is the wave number}
\end{cases}
$$

Γ is a force which appears in the resonant frame. A non-Galilean change of frame has to be done, so that to remain in that frame (where particle motion is given by Equa-tion (1). Γ can be expressed as follows:

$$\Gamma = V_R \frac{dV_R}{dt} + \mu \frac{\partial B_0}{\partial z} \tag{2}$$

where V_R is the resonant velocity, μ the first adiabatic invariant, B_0 the static terrestrial magnetic field, and z the distance along field lines, as measured from the equator.

In an homogeneous medium, $\Gamma = 0$; Equation (1) is then a pendulum equation describing trapping oscillations of particles in the potential wells of the wave.

Let us define $F_T = \omega_T^2/k_\parallel$ the trapping force.

As long as $F_T > |\Gamma|$, particles remain trapped, whereas when $F_T < |\Gamma|$, all the particles are detrapped. Then, in the following, Γ will be called the detrapping force.

2.2. Effect of trapping oscillations on the wave saturation level

Let us first discuss briefly the saturation mechanism by trapping in an *homogeneous medium*.

Trapped particles are oscillating back and forth in the potential troughs of the wave. Thus their pitch angles have the same probability either to increase or to decrease. Consequently, in a trapping dominated regime, the energy exchange tends to be null. Such a regime is reached, for an initially unstable wave, when the trapping frequency (which is proportional to the square root of the wave amplitude) reaches the following limit

$$\omega_T \sim \gamma_L \tag{3}$$

where γ_L is the linear growth rate of the wave.

In an *inhomogeneous medium*, as shown previously, trapping and inhomogeneity compete in order to govern particle motion. From the balance between these two forces, we will deduce an estimation of the saturation level relevant to an inhomogeneous medium.

In a recent work, relevant to electrostatic waves propagating at a fixed frequency but with a varying wave number, Asseo *et al.* (1972) studied in detail trapping effects in an inhomogeneous medium.

The transposition of their work to electromagnetic waves is straightforward. We will only summarize here their main conclusions, concerning the wave amplitude evolution.

Let us suppose that the medium is initially unstable and that the inhomogeneity is such that Γ is monotonous. As long as $|\Gamma| > F_T$, the wave grows exponentially at the rate fixed by the linear theory (despite the finite amplitude of the wave). When $F_T > |\Gamma|$, the wave still grows but very slowly and at a non exponential rate. There are two possible things that can happen:

First, when the resonant velocity variation is smaller than the trapping extension $\omega_T/k_{\parallel}$, the non-linear spatial increment is

$$K_{NL} = \tfrac{1}{2} \frac{k_{\parallel}}{B_1^2} \frac{\partial}{\partial z} \left[\frac{B_1^2}{k_{\parallel}} \right] = \frac{\gamma_L}{V_g} \left(\frac{\Gamma}{F_T} \right) \tag{4}$$

where V_g is the group velocity. If $\gamma_L = 0$, Equation (4) describes the WKB variation of the wave amplitude along field lines ($B_1 \propto \sqrt{k_{\parallel}}$). When $\gamma_L \neq 0$, this equation describes the averaged (over trapping oscillations) evolution of the wave amplitude. As $F_T \propto B_1$, the amplification is no more exponential in this regime, but wave amplitude increases linearly with distance.

When the resonant velocity variation is larger than the trapping extension ($\omega_T/k_{\parallel}$), a slightly different expression was obtained:

$$K_{NL} = \tfrac{1}{2} \frac{k_{\parallel}}{B_1^2} \frac{\partial}{\partial z} \left[\frac{B_1^2}{k_{\parallel}} \right] = \frac{\gamma}{V_g} \left(\frac{\Gamma}{F_T} \right)^2 \frac{\omega_r z}{V_R}. \tag{5}$$

Wave amplitude is now varying like the square root of the distance.

Similar expressions were obtained independently by Budko *et al.* (1971). The fact that the non-linear growth rate becomes proportional to distance was first mentioned by Nunn (1971) in a computational study.

These results are summarized in Figure 1, where, for the sake of simplicity, the inhomogeneity parameter Γ has been assumed to be constant. It is clear that in usual

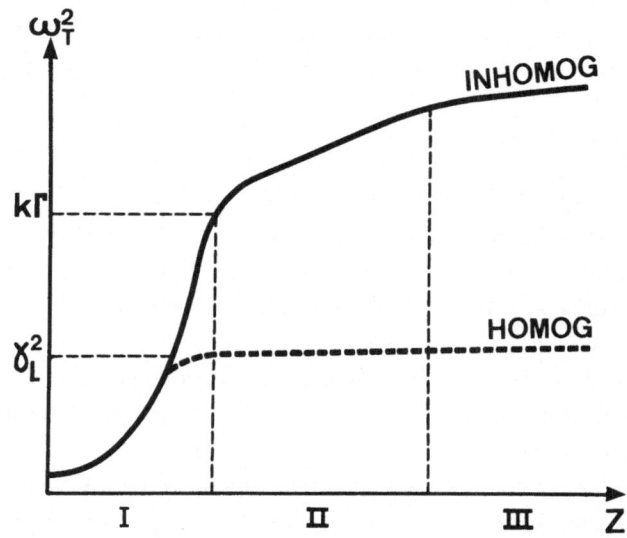

Fig. 1. Schematic plot of wave amplitude variation. The square of the trapping frequency (which is proportional to wave amplitude) is plotted vs. distance. In an homogeneous medium, saturation occurs when $\omega^2_T \sim \gamma^2_L$. Due to inhomogeneity, wave amplitude continues to grow exponentially at the rate fixed by the linear theory until $\omega_T{}^2 \sim k_\| \Gamma$ (Region I). In Region II ($\Delta V_R < \omega_T/k_\|$), wave amplitude increases linearly with distance. In Region III ($\Delta V_R > \omega_T/k_\|$), it still increases but as the square root of distances.

magnetospheric situations, Γ is not constant; furthermore, the residual non-linear growth rate (non-exponential when $|\Gamma| > F_T$) will generally not be strong enough to overcome wave energy losses (due, for instance, to bad ionospheric reflexion).

Let us study now in detail the appropriate consequences of this new saturation, in the particular geometry of terrestrial field lines.

All the quantities in the previous equations depend on the distance with respect to the equator. At the equator, trapping tends always to dominate, whereas far from the equator, no trapping can occur because $|\Gamma| \gg F_T$. The exact point where $|\Gamma| \sim F_T$ depends on the wave amplitude, but it should be noted that it does not depend very much on the pitch angle we are considering (Roux *et al.*, 1973).

Then one can expect that the instability easily saturates near the equator, whereas, exponential amplification still holds far from it. It should be stated that the preceding calculations are valid only if the number of trapping oscillations is large. Linear theory still holds over distances much smaller than the trapping length. Thus, the quasi-saturation in the equatorial region can be efficient only if particles suffer at least one trapping oscillation before being detrapped. This leads to the condition that saturation begins to be efficient when the trapping force dominates the inhomogeneity force over at least half a trapping length on each side of the equator. This can be written:

$$\Gamma\,(z = z_1) = \frac{\omega_{r_1}^2}{k_\parallel\,(z_1)} \quad \text{and} \quad z_1 = \frac{\mathscr{L}_\text{T}}{2} = \frac{\pi V_\text{R}}{\omega_{r_1}} \tag{6}$$

$\mathscr{L}_\text{T}$ being the trapping length.

Solving these equations, one obtains a minimum value of the wave amplitude B_1 in order that trapping becomes important. If $B > B_1$, wave amplification is largely reduced over the interval $(+z_1, -z_1)$. If exponential amplification remains effective out of this interval, then the wave can continue to grow for a while.

As shown by Liemohn (1967), the linear growth rate as a function of distance from the equator decreases rapidly. The quasi-saturation by trapping effects over the length $z > z_1$ provides a large decrease of the integrated growth rate along field lines, then marginal stability will be rapidly reached. The exact values of z_s and B_s, where marginal stability is reached depend on the ratio $z_s/\mathscr{L}$ ($\mathscr{L}$ being the length of the field lines), on the linear growth rate profile, and on the reflexion coefficient of the wave. An application of this saturation mechanism to the particular case of interaction of protons with left hand waves will be presented in the last part of this article.

It should be noted that this treatment is only valid for convective instabilities. The saturation condition $\gamma_\text{L} \sim \omega_\text{T}$ still holds for absolute instabilities.

3. Quasi Linear Like Diffusion by EM Waves Propagating in an Inhomogeneous Medium

Let us study now wave amplification and its effects on the particle distribution function, for amplitudes much smaller than B_1; thus no trapping effects can occur.

In that way, studying the frequency offset of VLF triggered emissions, Ashour-Abdalla (1972) has shown numerically that continuous low amplitude monochromatic waves can affect the electron distribution in such a way that a sustained growth rate is obtained for a frequency slightly different from the initial one. In a more recent work, Welti et al. (1973) studied analytically the corresponding problem for interactions between protons and ULF waves.

Let us discuss the physical mechanism.

Particles are moving back and forth along field lines. When their parallel velocity

$$V_\parallel = \{v^2 - 2\mu B_0\}^{1/2} \tag{7}$$

is equal to the resonant velocity:

$$V_\text{R} = \frac{\Omega_\text{c}(z) - \omega}{k_\parallel(z)} \quad \left(\text{where } \Omega_\text{c} = \frac{eB_0}{m}\right) \tag{8}$$

interaction can take place in two points on the field line, symmetric with respect to the equator. For a given model of the static magnetic field B_0 and a magnetic model for the cold plasma density $(N_0 \propto B_0)$, it can be easily seen that interacting particles should have a minimum parallel velocity, namely $V_\parallel > V_R(z=0)$.

Let us write

$$\psi = \int\limits_0^z (1 - V_R/V_\parallel)\, k_\parallel \, \mathrm{d}z$$

the relative phase between waves and particles. Resonance takes place when

$$\frac{\mathrm{d}\psi}{\mathrm{d}z} = 0 \quad \text{for} \quad z = z_R.$$

Near to the equator, ψ varies slowly, then $\cos\psi$ varies slowly too, whereas far from the equator, $\cos\psi$ varies very rapidly. The diffusion coefficient being proportional to $\{(\cos\psi \, \mathrm{d}z\}^2$, it is much larger near the equator than far away from it.

Let us call L_i the characteristic length over which ψ varies 1 rad. Expanding the phase around the equator, one finds that (for ULF left hand waves interacting with protons):

$$\frac{L_i}{z_0} = \left[\frac{6}{k_\parallel z_0 \left[\dfrac{3}{1-x} + \dfrac{V_{\perp 0}^2}{V_{\parallel 0}^2}\right]^2}\right]^{1/3} \tag{9}$$

where $X = \omega/\Omega_c (z=0)$ is the normalized frequency, and z_0 is defined by the parabolic expansion of the magnetic field near the equator.

$$B(z) = B(z = 0)\left[1 + (z/z_0)^2\right] \tag{10}$$

$V_{\perp 0}$ and $V_{\parallel 0}$ being perpendicular and parallel components of the velocity at the equator.

One can show easily that, due to the finite width of the resonance, a finite range or parallel velocities $\Delta V_\parallel$ (around the exact resonance velocity) are concerned by the diffusion process, with

$$\frac{\Delta V_\parallel}{V_{\parallel 0}} \simeq \frac{L_i}{2}\left[\frac{\partial^3 \psi}{\partial s^3}\right]_{\text{eq.}}. \tag{11}$$

This can be visualized in Figure 2; the distribution function tends towards diffusion curves over the width $\Delta V_\parallel$.
This can be visualized in Figure 2; the distribution function tends towards diffusion curves over the width $\Delta V_\parallel$.

At a fixed modulus of the velocity, a finite pitch angle $\Delta\alpha$ extension corresponds to $\Delta V_\parallel$ (with $\Delta\alpha = \Delta V_\parallel/V_\perp$)

For the sake of simplification, let us assume a pure pitch angle diffusion; over the width $\Delta\alpha$, the distribution tends towards isotropy. Out from the interval $\Delta\alpha$ there is no diffusion, thus, in order to ensure continuity, one has to increase the slope $\partial F/\partial\alpha$ at each side of this interval. Thus, $\partial F/\partial\alpha$ has to increase, for:

$$\alpha = \alpha_0 \pm \Delta\alpha/2.$$

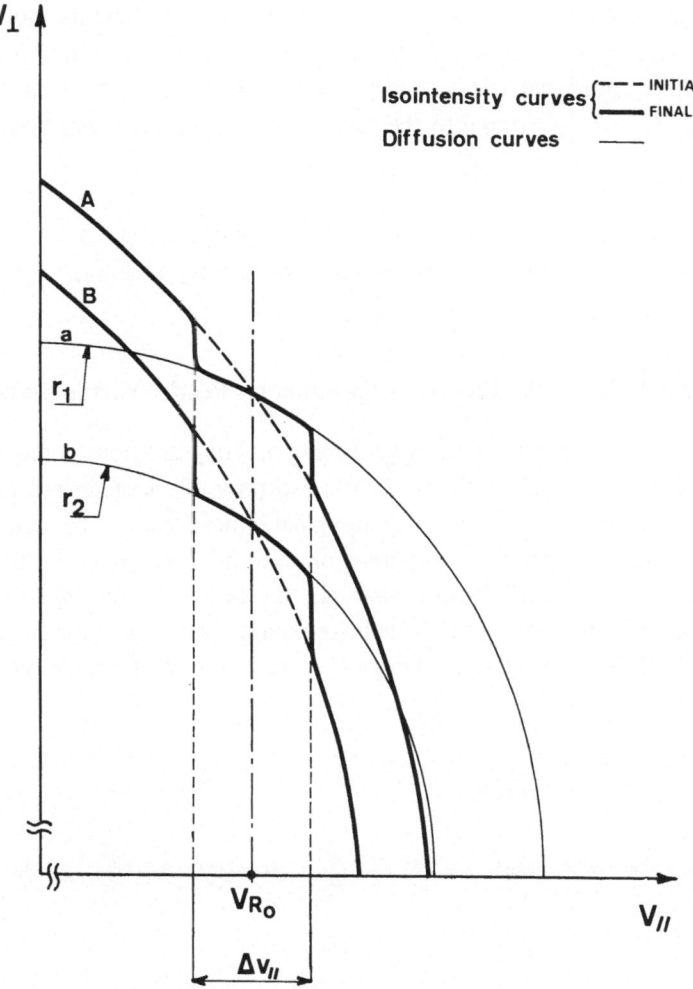

Fig. 2. Changes in the distribution function on an enlarged scale; $\partial F_0/\partial \alpha$ is greater in the vicinity of $V_{R_0} \pm \Delta V_\parallel/2$ (the subscript '0' stands for equatorial value of V_R) than it was before the diffusion took place. This figure is taken from the work of Roux *et al.* (1973).

Consequently, one can expect that sideband waves with frequencies

$$\omega = \omega_0 \pm \Delta\omega$$

are now preferentially amplified with

$$\frac{\Delta\omega}{\Omega_c} = c \, \frac{[4(1-x)]^{2/3}}{1 + 2/x} \, \frac{1}{(k_\parallel z_0)^{2/3}} \qquad (12)$$

C depending slightly on time; its exact evaluation can be obtained from Fokker-Planck equation (or quasi-linear theory). The validity conditions for such an equation are studied in the Appendix. Then the diffusion coefficient appears as the square of an Airy function (Ashour-Abdalla, 1972; Welti *et al.*, 1973).

It should be noted that $\Delta\omega$ essentially depends on the medium inhomogeneity along field lines, whereas it is quite independent on wave amplitude. In a trapping dominated regime, if the trapped particle instability could take place, the resulting frequency shift $\Delta\omega$ would be proportional to the square root of the wave amplitude:

$$\Delta\omega_{\text{Trap. part. inst.}} \propto \omega_{\text{T}}. \tag{13}$$

These different mechanisms, which can both provide sideband wave instabilities, will be discussed in detail in Section 4 where we will try to interpret Pc1 frequency spectrum.

4. Interpretation of the Frequency Spectrum of 'Pearls' Micropulsations

'Pearls' are regular micropulsations (Pc1 range). They are considered as left hand polarized waves, amplified by the thermal anisotropy of energetic protons (Gendrin *et al.*, 1971). They travel back and forth along field lines. This can be seen from Figure 3 where the repetition period is apparent on the chart record or on the integrated amplitude. It appears clearly from sonagrams that pearls are very monochromatic at the beginning of each event, but that their frequency spectrum seems to broaden after a while. Nevertheless, in some occasions (Figure 4) the frequency spectrum of pearl

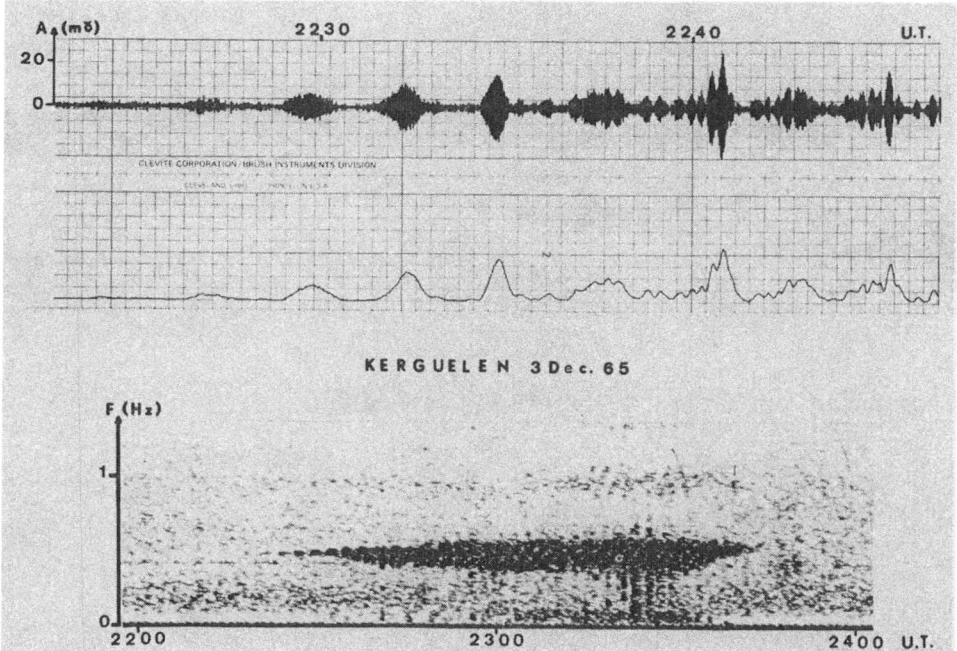

Fig. 3.   Usual representations of Pc1 events. At the top, a chart (amplitude vs. time) shows the repetition period of 'pearls'. This repetition period is still clearer on the integrated amplitude. Finally, in the lower part, a sonagram of the same event is presented, darkness being an indication of the wave amplitude. On this sonagram, dispersive properties are rather difficult to be seen.

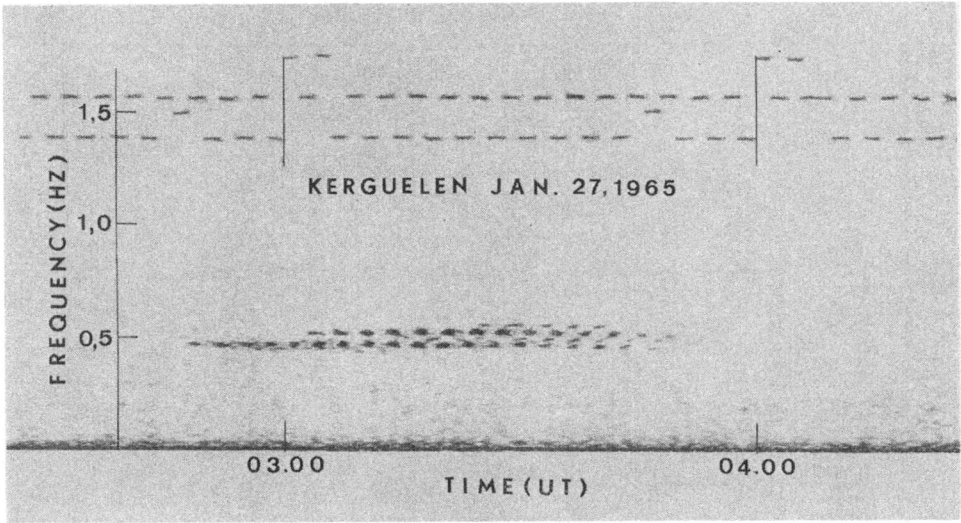

Fig. 4. A well defined structure can be seen in this particular sonagram, including well defined mother and daughter waves. Two granddaughter waves are also present.

events seems to be composed of two or more well defined components at slightly different frequencies.

As sonagraphs have a rather bad frequency and amplitude resolution, Wehrlin *et al.* (1973) made a more careful analysis, using a SAICOR apparatus which gives a three-dimensional plot (frequency-time-amplitude) with a good resolution. The result of such an analysis is presented on Figure 5. The frequency spectrum clearly appears on this figure. This well defined sideband structure can be found for all events which were analyzed by Wehrlin *et al.* This frequency structure is still clearer at the lower part of Figure 5 where the wave spectrum has been integrated over the whole pearl duration. This can be done for events having a constant mother wave frequency. The frequency offset between mother and daughter waves was always constant and of the order of 5×10^{-2}. Furthermore, on some occasions, granddaughter waves were observed to be triggered by daughter waves with a frequency offset (relative to the daughter wave) of the same order of magnitude. These frequency offsets exceed frequency resolution of the apparatus; furthermore, Wehrlin *et al.* have carefully checked their analysis in order to avoid parasitic effects which could have been due to badly analyzed gliding tones.

From these measurements, it is possible to follow the amplitude of each component of the spectrum. This is done in Figure 6. In the case presented here, there are two daughter waves, one at a higher frequency (higher than the mother) and one at a lower frequency; furthermore, a granddaughter wave is observed. From this figure (which is characteristic of what was always observed), one can conclude that:

(a) When daughter wave(s) appear, the mother wave amplitude decreases for a while (similarly, daughter wave amplitude decreases when granddaughter waves are observed).

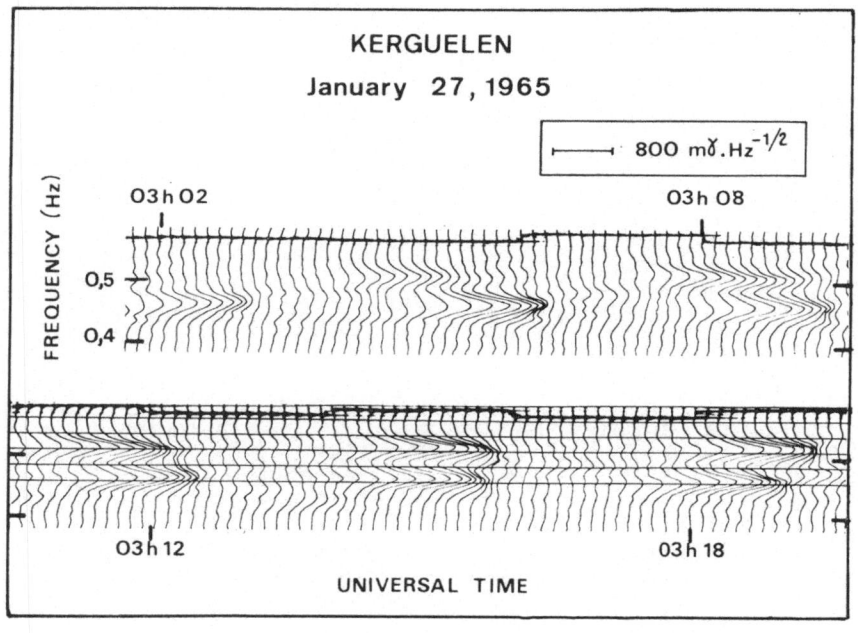

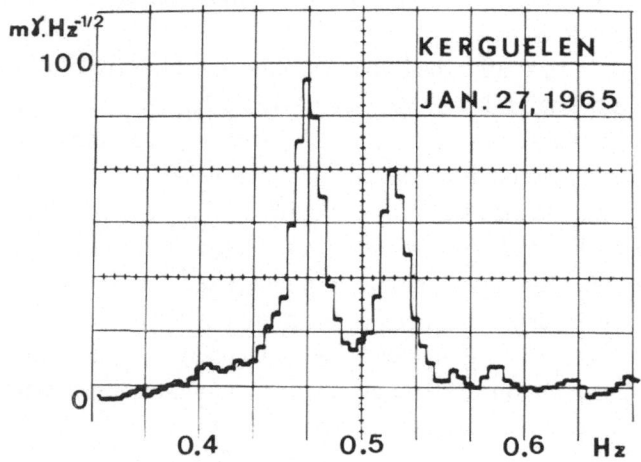

Fig. 5.   Result of pearl events analysis with a SAICOR analysis. At the top, a three dimensional plot; each line corresponds to a particular time, the displacement from a straight line indicates amplitude for each frequency. The dispersive properties of the pearls are easily observed if one compares the time evolution of each component of the spectrum. The frequency structure is still clearer on the integrated spectrum at the lower part of the figure.

(b) When daughter wave(s) appear, the mother wave amplitude is not saturated; after 'delivery', it continues to grow until a significantly higher amplitude level (usually three to five times higher). This real saturation level is approximately the same for all the components of the spectrum (if they reach saturation).

(c) All the frequency offsets are constant during each event duration, and have the same order of magnitude.

How can these features be explained?

An attempt was made by Budko *et al.* (1971). They interpret these sideband waves in terms of trapped particle instability. In their theory sidebands are due to a coupling between trapped particle motion and the initial wave. In the resonant frame, the

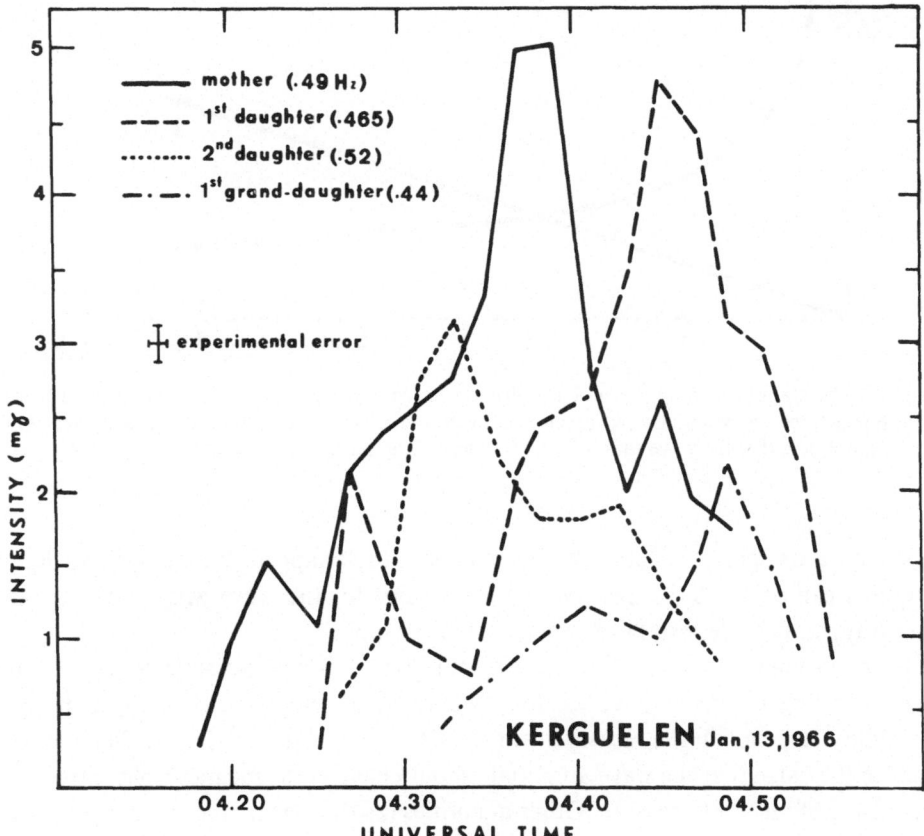

Fig. 6. Time evolution of the amplitudes of each component of the spectrum (in the present case it is composed of four sidebands). This figure was taken from Wehrlin *et al.* (1973).

frequency offset they found was proportional to the trapping frequency. Their calculations were made in an homogeneous medium.

In order that such a mechanism holds in an inhomogeneous medium, it is necessary that particles suffer several trapping oscillations before being detrapped.

We have written that the trapping force dominates the detrapping force over at least one trapping length (see Equation (5)). This can be visualized in Figure 7, where we plotted L_T the trapping length and l the distance where the trapping force is equal to the detrapping force as a function of wave amplitude. One sees that, if $B > B_s$, there

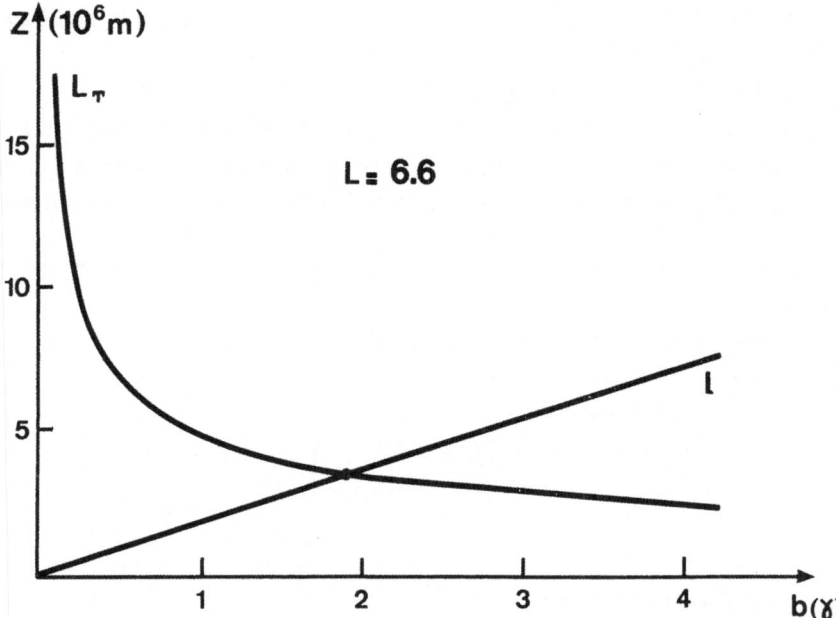

Fig. 7. The trapping length L_T and the distance l from the equator where trapping force and detrapping force are equal, are represented as functions of the wave amplitude. B_s is the minimum amplitude that the wave should have in order to trap particles over one trapping length.

is at least a trapping oscillation before particles are detrapped. Whereas, if $B < B_s$, no trapping can occur. B_s is then the minimum value for the wave amplitude in order that trapped particle instability theory could be applied.

From ground based amplitude measurements, it is not possible to deduce the corresponding *in situ* amplitudes. Meanwhile, let us suppose that trapped particle instability can explain the measurements made by Wehrlin *et al.* (1973). Then we can deduce the value that the wave amplitude should have in the magnetosphere from the frequency offsets they measure. Other quantities (as the gyrofrequency and the group velocity) can be deduced from the dispersive properties of pearls (Gendrin *et al.*, 1971). The wave amplitudes obtained in this way are always much smaller than B_s. Thus, trapped particle instability is unlikely to explain measurements made by Wehrlin *et al.*

We have already discussed in Section 2 another mechanism which explains, for lower amplitudes, sideband generation by inhomogeneity effects. Comparing the frequency offsets measured by Wehrlin *et al.* to the ones obtained theoretically by Welti *et al.* (1973) (see Equation (12)), Roux *et al.* (1973) have shown that the agreement is pretty good (see Figure 8). Computed Δf were always to be of the order of the observed Δf. It should be mentioned that this comparison has been made easier by the fact that this frequency offset depends a little on wave amplitude.

Meanwhile, in order to obtain large enough sideband growth rates, it is necessary that the initial wave amplitude overcomes a certain threshold. Welti *et al.* (1973) have shown that, depending on the L values, it varies between 0.3 γ and 1 γ .This was ob-

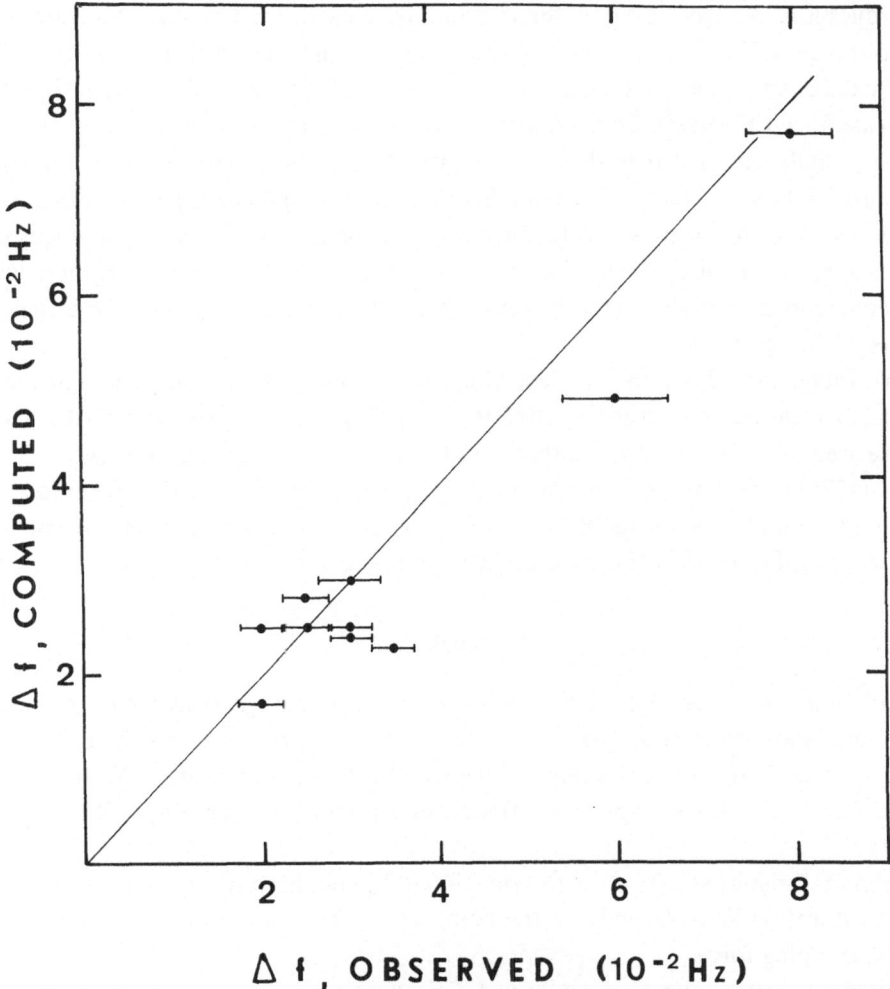

Fig. 8.   Theoretical value of the frequency offset (as given by Equation (12)) versus experimental ones. Taken from Roux *et al.* (1973).

tained by solving the diffusion equation. We will call this level B_i, the subscript 'i' standing for inhomogeneity.

One must keep in mind that this is not an absolute saturation level. After sideband generation the mother wave continues to grow until a significantly higher saturation level; this further growth of the mother wave being due to anisotropy restoration by azimuthal drift and to the diffusion induced by sideband waves. Then the mother wave (or the daughter wave) continues to grow. Thus they can now reach a trapping dominated regime. We have shown in Section 2 that, when the trapping force overcomes the detrapping force over at least one trapping length, the amplification is largely reduced over this length (see Equation (6)).

Does the residual amplification out of this interval (where trapping dominates) remain strong enough to overcome wave energy losses (in the ionosphere)? To answer this question, we need to know the reflexion coefficient, the linear growth rate, and the distance z_1. The first being almost unknown, we will focus on z_1. Roux *et al.* (1973) evaluated z_1 and found that it is an appreciable part of the line of force, say $z_1/R_E \sim 0.5$ (at least for ULF waves). Furthermore, we know from measurements that the mother wave initially grows rather slowly. Thus we should not be so far from marginal stability. A large decrease of the amplification over a significant part of the line of force will then stop wave growing. Thus $B \sim B$ can be considered for Pc1 monochromatic waves as a realistic evaluation of the saturation level B_s. This saturation level depending slightly on frequency, it is clear that it should be the same for all the components of the spectrum.

We found that $B_s \sim 1$ to $3\,\gamma$ (depending on L values). This is in good agreement with Pc1 measurements made by McPherron (1973) and coworkers onboard ATS 1.

We are now able to compute a theoretical ratio B_s/B_i for each analyzed event. Roux *et al.* (1973) have shown that it is in good agreement with the ratio that Wehrlin *et al.* deduced from measurements. Both are of the order of three to five. This agreement is indirect proof of the fact that saturation by trapping does occur in most of the cases.

5. Conclusion

In summary, we have first studied diffusion in an inhomogeneous medium by low amplitude monochromatic waves and it seems that the process is repetitive. The frequency offset arising from this theory agrees well with measurements by Wehrlin *et al.* (1973), and then, it can explain the frequency structure they observed (Roux *et al.*, 1973).

Second, if the wave continues to grow, it will be stabilized by trapping effects. The exact saturation level depends on the point where the inhomogeneity force is equal to the trapping force.

Third, we have shown that, as far as ULF waves are concerned, the distance from this point to the equator should be of the order of one trapping length. From this evaluation, we deduced *in situ* Pc1 wave amplitude at saturation. Whereas for VLF waves, it is easy to show that saturation occurs when particles have suffered many trapping oscillations. Thus the trapped particle instability, which was unlikely to occur for ULF waves, can reasonably be concerned in VLF waves.

Appendix

Validity Conditions of Quasi Linear-Like Diffusion in an Inhomogeneous Medium

We shall justify the statement that the diffusion in an inhomogeneous medium was a quasi-linear-like diffusion.

The Fokker-Planck equation for pure pitch angle diffusion can be written:

$$\frac{\partial F_0}{\partial t} = \frac{1}{\sin\alpha}\frac{\partial}{\partial\alpha}\left\{ -\sin\alpha\,\frac{\langle\Delta\alpha\rangle}{\Delta t}F_0 + \frac{\sin\alpha}{2}\frac{\partial}{\partial\alpha}\left[\frac{\langle(\Delta\alpha)^2\rangle}{\Delta t}F_0\right]\right\} \qquad (14)$$

$\langle \rangle$ denotes an average over phases;

Δt being averaged time between two collisions;

F_0 is the space averaged distribution function.

In an inhomogeneous medium, we have to work with μ as a new variable, or with equatorial pitch angle α_E (which is equivalent, because $\mu = v^2 \sin^2 \alpha_E / 2B(0)$).

It is easy to show that $\langle \Delta \mu^{(1)} \rangle$ or $\langle \Delta \alpha_E^{(1)} \rangle = 0$.

Laval and Pellat (1973) have shown that, as long as one works with variables ϕ and μ: *cyclic* variable and *constant* of the motion which are canonically conjugate, one obtains:

$$\frac{\langle \Delta \alpha^{(2)} \rangle}{\Delta t} = \frac{1}{2} \frac{\partial}{\partial \alpha_E} \left\{ \frac{\langle (\Delta \alpha^{(1)})^2 \rangle}{\Delta t} \right\}. \tag{15}$$

(1) and (2) stand for first and second order quantities.

Then the Fokker-Planck equation can be written:

$$\frac{\partial F_0}{\partial t} = \frac{1}{2} \frac{1}{\sin \alpha} \frac{\partial}{\partial \alpha_E} \left\{ \sin \alpha_E \frac{\langle (\Delta \alpha^{(1)})^2 \rangle}{\Delta t} \frac{\partial F_0}{\partial \alpha} \right\} \tag{16}$$

which is the usual quasi-linear equation for pure pitch angle diffusion. Then, at least in our case, the Fokker-Planck equation is equivalent to the quasi-linear diffusion equation.

Let us now study the validity conditions of this diffusion equation. In the usual quasi-linear theory (that is, for interaction of energetic particles with broad band noise), the diffusion process is Markovian, providing that the following inequalities are fulfilled (Roux and Solomon, 1970):

$$\tau_E, \tau_{F_0} \gg \frac{1}{V_R \Delta k_{\parallel}} \tag{17}$$

where τ_E, τ_{F_0} are characteristic times for the wave amplitude and averaged distribution function, and Δk the characteristic width of the spectrum.

Then, usual QLA holds when Δk is large enough.

For the monochromatic waves, we are considering $\Delta k \to 0$. But, as shown previously, the inhomogeneity of the medium induces a broadening of the resonance. Then, Welti et al. (1973) have shown that the validity conditions for QLA or the Fokker-Planck equation for interactions with monochromatic waves in an inhomogeneous medium are for independent collisions as follows:

$$\tau_E, \tau_{F_0} \gg \frac{L_i}{V_R}. \tag{18}$$

In that sense, one can speak of quasi-linear like diffusion by monochromatic waves, providing that the medium is inhomogeneous enough and that collisions are randomly distributed in phase.

Acknowledgments

Most of this work was done in collaboration with Drs R. Gendrin, R. Pellat, N. Wehrlin and R. Welti; it is a pleasure to acknowledge them here.

References

Ashour-Abdalla, M.: 1972, *Planetary Space Sci.* **20**, 639.

Asseo, E., Laval, G., Pellat, R., Welti, R., and Roux, A.: 1972, *J. Plasma. Phys.* **8**, 341.

Budko, N. I., Karpman, V. I., and Pokhotelov, O. A.: 1971, *J. Exp. Theoret. Phys. Letters* **14**, 320.

Dysthe, K. B.: 1971, *J. Geophys. Res.* **76**, 6915.

Etcheto, J., Gendrin, R., Solomon, J., and Roux, A.: 1973, *J. Geophys. Res.* **78**, 8150.

Gary, S. P., Montgomery, D., and Swift, D. W.: 1968, *J. Geophys. Res.* **73**, 7524.

Gendrin, R.: 1972, in B. M. McCormac (ed.) *Earth's Magnetospheric Processes*, D. Reidel Publishing Co., Dordrecht, Holland, pp. 311–328.

Kennel, C. F. and Petschek, H. E.: 1966, *J. Geophys. Res.* **71**, 1.

Laval, G. and Pellat, R.: 1970, *J. Geophys. Res.* **75**, 3255.

Laval, G. and Pellat, R.: 1972, *Cours des Houches*, to be published.

Liemohn, H. B.: 1967, *J. Geophys. Res.* **72**, 39.

McPherron, R. L.: 1973 private communication.

Nunn, D.: 1971, *J. Plasma Phys* **8**, 291.

Roux, A. and Solomon, J.: *Ann. Géophys.* **26**, 279.

Roux, A., Gendrin, R., Wehrlin, N., Pellat, R., and Welti, R.: 1973, *J. Geophys. Res.* **78**, 3176.

Wehrlin, N., Gendrin, R., Roux, A., and Welti, R.: 1973, *J. Geophys. Res.* **78**, 763.

Welti, R., Pellat, R., and A. Roux.: 1973, *Ann. Geophys.* **29**, 263.

A SELF-CONSISTENT THEORY OF
TRIGGERED VLF EMISSIONS

DAVID NUNN

Department of Mathematics, Kings College, London W.C. 2, England

1. Introduction

This work attempts a theoretical self-consistent description of the phenomenon of artificially triggered VLF emissions. The observations of this process are well known (Helliwell, 1965). High power VLF transmitters (e.g. NAA, NPG) transmit monochromatic morse pulses at ~ 14 kHz. These signals parallel propagate along a magnetospheric field line in a ducted mode. As a result of interaction with cyclotron resonant electrons such pulses are often observed to 'trigger' large amplitude narrow band emissions with sweeping frequencies.

2. Basic Assumptions

A number of important points need to be emphasized before going into theoretical detail. First, as far as resonant particle behavior is concerned inhomogeneity is most important, and the parabolic variation in background magnetic field has to be included from the start. Second, resonant particle motion is totally nonlinear and the application of quasi-linear theory to any aspect of this problem is quite wrong (Roux, 1974). Third, it is necessary to assume that the wave field is parallel propagating throughout the process, even though this ignores such difficult questions as Landau damping, duct leakage, etc. Fourth, it should be pointed out that except for very simple problems, nonlinear resonant particle behavior may not be tackled by analytic means, and accordingly the use of a high power digital computer is necessary if progress is to be made.

3. Resonant Particle Equations of Motion

In an inhomogeneous medium the equations of motion of a cyclotron resonant electron may be written in the form

$$\ddot{\psi} + \omega_{tr}^2 \left(\cos\psi - S(t) \right) = 0 \tag{1}$$

where all quantities are dedimensionalized with respect to $\bar{\omega} = \Omega_{EQ}/2$ and $\bar{K} = \Pi_{EQ}/c$. Here ω_{tr} is the trapping frequency

$$\omega_{tr} = K_0 \sqrt{R|V_\perp|/\omega_0}$$

and R is the dimensionless wave amplitude

$$R = e|E_\perp| \bar{K}/m\bar{\omega}^2 .$$

B. M. McCormac (ed.), Magnetospheric Physics, 313–321. All Rights Reserved.
Copyright © 1974 by D. Reidel Publishing Company, Dordrecht-Holland.

The quantity S is the total inhomogeneity divided by wave amplitude.

$$S = \frac{-\chi Z}{R} \left\{ \frac{2(2\beta - \omega_0)\,\omega_0}{K_0^3 |V_\perp|} + \frac{\omega_0 |V_\perp|}{2\beta K_0} \right\} - \frac{\ddot{\phi}}{R} \frac{\omega_0}{K_0^2 |V_\perp|} \tag{2}$$

where $\ddot{\phi}$ is the acceleration of wave phase

$$\ddot{\phi} \approx 2\,(\partial/\partial t + V_z\,\partial/\partial Z)\omega.$$

The angle ψ is that between a particle's perpendicular velocity vector and the electric field of the wave. Also we have

$$B_0 = \beta B_{EQ} = (1 + \tfrac{1}{2}\chi Z^2)\,B_{EQ}. \tag{3}$$

Now, if $|S| < 1$ nonlinear trapping may take place, and we may define a phase locking angle P_0

$$P_0 = \cos^{-1}(S); \quad 0 > P_0 > -\pi. \tag{4}$$

Assuming that R and P_0 vary slowly as compared to a trapping period, then we may write for a strongly trapped particle

$$\psi \cong P_0 + (\Delta\psi)\cos\left(\omega_{tr}\,|\sin P_0|\,t + \xi\right)$$

$$V^* = V_z - V_{RES} \cong -\frac{\Delta\psi\omega_{tr}}{K_0}\sin\left(\omega_{tr}\,|\sin P_0|\,t + \xi\right).$$

For trapped particles the phase ψ oscillates about the phase locking angle P_0. A detailed plot of resonant particle motion in the ψ, V^* plane is given in Nunn (1974). The most important aspect of trapping in inhomogeneous media is that stably trapped particles undergo a steady change in both energy and magnetic moment

$$\dot{W} = -R|V_\perp|\cos P_0$$

$$\dot{\mu} = \frac{d}{dt}(|V_\perp|^2/2\beta) \sim \frac{2}{\omega_0}\,\dot{W}.$$

Untrapped resonant particles pass straight through resonance and undergo relatively small energy changes.

4. The Resonant Particle Current

A vital and unavoidable intermediary in this problem is the small resonant particle current arising from the nonlinear interaction of electrons with the wave field. It is obtained in the normal way by integrating the resonant particle distribution function over three dimensional velocity space in the neighborhood of the resonance velocity:

$$J_R + iJ_I = \frac{-2\pi^2}{N_0} \int\limits_{|V_\perp|=0}^{\infty} |V_\perp|^2 \left\{ \frac{\partial F_0}{\partial W} + \frac{2}{\omega_0} \frac{\partial F_0}{\partial \mu} \right\}_{V_z = V_{RES}} \times$$

$$\times \int_{\psi=0}^{2\pi} \int_{V^*=-6R^{1/2}}^{6R^{1/2}} e^{i\psi} \Delta W\left(\psi, V^*, |V_\perp|, z, t\right) dV^* \, d\psi \, d|V_\perp| \qquad (5)$$

where ΔW is the energy charge of the particle in question due to interaction with the wave field. The current J_r is the component parallel to the wave electric field and J_I is the out of phase component parallel to the wave magnetic field. It transpires that for the nonlinear inhomogeneous problem the above integral is dominated by stably trapped particles for which ΔW is large. It is this fact coupled with the relatively simple and predictable behavior of trapped particles that renders the non-linear self-consistent problem tractable.

5. The Non Self-Consistent Approach

An obvious first step in tackling the problem is to specify various field configurations in advance and then rigorously compute the resonant particle current. This was done in Nunn (1973a) for a wide variety of cases and it was shown that invariably the non-linear current was dominated by stably trapped particles.

6. The Self-Consistent Model

6.1. THE FIELD EQUATIONS

We first require an equation giving the time development of the wave field resulting from the nonlinear resonant particle current. This is derived from Maxwell's equations and the linearized equation of motion of the cold plasma electrons, and is in c.g.s. units

$$\left[\left(\frac{\partial}{\partial t} - i\Omega(z)\right)\frac{\partial^2}{\partial z^2} - \frac{\Pi(z)^2}{c^2}\frac{\partial}{\partial t}\right]E_\perp = \frac{4\pi}{c^2}\left(\frac{\partial}{\partial t} - i\Omega\right)\frac{\partial}{\partial t}J_{RES} \qquad (6)$$

where

$$E_\perp = E_x + iE_y; \qquad J_{res} = J_{RES_x} + iJ_{RES_y}.$$

We now define a dimensionless amplitude R and an additional phase ϕ.

$$\left\{\frac{eE_\perp \bar{k}}{m\bar{\omega}^2}\right\} = R(z, t)\, e^{i(\omega_0 t - \int k_0\, dz + \phi(z,t))} \qquad (7)$$

$$\left\{\frac{2\pi e J_{RES}}{m\bar{\omega}\bar{k}c^2}\right\} = (J_r + iJ_1)\, e^{i(\omega_0 t - \int k_0\, dz + \phi)}$$

where ω_0, K_0 are in dimensionless units and satisfy the linear dispersion relation

$$K_0^2 = (2\beta - 1)\,\omega_0/(2\beta - \omega_0). \qquad (8)$$

We now apply the *narrow band* approximation and assume that R and ϕ are slowly varying functions of position and time:

$$\frac{\partial R}{\partial t} \ll \omega_0 R; \qquad \frac{\partial \phi}{\partial t} \ll \omega_0; \qquad \frac{\partial R}{\partial z} \ll K_0 R \qquad (9)$$

After some manipulation Equation (6) becomes

$$\left(\frac{\partial}{\partial t} + V_g \frac{\partial}{\partial z}\right) R = - AJ_R - \eta(R) R \qquad (10)$$

$$\left(\frac{\partial}{\partial t} + V_g \frac{\partial}{\partial z}\right) \phi = - AJ_I/R \qquad (11)$$

where

$$V_g = 2K_0(2\beta - \omega_0)/(K_0^2 + 2\beta - 1); \qquad A = V_g \omega_0/K_0$$

and an arbitrary factor η has been inserted to account for Landau damping and duct leakage. Differentiating Equation (11) we obtain

$$\frac{\partial}{\partial t}\left(-\frac{\partial \phi}{\partial z}\right) = \frac{\partial k}{\partial t} = V_g \frac{\partial^2 \phi}{\partial z^2} + A \frac{\partial}{\partial z}\left(\frac{J_I}{R}\right) \approx \frac{\partial \omega}{\partial t}/V_g. \qquad (12)$$

It is seen that the out-of-phase component of resonant particle current is able to cause the wave frequency to change.

6.2. MODEL EQUATIONS FOR THE NONLINEAR CURRENT

Ideally one should integrate Equations (10) and (11) using a self-consistent current J_{RES} that is rigorously computed. This is an impossible task computationally and we employ a model that assumes that J_{RES} is due to three bunches of stably trapped particles having different perpendicular velocities $|V_\perp|^j$. We consider events at half the equatorial gyrofrequency, whence

$$\omega_0 = A = V_g \cong 1.$$

The net inhomogeneity at a point is given by

$$S^j = -\frac{\chi z}{R}\left(\frac{2}{|V_\perp|^j} + \frac{|V_\perp|^j}{2}\right) - \ddot{\phi}/|V_\perp|^j R \qquad (13)$$

where

$$\ddot{\phi} = (\partial/\partial t - \partial/\partial z)^2 \phi$$

and

$$\chi = 2 \times 10^{-9} \quad \text{at} \quad L = 3.5.$$

Now if $|S^j| < 1$, then trapping is possible, and we define

$$P_0 = \cos^{-1} S^j; \quad 0 > P_0 > -\pi.$$

The centers of gravity of the trapped particle bunches are described by the variables $\bar{\psi}_j$ and $\bar{U}_j$

$$\bar{U}_j = \bar{V}_z^j - 1 + 2\beta = \bar{V}_z^j - V_{RES} + \dot{\phi}.$$

These quantities are advanced according to the particle equations of motion, with an extra term which moves the bunch continuously towards the center of the trap.

$$\left(\frac{\partial}{\partial t} - \frac{\partial}{\partial z}\right)\bar{U}_j = - R |V_\perp|_j \cos \bar{\psi}_i - \frac{\chi z |V_\perp|^{j^2}}{2} - 2\chi z +$$

$$- \frac{\alpha_j}{2\pi} \sqrt{R|V_\perp|_j |\sin P_0^j|} \, (\bar{U}_j - \dot{\phi}) \tag{14}$$

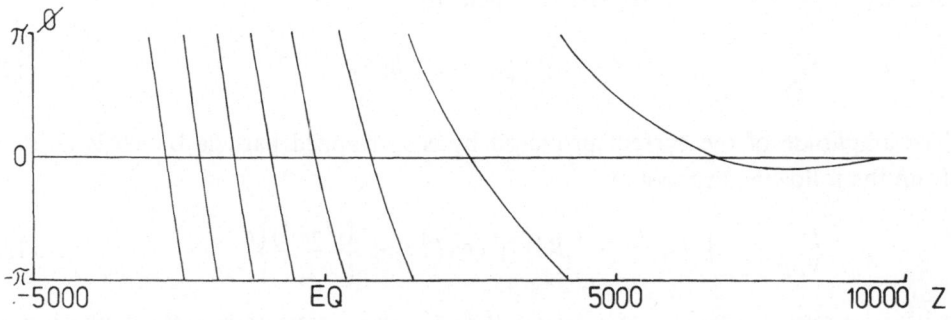

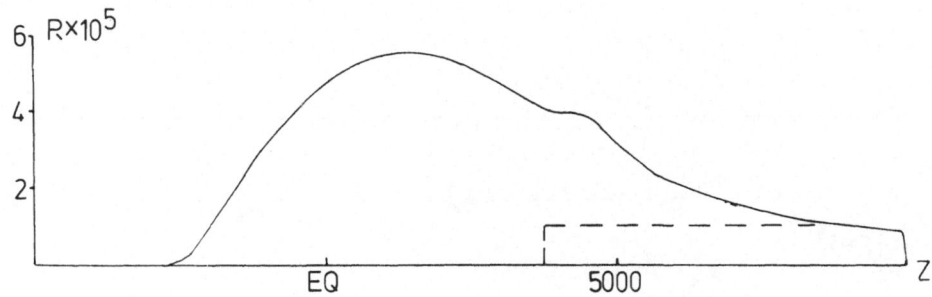

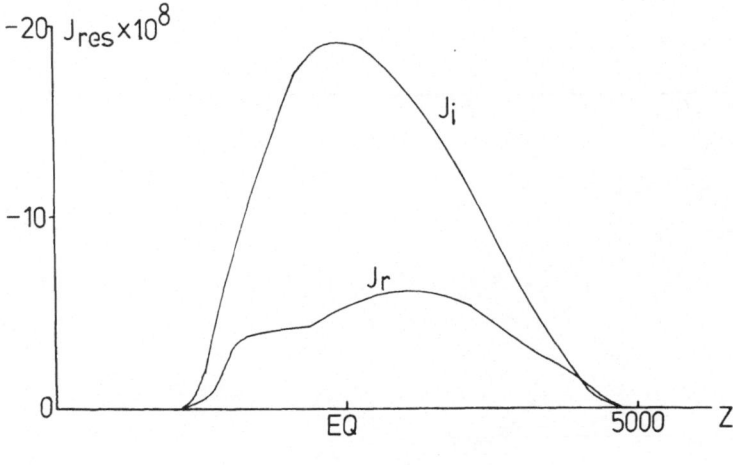

RUN A t = 15000

Fig. 1. Plots of amplitude R, additional phase ϕ, in-phase current J_r, and out-of-phase current J_i during the triggering of a riser. The diagram shows a quasi-static generation region of a riser of the kind described by Helliwell (1967).

$$\left(\frac{\partial}{\partial t} - \frac{\partial}{\partial z}\right)\overline{\psi}_j = \overline{U}_j - \phi - \frac{\alpha_j}{2\pi}\sqrt{R\,|V_\perp|_j}\,|\sin P_0^j|\,(\overline{\psi}_j - P_0^j) \tag{15}$$

If $|S^j| > |1$, $\alpha_j = 0$. Otherwise $\alpha_j = 2$.

We require a quantity proportional to the size of the resonant particle trap, τ_0^j.

$$\tau_0^j = R^{1/2}P_0^2 \quad \text{or} \quad R^{1/2}(\pi + P_0)^2 \quad \text{if} \quad S_j < 0 \tag{16}$$

and also τ^j, proportional to particle bunch size

$$\left(\frac{\partial}{\partial t} - \frac{\partial}{\partial z}\right)\tau_j = -(\tau_j - \tau_0^j)(R^{1/2} + 0.003). \tag{17}$$

The amplitude of the current produced by each trapped particle bunch is derived from the following expression

$$\left(\frac{\partial}{\partial t} - \frac{\partial}{\partial z}\right)|J_0|^j = -C_j R\,|V_\perp|^j \cos P_0^j \tau_0^j + \left\{\frac{|J_0|^j\,t_j}{\tau_j}\right\} \tag{18}$$

where the term in brackets only applies if $t_j < 0$, and corresponds to the dephasing of

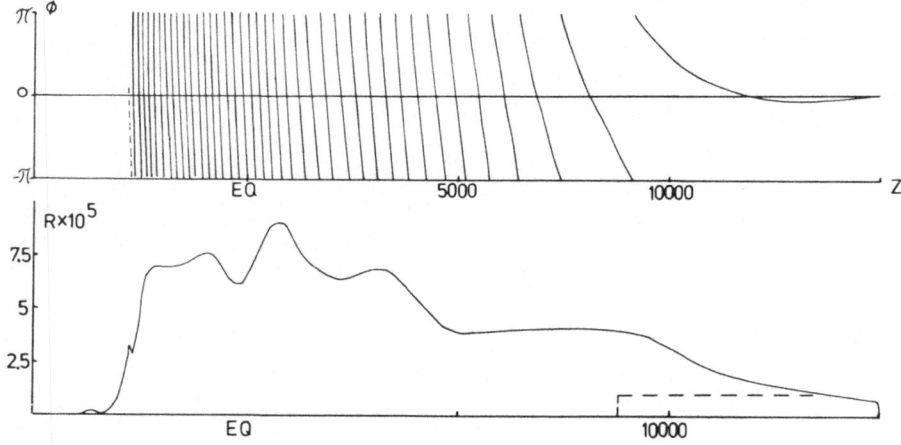

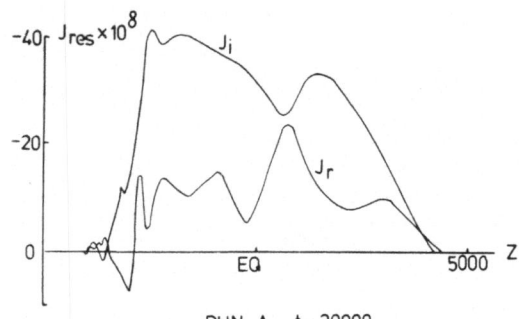

RUN A   t = 20000

Fig. 2.   The same emission at a later time showing the formation of a resonant upper sideband at + 50 Hz. This destroys the current model and terminates the computer run.

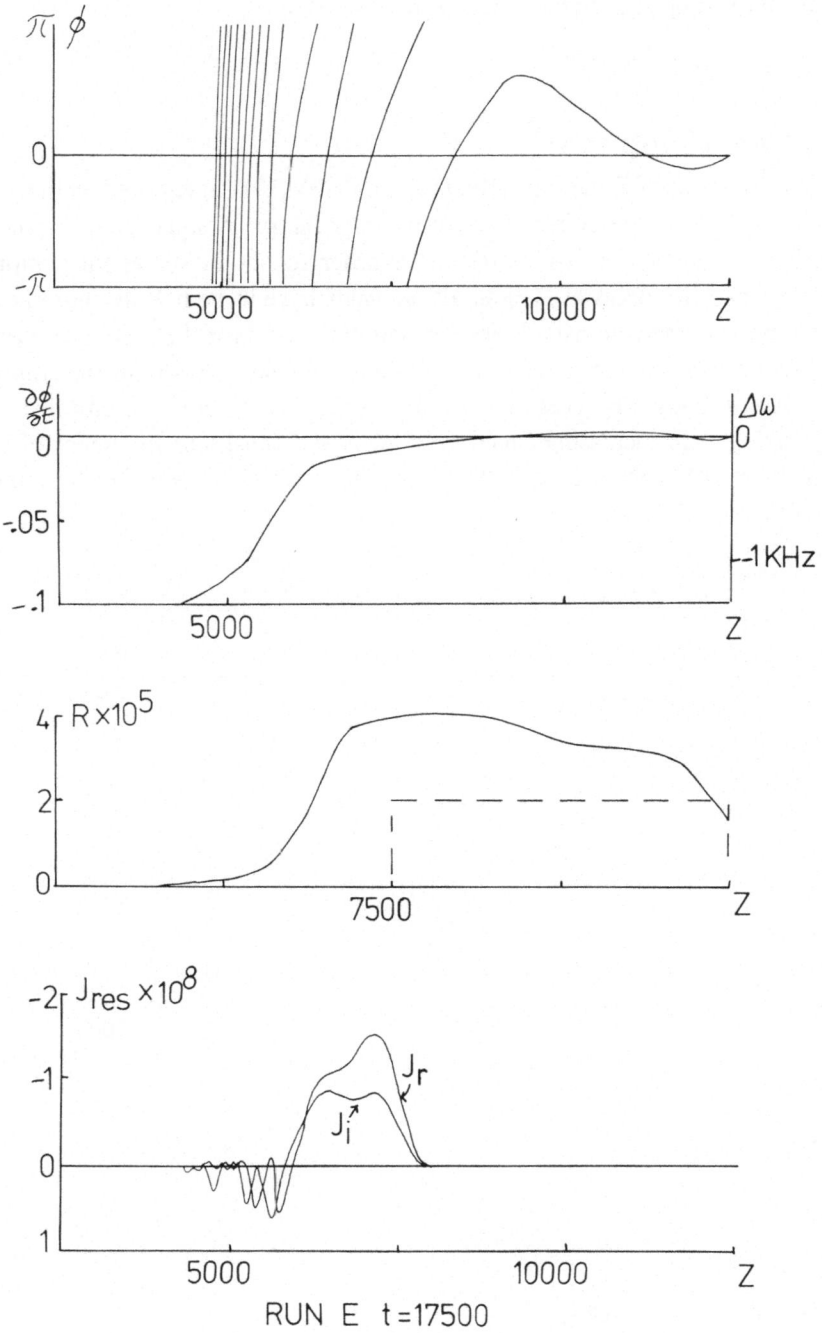

Fig. 3. A short steep faller triggered by the model.
The location is in accordance with the Helliwell theory.

a bunch of once trapped particles. Finally then we have

$$J_R + iJ_I = \sum_j |J_0|^j \, e^{i\bar{\psi}j}. \tag{19}$$

6.3. THE NUMERICAL INTEGRATION

The self-consistent set of Equations (10), (11), and (13)–(19) is integrated on the CDC 6600 using a Z grid of 800 points centered on the equator. A square morse pulse is introduced and triggered emissions result. In order to obtain successful computer runs two measures are necessary. First, all the fields have to be heavily smoothed in order to suppress numerical instabilities. Second, the model for the nonlinear current is only valid provided the spectral width of the wave field is less than the trapping frequency. Accordingly it is necessary to suppress the formation of the unstable sideband by making γ an increasing function of R. A self-consistent treatment of triggered emissions with sidebands is well beyond the reach of present day computers.

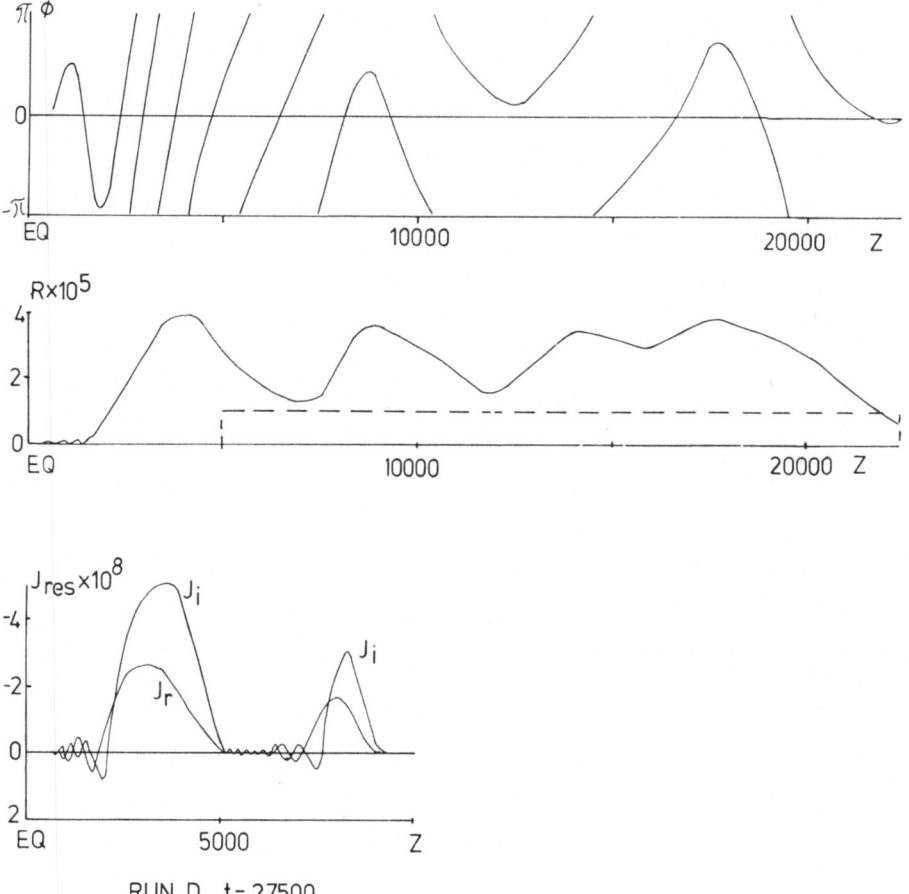

RUN D   t = 27500

Fig. 4.   Excitation of a long pulse, showing the formation of nodes corresponding to a double sideband formation at ± 15 Hz. Note the weak faller at the end of the pulse.

7. The Numerical Results

The first run shows the triggering of a riser by a morse pulse ~ 60 ms in length. Figure 1 shows the formation of a quasistatic generation region of a riser of the kind described by Helliwell (1967) in his well known theory. The frequency steadily rises, partly due to the advective term in Equation (12), and partly due to the effect of the out of phase current J_i. The in-phase current J_r maintains the wave profile in a constant position. The nonlinear current due to trapped particles is entirely reasonable, as may be checked by a rigorous computation in a model field. Note that the Helliwell criterion of zero inhomogeneity ($S = 0$) is satisfied at the low amplitude end of the generation region and is necessary if an emission is to be long enduring. Figure 2 shows the end of the emission when a large amplitude sideband wave appears at the resonant separation of $+50$ Hz. At this time the emission is about three times the length of the original pulse and rises to 1 kHz above the starting frequency. Figure 3 shows a triggered faller. The establishment of the generation region of a faller presents some difficulty because J_i tends to make a contribution to $\partial \omega / \partial t$ that is of the wrong sign. The faller in this figure is very short and steep and rather difficult to understand since appropriate variables are varying rather rapidly. However it is triggered in approximately the location required by the Helliwell theory. Figure 4 shows the case of a long pulse. It will be noticed that a succession of nodes is formed with trapping taking place at each nodal peak. This is in excellent agreement with Bell and Helliwell's (1971) NAA key down results. Also the weak faller at the end of the pulse has been observed in connection with Omega transmissions (Kimura, 1968).

8. Conclusion

There is little doubt that the full nonlinear theory is able to explain triggered VLF emissions, and is almost certainly the way to tackle the problem of banded chorus (Nunn, 1973).

References

Bell, T. F. and Helliwell, R. A.: 1971, *J. Geophys. Res.* **76**, 8414.
Helliwell, R. A.: 1965, *Whistlers and Related Ionospheric Phenomena*, Stanford University Press, Stanford, Calif.
Helliwell, R. A.: 1967, *J. Geophys. Res.* **72**, 4773.
Kimura, I: 1968, *J. Geophys. Res.* **73**, 445.
Nunn, D.: 1973, *J. Plasma Phys.*, in press.
Nunn, D.: 1974, *Planet. Space Sci.* **22**, 349.
Roux, A.: 1974, this volume, p. 297.

PART V

SUBSTORM PHENOMENA

GROUND BASED MAGNETIC SIGNATURES OF THE
PHASES OF MAGNETOSPHERIC SUBSTORMS

GORDON ROSTOKER

Institute of Earth and Planetary Physics, University of Alberta, Edmonton, Canada

1. Introduction

The term *magnetospheric substorm* was coined sometime in 1967 to describe the explosive dissipation of energy at high latitudes with auroral breakups. This made it possible to understand that the many modes of energy dissipation (e.g. VLF, ELF, ULF, current flow, etc.) really represent one phenomenon and allowed the study of substorms to progress through the understanding of the relationships among these various modes of energy dissipation.

The framework in which substorms are presently described was first proposed by Akasofu (1964) for auroral substorms. He broke down the phenomenon into an *expansion phase* followed by a *recovery phase*. The expansion phase involved brightening of auroral arcs in the midnight sector, the rapid poleward expansion of the region of brightened arcs, and the development of a westward traveling surge. The recovery phase was construed to start when the breakup arcs had reached their maximum poleward position and initially involved weak loop and surge activity at the poleward border of the auroral excited region followed by the equatorward drift of auroral forms and the eventual re-establishment of the pre-substorm quiet arc configuration. Accompanying these auroral events are intense magnetic perturbations associated with variations in the auroral electrojet system, and the growth and decay of strong field aligned currents. Each of the substorm phases has well defined magnetic signatures which have been studied in detail in the literature (Kisabeth and Rostoker, 1971, 1973; Rostoker and Kisabeth, 1973) and we will deal with some of these interpretations in this paper.

More recently, several workers have claimed that the expansion phase of a substorm is preceded by a period during which energy is stored and redistributed in the magnetosphere; this period has been termed the *growth* or *development* phase, and various signatures for this phase have been proposed in the literature (Belyakova *et al.*, 1968; McPherron, 1970, 1972; Mozer, 1971; Iijima and Nagata, 1972). However, there has been some difficulty in that each of the proposed signatures of the growth phase also appears to be the signature of either the expansion or recovery phase of the substorm. For example, while Mozer (1971) claims that equatorward drifting arcs are associated with a growth phase, Akasofu (1964) has clearly associated them with recovery phase activity. The source of the difficulty would appear to be that growth phase periods are defined by virtue of the existence of subsequent expansive phase activity. Since it is often difficult (if not impossible) to identify localized substorm activity solely through the use of expansion phase signatures, it is not immediately obvious if a given magnetic

B. M. McCormac (ed.), Magnetospheric Physics, 325–333. All Rights Reserved.

signature is that of growth or that of expansion seen at a distance. Wiens and Rostoker (1973) have attempted to resolve this controversy by noting the fact that expansion only takes place in a longitudinally localized sector of the nightside magnetosphere (see also Rostoker and Camidge, 1971). Accordingly they claimed that during the expansion phase of one substorm, energy was fed into an adjacent sector causing it to experience growth preparatory to becoming an expansion phase sector. This would imply that one would expect to see expansion phase and growth phase perturbations occur simultaneously in different regions of the nightside magnetosphere. This is found to be the case in the vast majority of cases studied to date, although there are exceptions to the rule. In this paper we will present one example of typical substorm morphology and one exception to the rule.

2. Development of a Typical Substorm Sequence

The substorm sequence we shall discuss occurred on Day 328, 1971. The auroral zone and low latitude H component magnetograms are shown in Figure 1. The top five traces in this Figure are from auroral zone observations arrayed from east to west down the Figure, while the bottom six traces are from low latitude observatories arrayed in a similar fashion. The low latitude signatures we shall look for are the depression of the H component (indicative of growth as suggested by McPherron (1972)) and a positive increase in the H component (indicative of expansion). The auroral zone signatures of interest are negative excursions in the H component (traditionally considered to be associated with the substorm expansion phase) and positive excursions in the H component (thought to be associated with eastward electrojet enhancements).

From Figure 1 we see that shortly after 0430 UT (line A) there is a clear onset of an expansion phase as observed in the auroral zone magnetograms at WHAL and CHUR. Further to the west, SMIT experiences a 'sloping negative H-bay' (often sited as a signature of growth, although it is always found at auroral zone observatories to the west of an expansion phase sector). Further into the evening sector COLL sees a +H perturbation indicative of the growth of an eastward electrojet. At the same time as the onset of the expansion phase, the low latitude stations of JUAN, FRED and DALL experience +H increases indicative of expansion phase activity. The +H perturbation is weaker at TUCS, and at VICT and HONL the response is a clear −H depression indicative of growth. *This appears to indicate that expansion and growth occur synchronously in adjacent sectors in the nightside magnetosphere.* The data from COLL, VICT and HONL indicate the growth of the asymmetric ring current while all observatories east of ∼295° E geomagnetic have signatures consistent with the development of a substorm-associated Birkeland current loop (Bonnevier *et al.*, 1970). This is the pattern we have noted for almost all substorm disturbances we have studied thus far.

We may look in more detail at the development of the expansion phase current system by noting that, after the onset of the above mentioned event, the auroral zone magnetograms exhibit a rather jagged structure particularly at WHAL. We shall look

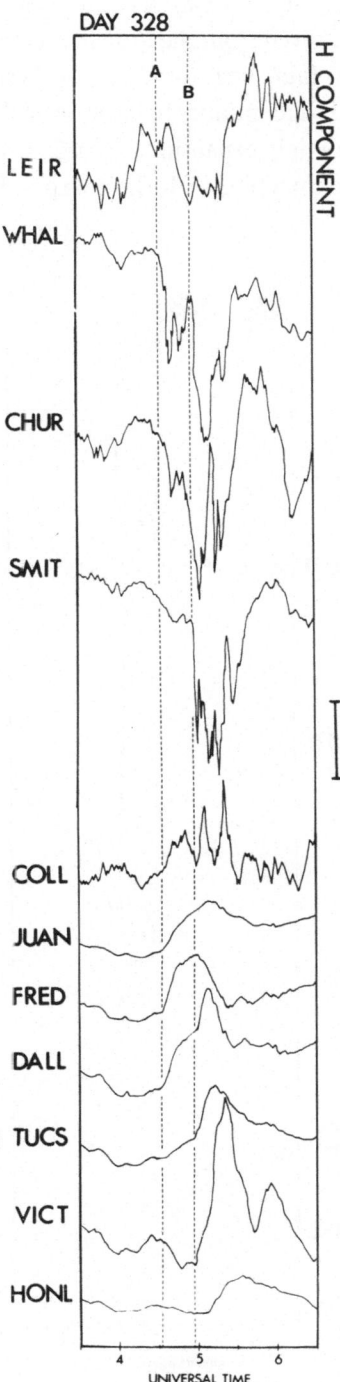

Fig. 1. H-component magnetograms from a group of average auroral zone and low latitude observatories for Day 328, 1971 (Nov. 24). The coordinates and code names of the observatories are given in the Appendix.

in detail at a portion of this structure starting ~0558 UT (line B on Figure 1). Prior to this time, expansion phase activity had been in progress since the 0430 UT onset and by 0445 UT VICT had moved from the growth sector to the region lying between growth and expansion (the 'fault line' discussed by Lesniak and Winckler (1970)). Note that west of VICT, HONL continues to imply growth, while east of VICT all low latitude records continue to imply expansion. At 0558 UT, DALL, TUCS and VICT all indicate expansion in conjunction with the sharp −H excursion at SMIT. It is

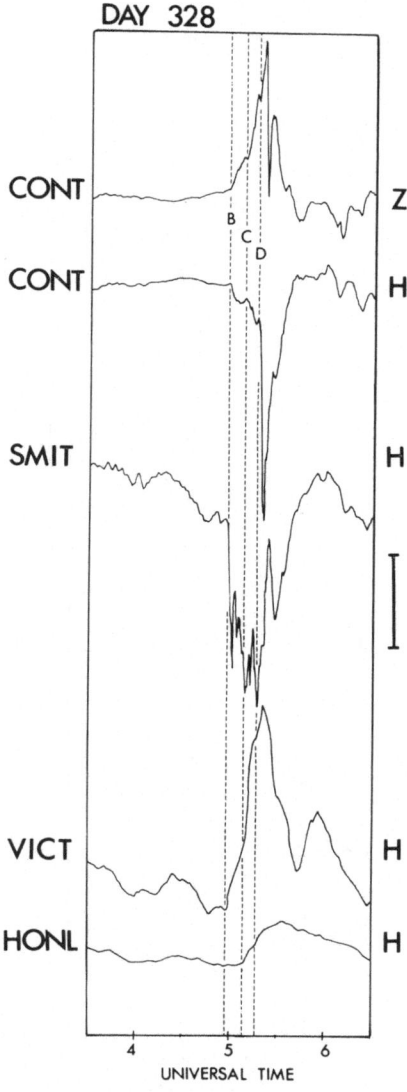

Fig. 2. H-component magnetograms displayed on an expanded scale to demonstrate the development of the poleward border of the auroral electrojets through the impulsive appearance of ionospheric current elements. The western end of each of the current elements is to the northwest of the preceding one.

interesting that HONL did not respond, implying it now lies at the boundary between growth and expansion phase sectors. The sequential development of the substorm current system is best followed from the expanded magnetograms shown in Figure 2. Note that CONT (74.1° N) is due north of SMIT (68.1° N). From the behavior of H and Z at CONT, it can be inferred that substorm intensified current elements developed at the times marked by lines B, C, and D. Each current element developed further northward, with the element which developed at line D being almost overhead at CONT. It is interesting to note that HONL experienced expansion phase signatures in association with the second two substorm intensifications. Therefore the western end of the substorm ionospheric current flow must have expanded westward during this time. A schematic picture of the development of the ionospheric current elements during this period is shown in Figure 3. It indicates that, after the onset of a substorm,

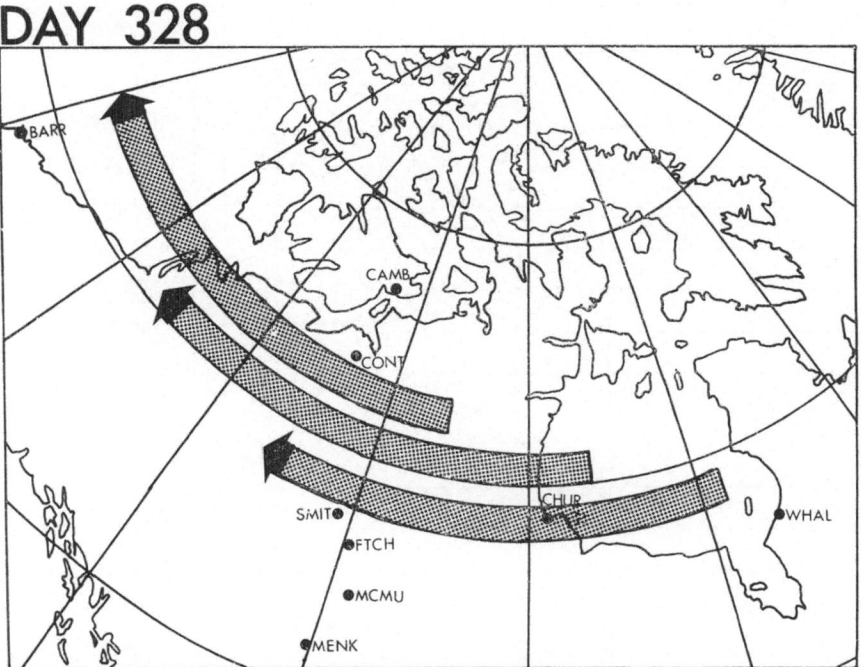

Fig. 3.   Schematic representation of the development of the ionospheric current elements which produce the magnetic perturbation pattern shown in Figure 2.

the substorm excited westward electrojet expands in a northwest direction in a series of impulsive jumps. Each of these jumps will have the same magnetospheric signature, although because of the localized nature of each perturbation current element, the appearance of the signature will depend on the placement of the monitoring observatories relative to the disturbed region.

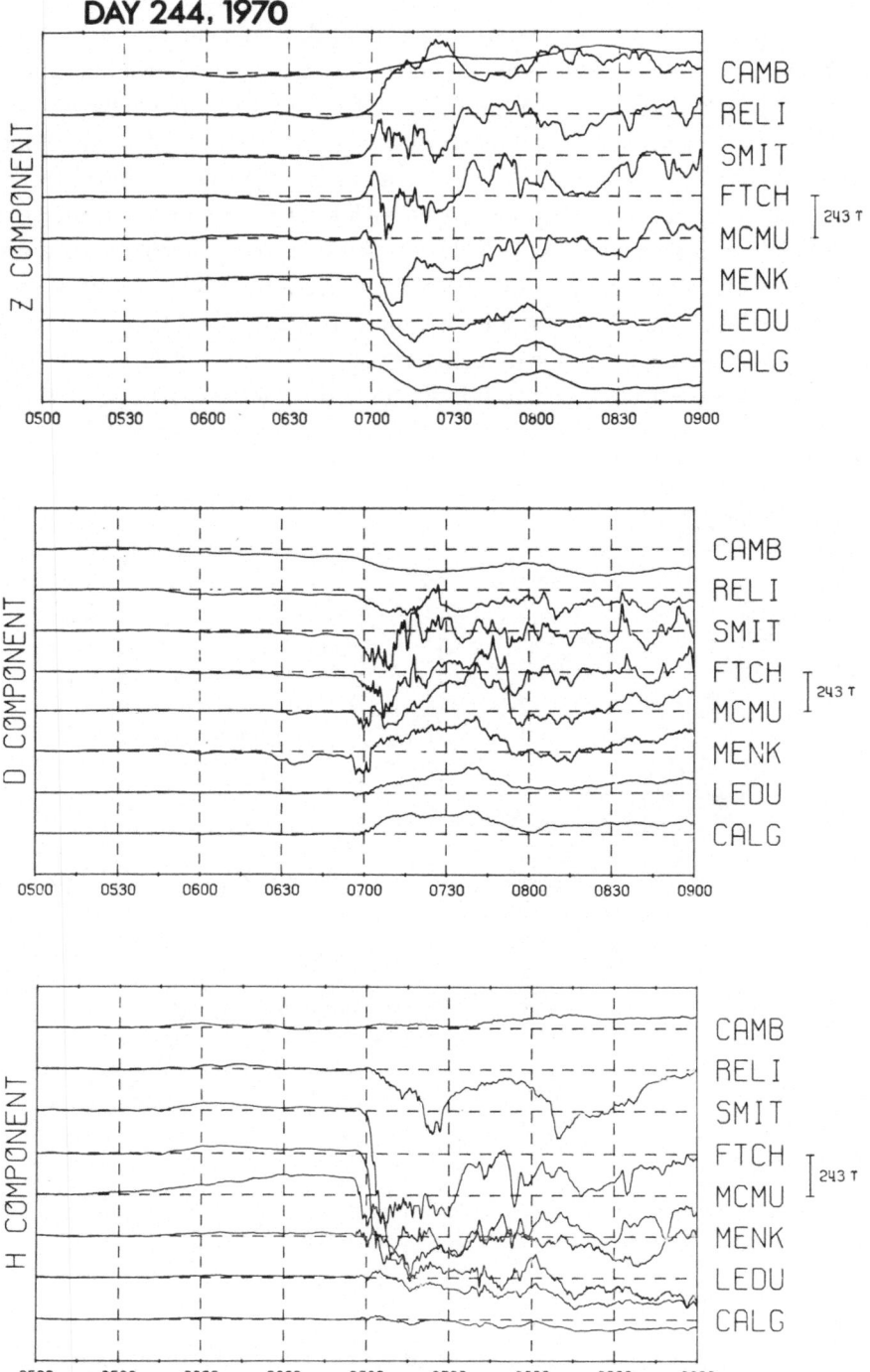

Fig. 4.   Magnetograms from the University of Alberta magnetometer line recorded on Day 244, 1970 (Sept. 1). Note the perturbations at the nominal polar cap station of CAMB and the development of the eastward electrojet as inferred from the increased +H at MCMU after ∼0540 UT.

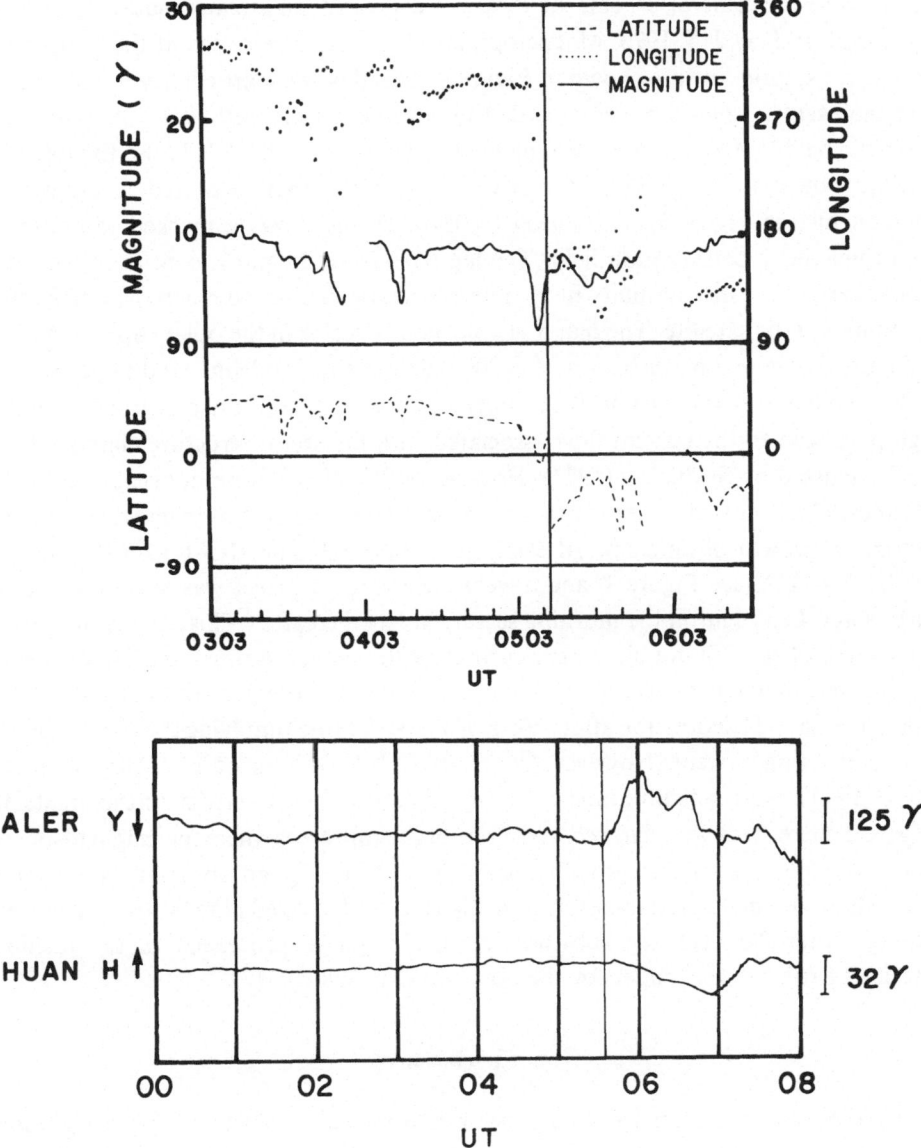

Fig. 5.   Parameters of the interplanetary magnetic field measured by Explorer 35 in lunar orbit and on the Sun-Earth line in front of the Earth. The solar wind velocity at this time was ∼ 348 km s^{-1}. Also shown are the recording of the H component of the geomagnetic field at the equatorial station of Huancayo and the recording of the Y component of the geomagnetic field at the polar cap station of Alert. The vertical line at ∼ 0515 UT on the top frame indicates the time at which the IMF turned southward and a sector structure change was experienced at the satellite. The vertical line at ∼ 0535 UT on the bottom frame indicates the onset of polar cap geomagnetic activity.

3. Development of an Isolated Substorm

In order to impress on the reader the fact that there are always a few events which do not fit into the general scheme of things, we present an isolated substorm which developed on Day 244, 1970. Magnetograms of the event recorded at the University of Alberta station line are shown in Figure 4. This day was particularly interesting in that the event in question was preceded by an interval of northward interplanetary magnetic field (IMF) and weak geomagnetic activity. At $\sim$0515 UT, the Explorer 35 satellite noted a change in sector structure and the IMF acquired a southward component (see Figure 5). At $\sim$0535 to 0540 UT there was a marked response in the geomagnetic perturbation field. The lag of $\sim$20 to 25 min is consistent with the time taken for the discontinuity in the interplanetary medium to reach the Earth after encountering the satellite. The major geomagnetic effect is in the polar cap (see ALER in Figure 5) where a perturbation of $\sim$200 γ was noted. Modeling studies performed by Kisabeth and Rostoker (1974) attribute a major portion of the perturbation in the high polar cap to the current flow associated with the sector structure change in the IMF discussed by Svalgaard (1973). However, other effects were noticeable at lower latitudes which are consistent with enhanced convection in the magnetosphere. In particular, growth of an eastward electrojet is noticeable at MCMU on the station line by 0540 UT (see Figure 4) and a weak westward electrojet was also observed at Kiruna and Leirvogur in the morning sector shortly thereafter. This is consistent with the growth of the SD double vortex current system which Axford and Hines (1961) associated with magnetospheric convection. In addition, the equatorial station HUAN experiences a $-$H depression after $\sim$0540 UT (see Figure 5) implying the growth of the asymmetric ring current. However, from $\sim$0540 UT until the onset of the substorm at $\sim$0655 UT there was no evidence of any expansion phase activity in the nightside magnetosphere. This would *appear* to be one of the rare cases where the magnetosphere as a whole was in a state of growth, although the lack of good observatory coverage at high latitudes makes it impossible to really rule out localized high latitude substorm activity. After 0655 UT the substorm expansion phase proceeded in the fashion described earlier in this paper for the event on Day 328, 1971.

4. Conclusions

We have presented material in this paper which indicates that typically, the ground based signatures of the growth phase of substorms occur concurrently with the expansion phase of substorms. This indicates that one sector of the nightside magnetosphere may be experiencing growth at the same time as an adjacent sector is experiencing expansion. In fact, it is highly probable that some of the energy transported from the magnetotail into the nightside magnetosphere during a substorm expansion phase is stored in the evening sector where it triggers subsequent substorm expansive phase activity. However, on rare occasions, one can find periods where no substorm expansion phase activity is in progress and yet the magnetosphere is in a state of growth. Such cases are the exception, rather than the rule, and in general the expansion phase of one substorm is responsible for the input of energy which causes ensuing expansion phase activity.

Acknowledgments

The work described in this paper has been carried out in association with R. G. Wiens. The material involving Day 244, 1970 was analyzed in collaboration with Dr J. L. Kisabeth. This research was supported by the National Research Council of Canada.

References

Akasofu, S.-I.: 1964, *Planetary Space Sci.* **12**, 273.
Axford, W. I. and Hines, C. O.: 1961, *Can. J. Phys.* **39**, 1433.
Belyakova, S. I., Zaytseva, S. A., and Pudovkin, M. I.: 1968, *Geomag. Aeron.* **8**, 569.
Bonnevier, B., Boström, R., and Rostoker, G.: 1970, *J. Geophys. Res.* **75**, 107.
Iijima, T. and Nagata, T.: 1972, *Planetary Space Sci.* **20**, 1095.
Kisabeth, J. L. and Rostoker, G.: 1971, *J. Geophys. Res.* **76**, 6815.
Kisabeth, J. L. and Rostoker, G.: 1973, *J. Geophys. Res.* **78**, in press.
Kisabeth, J. L. and Rostoker, G.: 1974, *J. Geophys. Res.*, to be published.
Lesniak, T. W. and Winckler, J. R.: 1970, *J. Geophys. Res.* **75**, 7075.
McPherron, R. L.: 1970, *J. Geophys. Res.* **75**, 5592.
McPherron, R. L.: 1972, *Planetary Space Sci.* **20**, 1521.
Mozer, F. S.: 1971, *J. Geophys. Res.* **76**, 7595.
Rostoker, G. and Camidge, F. P.: 1971, *J. Geophys. Res.* **76**, 6944.
Rostoker, G. and Kisabeth, J. L.: 1973, *J. Geophys. Res.* **78**, in press.
Svalgaard, L.: 1973, *J. Geophys. Res.* **78**, 2064.
Wiens, R. G. and Rostoker, G.: 1973, *EOS Trans. Amer. Geophys. Union* **54**, 312.

Appendix

Corrected geomagnetic coordinates and code names of magnetic observatories used in this study

Station	Code Name	Latitude (°N)	Longitude (°E)
Alert	ALER	86.5	122.6
Calgary	CALG	58.7	302.0
Cambridge Bay	CAMB	77.9	300.3
College	COLL	64.9	260.3
Contwoyto	CONT	74.2	297.0
Dallas	DALL	43.4	326.9
Fort Chipewyan	FTCH	67.0	301.3
Fort Churchill	CHUR	69.7	247.0
Fort McMurray	MCMU	65.0	302.7
Fort Reliance	RELI	71.4	300.0
Fort Smith	SMIT	68.2	299.8
Fredricksburg	FRED	51.8	352.2
Great Whale River	WHAL	68.2	353.8
Honolulu	HONL	21.1	266.5
Huancayo	HUAN	−0.6	353.8
Leduc	LEDU	61.2	301.5
Leirvogur	LEIR	66.3	72.0
Meanook	MENK	62.5	301.2
San Juan	JUAN	29.6	3.1
Tucson	TUCS	39.7	311.4
Victoria	VICT	53.9	292.6

CRITICAL PROBLEMS IN ESTABLISHING
THE MORPHOLOGY OF SUBSTORMS IN SPACE

R. L. McPHERRON

Department of Planetary and Space Science

and

Institute of Geophysics and Planetary Phyics, University of California, Los Angeles, Calif., U.S.A.

1. Introduction

For the past 6 yr many investigators have been attempting to establish the morphology of substorms in space. In our experimental work at UCLA we have been mainly concerned with the sequence of changes which occurs in the Earth's magnetic field. By using a combination of ground and satellite observations of the field, we have attempted to piece together a phenomenological model of the magnetospheric substorm. This model has been described at numerous meetings and also has been presented in detail in several recent publications (Russell and McPherron, 1973; McPherron *et al.*, 1973a; McPherron, 1972). To remind you of the observations, we will briefly review this model in the following section.

Theoretical work at UCLA has generally followed the same lines as the experimental work and has produced a theoretical model of the magnetospheric substorm. Various aspects of this theoretical model have been reported recently in several publications (Coroniti and Kennel, 1972a, b, 1973; Kennel *et al.*, 1973; Kennel, 1974).

Many investigators at other institutions working with different types of data have developed phenomenological models similar to ours as presented below. A fundamental characteristic of these models is a division of the substorm into three phases: a growth phase, an expansion phase, and a recovery phase. These models have been severely criticized (e.g., Akasofu, 1972) on the grounds that there is no evidence for the existence of the growth phase. Data presented by various investigators purporting to prove its existence have been rejected as inconclusive for a variety of reasons.

In our recent work, which is as yet unpublished, we have attempted to answer some of this criticism by increasingly sophisticated analysis. Much of this criticism has some justification, but in our opinion is not sufficiently persuasive to require radical changes in our phenomenological model. It is the purpose of this paper to discuss the various forms of criticism which our model has recently received. In so doing, we will review our recent work and describe changes in the model which are apparently required. In our conclusions, we discuss the requirements of future observations of the magnetic field necessary to further advance in our understanding of substorm morphology.

2. A Phenomenological Model of Magnetospheric Substorms

A schematic representation of the magnetic variations seen in different spatial regions

B. M. McCormac (ed.), Magnetospheric Physics, 335–347. All Rights Reserved.

during a magnetospheric substorm is presented in Figure. 1. We take as an example an idealized situation where the solar wind magnetic field suddenly turns southward after a long interval of northward field. For simplicity, no dynamic pressure change accompanies the discontinuity in field. The substorm growth phase begins when the discontinuity encounters the magnetopause. After a short delay, the magnetopause begins to move earthward due to erosion of magnetic fiux. This flux is transported by

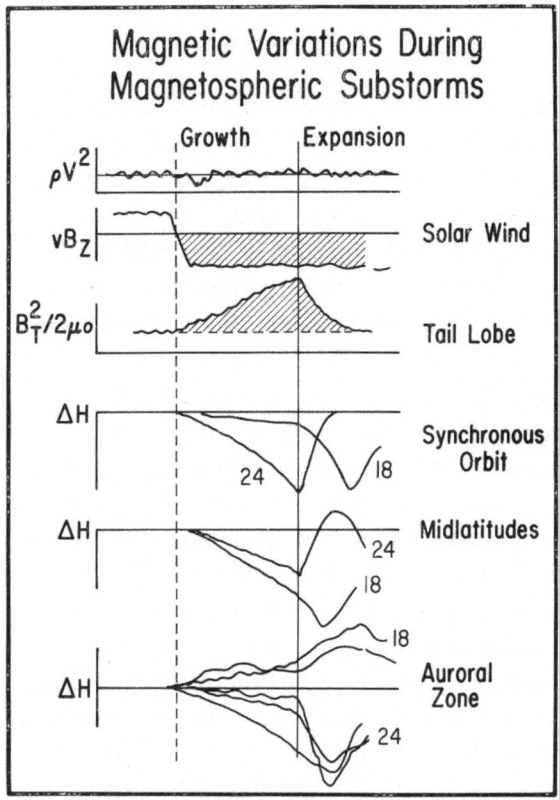

Fig. 1. Phenomenological model of magnetic variations during magnetospheric substorms. The recovery phase, not shown, begins at the time bays reach maximum development. Magnetopause erosion and near Earth plasma sheet thinning occur in the growth phase. Plasma sheet expansion begins in the expansion phase near the Earth and in the recovery phase far from the Earth.

the solar wind to the lobe of the tail. As flux is added to the lobe, the magnetic energy density begins to increase above the normal value. Simultaneously, the plasma sheet thins and the tail current moves earthward. Ionospheric currents driven by the earthward convection of plasma appear in the auroral zone as a weak westward electrojet. A partial ring current formed by injection of energized plasma close to the Earth closes through the ionosphere as an eastward electrojet. Effects of the partial ring current are seen at synchronous orbit and mid-latitudes as a depressed field.

The expansion phase begins when the near Earth plasma sheet becomes extremely thin and an X-type neutral point forms within the plasma sheet. The cross tail current is disrupted by the neutral point and diverted through the auroral ionosphere as another westward electrojet. Effects of this 'short circuit' are seen at synchronous orbit and mid-latitudes as an increase in the magnetic field. Magnetic flux previously stored in the tail lobe is annihilated at the neutral point, causing a decrease in the magnetic energy density of the tail lobe. Plasma energized at the neutral point is injected both earthward and tailward causing an increase in the partial ring current and its effects, and also an expansion of the plasma sheet.

The recovery phase begins when the additional energy in the lobe of the tail is exhausted through field annihilation. Field aligned currents, the westward electrojet, injection into the partial ring current, and the eastward electrojet begin to decay. Some time during the end of the expansion phase or the beginning of the recovery phase, the neutral point moves tailward returning the near Earth tail region to its presubstorm configuration.

3. Difficulties in Establishing a Phenomenological Model of Magnetospheric Substorms

By definition, a phenomenological model of a substorm is a description of the time sequence of events which occur in different regions of space throughout a substorm. Ideally, such a model should be devoid of observer bias and be free of interpretation. Perhaps the most serious difficulty in determining such a model is the fact that the model must be constructed from observations made on, at most, one or two satellites. To accomplish this, one must make several assumptions. First, one divides space into a number of regions wherein he believes similar behavior occurs. For example, if one believes that the tail lobe acts as an entity, then observations from any point in the lobe are adequate. Such a division is necessary to minimize the effort in determining the model and to maximize the number of observations available in any one region.

Unexpected spatial structure will clearly cause error in the resulting model. Such errors may not be easily recognized. Geophysical phenomena are highly variable and events which do not correspond to the usual situation may be attributed to such variability.

Substorm variability provides another source of difficulty. A second assumption one usually makes is that all substorms are similar. Thus, if a satellite repeatedly occupies the same spatial region during successive substorms, it should record similar observations. If substorms are highly variable, for example, if they may be localized at any local time in the night sector, one needs an independent determination that two substorms observed by a single satellite are, in fact, similar. Lack of consideration of such variability can easily lead one to conclude substorms show no systematic behavior in space.

Still another assumption one usually makes is that a time origin can be determined for each substorm. This question of substorm onset time is considered by many people to be the most serious difficulty in establishing a substorm model. Since satellites move through space as a function of time, it is crucial to know when the substorm starts because this determines the spatial region to which the subsequent observations will be attributed. We note that any substorm feature common to all substorms may be used as a time reference. Difficulty arises, first, as a consequence of substorm variability and lack of agreement between observers as to exactly when a given feature occurs. It arises, also, in a more serious way when the same name, i.e., expansion onset, is given different definitions by different investigators. Such a circumstance appears to be responsible for the present growth phase controversy. In some cases, the onset of mid-latitude and auroral zone bays near midnight is simultaneous. Because of this, many investigators have used data from either location interchangeably in timing substorms. However, as pointed out by Akasofu (1972), and as verified by McPherron *et al.* (1973b), this is not always true. Figure 2, taken from the recent paper by McPherron *et al.* (1973b) illustrates this point. Mid-latitude onsets, shown by vertical lines, are clearly delayed with respect to the onset of auroral electrojet activity. While it is argued by McPherron *et al.* (1973b) that the major substorm ex-

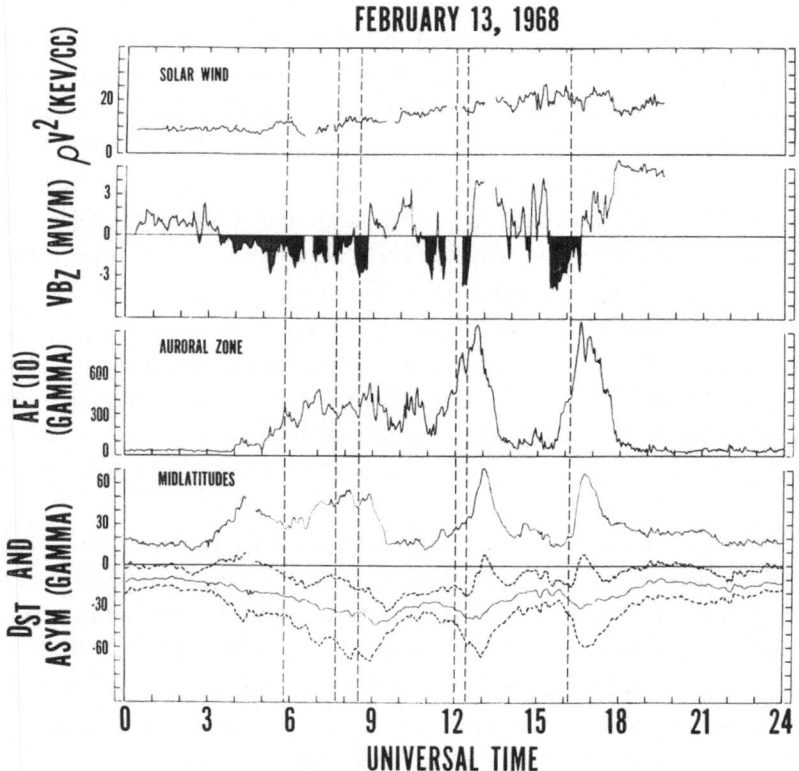

Fig. 2. Solar wind dynamic pressure and electric field at Explorer 33, AE(10) index, asymmetry, maximum, average (Dst), and minimum of local time profile of the H component at mid-latitudes on Feb. 13, 1968. Vertical dashed lines are mid-latitude determinations of substorm reference time.

pansion onsets in the auroral zone correspond to the mid-latitude onsets, it is undeniable that electrojet activity begins before the mid-latitude onsets.

Among the arguments raised against the phenomenological model discussed above is the fact that it is based, to a great extent, on the analysis of single events. This case history approach is said to be unscientific because unique events are chosen by the investigator to confirm his preconceived notion as to what substorm behavior ought to be. While this is, in fact, possible, the case history approach is a practical problem. To acquire and display the data relevant to a given event are time consuming and expensive tasks.

Some investigators have attempted to establish statistical pictures by choosing their events and displaying only one type of observation with a table summarizing the remaining observations. Such approaches have been criticized on the basis of lack of appropriate data or observer bias in selecting the events. For example, even in the most thorough studies it has been suggested that had another magnetic observatory or all-sky camera been in operation during a given substorm, it might have observed an earlier onset than seen at existing observatories. Such arguments cannot, of course, be refuted.

Many investigators have attempted to simplify their studies by choosing the first substorm after prolonged intervals of magnetic calm or by choosing relatively isolated substorms. The argument against this approach has been that such substorms are not typical and hence inappropriate objects of study. On the other hand, arguments have been made that substorms during disturbed times are too complicated to study.

Since our phenomenological model proposes that substorms are caused by the solar wind magnetic field, numerous arguments have been raised by others in attempt to refute demonstrations of this relation. For example, it is well known that solar wind parameters are highly correlated. Thus, a southward turning of the solar wind field is usually associated with a change in dynamic pressure. If there is subsequent substorm activity, it is difficult to prove that the pressure change did not have an important effect on the magnetosphere.

The actual spatial structure of discontinuities in the solar wind is not known. Thus, the fact that a southward turning is observed at a satellite in the solar wind may not imply that it will encounter the Earth. Furthermore, little is known about the way in which the discontinuities are spatially oriented or how they are propagated by the solar wind. Consequently, large errors are possible in the calculated arrival times of such discontinuities at the Earth. Finally, very little is known about how the bow shock distorts these discontinuities before they arrive at the magnetopause.

Even if one accepts the phenomenological model discussed above, it must be recognized that it does not make quantitative predictions. Some feel this is the most serious objection of all.

4. Emphasis of Our Current Research

In our current experimental work we are attempting to answer some of the criticism outlined above. In this section, we touch upon each major area of research in a brief review of recent results. The lack of appropriate data is a problem of great concern. To alleviate this problem, we have been accumulating a large file of digital data covering the years 1967–1969, during which extensive satellite data are available. The wide variety of data formats, time resolutions, etc., in this file provides a great obstacle to progress. We have undertaken a number of program developments designed to enable us to carry out statistical studies on this file.

Determination of substorm onset time is a problem we have investigated in some depth (Clauer and McPherron, 1973a). In an attempt to eliminate subjectivity in this definition, we have developed computer mapping procedures which automatically

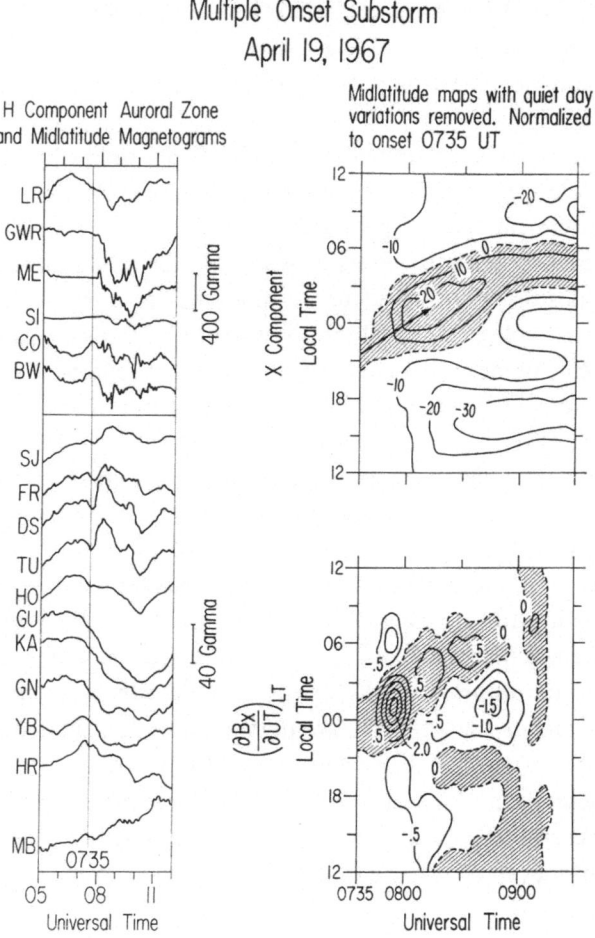

Fig. 3.   Ground observations of a substorm having an eastward sweeping expansion phase. Auroral zone and mid-latitude magnetograms appear at left. A vertical line marks substorm onset. The top contour map shows the mid-latitude magnetic field perturbations of the X component normalized to the onset. The bottom map contours the rate at which the field changed.

determine the time and location of the onset of mid-latitude positive bays. Figure 3 is an example of mid-latitude magnetic data processed by these procedures. A positive bay at U.S. stations (SS, FR, DS, TU) begins at 0735 UT, with a corresponding negative bay onset in the auroral zone at about this time. A map of the X component normalized to the beginning of the mid-latitude bay (magnetic N–S) shows this substorm begins premidnight and sweeps eastward. Ring current injection at dusk and in the late afternoon is also apparent. Maps of the slope of the X component (bottom right) show the exact onset times as contours of zero slope (dashed lines), corresponding to the minimum in X at the beginning of the bay. Similarly, the end of the expansion is shown by zero contours at the transition from positive to negative slope.

Similar maps created for many other substorms reveal there is a wide variety of substorms apparent at mid-latitude stations. Among the various types of onset are sharp onset, slow onset, multiple onset, eastward sweeping onsets, and westward sweeping onsets. Furthermore, as shown in Figure 4, the central meridian of maximum

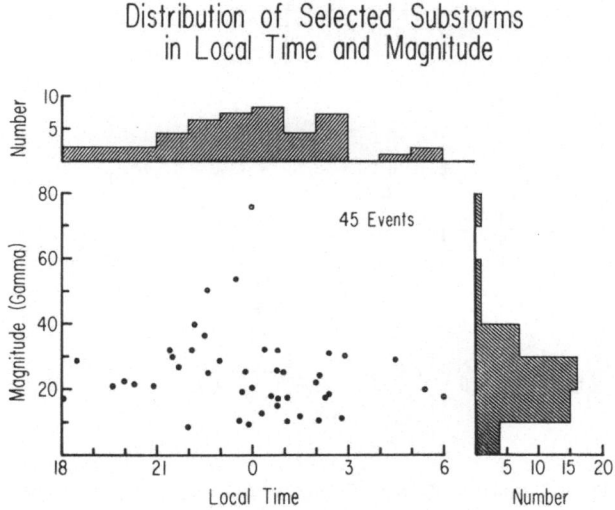

Fig. 4.   Local time and magnitude distributions of the central meridian of substorms at the end of the mid-latitude expansion phase. Some substorms are seen centered at any hour in the night hemisphere.

development of various substorms may be centered anywhere in the night hemisphere (Clauer and McPherron, 1973b). Clearly, there is far greater variability to substorms than has generally been appreciated.

Mid-latitude and satellite data of substorm magnetic perturbations have also been used by Horning *et al.* (1973) to determine parameters in simple models of substorm current systems. Figure 5 shows a current model of the expansion phase of a substorm determined by least squares inversion techniques. The model assumes that expansion

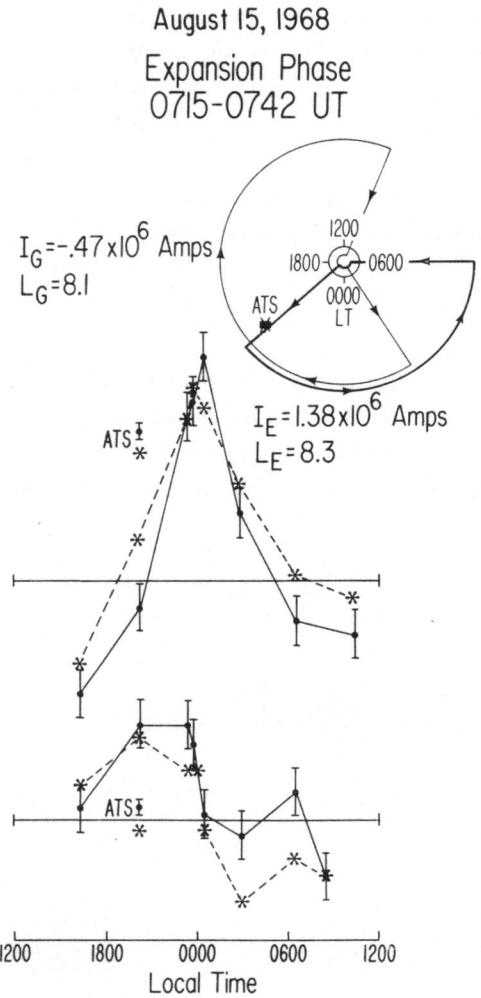

Fig. 5. Simple model of magnetospheric, field aligned and ionospheric currents flowing at end of substorm expansion on Aug. 15, 1968. A comparison of the observed (solid) and predicted (dashed) profiles of mid-latitude magnetic perturbation is made for the north and east component at the middle and bottom, respectively.

phase currents (heavy lines in equatorial view) are superimposed on the growth phase currents. Since the model is for perturbations in previously flowing currents, it implies the following. A partial ring current of wide angular extent and centered near dusk in the growth phase increases by $\simeq 0.5$ MA in the expansion phase. The tail current is short circuited through the ionosphere centered about 0200 LT with a strength of 1.4 MA. Agreement between observations (solid line) and predictions (dashed lines) is shown for the north component (middle) and the east component (bottom).

Empirical models of the dependence of the quiet field at synchronous orbit and in

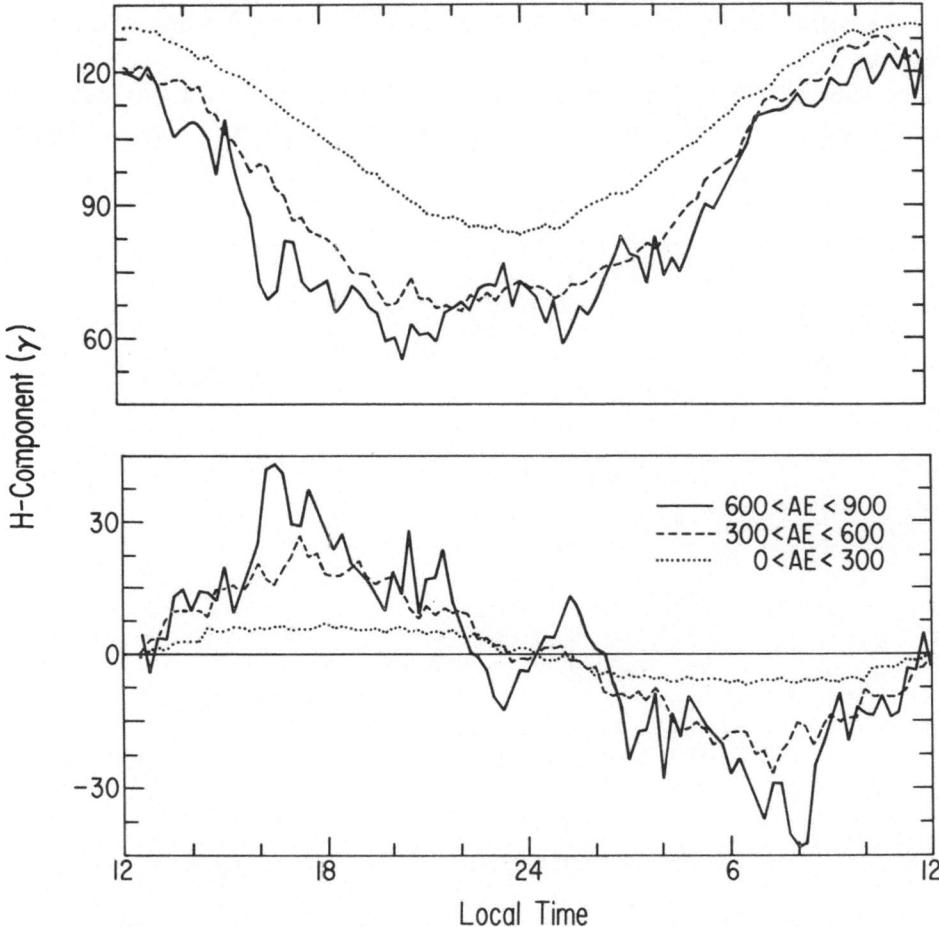

Mean Value of H and Asymmetry Around Midnight at ATS-1

Fig. 6.   Top panel shows the average value of the H component at synchronous orbit during 1968 for three different levels of the AE index. Bottom panel shows the asymmetry of average H around midnight. Asymmetry is defined as the difference in average H values at corresponding times on either side of midnight.

the tail lobe on solar wind dynamic pressure are being developed. Figure 6 shows initial results of a study of the H component at synchronous orbit (Subbarao *et al.*, 1973). One year of data has been sorted according to the level of auroral zone activity occurring at the time of the field observation. The H component is clearly depressed during auroral zone activity. Also, the depression becomes increasingly asymmetric with respect to midnight as activity increases. A widespread distribution of H values (not shown) about the mean curves is a function of solar wind dynamic pressure.

A similar model has been developed for the tail lobe, as shown in Figure 7. Shading between the two traces at OGO-5 emphasizes excess energy density, presumably the effect of southward solar wind magnetic field. In the model (Caan *et al.*, 1973a) the difference between the observed energy density on a day of high solar wind velocity and the energy density of a model field is assumed a linear function of dynamic pressure.

Also shown in Figure 7 are expansion onsets determined by methods described above. Boxed characters on the vertical lines indicate the local times of onset and types

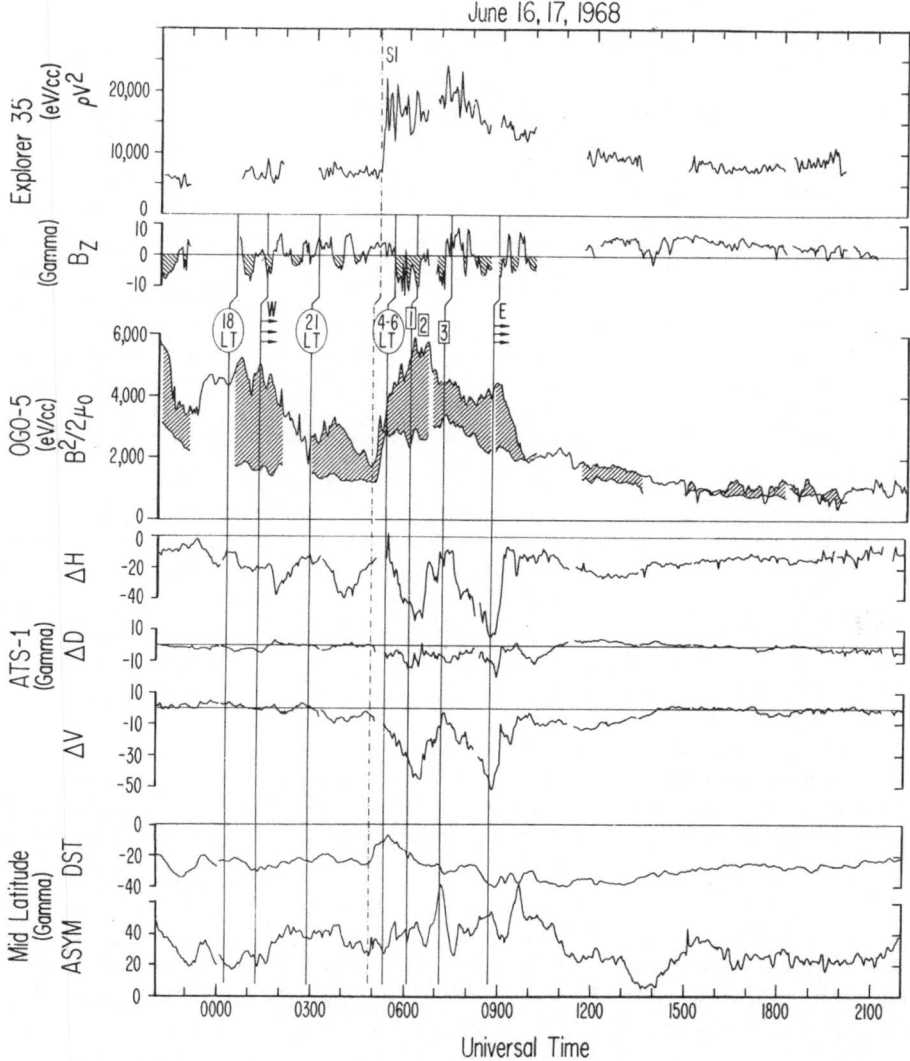

Fig. 7.   A correlation plot showing solar wind dynamic pressure and Bz (GSM) at Explorer 35, magnetic energy density at OGO-5 in tail lobe [both observed and predicted (lower curve)], perturbation field at ATS 1 and Dst, and its asymmetry for an interval of substorm activity on June 16, 17, 1968. Vertical lines are mid-latitude bay onsets with boxed characters showing substorm's central meridian and type of development.

TAIL LOBE MAGNETIC ENERGY DENSITY FLUCTUATIONS

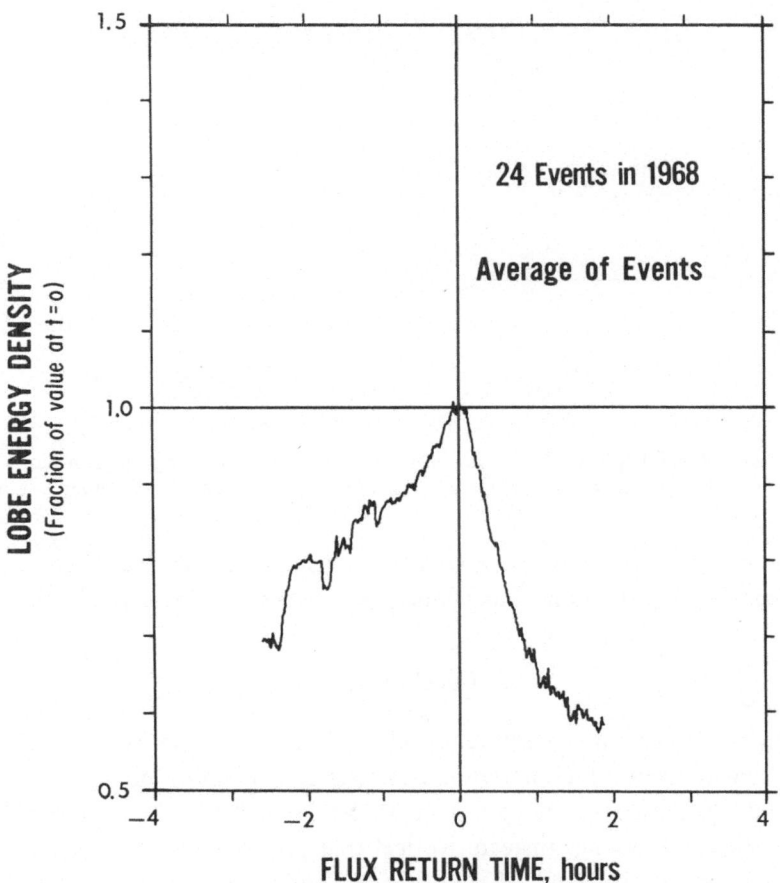

Fig. 8.   Results of superpositioning fractional fluctuations in tail lobe energy density during 24 isolated magnetospheric substorms in 1968. The time origin is defined as the beginning of midlatitude positive bays.

of substorm development. The lobe response is far more complex than suggested by our simple model outlined earlier.

Statistical studies of the lobe response during substorms have also been carried out by Caan *et al.* (1973b). Figure 8 shows superposed epoch analysis of 24 substorms organized by mid-latitude onset times. The characteristic increase before onset and decrease after onset are readily apparent for this group of selected substorms.

Empirical models of the solar wind control of Dst have been made quantitative by recent work of Burton *et al.* (1973). Figure 9 shows three successive days of magnetic activity and compares the observed and predicted Dst values. The empirical model assumes that the rate of change of Dst increases as a function of the square of the solar

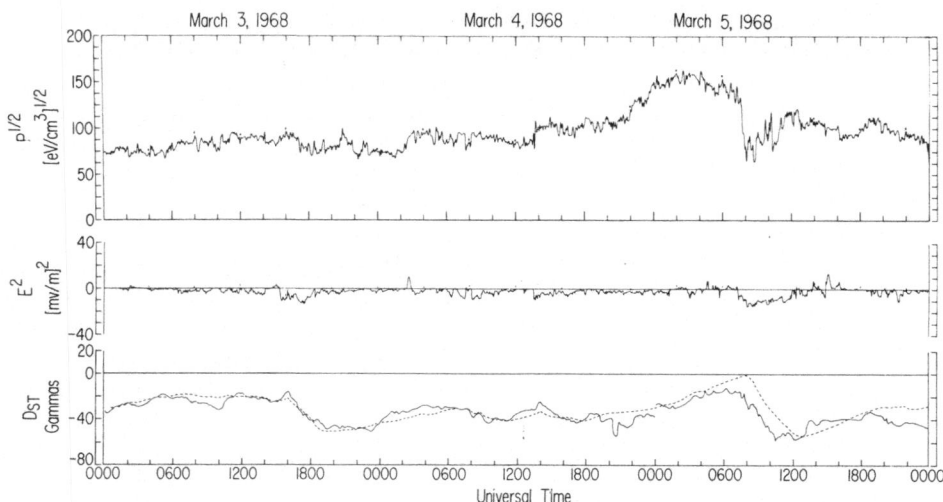

Fig. 9. Solar wind dynamic pressure and square of the electric field for a magnetic storm of March 3–5, 1968. Bottom curve shows comparison of observed and predicted (dashed line) Dst index.

wind electric field and decays proportionally to its instantaneous value. There is good agreement between observations and predictions, suggesting the general validity of the model.

5. Conclusions

The brief review of current research given above is the direction we have taken in attempts to answer the criticism our phenomenological model of substorms has received. While the criticism is, in part, justified, we do not find it persuasive evidence the model is radically wrong. Instead, we feel that previous research has not adequately considered the problems of substorm variability, errors in onset time, the number of distinct spatial regions, and dependence on past history of magnetic activity. Rather than reject the model entirely, we feel further research should include consideration of these problems.

As a final point, we note that careful consideration of previous criticism has revealed that our previous model is inadequate in several respects. The two most important are that the tail lobe and plasma sheet do not always react as single entities as previously supposed.

Future observations of substorm magnetic field changes must become increasingly sophisticated if further progress is to be made. Amplitude and time resolution of ground magnetometers must be improved. Digital data acquisition, real time transmission to central locations, and appropriate sorting and archiving techniques must be instituted. Better organized and more dense arrays of observatories must be set up. In space, simultaneous observations by several satellites must be performed to separate time and space variations. Particularly important are simultaneous observations of the magnetic field and low energy plasma.

Acknowledgments

The author wishes to express his gratitude for the helpful comments made by his students and co-workers during the preparation of this paper. The work reported herein has been carried out by several individuals, including R. Burton, M. Caan, N. Cline, R. Clauer, B. Horning, C. Russell, and S. Subbarao. The work and the author's time have been sponsored by several agencies, including the National Aeronautics and Space Administration (grant NGL 05-007-004), the National Science Foundation (grant GA-34148-X), and the Office of Naval Research (contract N00014-69-A-0200-4016).

References

Akasofu, S.-I.: 1972, *Scientific Design of a Shuttle Auroral Observatory System*, Vol. III, Report, Geophysical Inst., University of Alaska, Fairbanks.

Burton, R. K., McPherron, R. L., and Russell, C. T.: 1973, *Proceedings of the Chapman Memorial Symposium on Magnetospheric Motions*, p. 139.

Caan, M. N., McPherron, R. L., and Russell, C. T.: 1973a, *J. Geophys. Res.* 70, 8087.

Caan, M. N., McPherron, R. L., Russell, C. T., and Cline, N. E.: 1973b, *Proceedings of the Chapman Memorial Symposium on Magnetospheric Motions*, p. 103.

Clauer, C. R. and McPherron, R. L.: 1973a, *J. Geophys. Res.*, submitted.

Clauer, C. R. and McPherron, R. L.: 1973b, *J. Geophys. Res.*, submitted.

Coroniti, F. V. and Kennel, C. F.: 1972a, *J. Geophys. Res.* 77, 3361.

Coroniti, F. V. and Kennel, C. F.: 1972b, *J. Geophys. Res.* 77, 2835.

Coroniti, F. V. and Kennel, C. F.: 1973, *J. Geophys. Res.* 78, 2837.

Horning, B. L., McPherron, R. L., and Jackson, D. D.: 1973, *J. Geophys. Res.* submitted.

Kennel, C. F.: 1974, private communication.

Kennel, C. F., Coroniti, F. V., McPherron, R. L., and Russell, C. T.: 1973, *On the Initiation of Tail Reconnection During Magnetospheric Substorms*, in preparation.

McPherron, R. L.: 1972, *Planetary Space Sci.* 20, 1521.

McPherron, R. L., Russell, C. T., and Aubry, M. P.: 1973a, *J. Geophys. Res.* 78, 3131.

McPherron, R. L., Russell, C. T., Kivelson, M. G., and Coleman, P. J., Jr.: 1973b, *Radio Sci.*, in press.

Russell, C. T. and McPherron, R. L.: 1973, *Space Sci. Rev.* 15, 205.

Subbarao, S., McPherron, R. L., and Cline, N.: 1973, *Proceedings of the Chapman Memorial Symposium on Magnetospheric Motions*, p. 201.

THE DYNAMICS OF AURORAL ABSORPTION

J. K. HARGREAVES

Department of Environmental Sciences, University of Lancaster, Lancaster, England

1. The Global Picture of the Substorm in Auroral Absorption

1.1. AVERAGE PATTERNS OF MOVEMENT

The earliest riometer studies of auroral radio absorption paid little attention to movements. The first overt reference to them is probably a single example reported by Kavadas (1962). Evidence of movements can be seen in the diagrams of some other early papers (Little and Leinbach, 1958; Holt *et al.*, 1962) though the authors do not draw attention to them in the text. Thus, most of the earlier investigations of auroral absorption (AA) concentrated on questions such as the mechanism of radio absorption and the nature of the primary particles, the synoptics of time and place of its occurrence, and on relationships with other auroral and geophysical phenomena. Towards the middle of the 1960's there were further reports of AA movements, based on time differences between the appearance of bursts of activity at geographically separated stations (Eriksen *et al.*, 1964; Ansari, 1965; Little *et al.*, 1965), and there has been an increasing concentration on the dynamics of AA since that time. As a result, it is now realized that nearly all absorbing regions are in motion, that the motions are largely systematic, and that they occur on the global scale – that is, involve both night and day sectors and include the whole width of the auroral regions in both hemispheres. Clearly this has been a significant development and there can be little doubt that studies of the dynamics of auroral absorption represent one of the most important aspects of riometer work now and for the immediate future.

The substorm behavior of auroral absorption is important in relation to the auroral electrons of higher energy (say >15 keV) in the same way that the substorm in luminous aurora is important for particles of somewhat lower energy (1 to 10 keV). The movement of an absorption patch in the ionosphere indicates the motion of a disturbance in the magnetosphere; thus, observations of the velocity and direction of an absorption event should have something to say about the mechanism(s) by which the disturbance propagates through the magnetosphere. The systematic growth and decay of the absorption with local time and latitude must be related to the mechanisms of precipitation, energization, and loss of the auroral particles. An important difference between substorms in luminosity and in radio absorption is that on the dayside of the Earth the latter occur at a lower latitude and within the trapping region; indeed, in the daytime absorption region we are probably seeing the major dumping ground for excess outer-zone electrons, these having been introduced during substorms in the night sector where the radio absorption and auroral luminosity more or less coincide. Thus, it is now fairly clear that the AA substorm as observed from the ground is closely related to important fundamental processes occurring within the disturbed magnetosphere.

B. M. McCormac (ed.), Magnetospheric Physics, 349–356. All Rights Reserved.

The global pattern of movement of the AA substorm has been studied by several workers using spaced riometers (Jelly and Brice, 1967; Hargreaves, 1967, 1968; Jelly, 1968; Pudovkin *et al.*, 1968; Driatsky, 1969; Driatsky and Shumilov, 1970; Driatsky *et al.*, 1972; Theander, 1972). These results are reasonably consistent (though, as we shall see, they may in some respects be inadequate) and they lead to an average or median picture in which the substorm begins near the midnight meridian at an IN Lat near $65°$ ($L \sim 5.5$) and then spreads to higher and lower latitudes and eastward and westward from the origin. Generally the above studies used riometer records made at widely-spaced stations where the separations are typically several hours of local time or several degrees of latitude. To work out a pattern of movement it is necessary to identify the same feature at each of the stations being compared, and both onsets and absorption peaks have been selected. The onset is defined as the time when activity is seen to begin at a given place, and is not to be confused with the origin or beginning of the substorm which is the first appearance of activity anywhere. The movement of the onset is clearly a characteristic of the substorm, considered as the whole absorption event between quiet periods. It is less obvious that the movement of a selected peak is typical of the substorm as a whole, but nevertheless most studies have shown similar velocities for peaks and for onsets. On the average, eastward velocities are 3 to $4°$ min^{-1} and westward velocities are 1.5 to $2.5°$ min^{-1}. These are with respect to a non-rotating coordinate system, where the Earth's rotation rate, $0.25°$ min^{-1}, has been taken into account. Observations in the Russian sector show peak velocities less than half the onset velocities (Driatsky, 1969; Driatsky *et al.*, 1972). All studies report westward velocities smaller than eastward, and the two branches meet in the afternoon sector where AA is weak. The velocity of the eastward branch decreases with time from the substorm beginning, and that of the westward branch increases with the intensity of the event (Theander, 1972). The zonal velocity is a function of latitude, but K_p seems to have no effect.

For meridional movements in the night sector, average velocities range over 0.5 to 0.9 km s^{-1}, or 0.3 to $0.5°$ min^{-1}, with no significant differences reported between onsets and peaks, or between poleward and equatorward movements. The latitude of the origin is a function of K_p, and there is a change in the character of the onset at this latitude, the poleward section having a sharp onset, and the equatorward part beginning more gradually (Jelly, 1970). Jelly suggested that the change of character and the origin of the substorm occurred near the trapping boundary.

Stated thus, the pattern of movement of the AA substorm appears quite simple. Certainly, the average picture is useful as a summary of typical behavior, but in what follows we will see how detailed considerations lead to indications of greater complexity.

1.2. MECHANISMS OF ZONAL MOVEMENT

The median global picture, in setting the order of magnitude of the dominant motions, also limits the range of candidate mechanisms for substorm propagation. The various suggestions put forward include gradient-curvature drift of trapped particles, hydro-

magnetic waves, and electric fields. Most attention has been given to the zonal movements, eastward and westward; the meridional movements in the midnight sector
could be somewhat different and more complicated in that they may also be affected
by movement of an injection region. The most attractive of the theories of zonal
movement is probably that of Pudovkin and Shumilov (1969), which includes both
electric fields (i.e., the magnetospheric circulation) and gradient-curvature drift of
energetic electrons, and can accomodate the observed velocities of both the eastward
and the westward branches with reasonable parameters. However, it should be pointed
out that, although both mechanisms presumably operate, either gradient-curvature
drift or electric field drift alone can explain aspects of the movements with parameters
that are not too unreasonable. For instance, the eastward drift quoted above could be
due to electrons of energy ~ 70 keV or to a magnetospheric electric field of 4 mV m^{-1},
or to an indefinite number of combinations of the two. A coincidence of numbers
does not prove a mechanism, and better direct evidence is required before the relative
importance of these or other mechanisms can be established.

1.3. Variations about the average

Satisfactory theories must explain not only the average pattern but also variations
about the average. For the AA substorm these variations are large, and to neglect
them brings a danger of serious over simplification. Figure 1 shows individual measurements of the time difference between Great Whale River and Reykjavik, stations
separated by $5\frac{1}{2}$ h of MLT and 82.5° of longitude. These are the same data that were
previously used to compile average zonal velocities (Hargreaves, 1967), but are here
divided into onsets and peaks. If such a distribution of time differences is averaged
in the night and the afternoon sectors, where the movement may be eastward on one
occasion and westward on another, the average leads to a velocity much higher than
those actually observed. (When small time differences were observed this was probably
because the origin fell between the stations.) If eastward and westward movements
are regarded separately, velocities come out somewhat lower – for instance about
3° min^{-1} eastward (with respect to a non-rotating frame) instead of 4° min^{-1} reported
previously (Hargreaves, 1967). Where averaging is not complicated by reversals of
direction, such as for onsets during 0900 to 1500 UT one may see the amount of variation in the velocity. A factor of 3 with regard to the median is required if 90% of the
observations are to be included; thus, velocities between the slowest and the fastest
vary by about an order of magnitude. The range is similar in other parts of the day for
which there are enough observations for an assessment. It is important to establish
experimentally the reasons for these variations of velocity between one substorm and
another, and theories must be able to accomodate the full range as well as the average.

2. Dynamics of Structure Within the Substorm

2.1. Comparison of observations over wide and narrow baselines

It is not obvious that measurements on features, such as peaks, within the pattern of

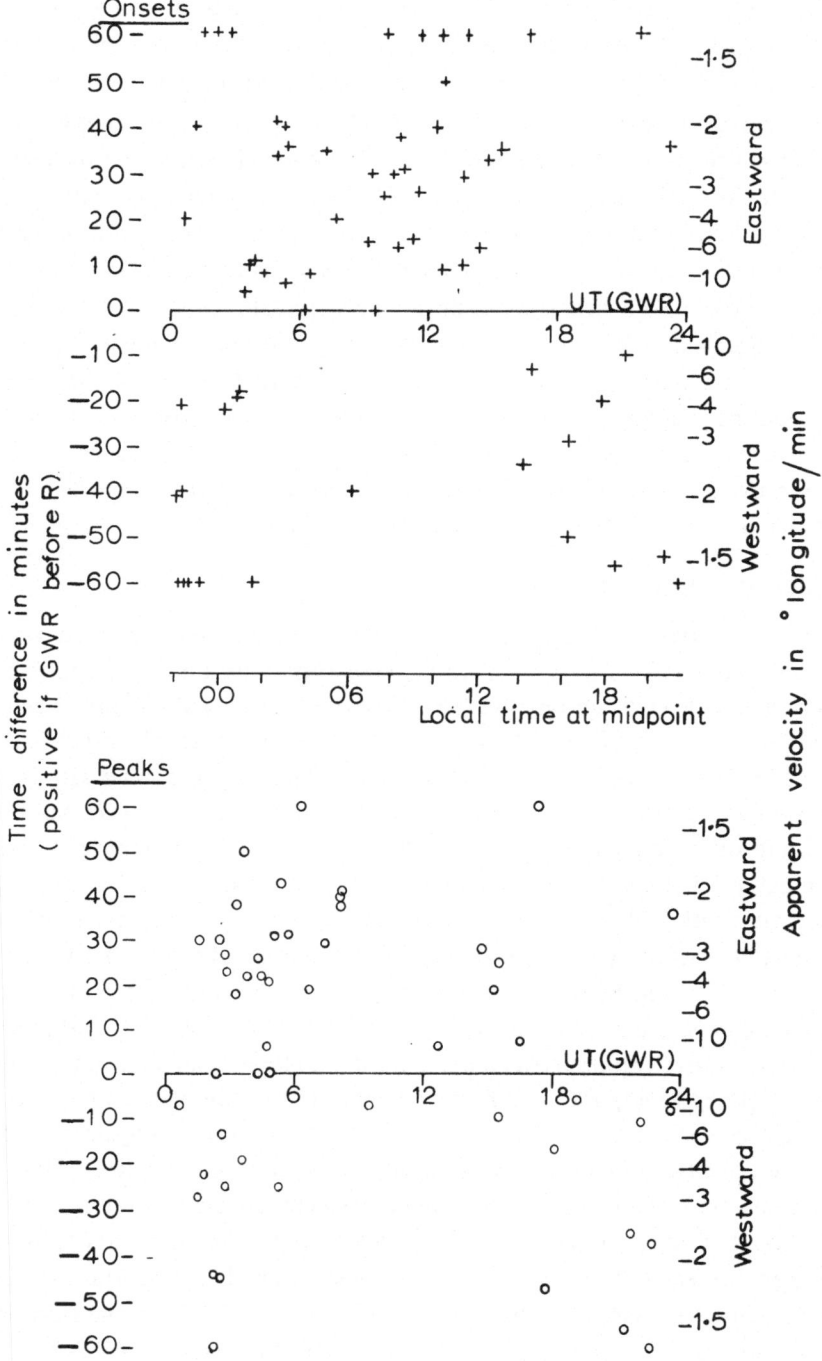

Fig. 1.   Time differences of AA features between Great Whale River and Reykjavik. Only time differences up to 2 h were recorded, and values between 1 and 2 h are plotted at 60 min. Velocities are with respect to the rotating Earth. The central time axes give the UT at which the feature appeared at G.W.R., where local magnetic midnight is at 05 UT. The observations were made in Dec. 1965, March–Apr. 1966, and June 1966.

absorption should be applicable to the substorm as a whole. It is well known that the typical fine structure of AA differs considerably between night and day (Hargreaves, 1969), and therefore it must evolve as the substorm propagates. Also, it is quite conceivable that the fine structure could be a modulation imposed by a different process from that controlling the form of the substorm as a whole.

In a study of movements over a short baseline of 250 km (Hargreaves, 1970), zonal velocities were found to be much smaller than those usually reported from wide baselines. We will confine our discussion at this point to eastward movements in the interval 0300 to 0700 LT at the middle of the baseline. Figure 2 shows the distribution of

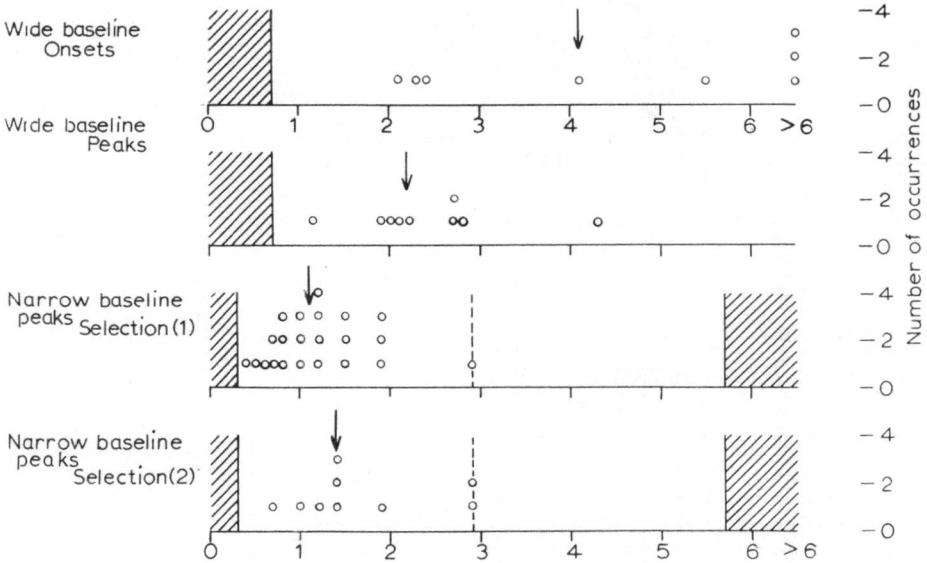

Fig. 2. Comparison of zonal velocities measured over wide and narrow baselines. The time interval is 03–07 LT at the midpoint of the baseline. Wide baseline observations are from Hargreaves (1967) and narrow baseline selection (1) from Hargreaves (1970). Selection (2) comes from the same set of observations but was based on criteria other than the presence of obvious movements. The similarity of results of selections (1) and (2) shows that selection (1) was not seriously biased by the selection criteria used. Observational limits are shown as shaded areas. The unit of velocity is degree min^{-1} and medians are marked by arrows. The high-velocity limit in narrow-baseline observations could be as low as 2.9° min^{-1} corresponding to $\Delta t = 2$ min.

measured velocities over 82.5° of longitude (Hargreaves, 1967) and over 250 km (which is 5.75° of longitude). The limits imposed by observational circumstances, such as timing accuracy and the range of time differences sought, are indicated. It appears that the velocities of peaks over the wide baseline seldom fall below 2° min^{-1}, and over the short baseline seldom exceed 2° min^{-1} (with respect to the rotating Earth) and that these limits are not determined by the timing restrictions. This may be explained if the peaks have a limited lifetime, so that only the faster ones survive unchanged over the wide baseline. Following a previous analysis (Hargreaves, 1970), if

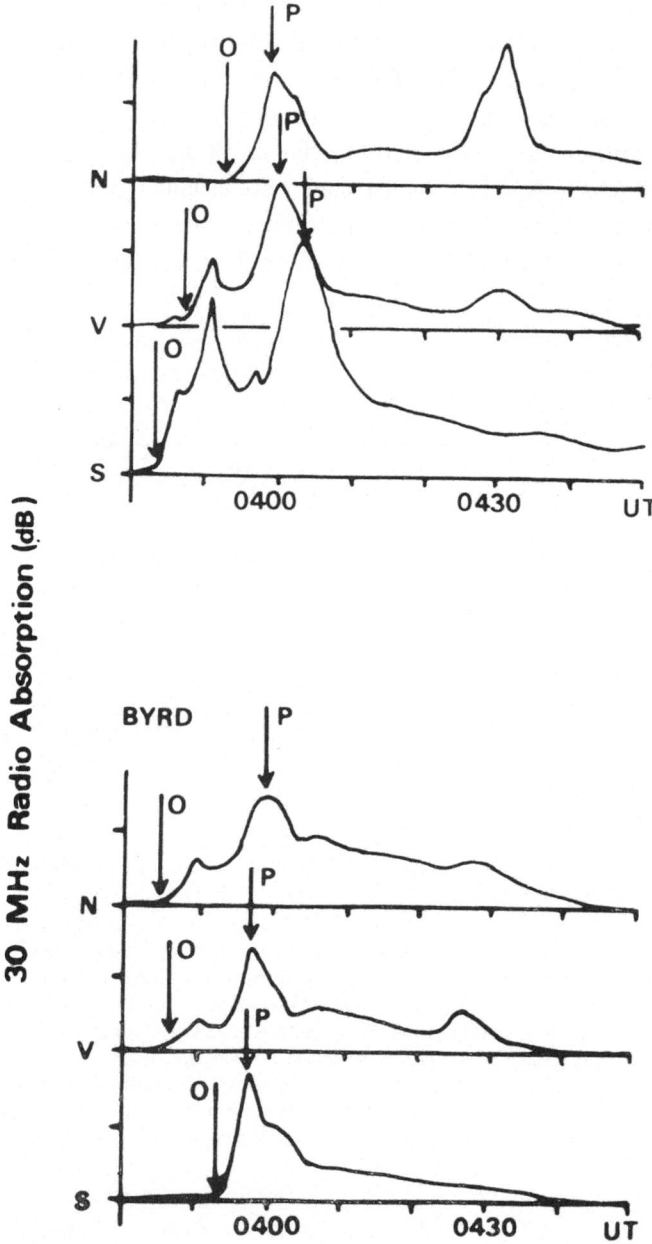

Fig. 3.   Moving onset and peak, April 7, 1966. Great Whale River is in the northern hemisphere and Byrd is conjugate to it in the southern hemisphere. Channels are separated about 125 km in a north-south direction. At each station the onest (O) moves poleward but the first major peak (P) moves equatorward. The first peak of the event is not seen at either of the poleward stations.

Δd is the station separation, d_0 the width of the peak, t_0 its duration as seen by an observer moving with the peak at velocity $V = \Delta d / \Delta t$, and $V_c = d_0 / t_0$ by definition, then a peak is identified at both stations if $\Delta t = \Delta d / V < t_0$. Hence, the requirement is for $\Delta d / V\, d_0 < t_0 / d_0$; $V / V_c > \Delta d / d_0$.

To give some examples, if the lifetime of a peak, t_0, is 20 min, $d_0 = 250$ km $= 5.75°$ (Hargreaves, 1970) and $V = 2°$ min^{-1}, then $V_c = 12.5$ km min^{-1} or $0.29°$ min^{-1}. The event would cover the narrow baseline in only 3 min, but would take 40 min to cross the wide baseline during which time it would have changed form sufficiently to prevent recognition at the second station. (In this case, $V / V_c = 7$, which is less than $\Delta d / d_0 = 14$.) Naturally some variations of d_0 and t_0 must occur, and slowly moving events are observed over wide baselines very occasionally: for instance, a 2 h delay over 5 h LT, giving $V = 0.68°$ min^{-1} and hence $V_c < 0.05°$ min^{-1}, is not unknown. Some peaks take as long as 30 min to cross the 250 km narrow baseline. For these, $V_c < 0.19°$ min^{-1}, of the same order as the value deduced above $(0.29°$ min$^{-1})$; such slow events will have decayed long before the second station of a wide baseline is reached.

We conclude from this comparison that wide-baseline observations select the faster peaks at the expense of the slower ones, and that the eastward median velocity of absorption peaks is nearer to $1°$ min^{-1}, as determined from short-baseline studies, than to $3°$ min^{-1} as is usually quoted. The parameters required in the theories based on gradient-curvature drift and electric fields would be correspondingly smaller than those quoted in Section 1.2.

2.2. MERIDIONAL MOVEMENTS IN THE MIDNIGHT SECTOR

Near midnight the dominant movements are meridional rather than zonal. Since the onset originates at about $L = 5.5$, it is expected that at $L = 7$ the onset will be moving poleward. Observations with the multiple riometer systems at Great Whale River and Byrd confirm this expectation: during February 1966 to January 1967, only 4% of the onsets during 2100 to 0100 LT showed equatorward movement, and the median time difference over 250 km was 3.3 min giving a median poleward velocity of 1.3 km s^{-1} $(0.7°$ lat min$^{-1})$. However, study of a number of particular events shows that the direction of movement of individual peaks within the event bears no relation to the direction of movement of the onset, although the velocities are of similar magnitude. In fact, near midnight the peaks tend to move towards the equator (at $L = 7$) – that is, in the opposite direction to the moving onset. A clear example is shown in Figure 3. It is fairly clear from this and other events in the night sector that successive peaks within an event are not just repeated substorms but reflect a dynamic structure within the substorm and it appears that at a given place the main function of the onset is to connect the ionosphere to some region of the magnetosphere which is by this time already active.

3. Conclusions

Although the basic pattern of motion of the substorm in AA is fairly well established in terms of the existing networks of riometer stations, in matters of detail many questions remain. The variability of velocity from event to event creates some difficulties when an average or median picture is being derived, and is important in regard to the theory of substorm propagation since it is necessary to explain the range of observed values as well as the median. There are indications that a serious observational bias may be introduced in wide-baseline studies of absorption peaks, and that the slower features tend to be neglected because of their limited lifetime. Finally, it is clear that the structure within the substorm is complex and that its dynamics are not necessarily related to those of the substorm as a whole.

References

Ansari, Z. A.: 1965, *J. Geophys. Res.* **70**, 3117.
Driatsky, V. M.: 1969, *Geomag. Aeron.* **9**, 398.
Driatsky, V. M. and Shumilov, O. I: 1970, *Geomag. Aeron.* **10**, 235.
Driatsky, V. M., Shumilov, O. I., and Frank-Kamenetsky, A.I.: 1972, *Geomag. Aeron.* **12**, 396.
Eriksen, K. W., Gilmor, C. S., and Hargreaves, J. K.: 1964, *J. Atmospheric Terrest. Phys.* **26**, 77.
Hargreaves, J. K.: 1967, *J. Atmospheric Terrest. Phys.* **29**, 1159.
Hargreaves, J. K.: 1968, *J. Atmospheric Terrest. Phys.* **30**, 1461.
Hargreaves, J. K.: 1969, *Proc. IEEE* **57**, 1348.
Hargreaves, J. K.: 1970, *Planetary Space Sci.* **18**, 1691.
Holt, O., Landmark, B., and Lied, F.: 1962, *J. Atmospheric Terrest. Phys.* **23**, 229.
Jelly, D. H.: 1968, *Can. J. Phys.* **46**, 33.
Jelly, D. H.: 1970, *Can. J. Phys.* **48**, 335.
Jelly, D. H. and Brice, N. M.: 1967, *J. Geophys. Res.* **72**, 5919.
Kavadas, A.: 1962, *J. Atmospheric Terrest. Phys.* **23**, 170.
Little, C. G. and Leinbach, H.: 1958, *Proc. IRE* **46**, 334.
Little, C. G., Schiffmacher, E. R., Chivers, H. J. A., and Sullivan, K. W.: 1965, *J. Geophys. Res.* **70**, 639.
Pudovkin, M. I. and Shumilov, O. I.: 1969, *Ann. Geophys* **25**, 125.
Pudovkin, M. I., Shumilov, O. I., and Zaitzeva, S. A.: 1968, *Planetary Space Sci.* **16**, 881.
Theander, A.: 1972, *Kiruna Geophysical Observatory Report* 723.

POLAR CAP OPTICAL AURORA SEEN FROM ISIS-2

C. D. ANGER and W. SAWCHUK

Department of Physics, The University of Calgary, Calgary, Alberta, Canada

and

G. G. SHEPHERD

Centre for Research in Experimental Space Science, York University, Toronto, Ontario, Canada

1. Introduction

Although the optical aurora represents one of the least direct ways of observing magnetospheric particles and fields it is nevertheless one of the most powerful. The polar atmosphere acts as a giant scintillator, reproducing by virtue of its connection to magnetospheric field lines the complex fields, patterns, and motions of the magnetospheric plasma. Remote sensing of these optical emissions from a high altitude satellite can provide a detailed snapshot of the magnetosphere as it projects into the polar atmosphere.

Although the details of the resulting pattern are clearly complex, as is the mapping itself, the aurora in one sense can be regarded as very simple. It consists of arcs and regions of diffuse aurora. The correspondence between these auroral features and their counterparts in the magnetosphere is of the greatest importance. The broad extent and importance of the diffuse aurora was first recognized by Sandford (1963). This 'mantle aurora' accounts, at least in part, for the 'diffuse auroral belt' which appears as a dominant feature in satellite pictures of the aurora (Lui and Anger, 1973). It is now fairly well established that the magnetospheric counterpart of the diffuse aurora is the plasma sheet, with the possible exceptions of the noon sector (Lui *et al.*, 1973) and the diffuse polar cap auroras described below.

The magnetospheric counterpart of the auroral arcs has not yet been identified, except perhaps in relation to regions of electric field reversal.

In this paper we shall describe the variety of different auroral features which has been observed inside the auroral oval with the two photometers on the ISIS-2 satellite. The morphology of auroras in this region places significant constraints on any theory of magnetic field configuration and magnetospheric convection over the polar cap. As well, the simple fact that in this region auroras occur which are observationally similar to those which occur within the auroral oval is profoundly significant.

The two instruments on the ISIS spacecraft scan the Earth from horizon to horizon at wavelengths of 3914, 5577 and 6300 Å, providing a complete mapping of a large part of the auroral zone during one favorable pass. Details of the instruments are given in Anger *et al.* (1973) and Shepherd *et al.* (1973). For purposes of this paper the data will be presented in pictorial format, similar to that described in Lui and Anger (1973). We shall consider the polar cap auroras in four categories, two of which have not been previously identified in ground-based and aircraft studies. An additional

B. M. McCormac (ed.), Magnetospheric Physics, 357–366. All Rights Reserved.
Copyright © 1974 by D. Reidel Publishing Company, Dordrecht-Holland.

category, the dayside cusp aurora, is described elsewhere (Shepherd and Thirkettle, 1973) and will not be considered here.

1.1. MIDNIGHT POLEWARD-EXPANDED AURORA

Akasofu (1968) noted the poleward expansion of auroras in the midnight sector as an essential part of his description of the 'auroral substorm'. The amount of expansion which can occur is dramatically illustrated by a substorm seen by the ISIS instruments on December 22, 1971 and reproduced in Figure 1, along with a transformation at reduced resolution onto a corrected geomagnetic coordinate grid. The superimposed cg latitude and longitude lines can be used to find the coordinates of any region of interest. The four upper pictures show the 5577 Å data on the left and the 3914 on the right. Different threshold intensities have been used in the four pictures in order to bring out details in both weak and intense features. The bright sloping lines running down the pictures are the Earth's limbs as seen during each spin of the spacecraft in the 5577 Å channel. The picture is made up of individual spin strips, each 100 to 200 km wide at the center of the field of view. The fact that features in some regions of the field of view (e.g., the limbs) present a zig-zag appearance is due to the fact that distant regions are seen for more than one spin. Generally in these pictures noon is at the top, midnight at the bottom, dawn at the right and dusk at the left. Time runs from left to right and down. The bright stripe across the third strip from the top is an internal calibration. The bright oval-shaped region surrounding the polar cap is the region of diffuse aurora mentioned earlier. Inside (poleward of) the main part of the diffuse aurora are several discrete auroral arcs or bands exhibiting characteristics typical of auroral arcs during substorms.

Although some of the apparent spatial features in the picture may actually be due to time variations, since a total of about 13 min was required to scan across the oval, it is very unlikely that such effects could account for the main features described below.

(1) There is an abrupt transition in intensity in the arc at A, but note also that the arc is continuous across A. Another abrupt drop followed by a rise in intensity occurs at B. This behavior is frequently observed in auroral arcs during substorms. Localized bright arc segments or sharply defined bright regions within arcs are much more common than long arcs of uniform intensity.

(2) At C the arc coming from the morning sector makes a very sharp bend and

---

$\rightarrow$

Fig. 1. (a) The top four pictures show 5577 Å (left) and 3914 Å (right) data from the ISIS auroral scanner reproduced in pictorial form for the pass on Dec. 22 1971 at 0440 UT. Letters indicate features described in text. Lower threshold intensities for the four pictures are, respectively, about 0, 0.2, 1, and 3 kR. Intensities more than 1 kR above the lower threshold are displayed at maximum brightness. (b) The bottom pair of pictures, one with coordinate lines and the other without are a transformation of the 5577Å data from the same pass onto a polar projection in corrected geomagnetic coordinates. Geomagnetic midnight is indicated by the letter M (at corrected geomagnetic longitude 5° E). Latitude lines are shown at 10° intervals and longitude lines at 30° intervals.

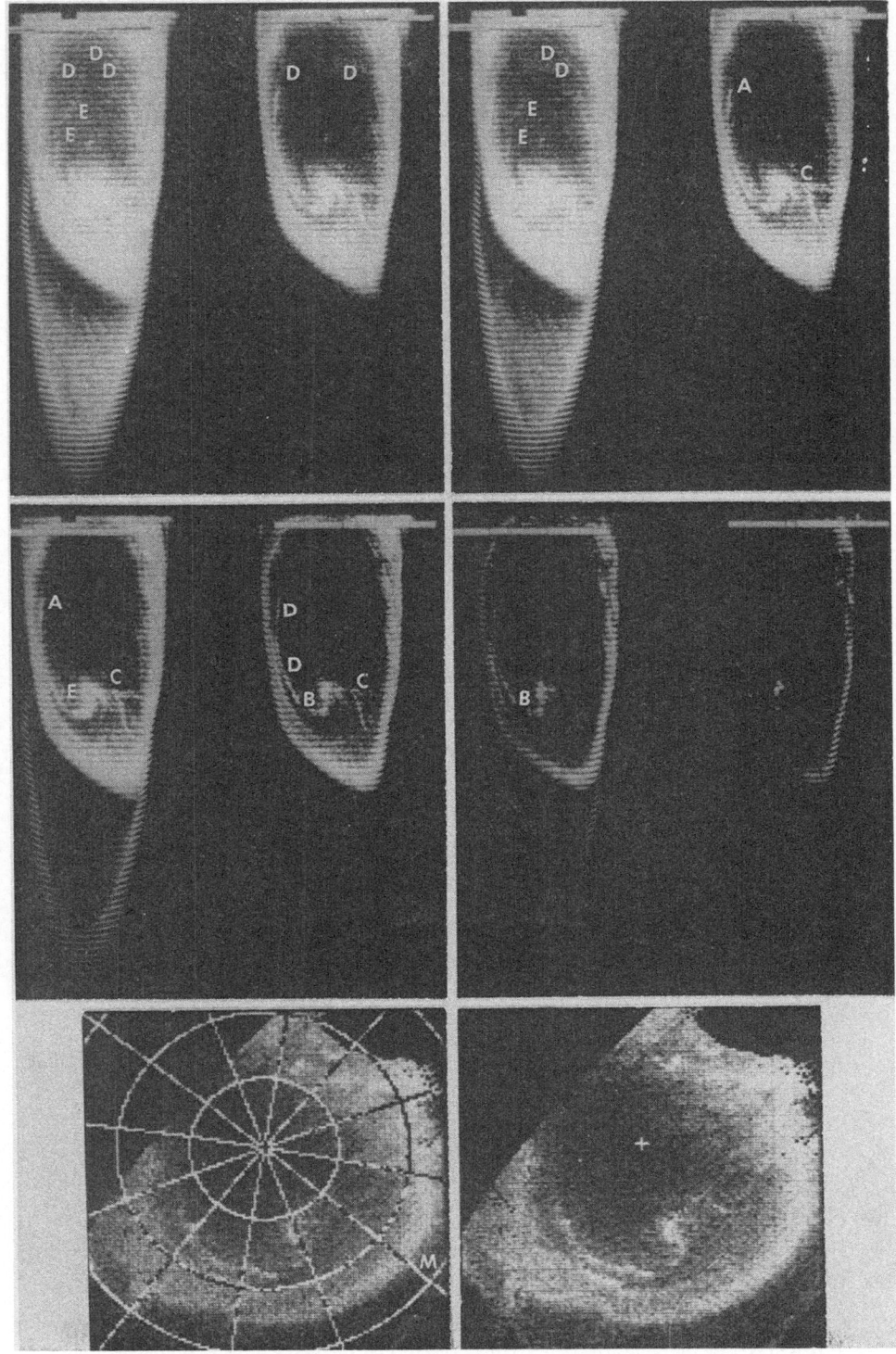

then continues on, producing a discontinuity at midnight. Similar discontinuities are usually observed in the midnight region during substorms.

(3) In this pass we note a special feature which we have not observed before. It is the faint extension of two arcs which originates in the midnight sector and extends right into the central polar cap, the outermost forming a continuous circle around the polar cap. These are indicated by D and E in Figure 1a.

1.2. SUN-ALIGNED 6300 Å ARCS

Studies based on all-sky camera data have led to the conclusion that auroral arcs in the polar cap exhibit a negative correlation with magnetic activity (Davis, 1963), and that the forms tend to be aligned so that they point toward the Sun, rather than being aligned with the auroral oval (Gustaffson, 1967). Although there is room for some ambiguity here, since arcs which lie along the oval in the evening and morning sectors are also aligned toward the Sun, the evidence seems compelling that the Sun-aligned arcs must be a distinct polar cap phenomenon. Eather and Mende (1971) have shown by means of airborne photometric observations that polar cap auroral forms have high ratios of 6300 Å to 5577 Å output. Three examples of 6300 Å arcs, all during magnetically quiet times, are shown in Figure 2. These arcs are not visible in the 5577 Å data. In one case the arc starts in the vicinity of the dayside cleft and extends almost to the midnight aurora. The other two cases are not as well defined but would probably be classified as Sun-aligned and in the polar cap if seen on an all-sky camera. K_p values for these three events were low $(1-, 0+, 0+)$. December 14 was a QQ day $(\sum K_p = 5)$, and $\sum K_p$ for December 16 was 10.

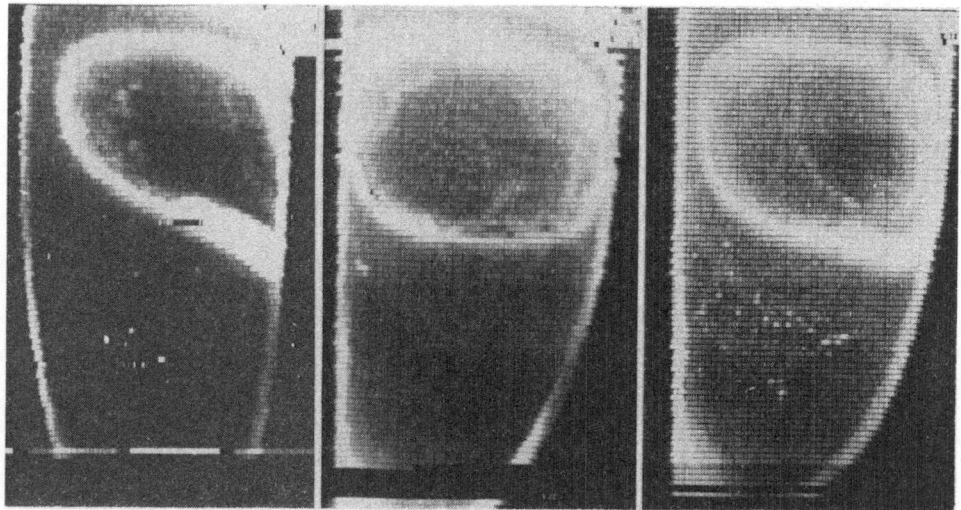

Fig. 2.   Three cases of polar cap Sun-aligned 6300 Å arcs on Dec. 14 1971 at 0715 UT (left) Dec. 16 1971 at 0245 UT (middle) and 0445 UT (right). The format of the pictures is similar to Figure 1a, but there is no resolution within the individual strips. Noon, evening, midnight, and morning are, respectively, at the top, left, bottom, and right of the pictures.

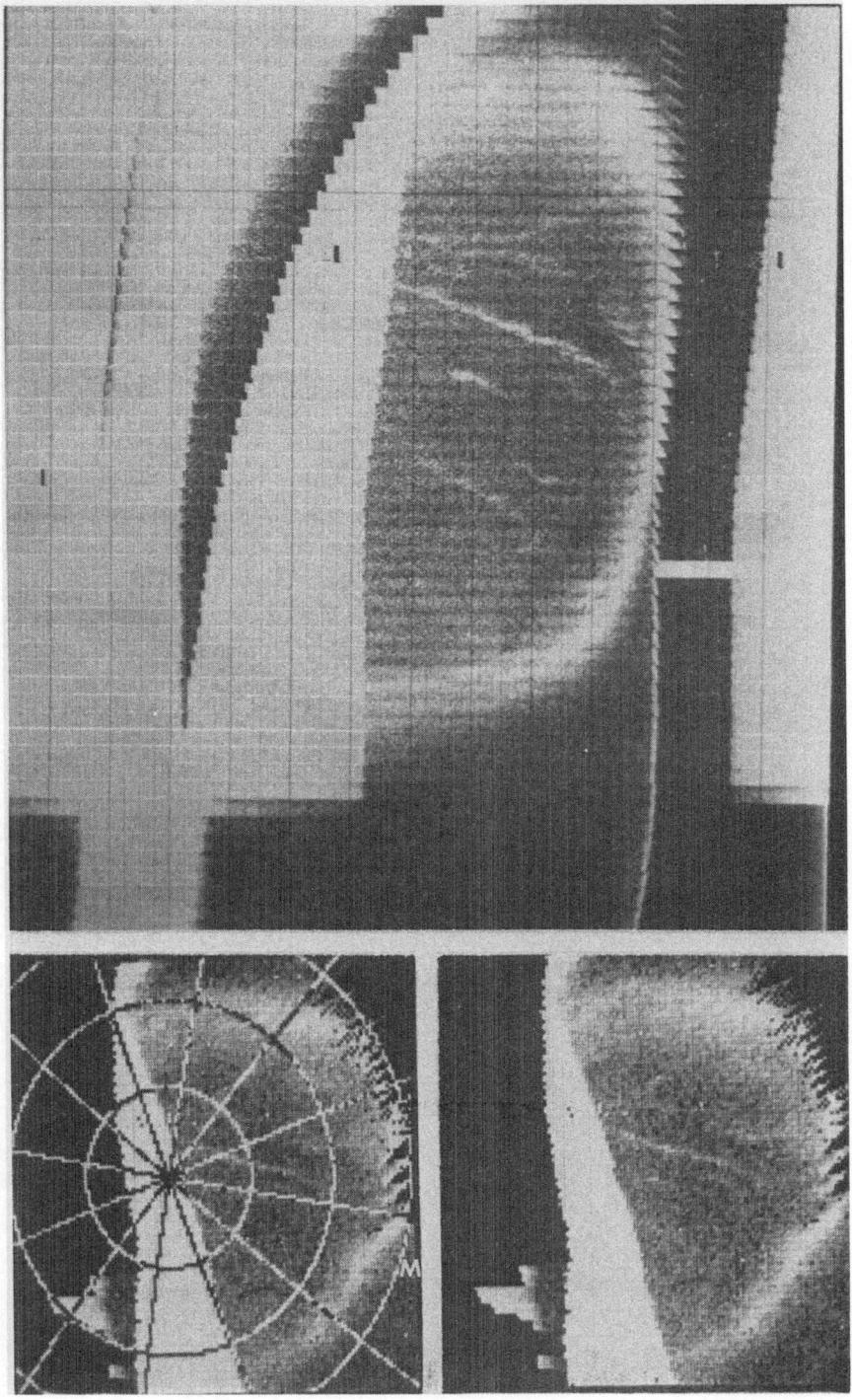

Fig. 3.   Sun-aligned arcs in the polar cap on Jan. 21, 1972 at 2330 UT. Format is similar to Figure 1 except that only one high resolution picture at 5577 Å is shown and the local time of the orbit plane has shifted. The midnight auroral zone lies off the right hand limb. Midnight is at 81 °E (see Figure 1).

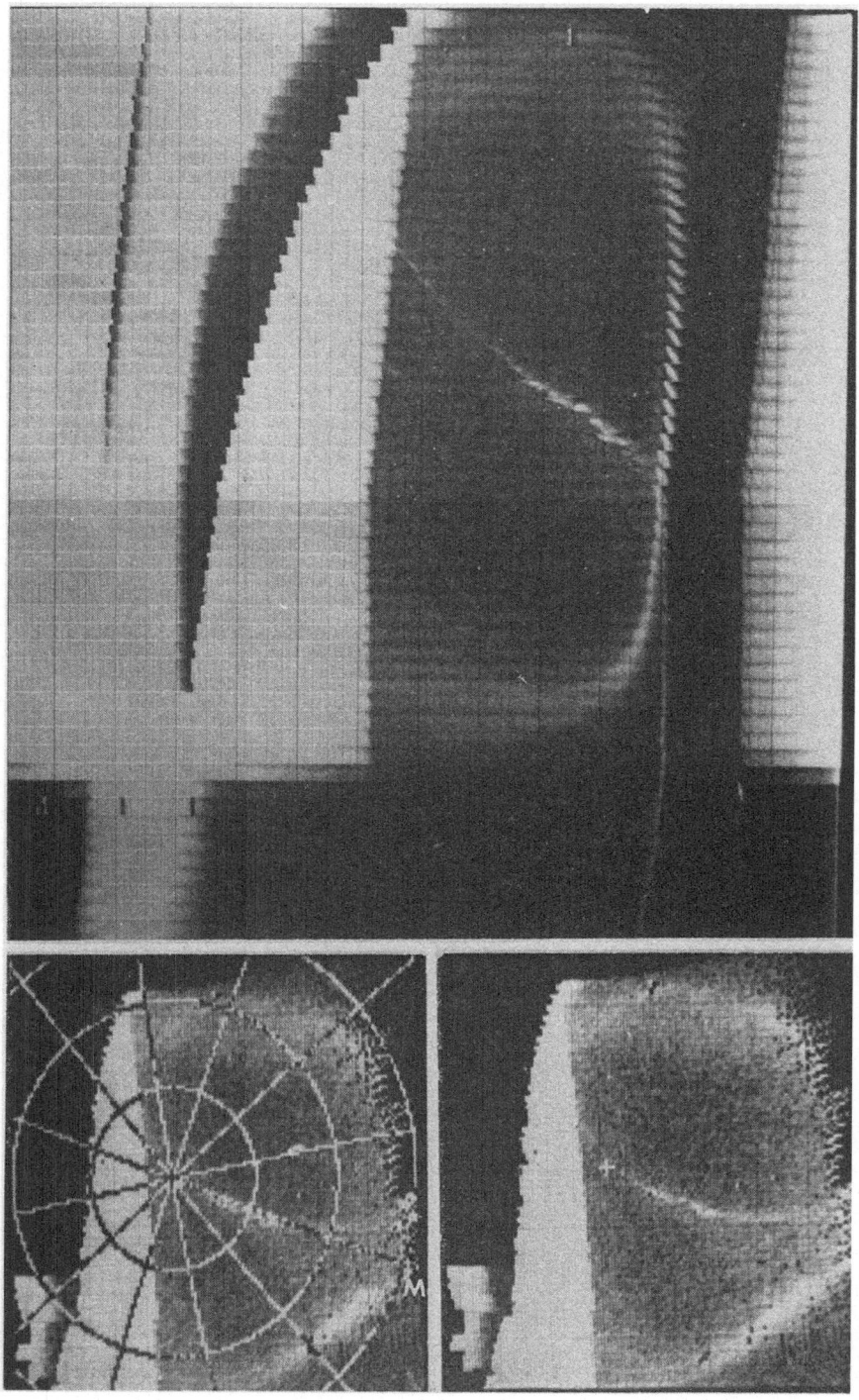

Fig. 4.   Same as in Figure (3), except for the immediately following orbit at 0130 UT. (Jan. 22, 1972).
Midnight is at 56°E (see Figure 1).

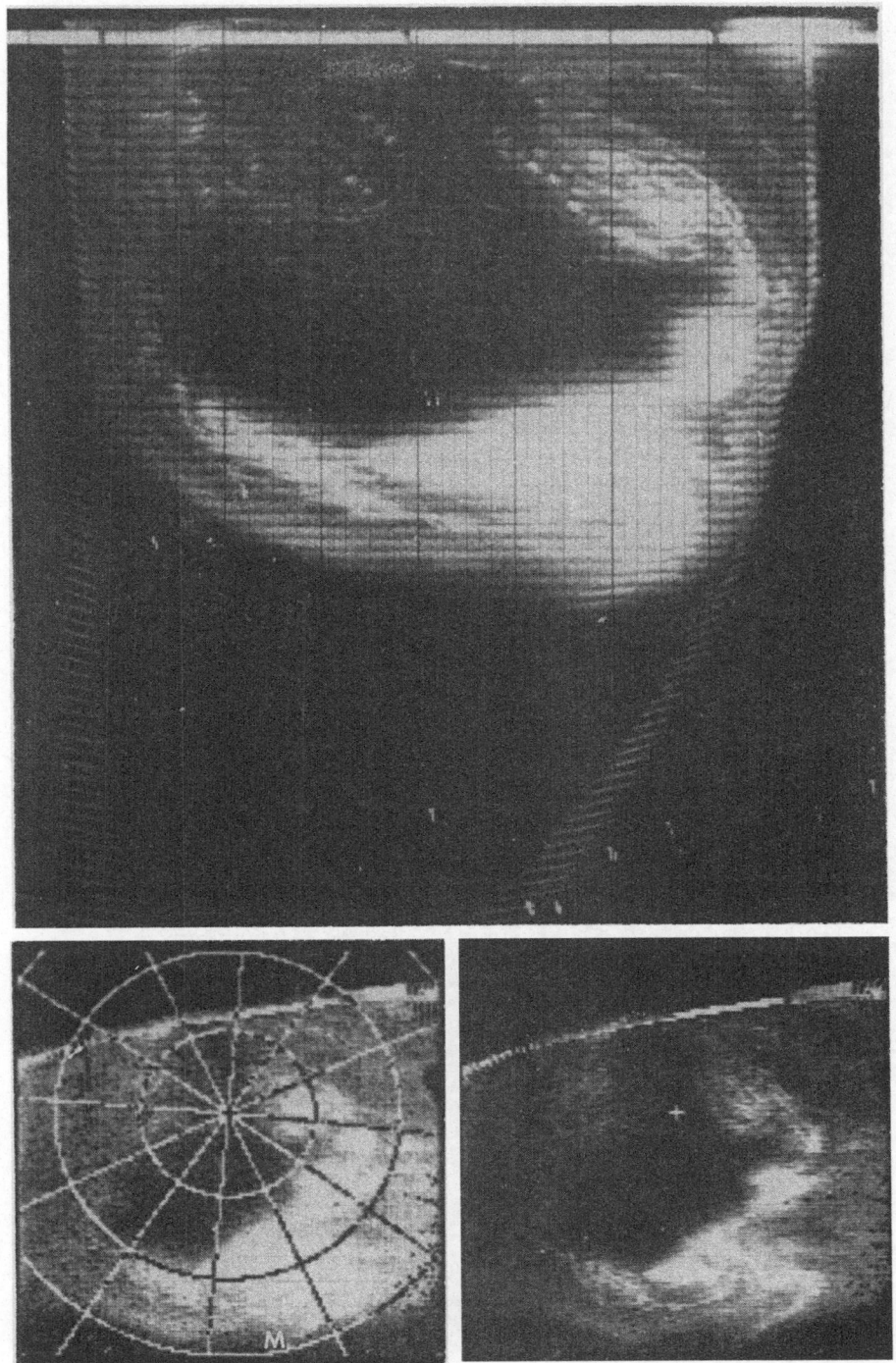

Fig. 5.   A complex spiral arc structure at very high latitudes in the late morning sector at 0410 UT during the intense magnetic storm of Dec. 18, 1971. The same spiral form is seen during passes 2 h before and 2 h after. Format is similar to Figure 1 except that only one high resolution 5577 Å picture is shown. Midnight is at 15°E (see Figure 1).

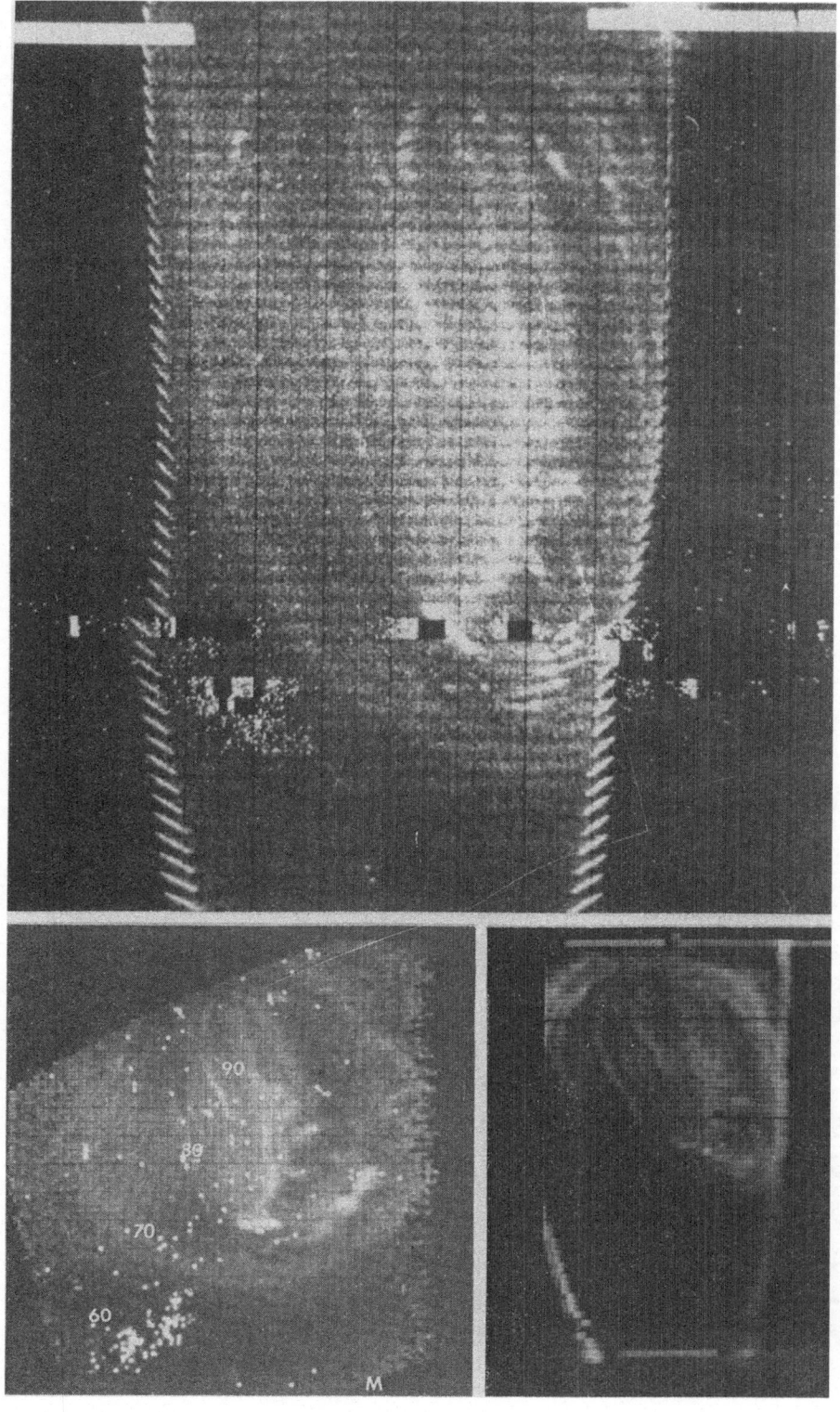

Fig. 6. A representative pass (0520 UT Dec. 20, 1971) showing the widespread Sun-aligned discrete and diffuse aurora in the polar cap following the intense magnetic storm of Dec. 17–18, 1971. Midnight is at 8°W (see Figure 1). Format is the same as that for Figure 5 except that the lower right picture shows the corresponding 6300 Å data (similar to Figure 2)

1.3. SUN-ALIGNED 5577 ARCS

On rare occasions we have observed Sun-aligned arcs in the polar cap which are prominent in the 5577 Å/3914 Å data. Observations of one such occurrence spanning at least two consecutive passes are given in Figures 3 and 4 (5577 Å). The local time for the center line of these pictures is different from that of the previous passes, which accounts for the alignment of these arcs in the picture. They are Sun-aligned, and the most intense region in the arc in Figure 4 exceeds 10 kR in intensity.

1.4. DIFFUSE AND DISCRETE AURORAS IN THE POLAR CAP ASSOCIATED WITH A MAJOR MAGNETIC STORM

The major geomagnetic storm of December 17–18, 1971, and its aftermath extending through December 21, produced dramatic auroras in the polar cap. These are described in detail elsewhere (Anger and Lui, 1973; Anger, 1973; Bunn *et al.*, 1973) and will only be summarized here. The polar cap during three successive passes on December 18 was dominated by a complex spiral auroral arc structure connected to the morning arcs along the oval and extending up to very high geomagnetic latitude. The 5577 Å data for the pass at 0410 is shown in Figure 5.

For three days following the storm, auroras in the form of Sun-aligned discrete and diffuse forms were abundant in the polar cap during each pass of the satellite for which data are available (one or two passes per day). Substorm activity in the midnight sector appears to have been absent during these times. The 5577 Å data for the pass starting at 0525 UT on December 20 is shown in Figure 6 along with the corresponding 6300 Å data in a slightly different format. The same features are present in the 6300 Å data, but they are less intense than the 5577 Å emissions. We do not know how rare such an event as this is. It clearly does not fit very well with ground-based descriptions of polar cap auroras, although it is quite consistent with the polar cap 'squalls' observed by Winningham and Heikkila (1974) in their particle data. Threshold and contrast effects in all-sky cameras could easily account for the fact that there have not been previous reports of the polar cap diffuse aurora, and any such discrete features would probably have been combined statistically with the quiet-time Sun-aligned arcs which seem on the basis of our results to be primarily a 6300 Å feature. Lassen (1973) has reported the presence of arcs in the polar cap at all K_p values.

2. Discussion

As is amply evident from the data presented here, auroral displays are highly varied. However, observations, whether from above or below, indicate a striking similarity between auroral arcs wherever they are seen – in the middle of the polar cap or at the equatorward boundary of the oval, at noon or midnight, morning or evening. It is apparent that arcs can occur anywhere in the polar cap. The fact that they do occur in regions that map into totally different parts of the magnetosphere is profoundly significant. It indicates that these disparate regions possess in common some process that leads to the formation of linear regions of particle energization or pitch angle scattering within which abrupt spatial increases and decreases of intensity occur. The fact that arcs often fold back on themselves and exhibit very complex small-scale structures is probably best viewed as a secondary process, occurring while the particles are in transit to the atmosphere (Davis and Hallinan, 1973; Hallinan and Davis, 1970). However, the frequently observed abrupt change in the over-all direction of an arc (observed mainly in the midnight sector) may provide a significant clue as to the production mechanism. If the explanation of the configurations and motions of auroral arcs in relation to magnetospheric convection is regarded as a first-order problem – one which is now receiving much attention – then the 'zero-order' problem of explaining the actual existence of auroral arcs themselves (Atkinson, 1970) ought to be considered of equal or greater importance. It seems reasonable to associate the arc with some form of discontinuity in magnetospheric convection. If this is the case, then two or more different mechanisms must operate to energize particles at such a boundary, one strong and one weak, in order to account for the abrupt intensity changes observed along some arcs.

The different large scale patterns of auroras in the polar cap appear difficult to reconcile with any one model of magnetospheric convection. In one case we have an arc extending from the midnight sector and forming a circle across the polar cap; in another, extended Sun-aligned discrete and diffuse forms; and yet a third configuration in which a complex spiral extends well up into the polar cap from the morning hours.

Acknowledgments

The authors wish to acknowledge the assistance of B. Wray, R. Sidebotham and J. Gosling in producing the scanner pictures.

References

Akasofu, S.-I.: 1968, *Polar and Magnetospheric Substorms*, D. Reidel Publishing Company, Dordrecht-Holland.
Anger, C. D.: 1973, in D. Venkatesan (ed.) *Solar Terrestrial Relations*, The University of Calgary, Calgary, Canada, p. 617.
Anger, C. D. and Lui, A. T. Y.: 1973, *J. Geophys. Res.* **78**, 3020.
Anger, C. D., Fancott, T., McNally, J., and Kerr, H. S.: 1973, *Appl. Opt.* **12**, 1753.
Atkinson, G.: 1970, *J. Geophys. Res.* **75**, 4746.
Bunn, F. E., Gordon, K. S., and Shepherd, G. G.: 1973, *Trans. Am. Geophys. Union* **54**, 418.
Davis, T. N.: 1963, *J. Geophys. Res.* **68**, 4447.
Davis, T. N. and Hallinan, T. J.: 1973, IAGA Bulletin, No. 34, Programs and Abstracts for the Second General Scientific Assembly – Kyoto, p. 434.
Eather, R. H. and Mende, S. B.: 1971, in B. M. McCormac (ed.), *The Radiating Atmosphere*, D. Reidel Publishing Company, Dordrecht-Holland, p. 255.
Gustaffson, G.: 1967, *Planetary Space Sci.* **15**, 277.
Hallinan, T. J. and Davis, T. N.: 1970, *Planetary Space Sci.* **18**, 1735.
Lassen, K.: 1973, Danish Meteorological Institute, Geophysical Papers R-33.
Lui, A. T. Y. and Anger, C. D.: 1973, *Planetary Space Sci.* **21**, 799.
Lui, A. T. Y., Anger, C. D., Perreault, P., and Akasofu, S.-I.: 1973, *Planetary Space Sci.* **21**, 857.
Sandford, B. P.: 1963, *J. Atmospheric Terrest. Phys.* **26**, 749.
Shepherd, G. G. and Thirkettle, F. W.: 1973, *Science* **180**, 737.
Shepherd, G. G., Fancott, T., McNally, J., and Kerr, H. S.: 1973, *Appl. Opt.* **12**, 1767.
Winningham, J. D. and Heikkila, W. J.: 1974, *J. Geophys. Res.* **79**, 949.

REMARKS ON THE GROWTH PHASE OF SUBSTORMS

L. ROSSBERG

Max-Planck-Institut für Aeronomie, Institut für Stratosphären-Physik, 3411 Lindau/Harz, Germany

Abstract. A pre-bay poleward expansion of > 30 keV electron intensities near midnight is shown to be coincident with the maximum phase of a plasma sheet expansion in the predawn sector of the Vela orbit. A study of simultaneous particle observations by ATS 1 and 5, and of perturbations of the horizontal component of the magnetic field on ground in the auroral zone, shows that one can arrive at two different conclusions about the relevant substorm phase. Based on the results of a multi-satellite study it is suggested to supplement the search for growth phase phenomena by a more general consideration of the varying modes of energy release into the auroral zone.

1. Introduction

Using ground based observations of magnetic field variations and of the patterns of auroral luminosity it was possible quite early to define two phases of a magnetospheric substorm, e.g., the breakup phase and the recovery phase (Akasofu, 1968). For a better understanding of the whole phenomenon, however, it became necessary to also study the developmental phase in order to determine what triggers the disturbance. This aspect received increasing attention recently with the secondary goal of being able to predict the occurrence of the breakup phase. The study of this developmental phase, called the "growth phase" by McPherron (1970), appeared at first quite promising. Including observations in the magnetosphere and in the interplanetary medium the UCLA group developed a phenomenological model of the substorm and its growth phase (McPherron, 1974). A schematic representation of the magnetic variations during magnetospheric substorms is presented in Figure 1 of McPherron's paper. In short the expected sequence of events during a growth phase is as follows: southward turning interplanetary magnetic field; enhanced erosion of the magnetopause; increased transport of magnetic flux towards the tail; increasing magnetic energy density in the lobes of the tail; thinning of the plasma sheet associated with an earthward convection of plasma. Near the Earth a dawn-dusk asymmetry develops owing to the different directions of gradient drifting protons (westward) and electrons (eastward). This results in the pre-midnight area in the formation of a partial ring current closing through the ionosphere as an eastward electrojet. Effects of the partial ring current are seen at synchronous orbit and mid-latitudes as a depressed field.

While all these phenomena are considered to represent the growth phase, the expansion phase begins when the near-Earth plasma sheet becomes extremely thin and an X-type neutral point forms within the plasma sheet. The cross tail current is disrupted by the neutral point and diverted through the ionosphere as a suddenly enhanced westward electrojet.

In discussing the criticism which the UCLA phenomenological model of substorms has received McPherron pointed out that the model is inadequate in several respects.

B. M. McCormac (ed.), Magnetospheric Physics, 367–376. All Rights Reserved.

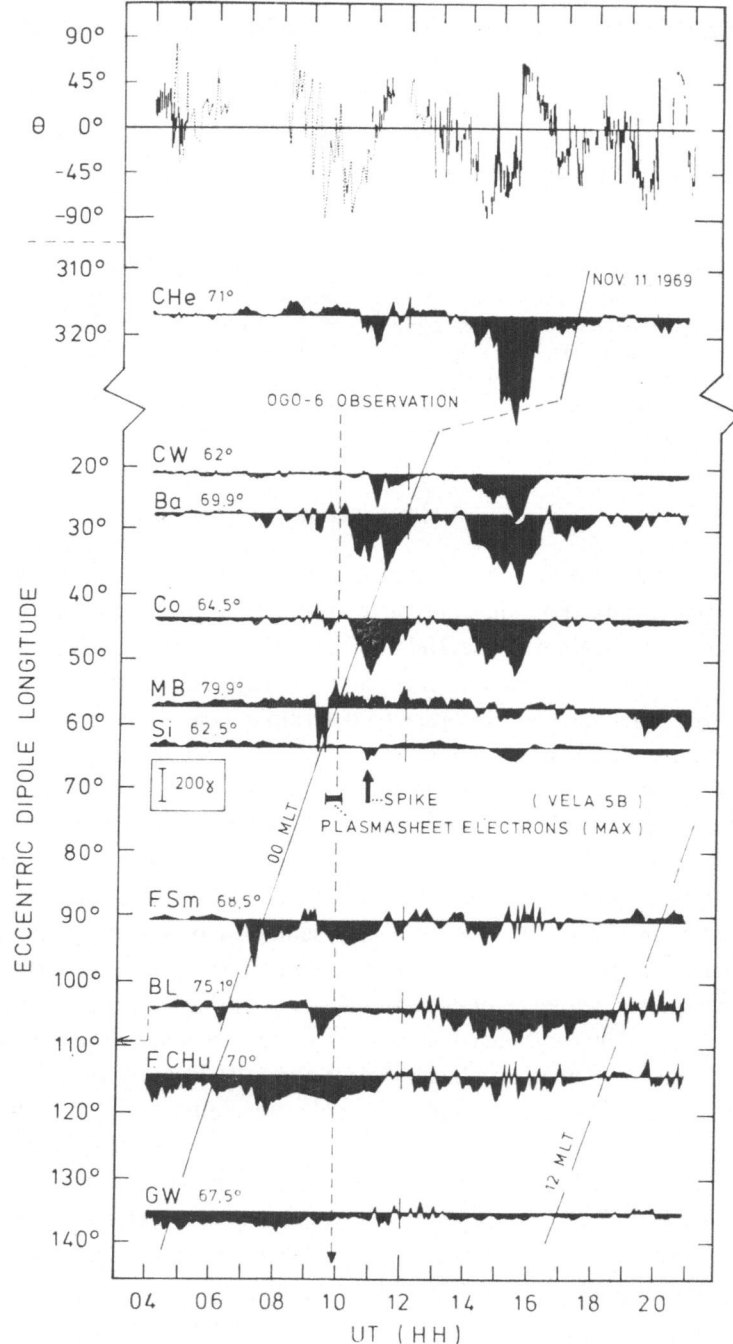

Fig. 1. Latitudinal angle θ of the interplanetary magnetic field is shown on top of the figure. The solid line is used for data from HEOS-1; the dotted line for data from EXPLORER-35. H traces observed at several auroral zone and polar cap observatories drawn to a scale of 200 γ per unit length are shown in the lower part of the figure. The base lines are drawn at the eccentric dipole longitude (Matsushita and Campbell, 1967) of the observatory, and its IN Lat. coordinate is given after the abbreviation of the observatory's name. A 1 h per division UT scale is given at both the top and bottom of the figure. Lines of constant MLT (eccentric dipole) are shown for midnight (solid line) and noon (broken line.) The vertical broken line is inserted at the time of the OGO-6 observation discussed in the text. The bar and the arrow refer to the Vela observation shown in Figure 5.

The most important is that the tail lobe and plasma sheet do not always react as single entities as previously supposed.

The observations reported here lend support to this hypothesis.

2. Observations

The observations were made on November 11, 1969, at the end of a 3 h period during which repetitive high latitude pre-midnight injections of 40 keV electrons occurred and prior to the onset of a magnetic substorm (Rossberg *et al.*, 1973a).

Figure 1 shows the variation of the latitudinal angle θ of the interplanetary magnetic field and a compilation of the H-perturbations from 04 to 20 UT. The magnetic

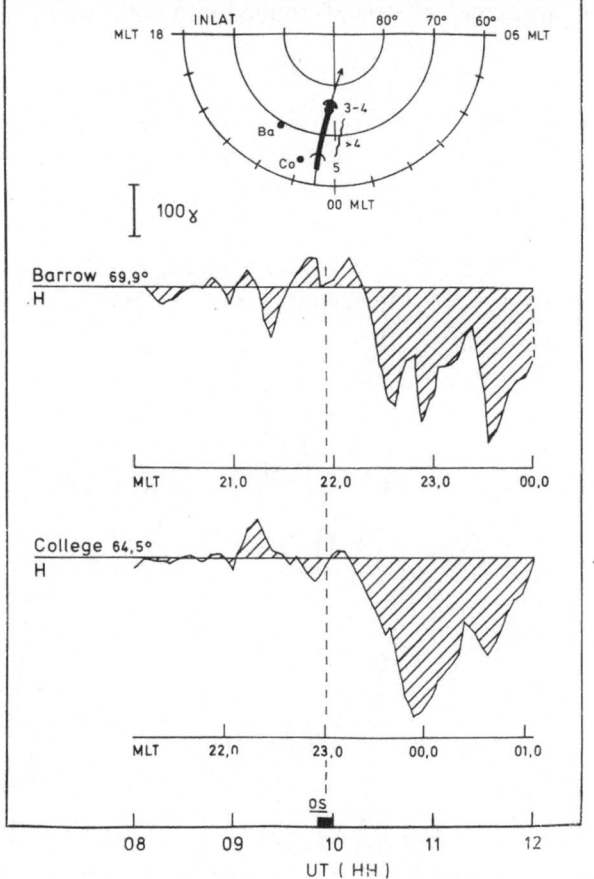

Fig. 2.   Expanded view of the H traces observed at College and Barrow shown in Figure 1. Separate MLT scales are given for each magnetogram. The inset shows in IN Lat–MLT coordinates the trajectory of a southern hemisphere pass of OGO-6 during the time interval given by a black bar in the UT scale. The umbrella-like symbol gives the position of the background boundary and the semicircles give intensity levels of 30 keV electrons in powers of 10. Enhanced precipitation is indicated by a heavy line along the trajectory. Also shown is the position of the two observatories at 09:56 UT. The vertical broken line is drawn at the midpoint of the time interval of the satellite observation.

L. ROSSBERG

substorm in question is observed near midnight by the College (Co) and Barrow (Ba) observatories from 10 UT to 12 UT. The vertical broken line indicates the time (09:56 UT) when, at a near conjugate point of the southern hemisphere, OGO-6 observed a displacement of the 40 keV electron boundary considerably poleward of the average boundary position which is near 71° (Mc Diarmid and Burrows, 1968). Figure 2 shows information on the OGO-6. The symbols used in the figure are explained in the figure caption. Also shown are the locations of the magnetic observatories (College and Barrow) at the time of the satellite observation and an enlarged view of the H-perturbations observed at the two stations. The detailed latitudinal intensity profiles of > 30 keV, > 100 keV and > 300 keV electrons is shown in Figure 3. At the time of this observation Vela 5B observed maximum intensities of 30 keV electrons in the dawn sector of its orbit, 10 R_E below the neutral sheet. Figure 4 shows the trajectory of Vela 5B rotated into both the equatorial plane and

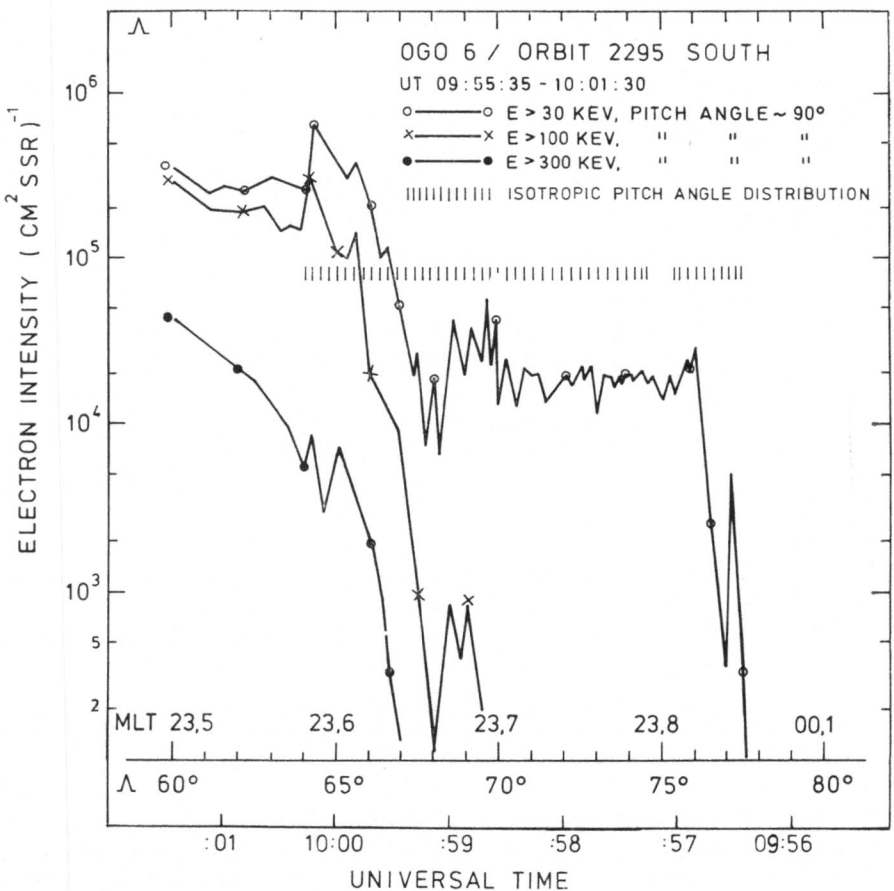

Fig. 3.   Intensity vs. latitude profile of > 30, > 100 and > 300 keV electrons at 90° pitch angle as observed by OGO-6. The hatching indicates the range of an isotropic pitch angle distribution of the 30 keV electrons.

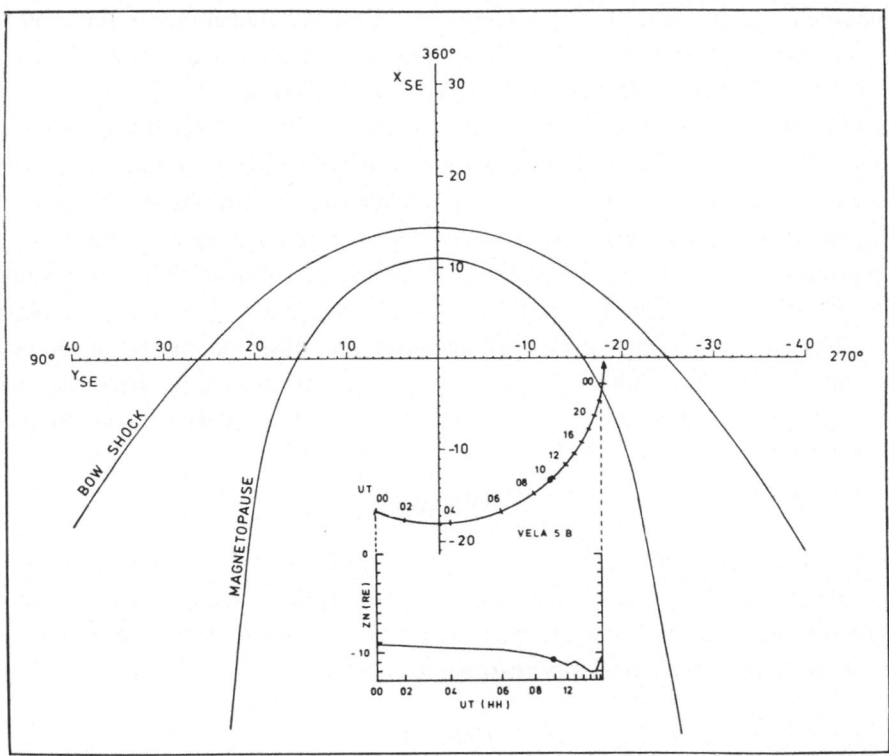

Fig. 4. Trajectory of Vela 5B on November 11, 1969, rotated into the XY$_{SE}$ plane. UT is given along the trajectory. The inset shows the distance from the neutral sheet. The two dots indicate the position at the OGO-6 observation.

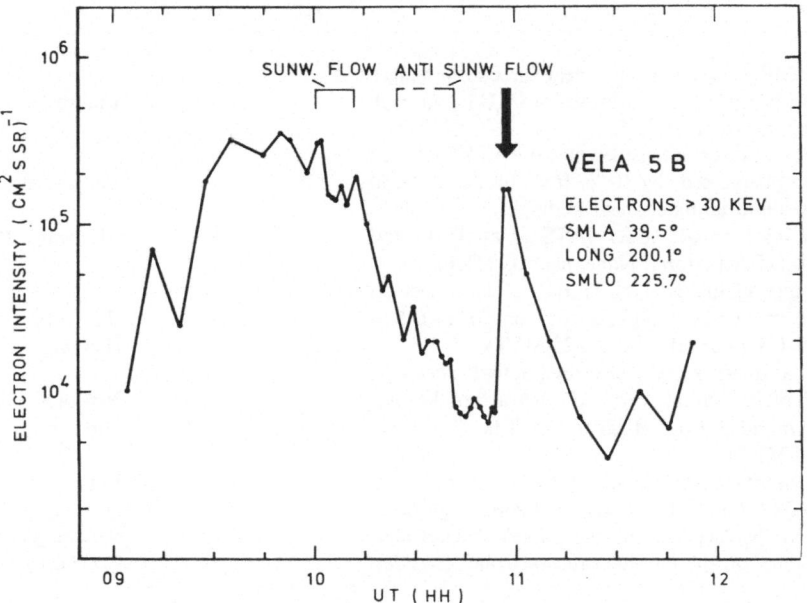

Fig. 5. 30 keV electron observation by Vela 5B. Periods of sunward and anti-sunward flow of low energy protons are given above the intensity curve. The UT position of the arrow is the same as in Figure 1.

projected onto a plane perpendicular to the Sun-Earth line. Dots indicate the position of the satellite at the time of the OGO-6 observation. A detailed plot of the > 30 keV electron intensity observed by Vela 5B is shown in Figure 5. The horizontal bar in the middle of Figure 1 refers to the time when Vela 5B observed maximum electron intensities. An inspection of Figures 1 and 2 shows that the two satellite observations are made at a time when the magnetic activity near midnight is at a minimum, shortly before the onset of the magnetic substorm. While the OGO-6 observation indicates an expansive phase, the Vela observation can be interpreted either as a 'thickening' of the plasma sheet near the end of the recovery phase of a substorm or as the beginning of the growth phase generally associated with the subsequent thinning of the plasma sheet. The slowly varying negative H-perturbations observed in the dawn sector (Fort Smith and Fort Churchill) and an associated small positive perturbation at Chelyuskin, in the dusk sector, is in accordance with McPherron's definition of the growth phase.

3. Discussion

It was shown that observations of magnetic perturbations on ground, satellite observations of 30 keV electron intensities over the auroral zone near midnight and in the plasma sheet lead to different conclusions about the relevant substorm phase. In view of the simple and physically reasonable substorm model and its inadequacies,

TABLE I

Satellite and ground based observations preceding the substorm onset at 10 UT, Nov. 11, 1969

Observation (Location)	Expected during a 'growth phase'		Reference
	yes	no	
The interplanetary magnetic field direction changes from northward to southward at 09 UT and back to northward at 10:30 UT.	X		Figure 1
Energetic electron intensities at 6.6 R_E (ATS 1) are slightly depressed (by 50 %) from 08:30 to 09:50 UT (equatorial plane, near midnight).	X		Lanzerotti (1972)
Plasma injection at 6.6 R_E (ATS 5) at 07:30 and 09:50 UT (equatorial plane, near 03 MLT).		X[a]	McIlwain (1972)
Occurrence of intense high latitude 30 keV electron bursts in the premidnight sector from 07:14 UT to 09:21 UT (auroral zone, 21–23 MLT).		X	Rossberg et al. (1973a)
Continual quasi-periodic electrojet activity localized north of 70° IN Lat, near and westward of the midnight meridian from 04 UT to 10 UT (auroral zone, 21–24 MLT).		X	Rostoker (1972) and Figure 1
Slowly varying electrojet activity in the dawn sector from 08 UT to ~ 11 UT (auroral zone, nightside).	X		Figure 1
Poleward displacement of the 30 keV background boundary on the dayside (auroral zone, dayside).		X	Rossberg et al. (1973a, b)

[a] Associated with 'substorms' in general but more likely with the expansive phase.

as presented by McPherron (1974), an attempt is now made to supplement the model by a more general consideration of the varying modes of energy release into the auroral zone.

According to the model the development of a substorm from the beginning of the growth phase to the breakup phase may last 3 to 4 h. It is thus appropriate to include observations made before the bay onset. This information was given in a comprehensive multisatellite study of auroral-zone phenomena on November 11, 1969 (Rossberg et al., 1973a). The observations from 07 to 10 UT (before the bay onset) are compiled from this study in Table I together with an indication of whether the observation is expected during a 'growth phase' or not.

The first two observations clearly indicate a growth phase (growth of magnetic flux transport from the dayside magnetosphere tailward owing to the southward IMF direction and the development of a partial ring current). This phase starts between 08.30 and 09.00 UT and terminates between 09.50 (recovery of energetic electron intensities ATS 1) and 10.30 UT (rapid transition from southward to northward directed IMF). A termination of the growth phase is also indicated by the occurrence of hot plasma at ATS 5 at 09.50 UT and last but not least by the bay onset at College at 10 UT.

The main problem then is how to explain the plasma sheet expansion from 09.00 to 10.00 UT and the poleward displacement of 30 keV electrons both on the night- and dayside of the auroral zone (last item in Table I) observed by OGO-6 at 10.00 UT particularly with regard to the simultaneously southward directed IMF.

Recalling McPherron's statement that the plasma sheet may not always react as a single entity and further that plasma energized at a neutral point is injected both earthward *and tailward* causing an expansion of the plasma sheet, it is interesting to note the fourth observation quoted in Table I. The 30 keV electron burst observations were made during 6 satellite traversals of the auroral zone in the pre-midnight area, centered around 07.20 UT and 09.00 UT (Figure 6). Near 09.00 UT simultaneous observations before and after midnight show that the burst occurrence is confined to the pre-midnight area whereas a $\sim 7°$ wide precipitation zone prevails from midnight towards noon associated with slowly varying CNA on the ground (Rossberg et al., 1973a). Boundary observations change from the average position (Mc Diarmid and Burrows, 1968) near midnight to considerably poleward displaced positions near noon (last item in Table I).

The high intensity of the bursts and their occurrence poleward of the stable trapping region suggests a localized source region near the equatorial plane, capable of heating the ambient plasma and scattering it into the loss cone. The heating process may be physically similar to the one required by McIlwain (1974). Note that the first burst observed precedes the plasma injection quoted in Table I by only a few minutes.

The first Vela observation of 30 keV electrons in the post-midnight sector of its orbit at 09.00 UT occurs ~ 1.5 h after the first and almost time coincident with the simultaneous pre-post-midnight observations quoted above, showing that the occurrence of the second burst is confined to the pre-midnight sector. The appearance

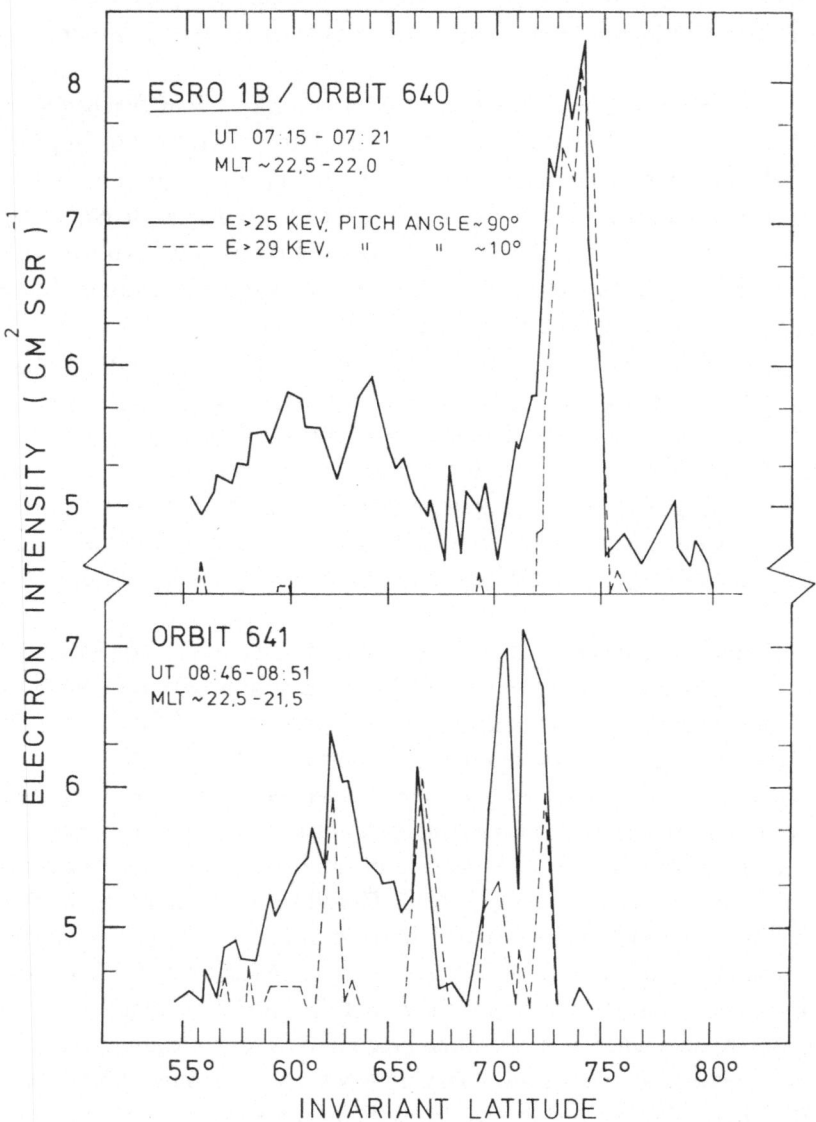

Fig. 6. Observation of high latitude ∼ 30 keV electron bursts observed by ESRO-1B near 07.20
and 09.00 UT.

of 30 keV electrons at Vela's position thus may be due to plasma, energized, injected tailward and drifting eastward from the source region of the high latitude bursts. If the source is an X-type neutral 'point' this implies an unusual Y-dependence of the plasma sheet thickness from a small value before midnight to the observed value after midnight. The possibility of assuming the existence of a neutral line extending across the tail but earthward of Vela's orbit thus avoiding the implication of the large dawn-dusk asymmetry of the plasma sheet thickness appears unlikely in view of the observed localization of the electron burst event at 09 UT in the pre-midnight sector.

The poleward displacement of 30 keV electrons near midnight and possibly also near noon (in the auroral zone) before the bay onset at College may be explained by projecting the high latitude edge of the expanded plasma sheet into the auroral zone (Akasofu *et al.*, 1973). It remains an open question, however, as to how to reconcile these high latitude phenomena observed from midnight towards noon (but not in the afternoon sector) (Rossberg *et al.*, 1973a) with the simultaneously increasing southward component of the interplanetary magnetic field.

Regardless of the fact that the interpretation given above may be wrong the mere occurrence of observations which contradict each other, if interpreted in terms of the established substorm model, requires at least some modifications. This is particularly true of the growth phase. Using observations of 40 keV electrons by Injun 3 O'Brien had already stated in 1964 that the particle source must always be operative. In other words, energy stored in the tail is always released towards the auroral zone. This loss must be balanced by continuous energy input from the solar wind. Both processes are likely to occur at different rates without being closely correlated. This follows simply from the fact that the rate of energy loss is dominated by characteristic time scales of magnetospheric phenomena. Any time interval when the rate of energy input exceeds the rate of energy loss is, in a physical sense, a growth phase. The resulting mode of energy loss (continuous and wide spread causing the mantle aurora, localized impulsive precipitation events or large scale events like typical substorms) depends almost certainly on the growth rate of the difference between energy loss and energy input. The unpredictable nature of this interplay may explain the impossibility of assigning to any substorm unique growth phase features and renders any such attempts useless.

Acknowledgements

The author gratefully acknowledges the encouragement given by Professor G. Pfotzer and thanks Drs D. Colburn, P. Hedgecock, E. W. Hones, C. McIlwain, L. J. Lanzerotti and D. J. Williams for making their satellite data available. Appreciation is expressed to Dr T. Fritz for many helpful comments.

References

Akasofu, S.-I.: 1968, *Polar and Magnetospheric Substorms*, D. Reidel Publishing Company, Dordrecht-Holland.
Akasofu, S.-I., Hones, E. W., Jr., Bame, S. J., Asbridge, J. R., and Lui, A. T. Y.: 1973, *J. Geophys. Res.*, to be published.
Lanzerotti, L. J.: 1972, private communication.
Matsushita, S. and Campbell, W. H.: 1967, *Physics of Geomagnetic Phenomena*, Academic Press, New York and London.
Mc Diarmid, I. B. and Burrows, J. R.: 1968, *Can. J. Phys.* **46**, 49.
Mc Ilwain, C.: 1972, private communication.
McIlwain, C.: 1974, this volume, p. 143.

McPherron, R. L.: 1970, *J. Geophys. Res.* **75**, 5592.

McPherron, R. L.: 1974, this volume, p. 335.

O'Brien, B. J.: 1964, *J. Geophys. Res.* **69**, 13.

Rossberg, L., Lammers, E., Riedler, W., Skovli, G., Søraas, F., Stauning, P., Theile, B., and Thomas, G. R.: 1973a, ESRO Scientific Report No. 23.

Rossberg, L., Riedler, W., Skovli, G., Søraas, F., Stauning, P., and Thomas, G. R. 1973b, Proc. VII. ESLAB Symp.

Rostoker, G.: 1972, private communication.

SPATIAL AND TEMPORAL VARIATIONS OF
ULF DISTURBANCES NEAR $L = 4$ DURING A MAGNETIC STORM

C. G. MACLENNAN, L. J. LANZEROTTI, and H. FUKUNISHI

Bell Laboratories, Murray Hill, N.J., U.S.A.

Abstract. Latitudinal and conjugate observations of ULF variations near $L = 4$ during the December 17, 1971, magnetic storm are discussed. The temporal variations in the latitude of the peak power in the 10 to 15 mHz band may be indicative of plasma density gradients in the magnetosphere. Diurnal variations are observed in the signal coherences between the observations at the several stations.

1. Introduction

Recent advances in magnetospheric plasma wave observations at high latitudes (Samson *et al.*, 1971; Rostoker *et al.*, 1972; Samson and Rostoker, 1972; Samson, 1972) and near the nominal plasmapause location (Lanzerotti *et al.*, 1972, 1973; Fukunishi and Lanzerotti, 1974), together with new theoretical interpretations of magnetospheric micropulsation phenomena (Chen and Hasegawa, 1974a, b) have given additional stimulation for the use of magnetospheric wave phenomena in the interpretation and understanding of physical processes in the magnetosphere. Furthermore, for the foreseeable future, only appropriately sited ground stations will be able to give the necessary information on the geomagnetic scale size of many magnetospheric phenomena.

This brief report presents results of observations of ULF wave phenomena at three magnetometer stations ($L \sim 4.4$, ~ 4.0, ~ 3.2) in the northern hemisphere ($\sim 3°$W geomagnetic) and at a conjugate station (Siple; $L \sim 4.0$) in Antarctica during the magnetic storm of December 17, 1971. The basic magnetometer instrumentation and data acquisition scheme at each station has been briefly outlined in Lanzerotti *et al.* (1972). The magnetic field values, measured for each of the three axes (H, D, Z) to a resolution of 60 mγ, are sampled at 2 s intervals and written in a computer-compatible format on magnetic tape. The results in this paper are derived from 2 h non-overlapping power spectra computed for each of the field components for December 16 to 20, 1971 (only December 18 to 20, 1971, data are available from Siple). Every fourth data point was used and a digital low-pass filter with a cutoff at 50 mHz and a Thomson data window (Thomson, 1971) were applied to the data before obtaining the Fourier coefficients of the time series with a fast Fourier transform algorithm.

2. Spectral Power Variations

The Fourier coefficients of the time series were used to obtain auto- and cross-spectra between the various stations. The variation in intensity of the geomagnetic power as a function of time during December 16 to 20, was determined from the power spectra

B. M. McCormac (ed.), Magnetospheric Physics, 377–383. All Rights Reserved.
Copyright © 1974 by D. Reidel Publishing Company, Dordrecht-Holland.

C. G. MACLENNAN ET AL.

of the four frequency bands 0.6 to 50 mHz, 2 to 5 mHz, 10 to 15 mHz, and 15 to
27 mHz. The latter three bands correspond roughly to the Pc-5, Pc-4, and Pc-3
frequency bands of the present micropulsation classification scheme (Jacobs *et al.*,
1964). For each of the frequency bands, the data at each 2 h interval was normalized
to the northern station with the maximum intensity. In this way, a contour plot, as

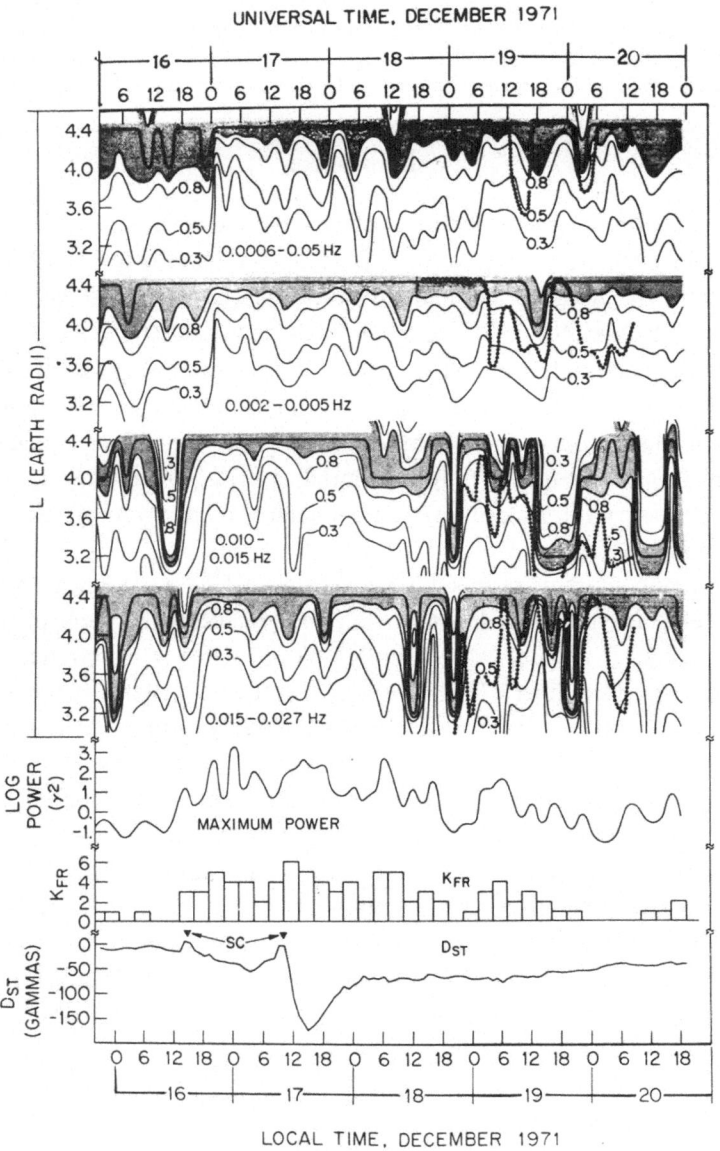

Fig. 1. Contour plots of ULF power in four frequency bands near $L = 4$ during the December 17,
1971, magnetic storm. The $\log_{10}$ maximum power in the 0.6 to 50 mHz band is plotted as a function
of time. Also shown are the Fredericksburg K-index and the equatorial Dst.

a function of time, is obtained of the ULF variations in a particular frequency band. The contours between stations are obtained by linear interpolations.

Plotted in Figure 1 are contour plots of the H-component magnetic power in each of the four frequency bands. The contour interval $\geqslant 0.9$ is indicated by shading in order to more strongly emphasize the temporal variation in the location of the peak power. Of course, because of the limited latitudinal extent of the present station array (which is designed to study the plasmapause), actual peak powers may occur at L values > 4.4 or < 3.2. The contours of the Siple power level (also normalized to the maximum northern power) are indicated by the closed circles. The open circles in these contours indicate periods when the normalized Siple power was > 1. Beneath the power level contours are plotted the time variation of the maximum amplitude power in the 0.6 to 50 mHz frequency band, the Fredericksburg K-index (an United States observatory near the longitude of the stations) and the Dst (Sugiura and Poros, 1973).

The important features evident in the spectral contours of Figure 1 are the following: (a) except for the 10 to 15 mHz frequency band, the maximum ULF power in the frequency bands tends to occur at the highest ($L \sim 4.4$) latitude; (b) the latitudinal gradient in the power in the 0.6 to 50 mHz band steepens sharply at ~ 2000 LT, December 16, and then slowly decreases over the next several days to the gradient existent prior to the storm; a similar behavior of the latitude gradient of the power is observed in the 2 to 5 mHz band; (c) during ~ 0700 to ~ 1800 LT on December 16, the peak power in the 10 to 15 mHz band was at $L < 4$, (d) during the periods ~ 1200 LT December 19 to ~ 0000 LT December 20 and ~ 0700 LT to ~ 1800 LT on December 20 the peak intensity in the 10 to 15 mHz frequency band was at $L < 4$; and (e) in terms of the latitude location of the peak power intensity, the location of the Siple conjugate area has an appreciable variation with time during the interval when data are available.

3. Spectral Coherence Variations

The coherence of the ULF variations between the several stations was calculated and examined as a function of time during the magnetic storm interval. Although not shown here because of lack of space, it was found that in general, as would be expected, the coherence was usually less for the higher frequency bands than for the lower and generally decreased with increasing separation between stations; *i.e.*, the coherence was generally lower for the $L = 4.4/3.2$ pair than for the $L = 4.4/4.0$ pair.

A somewhat surprising observation concerning the coherences was that some diurnal variations were apparent. This is evident in Figure 2 where the average coherence over the 5 day interval is plotted as a function of local time for two of the four frequency bands. For most components in both frequency bands, the coherence tends to be a maximum, on the average, during the local day. Such diurnal variations are also evident when the two days of the most intense portion of the magnetic storm (December 17 to 18) are omitted from the averages. The diurnal variations in the

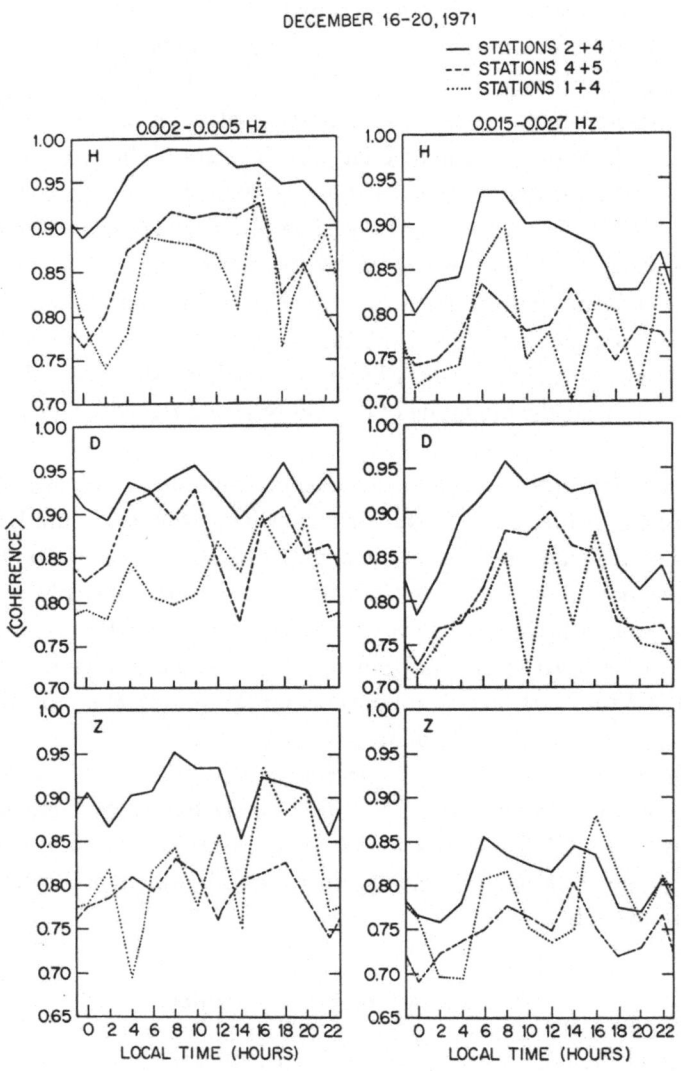

Fig. 2.   Local time variation of the 5 day average of the coherence between stations of the ULF
power in two of the four frequency bands shown in Figure 1.

coherences strongly suggest that the sources of the ULF power in both frequency bands are different at local day than at local night, even during a magnetic storm period.

4. Discussion: Latitude Gradient of Power

The increases in the maximum power (0.6 to 50 mHz band) and the K_{FR} index after the first SC on December 16 indicate that magnetic disturbances were occurring at the longitude of the stations during this local time period. The sharp break in the latitude gradients in all frequency bands at ~ 2000 LT on December 16 marks the time that the storm disturbance (in the form of ionospheric and field aligned current systems) noticeably modified the latitudinal distribution of ULF power levels. This break in the latitude gradients probably marks the onset of the major storm disturbances at $\sim 3°$W geomagnetic longitude. Although there are temporal variations lasting a few hours in the latitude gradients of the power in the 0.6 to 50 mHz and the 2 to 5 mHz bands after ~ 2000 LT, December 16, it takes 2 days or more for the latitude gradients to again reach the spatial distributions they had prior to the storm. This time dependence in regaining the equilibrium gradient distribution is probably a function of the magnetic storm intensity, a point that will need further examination.

5. Discussion: Time-Dependence of the 10 to 15 mHz Band

The intervals of intensity maxima at $L \sim 4.0$ and ~ 3.2 in the 10 to 15 mHz band are similar to the January 5, 1972 period studied in detail by Fukunishi and Lanzerotti (1974). These authors, using theoretical considerations of Chen and Hasegawa (1974a, b) and calculated eigen frequencies of toroidal and poloidal Alfvén waves from a simple model by Orr and Matthew (1971), concluded that the temporal 'movement' of the intensity maximum in the Pc4 band, particularly during the local day, was indicative of the location of the plasmapause at the local time of the station array. The maximum of the 15 to 27 mHz band was always at a higher latitude than the 10 to 15 mHz band. The situation during local night was more complex because of the probable existence of different ULF sources during local night than during local day.

In the data studied here, for the interval ~ 0700 to ~ 1300 LT during the quiet day of December 16, the intensity maximum of the 10 to 15 mHz band (if associated with the plasmapause) indicates that the plasmapause is at $L \lesssim 4$; after ~ 1400 LT, the plasmapause is at $L \gtrsim 4$ or 4.4, consistent with the nominal plasmasphere bulge region. During the first day of the magnetic storm, the maximum of intensity in the 10 to 15 mHz band was at $L \geq 4.4$. It is difficult to invoke field-line resonance models during storm periods to explain this observation in this disturbed period. However, beginning about ~ 0400 LT on December 18 there are considerable intervals when the intensity maximum in the 10 to 15 mHz band is at $L \leq 4.0$. For a continuous source spectrum, with the source at $L > 4.4$, such an intensity peak could only occur at plasma density gradients (possibly, but not necessarily, related to the plasmapause) in the magnetosphere (Chen and Hasegawa, 1974a).

Some equatorial plasmapause position data obtained from the electric field probe on the S^3 satellite exist for the local evening and local midnight regions of the magnetosphere during the storm period (Maynard and Cauffman, 1973). Comparisons of S^3 plasmapause positions and the peak intensity in the 10 to 15 mHz band are given in Table I for the intervals when the station array is passing near a satellite orbit. Out of the eight equatorial determinations of the plasmapause position by S^3 in the local evening/local night region, the 10 to 15 mHz peak intensity approximately corresponds

TABLE I

Comparison of S³ plasmapause location with ULF intensity

December	LT	PP: S³	10 to 15 mHz Peak	Agree
16	~ 18–20	~ 4.8	≥ 4.4	YES
16	~ 22–24	~ 3.5 (?)	≥ 4.4	(?)
17	~ 18–20	≥ ~ 4.0	≥ 4.0	YES
17	~ 22–24	~ 3.5	≥ 4.0	NO
18	~ 18–20	~ 3.6	≥ 3.2; ≤ 4.0	YES
18	~ 22–24	~ 4.0	≥ 4.0	YES
19	~ 18–20	~ 4.5	≥ 3.2; ≤ 4.0	NO
19	~ 22–24	~ 4.5	≥ 4.0	YES

to the plasmapause location on five occasions. The 10 to 15 mHz peak intensity does not seem to correspond on two occasions, and, on one occasion during the initial most intense part of the storm there is some question as to the possible motion of the plasmapause at the time of the satellite observation.

The results of Table I, combined with the detailed considerations of Fukunishi and Lanzerotti (1973), are very encouraging and strongly suggest that the variations in the 10 to 15 mHz peak location as a function of local time during the December 16 to 20 period indicate latitudinal changes with time in the density gradients of the mid-latitude magnetospheric plasma distribution.

6. Conclusions

The onset of magnetic storm disturbances at the longitude of the station array is marked by a sharp break in the latitude gradient of the magnetic power. The magnetic power in the 10 to 15 mHz frequency band often has a maximum intensity at $L \leqslant 4.4$. The theoretical considerations of Chen and Hasegawa (1974a, b) and the results of Fukunishi and Lanzerotti (1974) suggest that the intensity maximum in this frequency band may be indicative of plasma density gradients in the magnetosphere. Comparisons of the peak intensity of the 10 to 15 mHz frequency band with S³ determinations of the equatorial plasmapause position in the evening/night sector suggest such an interpretation may be correct. In terms of magnetic power, the conjugate point of Siple varies considerably as a function of time. The spatial coherence of ULF signals decreases with increasing station spacing. The average coherence over the time period studied here has a distinct diurnal variation, with smaller coherence near local night. It is important to further 'calibrate' these ULF measurements with satellite data in order to make the ULF data even more useful as a continuous monitor and probe of magnetospheric phenomena. Only with properly sited ground stations can the geographical extent of magnetospheric phenomena be properly studied and ultimately understood.

Acknowledgements

We thank D. J. Thomson and B. Kleiner for many helpful discussions on the spectral analyses and Liu Chen and A. Hasegawa for profitable theoretical consultations.

References

Chen, L. and Hasegawa, A.: 1973a, *J. Geophys. Res.* **79**, 1024.
Chen, L. and Hasegawa, A.: 1973b, *J. Geophys. Res.* **79**, 1033.
Fukunishi, H. and Lanzerotti, L. J.: 1973, *J. Geophys. Res.* **79**, 142.
Jacobs, J. A., Kato, Y., Matsushita, S., and Troitskaya, V. A.: 1964, *J. Geophys. Res.* **69**, 180.
Lanzerotti, L. J., Hasegawa, A., and Tartaglia, N. A.: 1972, *J. Geophys. Res.* **77**, 6731.
Lanzerotti, L. J., Fukunishi, H., Hasegawa, A., and Chen, L.: 1973, *Phys. Rev. Letters* **31**, 624.
Maynard, N. C. and Cauffman, D. P.: 1973, *J. Geophys. Res.* **78**, 4745.
Orr, D. and Matthew, J. A. D.: 1971, *Planetary Space Sci.* **19**, 897.
Rostoker, G., Samson, J. C., and Higuchi, Y.: 1972, *J. Geophys. Res.* **77**, 4700.
Samson, J. C.: 1972, *J. Geophys.* **77**, 6145.
Samson, J. C., and Rostoker, G.: 1972, *J. Geophys. Res.* **77**, 6133.
Samson, J. C., Jacobs, J. A., and Rostoker, G.: 1971, *J. Geophys. Res.* **76**, 3675.
Suguira, M. and Poros, D. J.: 1973, NASA Goddard Space Flight Center Report.
Thomson, D. J.: 1971, Ph. D. Dissertation, Polytech. Inst. of Brooklyn, New York.

PART VI

CONCLUSIONS

CONCLUSIONS

BILLY M. McCORMAC

Lockheed Palo Alto Research Laboratory, 3251 Hanover Street, Palo Alto, Calif. 94304, U.S.A.

1. Magnetospheric Structure and Characteristics

The magnetospheric structure is complicated and there are many dynamical processes occurring simultaneously, such that it is difficult to separate the various phenomena.

It is now identified that there are serious difficulties in relating solar wind variations to magnetospheric effects. Solar wind disturbances do not dominate the total mass and energy incident on the magnetosphere. There are no obvious solar cycle variations in solar wind properties. Thus, there is a problem relating observed solar cycle effects in the magnetosphere to solar wind properties. Several possibilities exist, for example, the energy dissipated in the magnetosphere is not derived from the solar wind; magnetospheric energy dissipation is not dominated by magnetic storms; or the efficiency of the solar wind-magnetospheric coupling changes with the solar cycle in some unknown manner.

It is now completely accepted that the magnetosphere is open, i.e., magnetic field lines from the Earth connect with interplanetary magnetic field lines. The concept of the magnetopause was developed as a boundary for the closed magnetosphere. One must now redefine the magnetopause for the open magnetosphere. The characteristics of the magnetosphere in terms of discontinuities, plasma, waves, and fields are yet to be resolved. Although magnetospheric phenomena are affected by the direction of the interplanetary magnetic field there is no agreement on the magnetospheric configuration when the interplanetary field is northward vs. southward.

An important process in an open magnetosphere is reconnection of magnetic field lines. Details of the reconnection process are poorly understood; for example, we do not know if the electric field across the magnetotail is the cause or effect of reconnection. The details of the types of processes occurring in the plasma flow and magnetic field reconnection are also poorly understood. Different processes may operate under different ambient conditions. Some believe that there is a series of standing MHD waves in the vicinity of the neutral line to produce the necessary plasma changes. Others prefer vacuum merging of field lines at a true neutral sheet. There is disagreement over the interaction layer, such as whether it can be as thin as the electron gyro-radius or not smaller than an ion gyro-radius. Some type of wave-particle interaction is required but is not understood yet.

There is also the problem of local vs. overall magnetospheric control for the rate of reconnection. Some argue that during substorms the neutral point is located $>15\,R_E$ and <25 to $35\,R_E$. Some argue that the substorms neutral point is part of the local structure and not a part of the magnetotail neutral point.

The plasmasphere is observed to have a bulge on the dusk side and its boundary, the

plasmapause, shrinks to lower L shells as K_p increases. It is suggested that the defini·
tion of the plasmapause should be 'the boundary where hot plasma becomes dominant over cold plasma'. This seems to make much more physical sense than trying to
identify the plasmapause from a cold plasma density profile. In the past it has been
argued that plasma is removed from the plasmasphere by convection, drifting out to
the magnetopause resulting in the plasmapause, although there are no data to prove
this contention. Synchronous orbit data are now interpreted to show that the plasma
is accelerated to high energies during a substorm to form the hot plasma sheet. The
plasmapause is thus the boundary beyond which all of the cool plasma is accelerated
to hot plasma. After a substorm the plasmapause expands to higher L values as the
plasmasphere is refurnished with plasma from the ionosphere. If the cold plasma
were detached and convected outward it could still be detected; however, data show
that it is not there.

The polar cleft region is particularly important as a source of particles for the
magnetosphere. While the connection between the cleft and the plasma sheet is still
obscure much useful data on cleft particle populations are now available. However,
the variation of cleft structure with solar changes is unknown. ISIS-1 data show that
the cleft particles have a spectra like those in the magnetosheath and extend about $3°$
in latitude at low altitudes.

The flow of plasma over the polar cap must be unravelled as it is critical to specifying
behavior in the magnetospheric tail. The plasma flow along the magnetopause seems
to be more or less uniform; therefore, in the past it has been assumed in the absence
of any other data that there was a uniform plasma flow over the polar caps, implying
a uniform electric field from dawn to dusk. These considerations are important to
help resolve the origin of the plasma sheet. Electric field measurements are now
interpreted to show that there is a concentration of plasma flow toward the edges of
the polar cap. On the other hand, particle data show that the plasma flows directly
over the polar cap.

Considerable detail on the inverted V's was presented. In the early evening the
inverted V seems to produce the visual arc. However, the arc is much narrower than
the 8 km optical resolution and the 15 km particle resolution of much of the data
now available. There are no protons in the inverted V. It is calculated that the electrons in the inverted V are accelerated between 2000 and 5000 km altitude. Particle
data show field aligned currents are narrower than the inverted V and at the edges
of the inverted V.

There is disagreement as to whether the inverted V is on open or closed field lines.
Some say that there are open magnetic field lines poleward of the ~ 40 keV trapping
boundary. However, others claim that the ISIS-1 data show that the intensity decrease in 40 keV electrons is caused by turbulence in the plasma sheet; therefore,
putting the trapping boundary at a lower latitude than the limit of closed field lines.
The trapping boundary should be the turbulence boundary not the open or closed
field line boundary.

The electric field imposed on the polar cap will extend into the ionosphere and form part of the magnetospheric convection system. More information is needed on the electric fields, magnetic fields, and conductivities if one is to use the available techniques for computing the magnetospheric current systems. Birkeland (field aligned) currents provide a coupling between the magnetosphere and the ionosphere at high latitudes. The coupling is important for many phenomena such as substorms, plasma convection, auroral particle behavior, instabilities producing double layers, etc. Magnetometer data for field aligned currents show field that aligned currents occur on all passes through the auroral oval, primarily in the form of east-west oriented sheets with current densities usually between 10^{-6} to 10^{-5} A m^{-2}. In particular, in the evening sector the current flow is outward at the poleward boundary of the field aligned current region with inward flowing currents at lower latitudes. The satellite magnetometer data are in complete disagreement with rocket experiments of field aligned currents. Rocket experiments have never sampled the entire field aligned current system. It cannot be assumed that the difference between the ionosphere and magnetosphere can be explained by parallel electric fields.

2. Wave-Particle Interactions

The research effort in wave-particle interactions has greatly increased in the last year or two. Enough data in the magnetosphere are now available to clearly demonstrate that nonlinear processes are involved and that linear theory is not applicable. However, there are insufficient data to uniquely determine the specific types of wave-particle interactions for the various parts of the magnetosphere.

The Earth's bow shock is always present and is one of the best regions in which to study nonlinear wave-particle interactions. This is the difficulty of not being able to resolve temporal and spatial effects from a single spacecraft. When the interplanetary magnetic field is parallel to the bow shock the shock front is sharp; however, when the field is perpendicular the shock is highly structured. The data mainly consist of electrostatic noise (leading to definition of fine scale structure) and the temperature of electrons and protons, both of which are observed to be heated in the bow shock. It is believed that cyclotron drift instability heats the electrons, but no mechanism is identified for the protons. Another poorly understood phenomena of the bow shock is that of reflected particles streaming back along magnetic field lines. These particles must be accounted for as a part of the dissipation mechanism and indicate the existence of an electric potential barrier at the shock front.

Wave-particle interactions near to the geostationary orbit can play an important role in various processes by: affecting the structure of the inner edge of the plasma sheet through strong pitch angle scattering, decay of the ring current, and radial diffusion of the hot plasma in the plasmasphere. Studies are underway to investigate the anisotropies induced in injected electrons. However, much effort remains to even get agreement on the initial conditions.

Studies of the electromagnetic whistler mode wave for gyroresonance with electrons show the thermal anisotropy is much more important than the loss cone anistropy inside of $L = 10$. Thus, in the outer zone only electrons well above the thermal energy are able to participate in unstable resonant interactions and hence whistlers cannot be expected to play an important role in establishing outer zone structure.

Much data now exist on the detection of VLF emissions from satellites. Strong high latitude emissions have a maximum $> 70°$ IN Lat. On most passes these emissions are poleward of the 35 keV trapping boundary. These are interpreted as arising from soft electrons in the cleft in the afternoon and the outer plasma sheet at midnight. The mechanism is believed to be due to Heaviside (Čerenkov) radiation, but does not adequately explain the observations at frequencies < 10 kHz.

The midlatitude VLF emissions tend to maximize between 50° and 65° IN Lat and show a strong dependence on K_p. The mechanism is believed to be cyclotron resonance generation with energetic elctrons in the equatorial region and they are ducted along the magnetic field lines. The emissions occur in longitudinally localized zones of intense emission which corotate with the Earth and show conjugate symmetry.

The low latitude VLF emissions are frequently observed $< 30°$ IN Lat. The most intense emissions agree well with the intense particle flux but show conjugate symmetry implying that the emissions are generated at relatively low altitudes, possibly by the Heaviside mechanism.

Diffusion in an inhomogeneous medium by low amplitude monochromatic waves leads to finite wave spread and some energy is radiated in the sidebands. The experimental and theoretical work on the generation of Pc-1 sidebands shows agreement on the frequency structure. This is mot important for the future study of non-linear proton interactions.

3. Magnetospheric Substorms

Much effort is being devoted to the study of magnetospheric substorms and morphological details are becoming better known than in the past, however, it is still difficult to define the primary features of a substorm because substorms are highly variable in time and space.

Initially substorms were divided into expansion and recovery phases. There is much interest in the initiation of the expansion phase to determine the triggering mechanism which is still unknown. This has led some to establish a growth phase prior to the expansion phase. Others argue that the growth phase is just expansion or recovery phase seen at a distance. There are definition difficulties on the expansion phase. It was originally defined in terms of auroral region observables. Now some want to define the expansion phase as the commencement of a midlatitude positive bay.

It is difficult to determine the onset time of a substorm. There is a wide variety of substorms. Even at midlatitudes where some tie the onset of the expansion phase to the midlatitude positive bay, the following types of onsets are found: sharp, slow, multiple, eastward sweeping, and westward sweeping. Also, the central meridian of maximum development of substorms may be centered anywhere in the night hemisphere.

It seems that the sequence of events with all of the cause and effect aspects of a substorm are yet to be agreed on. Observations of substorms become more sophisticated. More observing stations are required. Multidisciplinary observations are needed, such as optical, plasma, and magnetic field observations. Advanced data processing techniques are needed to organize and process the vast amount of information that is needed.

Polar cap pictures of optical emissions at 3914, 5577, and 6300 Å are proving to be very useful in studying the behavior and extent of arcs and diffuse aurora. This will eventually allow various optical features in the atmosphere to be correlated with regions and processes in the magnetosphere. Linear auroral arcs appear under a variety of conditions. They are seen in the polar cap, at any hour of the day, and down to the equatorward boundary of the oval. Thus, they map into many different regions of the magnetosphere. The plasma sheet maps into the diffuse aurora. The total energy input into the nightside diffuse zone is greater than the total input into the arcs. Auroral arcs may lie anywhere in the diffuse region but never equatorward. It is difficult with the available data to determine if there is a single auroral oval or if there are overlapping horseshoe structures. Does the auroral oval have any physical significance beyond a statistical one, or is it a composite of several distinct but overlapping regions of diffuse and discrete aurora? Discontinuities occur frequently.

The auroral absorption regions are in motion on a global scale. It is believed that the movement of an absorbing patch in the ionosphere indicates magnetospheric disturbance propagation. Dayside auroral absorption occurs within the trapping region as a result of the precipitation of outer zone electrons that were introduced by substorms in the night sector. Nightside auroral absorption indicates that the substorm begins near to the midnight meridian at $\sim 65°$ IN Lat then spreads to higher and lower latitudes and eastward and westward. The variation of velocities of absorbing regions is very large and is $\leqslant 3°$ to $4°$ min^{-1} eastward or westward.

GLOSSARY

Most of the abbreviations and terms utilized in this book are obvious. Several whose meaning may not be apparent are listed below.

AA. Auroral absorption. Measure of cosmic noise absorption through the ionosphere in the auroral oval.

AEJ. Auroral electrojet. Current system in the ionosphere of the auroral region.

AU. Astronomical unit.

Auroral Oval. Locus of auroras in latitude as a function of time, which has an oval shape.

Bow Shock. Collisionless shock set up by the interaction of the solar wind and the Earth's magnetosphere.

Closed Magnetic Field Line. Earth's magnetic field line which is continuous through space from one hemisphere to the other.

CNA. Cosmic noise absorption. Surface measurement of the absorption of cosmic noise, usually about 30 MHz, passing through the ionosphere.

Conjugate. Used herein for magnetic conjugacy and refers to the opposite ends of the same closed magnetic field line.

CRAND. Cosmic ray albedo neutron decay.

EDT. Eccentric dipole time.

ELF. Extremely low frequency and extends from 3 to 10^3 Hz.

Equatorial Electrojet. A current system in the ionosphere, flowing generally along the Earth's equator.

FD. Forbush decrease.

FP. Fabry-Pérot.

FWHM. Full width half maximum.

Geomagnetic Micropulsation. Magnetic field fluctuations in the period range of 0.2 s to 10 min.

Hall Currents. Current flow perpendicular to both the electric and magnetic fields.

IGY. International Geophysical Year.

IN Lat. Invariant latitude, Φ.

IN LT. Invariant local time.

IN Pole. Invariant pole, where $\Phi = 90°$.

Invariant Coordinate System. McIlwain's B, L space magnetic coordinates.

IQSY. International Year of the Quiet Sun.

IR. Infrared radiation covering from about 7800 Å to 1000 μm.

K_p. Quasi-logarithmic scale, from 0 to 9, measuring the range of activity of the most active component of the magnetic field within a 3 h interval.

L. McIlwain's invariant shell parameter, whose units are expressed in R_E at the magnetic equator.

B. M. McCormac (ed.), *Magnetospheric Physics*, 393–395. *All Rights Reserved.*
Copyright © 1974 *by D. Reidel Publishing Company, Dordrecht-Holland.*

LT. Local time.

Magnetic Bay. Positive or negative deviations from the normal magnetograms, having a characteristic shape of the shore line of a bay.

Magnetopause. Boundary of the Earth's magnetosphere.

Magnetosheath. Region between the magnetopause and the bow shock.

Magnetosphere. Region inside the magnetopause.

Magnetotail. Region of the magnetosphere extending in the antisolar direction beyond the trapping region.

MHD. Magnetohydrodynamics.

MLT. Magnetic local time.

M Substorm. Magnetospheric substorm.

Neutral Sheet. Narrow region about 1000 km thick in the middle of the tail of the plasma sheet where the magnetic field falls to a very low value.

Open Magnetic Field Line. One of the Earth's magnetic field lines which is connected to the interplanetary magnetic field.

PAD. Pitch angle distribution.

PCA Event. Polar cap absorption event. High energy proton precipitation in the polar cap producing high cosmic noise absorption.

Pedersen Current. Current flow along electric field which is perpendicular to the magnetic field.

PEJ. Polar electrojet. See auroral electrojet.

Pitch Angle. Angle between the instantaneous velocity vector of a charged particle and the direction of the magnetic field.

Plasmapause. Boundary at about L of 3.5 to 4 inside of which the plasma density is much higher.

Plasma Sheet. Thick slab of hot plasma in the magnetosphere.

Polar Cap. Region inside the auroral oval.

Pre-dawn Enhancement. Enhanced optical emission produced before normal sunrise behavior as a result of charge particles from the sunlite conjugate region.

QL. Quasi-linear.

R_E. Earth radius.

Ring Current. Current of trapped low energy protons at $L = 3$ to 6.

rms. Root mean square.

S. Siemens $= \text{ohm}^{-1}$.

SAR Arc. Stable auroral red arc.

SC. Sudden commencement.

SCA. Sudden commencement absorption.

SCNA. Sudden cosmic noise absorption.

SI. Sudden impulse.

Solar Wind. Electron, proton, α-particle and other charged particle emissions from the Sun.

SPAND. Solar proton albedo neutron decay.

Trapping Region. Region of closed lines wherein charged particles can bounce from one hemisphere to the other and can drift all of the way of the around Earth.

ULF. Ultra low frequency and is from 10^{-2} to 3 Hz.

UT. Universal time.

UV. Ultraviolet radiation and extends from 100 to 3800 Å.

VK. Vegard-Kaplan band system.

VLF. Very low frequency and is from 3 to 30 kHz.

VLF Chorus. Radiation consisting of a multiple of overlapping rising tones, usually in the band of 2 to 4 kHz, which peaks in the morning hours, sounding like those of a distant colony.

Whistler. Radio signals in the audio-frequency that 'whistle'.

WPI. Wave particle interaction.

INDEX OF SUBJECTS

ASTROPHYSICS AND SPACE SCIENCE LIBRARY

Edited by

J. E. Blamont, R. L. F. Boyd, L. Goldberg, C. de Jager, Z. Kopal, G. H. Ludwig, R. Lüst,
B. M. McCormac, H. E. Newell, L. I. Sedov, Z. Švestka, and W. de Graaff

1. C. de Jager (ed.), *The Solar Spectrum. Proceedings of the Symposium held at the University of Utrecht, 26–31 August, 1963.* 1965, XIV + 417 pp.
2. J. Ortner and H. Maseland (eds.), *Introduction to Solar Terrestrial Relations. Proceedings of the Summer School in Space Physics held in Alpbach, Austria, July 15–August 10, 1963 and Organized by the European Preparatory Commission for Space Research.* 1965, IX + 506 pp.
3. C. C. Chang and S. S. Huang (eds.), *Proceedings of the Plasma Space Science Symposium, held at the Catholic University of America, Washington, D.C., June 11–14, 1963.* 1965, IX + 377 pp.
4. Zdeněk Kopal, *An Introduction to the Study of the Moon.* 1966, XII + 464 pp.
5. B. M. McCormac (ed.), *Radiation Trapped in the Earth's Magnetic Field. Proceedings of the Advanced Study Institute, held at the Chr. Michelsen Institute, Bergen, Norway, August 16–September 3, 1965.* 1966, XII + 901 pp.
6. A. B. Underhill, *The Early Type Stars.* 1966, XII + 282 pp.
7. Jean Kovalevsky, *Introduction to Celestial Mechanics,* 1967, VIII + 427 pp.
8. Zdeněk Kopal and Constantine L. Goudas (eds.), *Measure of the Moon. Proceedings of the Second International Conference on Selenodesy and Lunar Topography, held in the University of Manchester, England, May 30–June 4, 1966.* 1967, XVIII + 479 pp.
9. J. G. Emming (ed.), *Electromagnetic Radiation in Space. Proceedings of the Third ESRO Summer School in Space Physics, held in Alpbach, Austria, from 19 July to 13 August, 1965.* 1968, VIII + 307 pp.
10. R. L. Carovillano, John F. McClay, and Henry R. Radoski (eds.), *Physics of the Magnetosphere, Based upon the Proceedings of the Conference held at Boston College, June 19–28, 1967.* 1968, X + 686 pp.
11. Syun-Ichi Akasofu, *Polar and Magnetospheric Substorms.* 1968, XVIII + 280 pp.
12. Peter M. Millman (ed.), *Meteorite Research. Proceedings of a Symposium on Meteorite Research, held in Vienna, Austria, 7–13 August, 1968.* 1969, XV + 941 pp.
13. Margherita Hack (ed.), *Mass Loss from Stars. Proceedings of the Second Trieste Colloquium on Astrophysics, 12–17 September, 1968.* 1969, XII + 345 pp.
14. N. D'Angelo (ed.), *Low-Frequency Waves and Irregularities in the Ionosphere. Proceedings of the 2nd ESRIN-ESLAB Symposium, held in Frascati, Italy, 23–27 September, 1968.* 1969, VII + 218 pp.
15. G. A. Partel (ed.), *Space Engineering. Proceedings of the Second International Conference on Space Engineering, held at the Fondazione Giorgio Cini, Isola di San Giorgio, Venice, Italy, May 7–10, 1969.* 1970, XI + 728 pp.
16. S. Fred Singer (ed.), *Manned Laboratories in Space. Second International Orbital Laboratory Symposium.* 1969, XIII + 133 pp.
17. B. M. McCormac (ed.), *Particles and Fields in the Magnetosphere. Symposium Organized by the Summer Advanced Study Institute, held at the University of California, Santa Barbara, Calif., August 4–15, 1969.* 1970, XI + 450 pp.
18. Jean-Claude Pecker, *Experimental Astronomy.* 1970, X + 105 pp.
19. V. Manno and D. E. Page (eds.), *Intercorrelated Satellite Observations Related to Solar Events. Proceedings of the Third ESLAB/ESRIN Symposium held in Noordwijk, The Netherlands, September 16–19, 1969.* 1970, XVI + 627 pp.
20. L. Mansinha, D. E. Smylie, and A. E. Beck, *Earthquake Displacement Fields and the Rotation of the Earth. A NATO Advances Study Institute Conference Organized by the Department of Geophysics, University of Western Ontario, London, Canada, June 22–28, 1969.* 1970, XI + 308 pp.
21. Jean-Claude Pecker, *Space Observatories.* 1970, XI + 120 pp.
22. L. N. Mavridis (ed.), *Structure and Evolution of the Galaxy, Proceedings of the NATO Advanced Study Institute, held in Athens, September 8–19, 1969.* 1971, VII + 312 pp.

23. A. Muller (ed.), *The Magellanic Clouds. A European Southern Observatory Presentation: Principal Prospects, Current Observational and Theoretical Approaches, and Prospects for Future Research. Based on the Symposium on the Magellanic Clouds, held in Santiago de Chile, March 1969, on the Occasion of the Dedication of the European Southern Observatory.* 1971, XII + 189 pp.

24. B. M. McCormac (ed.), *The Radiating Atmosphere. Proceedings of a Symposium Organized by the Summer Advanced Study Institute, held at Queen's University, Kingston, Ontario, August 3–14, 1970.* 1971, XI + 455 pp.

25. G. Fiocco (ed.), *Mesospheric Models and Related Experiments. Proceedings of the 4th ESRIN–ESLAB Symposium, held at Frascati, Italy, July 6–10, 1970.* 1971, VIII + 298 pp.

26. I. Atanasijević, *Selected Exercises in Galactic Astronomy.* 1971, XII + 144 pp.

27. C. J. Macris (ed.), *Physics of the Solar Corona. Proceedings of the NATO Advanced Study Institute on Physics of the Solar Corona, held at Cavouri-Vouliagmeni, Athens, Greece, 6–17 September 1970.* 1971, XII + 345 pp.

28. F. Delobeau, *The Environment of the Earth.* 1971, IX + 113 pp.

29. E. R. Dyer (general ed.), *Solar-Terrestrial Physics 1970. Proceedings of the International Symposium on Solar-Terrestrial Physics, held in Leningrad, U.S.S.R., 12–19 May 1970.* 1972, VIII + 938 pp.

30. V. Manno and J. Ring (eds.), *Infrared Detection Techniques for Space Research, Proceedings of the Fifth ESLAB-ESRIN Symposium, held in Noordwijk, The Netherlands, June 8–11, 1971.* 1972, XII + 344 pp.

31. M. Lecar (ed.), *Gravitational N-Body Problem, Proceedings of IAU Colloquium No. 10, held in Cambridge, England, August 12–15, 1970.* 1972, XI + 441 pp.

32. B. M. McCormac (ed.), *Earth's Magnetospheric Processes. Proceedings of a Symposium Organized by the Summer Advanced Study Institute and Ninth ESRO Summer School, held in Cortina, Italy, August 30–September 10, 1971.* 1972, VIII + 417 pp.

33. Antonin Rükl, *Maps of Lunar Hemispheres.* 1972, V + 24 pp.

34. V. Kourganoff, *Introduction to the Physics of Stellar Interiors.* 1973, XI + 115 pp.

35. B. M. McCormac (ed.), *Physics and Chemistry of Upper Atmospheres. Proceedings of Symposium Organized by the Summer Advanced Study Institute, held at the University of Orléans, France, July 31–August 11, 1972.* 1973, VIII + 389 pp.

36. J. D. Fernie (ed.), *Variable Stars in Globular Clusters and in Related Systems. Proceedings of the IAU Colloquim No. 21, held at the University of Toronto, Toronto, Canada, August 29–31, 1972.* 1973, IX + 234 pp.

37. R. J. L. Grard (ed.), *Photon and Particle Interaction with Surfaces in Space. Proceedings of the 6th ESLAB Symposium, held at Noordwijk, the Netherlands, 26–29 September, 1972.* 1973, XV + 577 pp.

38. Werner Israel (ed.), *Relativity, Astrophysics and Cosmology. Proceedings of the Summer School, held 14–26 August, 1972, at the BANFF Centre, BANFF, Alberta, Canada.* 1973, IX + 323 pp.

39. B. D. Tapley and V. Szebehely (eds.), *Recent Advances in Dynamical Astronomy. Proceedings of the NATO Advanced Study Institute in Dynamical Astronomy, held in Cortina d'Ampezzo, Italy, August 9–12, 1972.* 1973, XIII + 468 pp.

40. A. G. W. Cameron (ed.), *Cosmochemistry. Proceedings of the Symposium on Cosmochemistry, held at the Smithsonian Astrophysical Observatory, Cambridge, Mass., August 14–16, 1972.* 1973, X + 173 pp.

41. M. Golay, *Introduction to Astronomical Photometry.* 1974, approx. 570 pp.

42. D. E. Page (ed.), *Correlated Interplanetary and Magnetospheric Observations. Proceedings of the Seventh ESLAB Symposium, held at Saulgau, W. Germany, 22–25 May, 1973.* 1974, XIV + 662 pp.

45. C. B. Cosmovici (ed.), *Supernovae and Supernova Remnants. Proceedings of the International Conference on Supernovae, held in Lecce, Italy, May 7–11, 1973.* 1974, XVII + 387 pp.